3rd Annual International Workshop on Materials Science and Engineering (IWMSE 2017)

IOP Conference Series: Materials Science and Engineering
Volume 250

Guangzhou, China
8 - 10 September 2017

Editors:

G. Elatharasan
Chen Hsu

ISBN: 978-1-5108-5029-3
ISSN: 1757-8981

Printed from e-media with permission by:

Curran Associates, Inc.
57 Morehouse Lane
Red Hook, NY 12571

Some format issues inherent in the e-media version may also appear in this print version.

Printed by Curran Associates, Inc. (2017)

For permission requests, please contact the Institute of Physics
at the address below.

Institute of Physics
Dirac House, Temple Back
Bristol BS1 6BE UK

Phone: 44 1 17 929 7481
Fax: 44 1 17 920 0979

techtracking@iop.org

Additional copies of this publication are available from:

Curran Associates, Inc.
57 Morehouse Lane
Red Hook, NY 12571 USA
Phone: 845-758-0400
Fax: 845-758-2633
Email: curran@proceedings.com
Web: www.proceedings.com

Table of contents

Volume 250

3rd Annual International Workshop on Materials Science and Engineering (IWMSE2017)

8–10 September 2017, Guangzhou, Guangdong, China

Accepted papers received: 2 October 2017
Published online: 22 October 2017

Preface

Papers

IOP Conf. Series: Materials Science and Engineering **250** (2017) 011001 doi:10.1088/1757-899X/250/1/011001

Preface

"The 3rd annual international workshop on Materials science and Engineering, IWMSE2017 conference" was held on September 8~10 in Guangzhou, Guangdong, China. The conference aimed at bringing together the academics and practicing engineers with interests in Materials sciencc, tcchnology, and a broad variety of topics to promote the exchange of ideas and experiences among participants and consolidate the recent advances in the areas of materials.

Nearly 70 full papers were accepted by the Organizing Committees and presented in the 7 sessions of IWMSE2017. World-renowned experts presented these papers from different countries during the three-day conference including 22 keynote speakers. All accepted papers are published in the proceeding of IOP Publishing.

The conference committee, advisory board, and editors would like to thank to all the participating authors for their valuable contributions to the workshop and made a most successful IWMSE2017.

Prof. Chen Hsu
Semiconductor and Ceramic Material Laboratory, Taiwan

IWMSE2017 Committee List

The Advisory Committee

Dr. Timur Sh. Atabaev, Seoul National University, Seoul, South Korea

Dr. Shahrom Mahmud, UniversitiSains Malaysia, Malaysia

Editor

Dr. G.Elatharasan, University College of Engineering Pattukkottai, India

Prof. Chen Hsu, Semiconductor and Ceramic Material Laboratory, Taiwan

Keynote Speakers

Dr. L. Q. Wang, The University of Hong Kong, Hong Kong, China

Dr. Marcela Elisabeta BARBINTA- PATRASCU, University of Bucharest, Romanian

Prof. Sobolev Vladimir, South Dakota School of Mines and Technology, U.S.A.

Dr. Hafez Maghsoudi Sahand, University of Technology, Iran

Dr. NijhumaKayal, Central Glass & Ceramic Research Institute, Kolkata

Dr. Jiangwei Liu ational Institute for Materials Science (NIMS), Japan

Dr. MohiniSain, University of Toronto, Canada

Dr. Jingwei Zhao, University of Wollongong, Australia

Dr. Udo Erich Schwingenschl¨ogl King, Abdullah University OfScience And Technology(KAUST), Germany

Prof. Tarek Aly Elsayed, Helwan University, Egyptian

Dr. TarasKavetskyyDrohobych Ivan Franko, State Pedagogical University, Ukraine

Prof. Kung-Chung Hsu, National Taiwan Normal University, Taiwan

Prof. MALAY KUMAR BANERJEE, Calcutta University, India

Prof. Ramakrishna Reddy K. T. Sri, Venkateswara University, India

Prof. Malgorzata SOPICKA-LIZER, Silesian University of Technology, Poland

Prof. Sergiy N. Lavrynenko, National Technical University "Kharkov Polytechnic Institute, Ukraine

Prof. Jyh-Ping Chen, Chang Gung University, Taiwan

Prof. Núria Ferrer-Anglada, UniversitatPolitècnica de Catalunya, Spain

Prof. Arezki BENFDILA, University M. Mammeri, Algeria

Prof. Pierre Ruterana, Centre de Recherche sur les Matériaux, les Ions et la Photonique, France

Prof. Michel Mortier, EcoleNationaleSupérieure de Chimie de Paris, France

Dr. Hajime HIRAO, City University of Hong Kong, China

Honor chair

Prof. Dean Vucinic, VrijeUniversiteit Brussel (VUB)

Prof. Peng-Sheng Wei, National Sun Yat-Sen University, Taiwan

Dr. Hajime HIRAO, City University of Hong Kong, China

Dr. Ali Davoodi, Ferdowsi University of Mashhad, Iran

Dr. Garima Mahajan, Institute of Engineering & Technology, Iran

Dr. EbadBagherpour, Semnan University, Iran

Dr. Mohamed Ackiel Mohamed, University Kuala Lumpur, Malaysia

IOP Conf. Series: Materials Science and Engineering **250** (2017) 011002 doi:10.1088/1757-899X/250/1/011002

Dr. Zhen Liu , South China University of Technology, China

The Science committee

Dr. Yu-Lin (Eugene) Song, Asia University, Taiwan
Dr. OndrejSikula, Brno University of Technology, Czech Republic
Dr. Jitendra Kumar Katiyar, Indian Institute of Technology Kanpur, India
Dr. Michaela Kendall, University of Birmingham, UK
Dr. BalaramKundu, Jadavpur University, India
Dr. HYUNDOO HWANG, InstitutoTecnológico y de EstudiosSuperiores de Monterrey (ITESM), México
Dr. Lincan Yan, "University of New Mexico", USA
Dr. Sohail Ahmed, Hazara University, Pakistan
Dr. Shien-KueiLiaw, National Taiwan University of Science and Technology, Taiwan

Technical Program Committees

Dr. Xuebiao Yuchi, China Internet Network Information Center, China
Dr. Roberto Suárez Antola, Energy and Mines of Uruguay, Uruguay
Dr. Zhiyu Jiang, University of Chinese Academy of Sciences, China
Dr. Limin Cao, Capital Normal University, China
Dr. MariyaPetrovaAleksandrova, Technical University of Sofia, Bulgaria
Dr. Liyuan Sheng, Peking University, China
Dr. Moola Mohan Reddy, Curtin University, Malaysia
Dr. BonifacioLlamazares, University of Valladolid, Spain
Prof. ParthaPratim Das, Haldia Institute of Technology, India
Prof. Roman Szewczyk, Industrial Research Institute for Automation and Measurements, Polish
Dr. Sanwar Ali, Indiana University of Pennsylvania, U.S.A
Dr. TheofilosChrysikos, University of Patras, Hellenic
Prof. SeifedineKadry
Dr. Khian-Hooi CHEW, University of Malaya, Malaysia
Dr. Dariusz Jacek Jakóbczak, Koszalin University of Technology, Poland
Dr. BappaAcherjee, Kulti College More, India
Dr. Xiaoning Zhang, Paine College, Augusta
Dr. Krzysztof Kuś, University of Warmia and Mazury in Olsztyn, Poland
Dr. Madalina Dumitriu, University Politehnica of Bucharest, Romania
Dr. Yoji Yamato, NTT Corporation, Japan
Dr. TheanPheng Lim, Universiti Utara Malaysia, Malaysia
Dr. Alberto Cano, Virginia Commonwealth University, USA
Dr. Shweta S Dumbre, Nagpur University, USA
Dr. Ray-Hwa Wong, Hwa-Hsia University of Technology, Taiwan
Prof. ViorelNicolau, Dunarea de Jos University of Galati, Romanian
Prof. Antonio Gloria, National Research Council of Italy, Italy
Dr. Marc Porti, Autonomous University of Barcelona, Spain
Dr. Dr. HueiRuey Ong, Universiti Malaysia Pahang, Malaysia
Prof. Mu-Chun Wang, Minghsin University of Science and Technology, Taiwan

IOP Conf. Series: Materials Science and Engineering **250** (2017) 011002 doi:10.1088/1757-899X/250/1/011002

IOP Conf. Series: Materials Science and Engineering **250** (2017) 011002 doi:10.1088/1757-899X/250/1/011002

Dr. Ke Chang, Applied Materials Inc. USA
Prof. GagikShmavonyan, National Polytechnic University of Armenia, Armenian
Dr. UthumpornUtraSapina Abdullah, UniversitiSains Malaysia, Malaysia
Dr. MohdYuhazri Bin Yaakob, UniversitiTeknikal Malaysia Melaka, Malaysia
Dr. Tian Zhou, Private own high-tech company, China
Dr. Saqib Nasir, premier S&T organization in Pakistan (Ministry of Science & Technology), Pakistan
Dr. Waleed K. Ahmed, United Arab Emirates University, Iraqi
Dr. Shaoping Xiao, The University of Iowa, USA
Dr.Almerich-Chulia, Ana UniversitatPolitecnica de Valencia, Spain
Dr. CH.V.V.Ramana, Chirala Engineering College, India
Dr. Jin Xin, Yunnan University, China
Dr. Xiang Ma, SINTEF Materials and Chemisty , Australia
Dr. Ibrahim khiderEltahir , Sudan University of Science and Technology, Sudan
Dr. Murat ÖZEN, University of Antwerp, Belgium
Dr. SankarRamu, National Taiwan University, Taiwan
Prof. Suresh Kumar, Kurukshetra University, India
Dr. Angel A. San-Blas, Miguel Hernandez University of Elche, Spain
Dr. Prashant G. Khakse, Air India Engineering Services Limited, Mumbai Indian
Dr. Mihai IORDOC, National Institute for Research and Development in Electrical Engineering, ROMANIA
Dr. Hisham, National Research Centre, Egypt
Dr. Mariusz R. Rzasa Opole, University of Technology, Poland
Dr. Ke Wang, Huaqiao University, China
Dr. Ngee, University Malaysia Perlis, Malaysia
Dr. Valery Borovik, National Academy of Sciences of Ukraine I. M. Frantsevich, Ukraine
Dr. Chi-Wai Chow, National Chiao Tung University, Taiwan
Dr. Raju KLE, Technological University, India
Dr. Zhenbo Xu, South China University of Technology, China
Dr. Angloria, National Research Council of Italy, Italy
Dr. Ghadir, Regional Center for Food and Feed (RCFF), Cairo
Dr. Mahmoud Ebrahimi, University of Maragheh, Iran
Dr. Paulo NB, ReisUniversity of Beira Interior, Portugal
Dr. Vahid, iversity of Kashan, USA
Prof. N. Kanthavelkumaran, Arunachala College of Engineering for Women, India
Dr.Liang Deng, Shanghai University of Electric Power, China
Dr.Arnulfo Luévanos Rojas, Autonomous University of Coahuila, México
Dr.Weiwei, Xi'an University of Technology, China
Dr.Jae-Ho Jeon, Principal Research Scientist, Korea
Dr.Sergo Sh. Rekhviashvili, Institute of Applied Mathematics & Automation, Russia
Dr.Sabyasachi Sarkar, Indian Institute of Engineering Science and Technology-Shibpur, India
Dr.MehrdadShaygan, Advanced Microelectronic Center Aachen (AMICA), Germany
Dr.Long-yuan LI, University of Plymouth, UK
Dr.GianlucaLandi PhD Research Institute on Combustion-CNR, ITALY

IOP Conf. Series: Materials Science and Engineering **250** (2017) 011002 doi:10.1088/1757-899X/250/1/011002

IOP Conf. Series: Materials Science and Engineering **250** (2017) 011003 doi:10.1088/1757-899X/250/1/011003

Keynote 1

Dr. Udo Schwingenschlogl, King Abdullah University Of Science And Technology(KAUST)

Speech Title:

First-principles predictions of substrate effects on silicene

Abstract:

Silicene is the Si analogue of graphene with the same honeycomb structure and linear dispersions of the π and π^* bands at the K point of the Brillouin zone. It realizes a buckled structure, due to sp^2-sp^3 hybridization, and is compatible with current Si-based nano-electronics. Silicene yet has not been achieved by mechanical exfoliation, but can be deposited on metallic substrates such as Ag(111), Ir(111), and ZrB_2(0001). Regrettably, strong interaction to these substrates destroys the Dirac physics. For this reason, semiconducting substrates have been explored in the literature, including Si(111) and SiC(0001). However, surface passivation is inevitable for these and similar substrates due to dangling bonds. Layered materials such as $MgBr_2$(0001), MoX_2, and GaX_2 (X = S, Se, and Te), on the other hand, simplify the preparation procedure (as passivation is not required), while preserving silicene's characteristic electronic states. Effects of different substrates on silicene will be compared and evaluated with respect to technological requirements.

IOP Conf. Series: Materials Science and Engineering **250** (2017) 011003 doi:10.1088/1757-899X/250/1/011003

Keynote 2

Prof. Pierre Ruterana, Centre de Recherche sur les Matériaux, les Ions et la Photonique

Speech Title:
The structural properties of strained heterostructures in hexagonal symmetry semiconducting materials at atomic level

Abstract:
The hexagonal semiconductors crystallize in the wurtzite structure; they include the nitride (Al,Ga,In)N family and ZnO, and they have undergone extensive investigation for the last 20 years. These materials exhibit physical properties which are subject to highly strategic applications; for instance, the nitride semiconductors constitute the unique family whose alloys direct band gaps allow to cover the largest wavelength span for emission and detection. Moreover, they are ceramics and polar especially along the main growth axis [0001]; this leads to spontaneous piezoelectricity which is now intensively applied to the fabrication of high power, high frequency transistors. For ZnO, its large exciton energy means that it should be at the origin of highest performance emitters in the blue range. Still, the active devices are made of heterostructures which involve various alloy compositions and, it is still a challenge to grow optimal quality devices even on free standing substrates, as for instance, the elemental compounds exhibit large lattice, as well as, thermal mismatch, moreover they grow at largely different temperatures. Therefore, it is important to investigate the extended defects in these heterostructures and at their interface, and to determine their formation mechanisms with the objective of contributing to the optimization of the layers structural quality and performances for the fabricated devices. In our work, using high resolution electron microscopy, atomistic modeling and image simulations, we have identified the atomic structure of **a** type threading dislocations. In the nitride layers, they typically exhibit three different cores which are found inside the high quality layers as well as in grain boundaries. In ZnO, it was possible to point out, for the first time, inside grain boundaries, depending on the tilt angles, the [10-10] dislocation which has only been suggested to exist in the wurtzite system through a theoretical report. In these materials, it was shown that stacking faults not only occur in the basal planes, but also inside the prismatic {11-20} planes with two displacement vectors. Due to the polarity of the wurtzite structure, other typical defects are inversion domains, such defects habit planes are usually {10-10}. They form in order to relieve residual strain at surface steps and their formation inside the basal plane has been shown to be connected to a precipitation of impurities such as Mg from heavy p-doping of the nitride layers. Very recently, we showed that such a (0001) polarity inversion can be driven to take place upon growth of ZnO/GaN heterostructure and within only one monolayer.

IOP Conf. Series: Materials Science and Engineering **250** (2017) 011003 doi:10.1088/1757-899X/250/1/011003

Keynote 3

Dr. Jiangwei Liu, National Institute for Materials Science (NIMS)

Speech Title:

Diamond field-effect transistors

Abstract:

It is well-known that wide bandgap semiconductors such as GaN, SiC, and diamond are suitable to replace silicon partly for fabrication of high-power and high-frequency electronic devices because of their large band-gap energies, high carrier mobility, and high breakdown field. According to figure of merit, diamond-based electronic devices have the largest power-frequency product, the highest thermal limitation, and the lowest power-loss at high-frequency. Thus, diamond semiconductor devices are expected to be of very importance for the future practice applications.

Recently, diamond-based metal-oxide-semiconductor (MOS) capacitors and MOS field-effect transistors (MOSFETs) have developed greatly. They were fabricated on *p*-type boron-doped oxygenated diamond (O-diamond) and hydrogenated diamond (H-diamond) channel layers.[1,2] Although thermal stability of the O-diamond is believed better than that of the H-diamond, trap charge density at insulator/O-diamond interfaces and leakage current density of the O-diamond based MOS capacitors are very high. It is still difficult to fabricate high-performance O-diamond based MOSFETs. On the other hand, many successful diamond MOSFETs were fabricated on the H-diamond channel layers. The H-diamond epitaxial layer has a high surface conductivity. The holes are accumulated on the surface with the sheet hole density of 10^{12}~10^{13} cm^{-2}. Notably, it was reported that the exposure of the H-diamond in NO_2 gas could increase the sheet hole density of H-diamond significantly to be as high as 1×10^{14} cm^{-2}.[3] Therefore, it is promising to fabricate high performance H-diamond based electronic devices.

In this talk, the fabrication of H-diamond-based electronic devices such as MOS capacitors, MOSFETs, and MOSFET logic circuits will be demonstrated and discussed.[4-11]

IOP Conf. Series: Materials Science and Engineering **250** (2017) 011003 doi:10.1088/1757-899X/250/1/011003

Keynote 4

Prof. Malgorzata Sopicka-Lizer, Silesian University of Technology

Speech Title:

Eu/Ce:oxynitride phosphors: advantages and drawbacks

Abstract:

Some ceramic compounds can be used as a solid matrix for accommodation of rare earth elements playing the role of the optical activators. The optical properties of these compounds are extremely related to their electronic band structure and can be tailored by changing the chemical composition and processing methods. Oxynitride compounds with ionic-covalent bonds show interesting physical and chemical properties due to their substantial structural diversity and possibility of solid solution formation with various nitrogen to oxygen ratio. However, synthesis of those compounds requires very high temperature and careful control of ambient atmosphere. This paper presents the optical properties of three main groups of RE:oxynitrides as prepared in Silesian University of Technology. The solid state method has been chosen for synthesis of the chosen phosphors. The efforts in the presented studies were oriented for the launching of nitrogen-silicon into Ce:$Al_5Y_3O_{12}$ garnet, for controlling an oxygen amount in $SrSi_2N_2O_2$:Eu^{2+} and for tailoring the range of solid solution in Ca,Eu-α-sialon powdered phosphors. As a result of the performed research, the green, yellow and orange phosphor powders with the excellent emission intensity and the high quantum yield after excitation under the blue light have been obtained and are presented.

IOP Conf. Series: Materials Science and Engineering **250** (2017) 011003 doi:10.1088/1757-899X/250/1/011003

Keynote 5

Dr. Hajime Hirao, City University of Hong Kong

Speech Title:

Computationally Looking into Complex Metal-Organic Frameworks and Nanomaterials

Abstract:

The physical principles used in computational chemistry underlie all branches of chemistry; as such, computational chemistry has unlimited potential to contribute to the advancement of fundamental chemistry in every different subdiscipline as well as to finding solutions to critical challenges that humankind faces today, such as healthcare and energy/environmental issues. With this in mind, our computational exploration of chemistry applies quantum chemistry, multiscale QM/MM and QM/QM approaches, and many other advanced computational chemistry techniques to a broad range of complex molecular systems such as metalloenzymes, transition-metal catalysts, drugs/drug targets, metal-organic frameworks (MOFs), and nanomaterials. In particular, using computational approaches and often with experimental collaborators, we seek to derive information about chemical reaction mechanisms and bonding patterns of these complex molecules. We are also developing efficient computational methods and algorithms, in the hope that our new computational methods will expand the capability of computational chemistry and thereby enable one to simulate the behavior of complex molecular systems with higher reliability and predictability in the future.

IOP Conf. Series: Materials Science and Engineering **250** (2017) 011004 doi:10.1088/1757-899X/250/1/011004

Peer review statement

All papers published in this volume of *IOP Conference Series: Materials Science and Engineering* have been peer reviewed through processes administered by the proceedings Editors. Reviews were conducted by expert referees to the professional and scientific standards expected of a proceedings journal published by IOP Publishing.

IOP Conf. Series: Materials Science and Engineering **250** (2017) 012001 doi:10.1088/1757-899X/250/1/012001

Selective Method for the Determination of Manganese in End-fitting of Spoolable Reinforced Plastic Line Pipe for Petroleum Industries

Xiaodong Shao[1,2,*], Dongna Zhang[1,2], Houbu Li[1,2]and Xuehua Cai[1,2]

[1]State Key Laboratory for Performance and Structure Safety of Petroleum Tubular Goods and Equipment Materials, No. 89 Jinyeer Road, Xi'an 710077, China
[2]CNPC Tubular Goods Research Institute, No. 89 Jinyeer Road, Xi'an 710077, China
[*]E-mail: shaoxd@163.com

Abstract. The fact that spoolable reinforced plastic line pipe is more flexible and spoolable than steel, and is also much lighter, means that it can becarried and deployedfrom smaller vessels and managed more easily. It was well known that manganese is an important element in end-fitting of spoolable reinforced plastic line pipe. In this paper, a simple spectrophotometric method was described for the determination of manganese in end-fitting of spoolable reinforced plastic line pipe. The method was based on the oxidation-reduction reaction between ammonium persulfate and manganese(II) producing manganese(VII) in the presence of silver nitrate as a catalyst. The characteristic wavelength of maximum absorption of manganese(VII) was obtained locating at 530 nm. Under the optimum reaction conditions the absorption value was proportional to the concentration of manganese in the range of 0.50%~1.80% (R^2 = 0.9997), and the relative standard deviation was less than 3.0% (n=5). The proposed method was applied successfully to determine manganese in end-fitting of spoolable reinforced plastic line pipe samples.

1. Introduction

Humanity's insatiable demand for oil and gas is driving the exploitation of offshore reserves that are ever deeper and harder to bringto the surface. This is subjecting the risers, flowlines, tubulars and other pipe work used to access and work sub-sea locations to greater loads, pressures and environmental challenges. At the same time, pipes need to be as light as possible in order to stay within the payload capacity of floating rigsand storage platforms. The condition of gas and oil wells in the west of China has complex characteristics such as deep, high pressure, and serious corrosive that only corrosion-resistant pipes can be applied to explore. One key feature there is obvious existence of corrosive hydrogen sulfide medium. In order to meet the exploitation requirements of these oil-gas wells with high pressure and high partial pressure of H_2S and CO_2, the pipes must adopt new materials with excellent resistance to corrosion. Spoolable reinforced plastic line pipe (Figure 1), also known as flexible composite pipes or reinforced thermoplastic pipes (RTP), contain three layers, among which, neck bush and reinforcement layer are keys to determine mechanical and thermal properties of these pipes. These composite pipesare key to enabling the industry to meet some of the energy security challenges that nations face today. The ability to combine tailored strength with flexibility, low weight, high thermaland pressure tolerance, minimal flow resistance, high resilience and durability in large-borepipes enables them to surpass thecapabilities of current predominantly metalpiping - albeit often polymer lined. The Spoolable reinforced plastic line pipe has excellent resistance to corrosion. For example, glass fibre reinforced polyoxymethylene (POM) and polypropylene (PP) performed well in saltwater, while carbon fibre reinforced polyetheretherketone (PEEK) proved well suited to a sour oil environment. Carbon in polyphenylenesulphide (PPS) behaved well in a sour oil environment and in hydrocarbon gas condensate at elevated temperature [1].

Figure 1 Spoolable reinforced plastic line pipe

The fact that spoolable reinforced plastic line pipeis more flexible andspoolable than steel, and is also much lighter, means that it can becarried and deployedfrom smaller vessels andmanaged more easily.Thus, spoolable reinforced plastic line pipe will be widely used in the exploitation of acid gas and oil wells. End-fitting is a mechanical device that forms the transition from the spoolable reinforced plastic line pipe to the connector. The content of elements in end-fittingof spoolable reinforced plastic line pipe directly influences the properties of materials. In order to control the quality of end-fitting products, the contents of trace impurities and alloy elements in end-fitting should be determined. In making processes, the content of alloyed elements such as nickel,manganese, chromium, molybdenum, vanadium, and so on, is required to be strictly controlled because it fundamentally determines the performance of end-fitting.It is well known that manganese is an important element in end-fitting. Therefore, a rapid and precise analytical method for the determination of manganese is essentially required for the production control ofspoolable reinforced plastic line pipe.

Manganeseisvaluable alloying agents and coexists in a number of industriallyimportant alloys and steels.Owing to their increasing industrialapplications of manganese, the determination of trace amounts of manganeseis of great interest.Therefore, it was important from the analyticalpoint of view to develop simple, sensitive and reliable methodsfor the determination of manganesein various complex matrixes.Numerous analytical methods were employed for the determinationof manganese in different types ofmatrices includingatomic emission spectrophotometry, atomic absorption spectrometry [2, 3], laser induced breakdown spectroscopy[4], ion chromatography[5],electroanalyticalmethod [6]and spectrophotometry[7, 8]. Spectrophotometric methods occupy special positiondue to their simplicity, less expensive instrumentationand high sensitivity.Similarly, spectrophotometric methods also play an important role in manganese determination. The literatures shows that aromatic amines, azo dyes, oximes, porphyrins and other reagents have been used for the colorimetric determination of manganese. The most significant reagents are permanganate, formaldoxime and 1-(2-pyridylazonaphthol),which is far more sensitive than the other two reagents. This present method was based on theoxidation-reductionreaction between ammonium persulfate and manganese(II). Employing ammonium persulfate as an oxidant, manganese(II) was oxidized to manganese(VII) in the presence of silver nitrate as a catalyst at the room temperature.The characteristic wavelength of maximum absorption was measured at 530 nm. Under the optimum reaction conditions the absorption value (A) was proportional to the concentration of manganese in the range of 0.50%~1.80%(mass fraction), and the relative standard deviation (RSD) was less than 3.0% (n=5). The proposed method was free from the interference from a large number of analytical important elements and has been applied satisfactorily to the determination of manganesein end-fitting of spoolable reinforced plastic line pipe samples with the improved accuracy and precision.Results were in agreement with those attained by inductivelycoupled plasma atomic emission spectrophotometry at the 95% confidence level. The satisfactory performance in the determination of manganese in the end-fittingof spoolable reinforced plastic line pipe demonstrated that the method was practical and suitable not only for quality control analysis but also for product analysis, confirming the promise for spoolable reinforced plastic line pipe research.

2. Experimental

2.1. Apparatus
A UV-visible spectrophotometer (Model TU-1901, Beijing Persee General Instrument Co. Ltd.) with 1.0-cm quartz cells was employed for all absorbance measurements.

2.2. General Procedures
Weigh accurately 0.1000 g of drilled low alloy steel samples and dissolve the samples in 20 mLmixture acid solution of sulphuric acid 15% (V/V) and phosphoric acid 15% (V/V)in a 150-mL taper bottle. Continue the treatment with mixture acid till the sample goes into solution and heat to fumes of oxides of nitrogen, sulphuric acid and evaporate almost to dryness. Cool the solution, add 50 mLdeionised water to dissolve some materials. Then, add 5 mL of 5.0 mg·mL^{-1}solution of silver nitrate, 10 mL of 200.0 mg·mL^{-1} solution of ammonium persulfate in the taper bottle to oxidate manganese(II) to manganese(VII) oxidation states.Boil the solution to no bubble and cool the solution at room temperature. Finally, dilute the solution ina 100-mL volumetric flask, followed make up to the mark with deionised water, mix well and let stand for 10 min. Then, transfer the above mulberry stock solution in 1.0-cm quartz cells. Add 5.0 mg·mL^{-1}solution of sodium nitrite to the remaining solution reduce manganese(VII) to the lower oxidation states, this transparent solution was employed as the blank solution in subsequent experiments. Measure the absorbance at 530 nm using a UV-visible spectrophotometer with 1.0-cm quartz cells against a reagent blank prepared similarly. Plot the amount of manganese in the sample solution against absorbance to obtain the calibration graph.

3. Results and Discussion

3.1. Spectral Characteristics
The absorption spectra of the reagents blank, manganese(VII) against the reagent blank were tested from 400 nm to 700 nm. It could be seen from Figure 2, the manganese(VII) yields the characteristic wavelength of maximum absorption at 530 nm wavelength. However, the absorption spectrum of the reagents blank under similar conditions shown that it does not absorb at 530 nm. And the experimentation also demonstrated that manganese(VII) increases the absorbance value considerably, resulting in increased sensitivity of the method. Therefore, measure the absorbance of manganese(VII) at 530 nm.

3.2. Select Dissolved Acid for Low Alloy Steel Samples
The effect of the different acids for dissolve the low alloy steel samples was examined. It was found that a solution mixed with 15% (V/V) sulphuric acid and 15% (V/V) phosphoric acid has a good effect for dissolve the low alloy steel samples. Thus, an acid solution mixed with 15% (V/V) sulphuric acid and 15% (V/V) phosphoric acid was recommended for dissolve the low alloy steel samples in subsequent studies.

3.3. Influence of the Amounts of Oxidant
Ammonium persulfate was selected as the oxidant in the proposed method. The effect of the oxidant was examined by measuring the absorbance of solution containing certain amounts of manganese(II) and variable amounts of ammonium persulfate. It was found that 10 mL of 200.0 mg·mL^{-1} solution of ammonium persulfate sufficed to oxidate the amounts of manganese(II) taken, with higher concentrations the absorbance was essentially constant. Ten millilitres of 200.0 mg·mL^{-1}ammonium

persulfate solution were recommended as a suitable amount of oxidant.

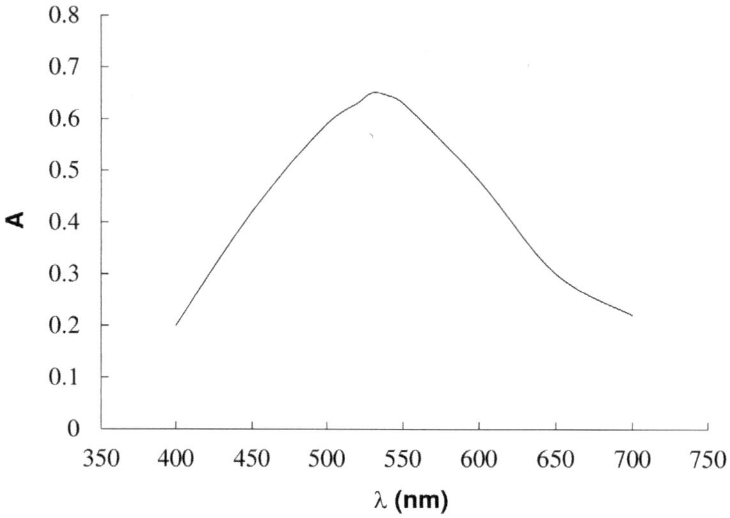

Figure 2 Absorbance-wavelength profile

3.4. Influence of the Amounts of Silver Nitrate
The oxidation-reduction reaction between manganese(II) and ammonium persulfate withouta catalyst was slowly and not fully. Selected the silver nitrate as a catalyst for the oxidation-reduction reaction between manganese(II) and ammonium persulfate. It has been further observed that the oxidation-reduction reaction between manganese(II) and ammonium persulfate occurs immediately even at room temperature if silver nitratewas used. Therefore, 5 mL of 5.0 mg·mL^{-1} solution of silver nitrate was employed in subsequent experiments.

3.5. Influence of the Reductant
Sodium nitrite was selected as the reductant in the proposed method. 5.0 mg·mL^{-1}solution of sodium nitrite reduce manganese(VII) to the lower oxidation states manganese(II). The transparent solution was employed as the blank solution in subsequent experiments.

3.6. Sequence of Addition of Reagents
From experiments in which sulphuric acid, phosphoric acid, ammonium persulfate and silver nitrate were added in all possible orders. It was found that there was no appreciable change in the colour of the solution and the variations in the absorbance values. These phenomenons demonstrate that the oxidation-reductionreaction between manganese(II) and ammonium persulfate was independent of the order of addition.

3.7. Preparation of the Calibration Graph
Under the optimal conditions, a series of standard solutions of manganese were prepared by the different steel certified reference materials in a 100-mL volumetric flask. Add 5 mL of 5.0 mg·mL^{-1} solution of silver nitrate, 10 mL of 200.0 mg·mL^{-1} solution of ammonium persulfate followed make up to the mark with deionised water, mix well and let stand for 10 min. Measure the absorbance at 530 nm using a UV-visible spectrophotometer with 1.0-cm quartz cells against a reagent blank. The absorbance of the solution was proportional to the concentration of manganese in the range of 0.50%~1.80% (mass fraction). Plot the amount of manganese in the sample solution against absorbance to obtain the calibration graph. The linear regression equation for manganese was $A = 0.41837C - 0.01073$, $R^2 = 0.9997$, and the relative standard deviation (RSD) was less than 3.0% (n=5).

3.8. Interference Studies

The extent of interference by diverse ions was determined by measuring the absorbance of solutions containing manganese and various amounts of diverse ions. The criterion for interference was an absorbance value varying by more than +2% from the expected value of manganese alone. The results show that a large excess of cations and anions which are usually associated in the determination of manganese do not interfere. The colorless metal ions do not interfere. The tolerable amounts of foreign species with respect to manganese for interference at +2% level were over 200~300 for the colored metal ions. When the colored metal ions contents of not more than 200~300 times manganese, it can be in some color solution adding a few drops of EDTA solution or sodium nitrite solution to fade the color, this solution as the measured reference solution to eliminate the interference. There was not obviousinterference ions exist in end-fittingof spoolable reinforced plastic line pipesamples.

4. Applications

4.1. Determination of Manganese in End-Fitting

The recommended procedure has been applied satisfactorily to the determination of manganese in end-fittingof spoolable reinforced plastic line pipe samples. The samples were dissolved according to the following procedure: Dissolve 0.1000 g of drilled end-fittingof spoolable reinforced plastic line pipe samplesin 20 mL mixture acid solution of sulphuric acid 15% (V/V) and phosphoric acid 15% (V/V) in a 150-mL taper bottle. Continue the treatment with mixture acid till the sample goes into solution and heat to fumes of oxides of nitrogen, sulphuric acid and evaporate almost to dryness. Cool the solution, add 50 mL deionised water to dissolve some materials. Then, add 5 mL of 5.0 $mg \cdot mL^{-1}$ solution of silver nitrate, 10 mL of 200.0 $mg \cdot mL^{-1}$ solution of ammonium persulfate in the taper bottle tooxidate manganese(II) to manganese(VII)oxidation states. Boil the solution to no bubble and cool the solution at room temperature.Finally, dilute the solution in a 100-mL volumetric flask, followed make up to the mark with deionised water, mix well and let stand for 10 min. Then, transfer the above mulberry stock solution in 1.0-cm quartz cells. Add 5.0 $mg \cdot mL^{-1}$solution of sodium nitrite to the remaining solution reduce manganese(VII) to the lower oxidation states, this transparent solution was employed as the blank solution in subsequent experiments. Measure the absorbance at 530 nm using a UV-visible spectrophotometer with 1.0-cm quartz cells against a reagent blank.The results were summarized in Table 1 with the good accuracy and precision. To verify the results obtained by the proposed method, inductivelycoupled plasma atomic emission spectrophotometry method was applied to determine the samples according to the literature and the results obtainedby the proposed method agreed well with thatfrom the reference method.

5. Conclusions

Based on the oxidation-reduction reaction between manganese(II) and ammonium persulfate, a simple and accurate spectrophotometric method for determination of manganese in the end-fittingof spoolable reinforced plastic line pipe was proposed. The purple-red colored manganese(VII) solution exhibits absorption maximum at 530 nm. The proposed method was free from the interference from a large number of analytical important elements and has been applied satisfactorily to the determination of manganese in end-fittingof spoolable reinforced plastic line pipe samples. Compared with other methods for the determination of manganese, this method offers advantages of simplicity, less expensive instrumentationand high efficient. The satisfactory performance in the determination of manganese in end-fittingof spoolable reinforced plastic line pipedemonstrated that the method was practical and suitable not only for quality control analysis but also for product analysis, confirming the promise for spoolable reinforced plastic line pipe research.

6. Acknowledgments

The authors gratefully acknowledge the *State Key Laboratory for Performance and Structure Safety of Petroleum Tubular Goods and Equipment Materials*, and *CNPC Tubular Goods Research Institute*, and *China National Quality Supervision, Testing and Inspection Center of Oil Tubular Goods*.

Table 1 Results of determination of manganese in end-fitting[a].

Sample No.	Size (mm)	*A*	RSD (%)	Content (*ω*, %)	
				By the proposed method	By AES
1	DN75	0.4955	1.05	1.21 ± 0.01	1.23 ± 0.02
2	DN75	0.4997	1.88	1.22 ± 0.01	1.22 ± 0.04
3	DN75	0.4997	1.65	1.22 ± 0.03	1.20 ± 0.03
4	DN75	0.4955	1.79	1.21 ± 0.01	1.21 ± 0.01
5	DN75	0.5039	2.08	1.23 ± 0.02	1.23 ± 0.03
6	DN65	0.4327	1.04	1.06 ± 0.02	1.06 ± 0.02
7	DN65	0.4286	0.57	1.05 ± 0.01	1.07 ± 0.03
8	DN65	0.4327	0.67	1.06 ± 0.02	1.10 ± 0.02
9	DN65	0.4369	2.32	1.07 ± 0.03	1.06 ± 0.01
10	DN65	0.4327	1.43	1.06 ± 0.02	1.05 ± 0.03
11	DN90	0.6043	2.54	1.47 ± 0.03	1.47 ± 0.02
12	DN90	0.6001	2.86	1.46 ± 0.02	1.44 ± 0.04
13	DN90	0.6085	0.49	1.48 ± 0.01	1.50 ± 0.01
14	DN90	0.6085	1.25	1.48 ± 0.02	1.48 ± 0.03
15	DN90	0.6126	1.63	1.49 ± 0.03	1.48 ± 0.01

Notes: [a]The average of five determinations.

7.Reference

[1] George M 2013 *Reinforced Plastics***57**26
[2] SahinC A, EfecınarM, NurayS 2010 *Journal of Hazardous Materials***176**672
[3] Ricardo J C, Luis G T, Ricardo ES, Eliane PO 2011 *Talanta***85**415
[4] Gonzaga FB,Pasquinia C2008*SpectrochimicaActa Part B: Atomic Spectroscopy***63**56
[5] Anna B, Wojciech D, Suceil S, Apeksha D, Rajinder R, Grażyna O, Remigiusz B 2010*Journal of Chromatography B***878**34
[6] Piech R, Baśa B,Kubiaka W W2008*Journal of Electroanalytical Chemistry***621**43
[7] Gouda AA 2014 *SpectrochimicaActa Part A: Molecular and Biomolecular Spectroscopy***131**138
[8] Andrew SM, Bradley MT,George W 2011*Talanta***84**374

IOP Conf. Series: Materials Science and Engineering **250** (2017) 012002 doi:10.1088/1757-899X/250/1/012002

First-Principles Study on Solution Strengthening Effect of Cu and Zn in Ag Alloy

Yongtai Chen[1,2], Jieqiong Hu[1,2], Youcai Yang [3], Jiming Zhang [3], Song Chen [1,2], Saibei Wang [1,2] and Song Wang[2], Ming Xie [1,2,4]

[1] School of Material Science and Engineering, Kunming University of Science and Technology, Kunming, Wenchang road 68, China.
[2] Kunming Institute of Precious Metals, Kunming, Keji Road 988, China.
[3] State Key Laboratory of Advanced Technologies for Comprehensive Utilization of Platinum Metals, Sino-platinum Metals Co. Ltd., Kunming, Keji Road 988,China.
[4] Corresponding author, email: cyt@ipm.com.cn

Abstract: A first-principles plane-wave pseudopotential method based on the density functional theory was used to investigate the mechanical properties and electronic properties of AgCu and AgZn solid solution. The VCA models have been set up for solid solution, and elastic constants, bulk modulus and shear modulus, Young modulus and density of states of Cu or Zn doped Ag system were also analysed. The results show that Cu and Zn in Ag alloy play a role of solution strengthening, the solid solution strengthening effect was obviously improved by adding a small amount of Cu(less than 2at.%) or Zn(less than 1at.%). The total DOS of AgCu(Zn) Solid solution is mainly constituted by the Ag 4d and Cu(Zn) 4d states. The AgZn Solid solution has a more stable structure than AgCu solid solution.

1.Introduction

Ag-Cu-Zn alloys are widely used for sliding electrical contact materials, jewelry, coins, silverware[1-3]due to their high hardness and strength and good performances of electrical conductivity, thermal conductivity and wear resistance. Nowadays, Ag-Cu-Zn alloys researches are mainly focused on the effects of Zn on microstructure and mechanical properties of materials, corrosion resistance, anti-tarnishing and strengthening mechanism [4-8] etc. However, many properties (including elastic properties, as well as electronic properties) of AgCu and AgZn solid solution still need to be understood at an atomic level. The maximum solid solubility of Cu and Zn in Ag is about 14 and 30wt, respectively. In the practical application, the content of Cu is 3-6wt. % and Zn is 1-2wt. % while the Ag-Cu-Zn alloys used for electrical contact materials, jewelry and coins. Here, we performed first principles calculations to investigate the elastic properties and electronic properties for AgCu and AgZn solid solution added less than 10at. % Cu and 10at. % Zn, in order to understand the solution strengthening effect of Cu and Zn on Ag Alloy.

2. Details of the Calculations

2.1 Computational Method

The crystal structure of metal Ag and Cu is the face-centered (FCC), and Zn is the hexagonal close packed (hcp),The cell model of face centered cubic crystals were build based on the virtual crystal approximation (VCA), the adding atoms (Cu and Zn) and Ag atoms are indefinite position, each position can be viewed as a mixture of two kind atoms, the atomic percentage show the probability of one kind atom appearing in that position .The VCA method was widely used to calculate the structure properties of disordered solid solution and ordered solid solution. The mass fraction of Cu and Zn added in Ag alloy are from 0 to 10at. %, the crystal cell of Ag-Cu and Ag-Zn alloy as shown in Figure 1.

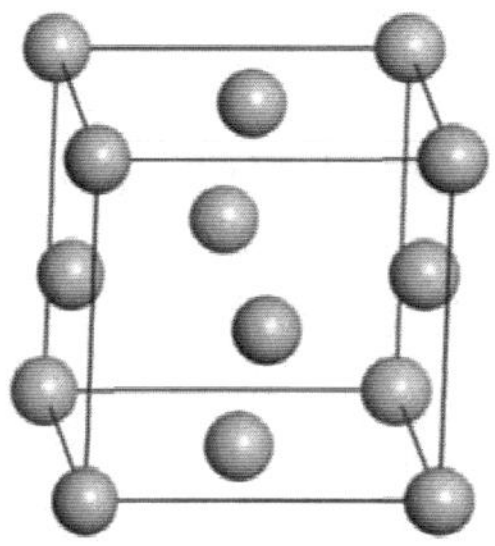

Figure 1. Crystal cell of Ag-Cu and Ag-Zn alloy

In order to research the electronic properties, the 2X2X2 supercell of substitution solid solution of Cu and Zn in Ag was structured, as shown in Figure 2.

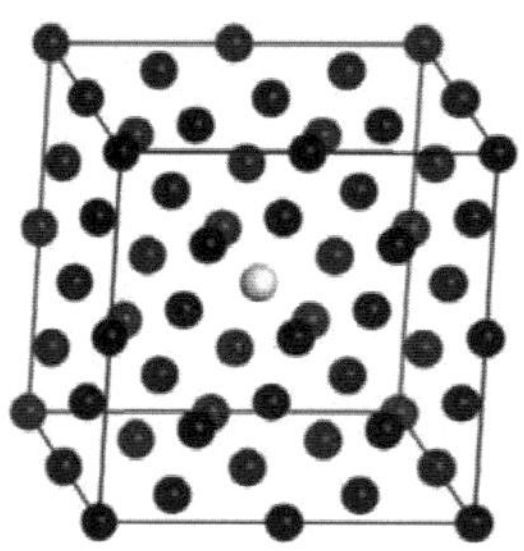

Figure 2. Crystal cell of substitution solid solution of Cu and Zn in Ag

2.2 Calculation Methods

All calculations were performed with Cambridge serial total energy package (CASTEP) code based on the first-principles method within the framework of density function theory (DFT).Through calculating in many times and the calculation results were compared with experimental data (such as lattice), Finally, the calculation parameters were determined as follows: the periodic boundary condition was used in the whole process. Interactions of electrons with an ion core were represented by the Vanderbilt-type ultrasoft pseudopotential. The electronic exchange-correlation energy was treated under the generalized gradient approximation (GGA-PBE). The plane-wave basis set cutoff was 460 eV for all calculations. The special k-points sampling integration over the Brillouin zone was employed by using the Monkhorst-Pack method with a $10\times10\times10$ special k-point mesh. The energy tolerance is 5×10^{-7}eV/atom. These parameters were sufficient to lead to convergences of total energy and geometrical configuration. Before calculation, the Cell Geometry was optimized by Broyden-Fletcher-Goldfarb-Shanno (BFGS) method to obtain the local stable structure.

3. Results and Discussion

3.1Mechanical Properties

The elastic constants of solids are closely related to its mechanical properties. It can be used to characterize the response of crystals to external stresses and as the interaction between atoms standard. Therefore, to research the elastic constants of Ag-Cu and Ag-Zn solid solution alloys is of great importance to understanding the mechanical properties of alloys. In the case of cubic crystals, there are three independent elastic constants (C_{11}, C_{12} and C_{44}). According to the Voigt-Reuss-Hill arithmetic approximation, the bulk modulus (B), shear modulus (G) and Young's modulus (E) can be obtained from the following equation[9]:

$$B_0 = B_V = B_R = \frac{C_{11}+2C_{12}}{3} \quad (1)$$

$$G_V = \frac{C_{11}-C_{12}+3C_{44}}{5} \quad (2)$$

$$G_R = \frac{5C_{44}(C_{11}-C_{12})}{4C_{44}+3(C_{11}-C_{12})} \quad (3)$$

$$G = \frac{G_V+G_R}{2} \quad (4)$$

$$E = \frac{9GB}{3B+G} \quad (5)$$

Where B_V is Voigt elastic modulus, B_R is Reuss elastic modulus, G_V and $G_R$are Voigt shear modulus and Reuss shear modulus, respectively.

The calculation result of the elastic constants, bulk modulus, shear modulus and Young's modulus of Ag-Cu and Ag-Zn are showed in Table 1 and Table 2. It can be seen that the mechanical constants of pure Ag are close to the calculated value of ref.[10] and experimental values of ref.[11], so the calculation result is credible.

Table 1. Mechanical property parameters of Ag-Cu alloys (GPa)

Composition	C_{11}	C_{12}	C_{44}	B	G_V	G_R	G	E
Ag	131.67	80.43	58.41	97.51	45.29	38.63	41.96	110.09
Ag-1Cu	147.48	89.05	61.84	108.52	48.79	42.75	45.77	120.38
Ag-2Cu	157.66	95.29	59.13	116.08	47.95	43.53	45.74	121.29
Ag-3Cu	157.91	99.87	54.39	119.22	44.24	40.30	42.27	113.40
Ag-4Cu	157.39	103.20	50.72	121.26	41.27	37.60	39.44	106.74
Ag-5Cu	157.87	106.71	47.94	123.77	39.00	35.52	37.26	101.58
Ag-6Cu	158.60	108.44	46.57	125.16	37.97	34.68	36.33	99.37
Ag-7Cu	159.81	110.40	45.77	126.87	37.35	34.13	35.74	98.01
Ag-8Cu	160.69	110.62	45.57	127.31	37.35	34.31	35.83	98.28
Ag-9Cu	160.35	111.57	45.49	127.83	37.05	33.80	35.42	97.29
Ag-10Cu	160.11	111.92	45.87	127.98	37.16	33.69	35.43	97.30

Table 2. Mechanical property parameters of Ag-Zn alloys (GPa)

Composition	C_{11}	C_{12}	C_{44}	B	G_V	G_R	G	E
Ag	131.67	80.43	58.41	97.51	45.29	38.63	41.96	110.09
Ag-1Zn	144.78	84.02	59.71	104.27	47.98	43.08	45.53	119.23
Ag-2Zn	143.11	85.33	59.60	104.59	47.32	41.82	44.57	117.07
Ag-3Zn	135.80	83.47	57.97	100.91	45.25	39.01	42.13	110.95
Ag-4Zn	130.09	84.30	56.78	99.56	43.23	35.67	39.45	104.53

Ag-5Zn	130.23	83.04	51.08	98.77	40.09	34.84	37.47	99.78
Ag-6Zn	128.77	81.41	52.83	97.20	41.17	35.40	38.29	101.53
Ag-7Zn	113.25	85.07	47.59	94.46	34.19	24.39	29.29	79.64
Ag-8Zn	108.92	79.43	40.14	89.26	29.98	23.77	26.87	73.27
Ag-9Zn	99.38	81.39	36.74	87.39	25.64	16.45	21.04	58.44
Ag-10Zn	94.91	79.44	35.25	84.60	24.24	14.55	19.40	54.06

The bulk modulus is usually used to describe the ability of materials on resistance to volume deformation under external force. The larger bulk modulus means that the volumetric deformation resistance of crystal is stronger. Shear modulus is a measure of the resistance to form change of crystal. It can reflect the anti-plastic-deform capacity of materials. It was suggested that the Young's modulus E and shear modulus G have close relations with the hardness of materials because it has a strong correlation with the cohesion energy or binding energy of atoms in crystal. It indicates that the higher the Young's modulus E and shear modulus G, the higher the hardness of materials. The hardness and shear modulus between different materials are not the exactly same, but generally speaking, with B, G and E increase, the hardness of the material increase accordingly.

Figure 3 is curve which reflect the change of B, G and E with the copper content in Ag-Cu solid solution alloy,it can be seen from the graph that the bulk modulus increases with the increase of copper content, the bulk modulus increase sharply atfirst, when the copper content is above 8at.%, the bulk modulus still rise but the trend becomes very slow, it shows that the processing difficulty increases with the increase of copper content, as well as ,the higher the Cu content, the higher the resistivity, it seriously affect the performance of electrical contact materials. In practical application, in order to obtain higher conductivity,the content of added elements is minimized under the condition of guaranteeing strength. For AgCu electrical contact materials that we commonly used, such as AgCu4Ni0.5. AgCu5, AgCu6Zn, the Cu content of them are about 6wt. % or less than 6wt. %,it also shows that the calculated results are in good agreement with the practical application, the physical model and parameters of simulation is reasonable, the result is credible.

It also can be seen from Figure 3, that the Young's modulus and shear modulus first increase then decreasewith the increase of Cu content, when Cu content rise to 2%at. %, the two modulus reach a maximum value and then drops down sharply.It shows that adding a small amount of Cu (2at. %) in Ag can effectively improve the strength and hardness of Ag, increasing the amount of Cu results in performance decreased instead.

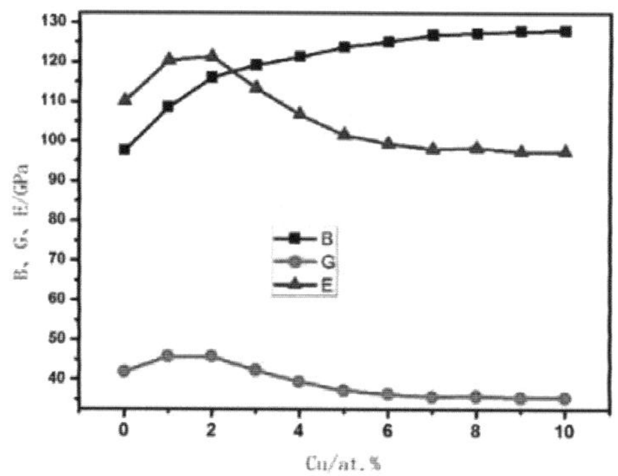

Figure 3. The relationship between Young's modulus (E), shear modulus (G) and bulk modulus (B) with the content of Cu.

The relationship between Young's modulus (E), shear modulus (G) and bulk modulus (B) with the content of Zn as shown in Figure 4.It can be seen that the three elastic modulus first increase then decreasewith the increase of Zn content, when Zn content rise to 1%at. %, the three elastic modulus reach a maximum value and then drops down sharply.It shows that adding a small amount of Zn

(1at. %) in Ag can effectively improve the strength and hardness of Ag, such as the Zn Content of wear-resisting silver based electrical contact material (AgCu6Zn1Ni0.5) and 925Ag are about 1wt. %, increasing the amount of Zn results in performance degradation. So the actual application is in good agreement with the calculation results.

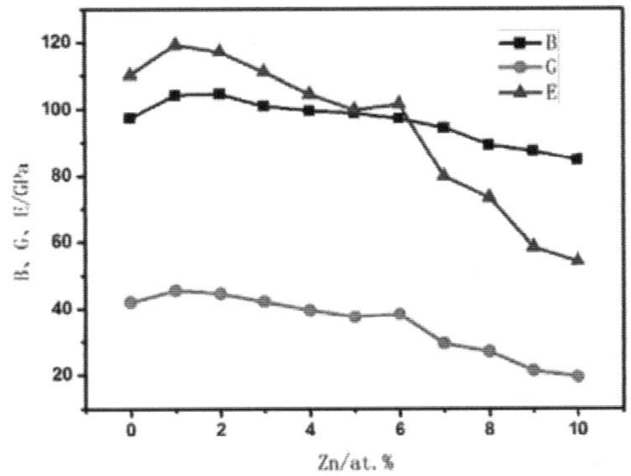

Figure 4. The relationship between Young's modulus (E), shear modulus (G) and bulk modulus (B) with the content of Zn.

3.2 Electronic Properties

Almost all the macro-properties of materials, such as hardness, elasticity, and conductivity, originate from their electronic structure properties as well as the nature of the chemical bonding. Therefore, it is necessary to perform the electronic structure analysis of AgCu Solid solution and AgZn Solid solution. The DOS is an important quantity for understanding the bonding in a compound. After geometry optimization, the total and partial DOS at equilibrium lattice constants forAgCu Solid solution and AgZn Solid solutionare calculated, as shown in Figure 5. In Figure 5(a) it is shown that the total DOS of AgCu Solid solutionis mainly constituted by the Ag 4d and Cu 4d states.In Figure 5(b) it is shown that the total DOS of AgZn Solid solutionis mainly constituted by the Ag 4d and Ag 4p and Zn 4d states.The density of states at the Fermi level, N(EF) is oftenused as a good indicator for the structural stability, and thevalues are 7.6 and 6.5 state·eV^{-1} perSupercellfor AgCu Solid solution and AgZn Solid solution, respectively, so the AgZn Solid solution has a more stable structure than AgCusolid solution ,thisconclusion is in agreement with the phase diagram of Ag-Cu and Ag-Zn, that the solid solubility of Zn in Ag is much larger than that of Cu in silver.

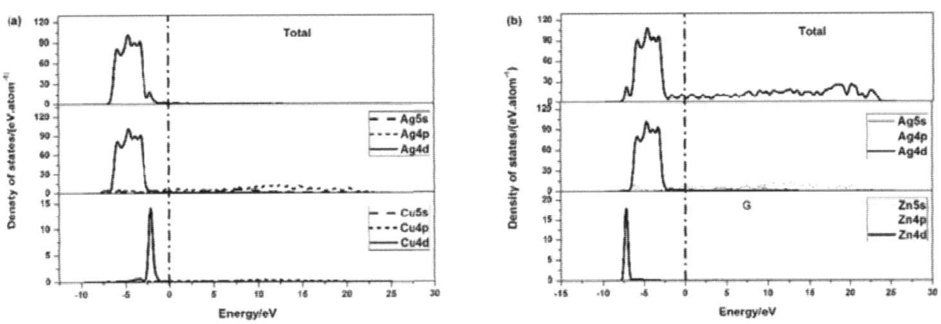

Figure 5. Calculated total and partial density of states for Solid solution(a)Ag31Cu,(b)Ag31Zn

4. Conclusions

(1)Cu and Zn in Ag alloy play a role of solution strengthening, the solid solution strengthening effect was obviously improved by adding a small amount of Cu(less than 2at.%) or Zn(less than 1at.%).

(2)The total DOS of AgCu(Zn) Solid solutionis mainly constituted by the Ag 4d and Cu(Zn) 4d states. The AgZn Solid solution has a more stable structure than AgCusolid solution.

5.Acknowledgement

The present work was supported by the National Natural Science Foundation of China (No.U1302272, U1602271, U1602275), Talent project of Yunnan (2016HB026), School science and technology cooperation project（2015IB012）

6. References

[1] Editor Team of Precious Metal Materials Process Handbook.Precious Metal Materials Machining Manual [M].Beijing:Metallurgical Industry Press, 1978: 14 (in Chinese).
[2]Yu Chaoqing, Wang Liqiong. *Small & Special Machines*[J], 2003(2): 31 (in Chinese).
[3]Lei Zhuo, Bai Xiaojun, Xiang Xiongzhi et al. *MaterialsReview* [J], 2006, 20(5): 434 (in Chinese)
[4]Bo Hairui, Yuan Junping, Zhou Yongheng et al. *Special Casting & Nonferrous Alloys* [J], 2011, 31(7): 680 (in Chinese).
[5] Yang Qingqing, XiongWeihao, Shen Dingjie et al. *Rare Metal Materials and Engineering* [J], 2008, 37(6): 947(in Chinese).
[6]YangQingqing, XiongWeihao, Yang Zhen et al. *Rare Metal Materials and Engineering* [J], 2010, 39(4): 578(in Chinese).
[7]Eccles P A. US Patent, US 6726877[P], 2004.
[8]Tang Youzheng, Xiang Xiongzhi, Bai Xiaojun et al. *RareMetal Materials and Engineering* [J], 2012, 41(5): 929 (in Chinese).
[9] Rong Yang,Bin Tang,Tao Gao. *Zeitschrift für Naturforschung A*[J],2017,72(6): 527-534.
[10]Wang Y， Aurtarolo S， Jiang C. *Ab initio lattice stabilityin comparison with CALPHAD lattice stability* [J].Calphad， 2004， 28: 79-90.
[11] Kittel C. Introduction to Solid State Physics [M]. 6th ed. New York: John Wiley&Sons Inc， 1986: 55-144.

Protection of 310l Stainless Steel from Wear at Elevated Temperatures using Conicraly Thermal Spray Coatings with and without Sic Addition

Yan Zhang[1,2], Tao Zhang[1,2], Kaiyang Li[2] and Dongyang Li[3]

[1]Jiangsu Key Laboratory of Advanced Structural Materials and Application Technology, Nanjing, 211167, China
[2]School of Materials Science and Engineering, Nanjing Institute of Technology, Nanjing, 211167, China
[3]Dept. of Chemical &Materials Engineering, University of Alberta, Edmonton, AB, Canada T6G 1H9
E-mail: Zhangyan3@njit.edu.cn

Abstracts: Due to its high oxidation resistance, 310L stainless steel is often used for thermal facilities working at high-temperatures. However, the steel may fail prematurely at elevated temperatures when encounter surface mechanical attacks such as wear. Thermal spray coatings have been demonstrated to be effective in protecting the steel from wear at elevated temperatures. In this study, we investigated the effectiveness of high velocity oxy-fuel(HVOF) spraying CoNiCrAlY/SiC coatings in resisting wear of 310L stainless steel at elevated temperature using a pin-on-disc wear tester. In order to further improve the performance of the coating, 5%SiC was added to the coating. It was demonstrated that the CoNiCrAlY/SiC coating after heat treatment markedly suppressed wear. However, the added SiC particles did not show benefits to the wear resistance of the coating. Microstructures of CoNiCrAlY coatings with and without the SiC addition were characterized in order to understand the mechanism responsible for the observed phenomena.

1. Introduction

310L stainless steel containing high contents of Cr and Ni, possesses many attractive properties, such as excellent oxidation resistance, corrosion resistance, and creep strength. It can be used as heat-resistant steel and can continuously operate at high temperature because of favorable high-temperature resistance. However, they may fail prematurely at elevated temperatures when encounter surface mechanical attacks such as hot crack, corrosion and wear [1, 2]. Considerable efforts have been made to improve the high-temperature performance of the steel. Various surface modification techniques have been conducted, such as aluminized coating, laser coating, nano-crystalline coating, and so on[3-5]. MCrAlY(M=Co、Ni or Co+Ni) coating was widely used for high-temperature oxidation resistance coating due to its favorable high-temperature oxidation resistance, hot-corrosion resistance and better associability with base metal[6]. The high content of Cr can provide hot corrosion resistance, the contents between 5~15wt.% of Al can provide oxidation resistance, and Y addition around 1wt.% maintain stability during cyclic oxidation[7]. As one of the best coatings for high temperature oxidation resistance, CoNiCrAlY alloy was selected. But its hardness is relatively lower, so is wear resistance at ambient and elevated temperatures [8-11]. As we all know, SiC powder has high hardness and excellent chemical stability, and it is often selected as reinforcement phase. In this study, a small content of nano SiCp has been added to coatings for wear and corrosion application. The coating is provided with the high toughness and ductility of metallic binder phase and the high hardness and wear resistance of ceramic-carbide.

Comparing with other coating techniques such as APS and LPPS, HVOF can deposite CoNiCrAlY bond coat with lower extent of oxide contaminants[12]. This work converted the conventional two step spray (to deposite bond layer and functional layer) to once, and the coating consists of CoNiCrAlY and SiC. We studied the microstructure and wear and friction properties of the CoNiCrAlY/SiC coat formed using HVOF.

2. Materials and Methods

2.1. Coating and Substrate Metal

Commercial 310L stainless steel (C 0.08%, Si 1.3%, Mn 1.6%, P 0.03%, S 0.025%, Ni 20%, Cr 25%) was used as the base metal in 3mm thickness. CoNiCrAlY and CoNiCrAlY–SiC powder were employed as the feedstock to produce the coatings respectively, supplied by Sulzer Metco Ltd and Van composites research Institute. The size distribution of CoNiCrAlY is -38+5.5μm, the chemical composition of CoNiCrAlY is Ni 32%，Cr21%，Al 8%，Y 0.5%，Bal Co. The average diameter of SiC is 50nm, purity ≥99.9%, apparent density≤0.09 g/cm3，density 3.2 g/cm3, specific surface area 60 g/cm2,cubic crystalline. The feedstock powder was the mixture of CoNiCrAlY and SiC,

2.2. Coating Procedures

The mixture of SiC and CoNiCrAlY were homogenized by Pulverisett-6 tank type high-energy ball mill. High performance stainless steel vials and balls were used as the milling components. The 310s substrates were roughed by a grit blasting machine, cleaned and dried. The CoNiCrAlY/ SiC powder was taken in a JP5000 high-velocity oxygen fuel spraying facility operated with a spray distance of 350- 380 mm, fuel pressure 120MPa, flow velocity of powder 50-60 g/min, N_2 pressure of powder feeder 10kPa, flow velocity of O_2 1800psi, and flow velocity of N_2 21scfh.

Sprayed coatings were heat-treated in a vacuum hot pressing furnace (ZRVS1500), and the parameters were: heating time 110min, at 1050°C for 2h, and vacuum degree 1×10-3Pa.

2.3. Coating Microstructure

X-ray diffraction analysis was made to determine phases on the coating using a Siemens diffractometer (D5005) with Cu Ka radiation. The depth of XRD analysis is decided by the diffraction angle and material. According to the 3~15μm of Cr_3C_2–CoNiCrAlY coatings[13], the consequently received results is about the phase structure of the superficial layers of the coatings.

The coatings' microstructure was characterized using SEM and EDS with a Vega-3 TESCAN Scanning Electron Microscope at 20kV voltage and an EDXS oxford instrument was employed to analyze local compositions.

2.4. Coating Hardness

Micro-hardness measurements were conducted on a polished cross-section of the coating under a load/force of 300gf for 15s using a FM-700 Micorhardness tester with a Vickers diamond indenter. Every location was performed five indents. 5 data points were measured and averaged for each hardness value. Hardness was tested from base metal to coating. The surface of the coated samples' cross-section was mechanically polished, finished with a 1-μm diamond paste, cleaned and dried.

2.5. Coating Wear Resistance

Dry Sliding wear tests were performed on a pin-on-disc tribometer (CSEM Instruments CH-2007, Neuchatel, Switzerland) at ambient with a relative humidity of approximately 50% and 500° C temperature. The pin was a silicon nitride (Si3N4) ball, with a diameter of 6mm and the disc was the sample under study. All tests were carried out at a sliding speed of 1cm/s along a circle path in diameter under a normal load 10N for 72m. The friction coefficient curves were recorded continuously during the test process. The wear track and volume loss of the samples was calculated by measuring crosssectional area of the wear scars with a surface profilometer (Form Talysurf PGI 1240, Taylor Hobson) Each reported value is an average of results received from at least three tests.

3. Results and Discussion

3.1. Morphology and Microstructure of the Powders

The morphology of the CoNiCrAlY and CoNiCrAlY+5wt.%SiC powders is shown in Fig. 1a and b, respectively. The powder is both predominantly spherical. The particle surface is smooth in CoNiCrAlY powders, and irregular and fuzzy in the SiC-CoNiCrAlY powders.

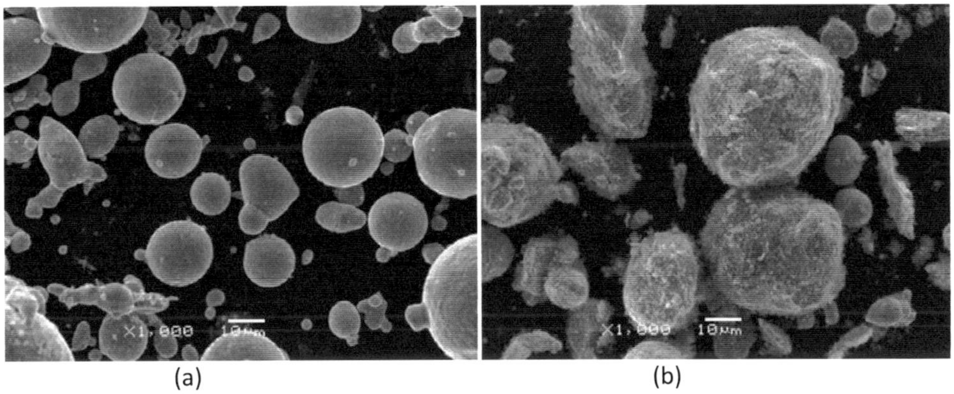

Figure 1. SEM micrograph of feedstock: (a) CoNiCrAlY and (b) CoNiCrAlY+5wt.%SiC

Fig. 2 shows the phase structures of the two powders, which exhibit the comparison of XRD patterns for the CoNiCrAlY and mixed CoNiCrAlY/SiC spraying powder. In the both powder, Co-Ni-Cr matrix and CoAl phase were the major phases. These results are in agreement with the previous results published by A.S. Khanna and W.S. Rathod in the study of CoNiCrAlY oxidation resistant hard coatings[14]. SiC was also detected in CoNiCrAlY/SiC powder.

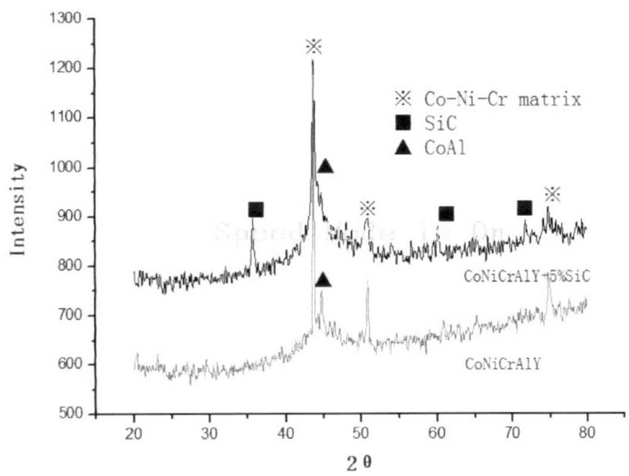

Figure 2. XRD pattern of the two powders

3.2. Microstructural Characterization of the Coatings

The BSE images of the cross-section of the as-sprayed HVOF CoNiCrAlY and CoNiCrAlY+5%SiC coating are shown in Fig. 3. Under the same spraying parameters, the coating of CoNiCrAlY+5%SiC coating is thinner than CoNiCrAlY coating. The most significant difference is a transition zone, about

IOP Conf. Series: Materials Science and Engineering **250** (2017) 012003 doi:10.1088/1757-899X/250/1/012003

50μm width, rich of Co, Ni, Cr and Al in the coating of CoNiCrAlY. Element Fe in the steel matrix diffused to the coating in both coatings.

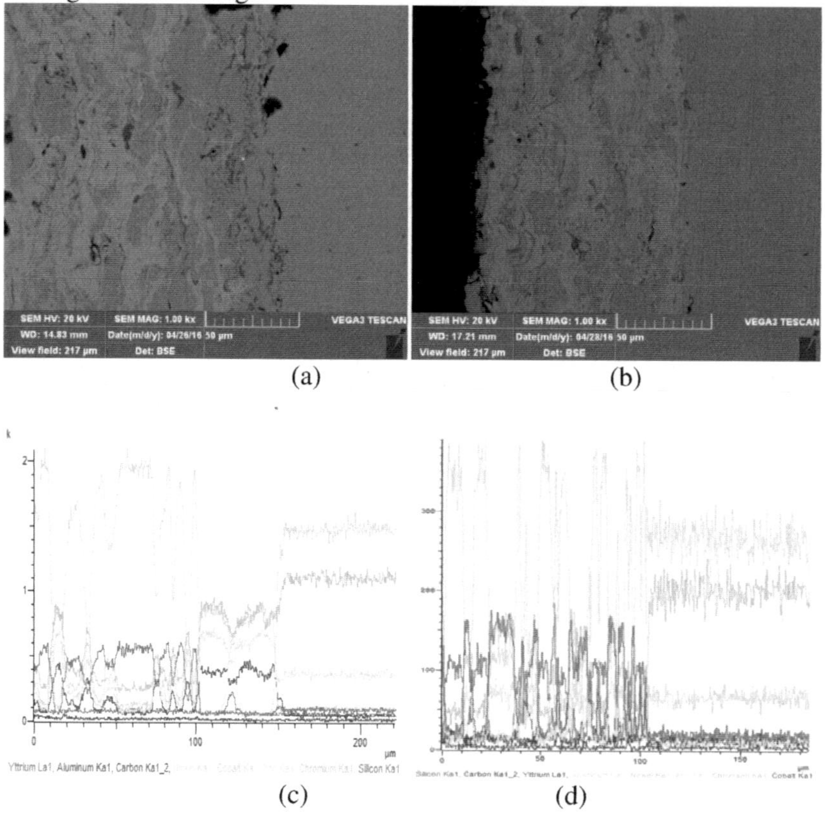

Figure 3. BSE image of the polished cross-section of the as-sprayed HVOF CoNiCrAlY(a) and 5%SiC+CoNiCrAlY (b),(c) and (d) are linar EDS of (a) and (b)

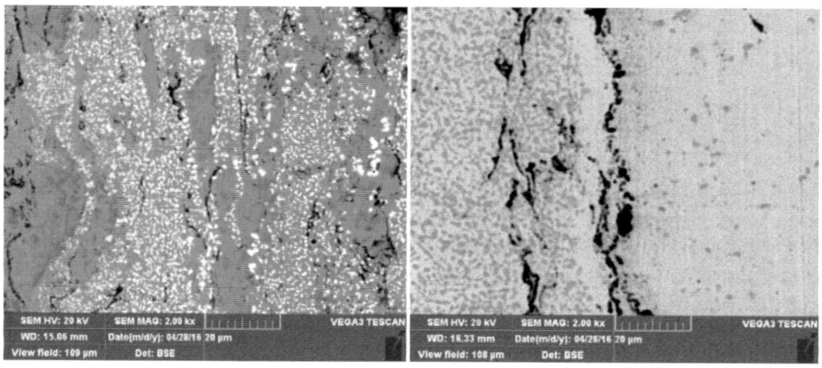

Figure 4. BSE image of the polished cross-section of the as-sprayed HVOF CoNiCrAlY(a) and 5%SiC+CoNiCrAlY (b) heated treatment

Fig. 4 shows the BSE images of cross-sections of as-sprayed HVOF CoNiCrAlY and CoNiCrAlY+5%SiC coating after heat treatment at 1050℃. There are many microcracks and white or grey plots in the both coatings.

Fig. 5 shows the comparison of XRD patterns for the spraying powder and for CoNiCrAlY /SiC coatings (as-sprayed and heat treated at 1050° C). In the both coatings, Ni_3Al and Ni_3Al_4 diffraction peaks were the major phases. With the addition of SiC, the Ni_3Al_4 content decreased.

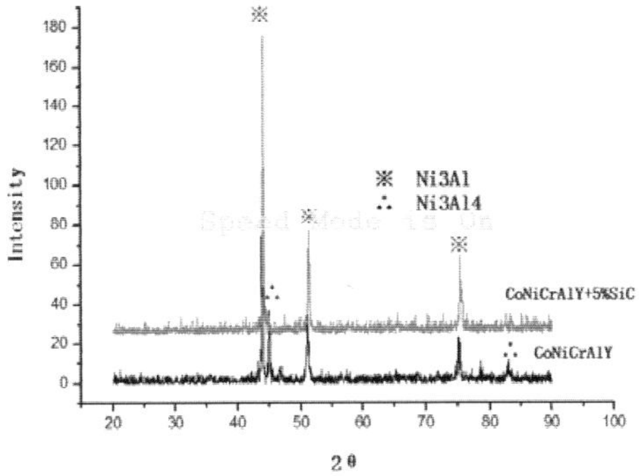

Figure 5. XRD patterns for the spraying powder and for CoNiCrAlY /SiC coatings (as-sprayed and heat treated at 1050°C).

3.3. Microhardness of the Coatings

The microhardness values were measured in the polished cross-section of the CoNiCrAlY and CoNiCrAlY/SiC coatings, shown in Table 2. Although there is some dispersion of the results, a trend in the evolution of microhardness values could be observed. The hardness of the coating is the maximum value of all the samples, perhaps double the matrix. This response was attributed to the precipitation of Ni3Al to strengthen the structure. The trend to softening of the near surface was resulted in the structure coarsening and non-homogeneous. The hardness of both contents coatings are higher than heat treated. At the transition area, the hardness of CoNiCrAlY+5wt.%SiC coating are higher than CoNiCrAlY coating.

Table 1 Microhardness evolution of coatings

	CoNiCrAlY	CoNiCrAlY (heat treated)	CoNiCrAlY+5wt.%SiC	CoNiCrAlY+5wt.%SiC (heat treated)
1	204.45	154.57	266.16	192.64
2	336.19	230.12	459.27	263.96
3	479.72	372.82	474.55	378.78
4	409.6	276.21	420.02	328.56

The number show the locations. 1 is the matrix, 2 is the transition zone, 3is the coating, 4 is the near surface.

3.4. Tribological Performance and Thermal Shock Resistance of the Coatings

In order to study the effect of SiC on the tribological performance of the coating, the wear performance of CoNiCrAlY coating on the steel substrate was compared with the wear behaviour of the CoNiCrAlY+5wt.%SiC coating. Based on the tribological testes results, the abrasion mark of CoNiCrAlY coating is narrow and deap, and that of CoNiCrAlY+5wt.%SiC coating is wide and shallow, as in Figure 6. And the specific volume wear loss of CoNiCrAlY coating is only 54.89% of CoNiCrAlY /SiC coating. The higher hardness of CoNiCrAlY /SiC coating can preserve the wear from deeper area.

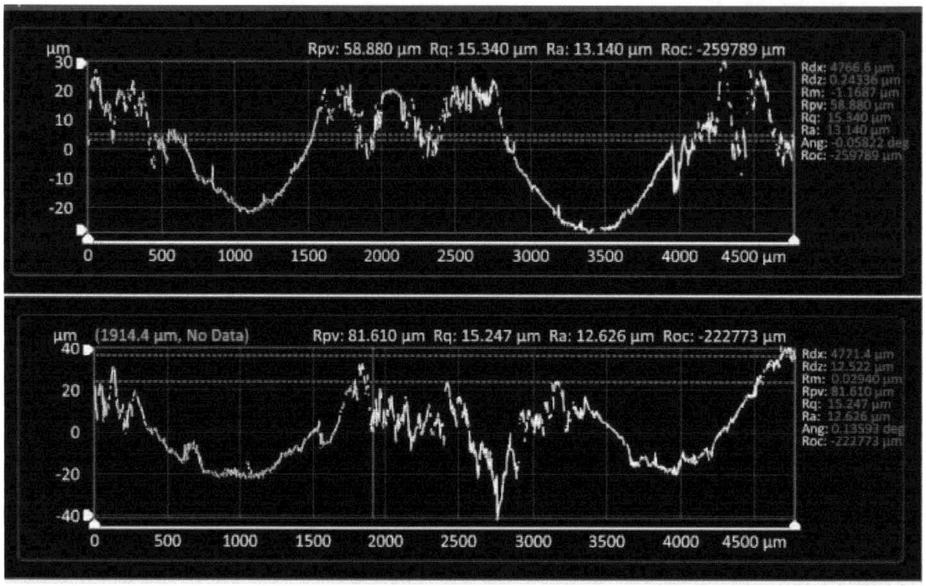

Fig. 6 Cross-sectional profiles of the worn scars made on CoNiCrAlY and CoNiCrAlY /SiC coatings

But there are some microcracks in the CoNiCrAlY+5wt.%SiC coating surface that involves an increase in the wear rate. The cracks is caused probably because of the significant differences between CoNiCrAlY and SiC thermal expansion coefficients, which increased the residual stresses and reduced the coating's integrity.

4. Conclusions

Effects of SiC and heat treatment on mechanical and tribological performances of HVOF sprayed CoNiCrAlY coating on 310L steel substrate has been investigated and the following conclusions are drawn:

(1) Co-Ni-Cr matrix and CoAl phase were the major phases, and SiC was only detected in CoNiCrAl/SiC powder. The most significant difference between the coatings is a transition zone, about 50µm width, rich of Co, Ni, Cr and Al in the coating of CoNiCrAlY. Element Fe in the steel matrix diffused to the coating in both coatings.

(2) The hardness of the coating is nearly double the matrix. At the transition area, the hardness of CoNiCrAlY+5wt.%SiC coating are higher. And at the near coating surface, they are almost the same of CoNiCrAlY and CoNiCrAlY/SiC coating as sprayed.

(3) The addition of SiC increases the residual stresses and reduced the integrity of the coating. The volume wear loss of CoNiCrAlY coating is less than CoNiCrAlY/SiC coating. And the abrasion mark of CoNiCrAlY coating is narrow and deap.

5. Acknowledgement

The authors are grateful to Nanjing Institute of Technology and University of Alberta for their support to this study. This study was supported by Innovation Foundation of Nanjing Institute of Technology(CKJB201501) and University student science and Technology Innovation Fund Project.

6. References

[1] K. Saida, Y. Okabe, K. Hata and et al. Hot cracking behaviour and susceptibility of extra high purity type 310 stainless steels[J]. Science and Technology of Welding and Joining, 2013 , 15 (1) :87-96

[2] B.T. Lu, J.Qiao, J.L. Luo, and et al. Role of hydrogen in stress corrosion cracking of austenitic stainless steels[J]. Philosophical Magazine, 2011 , 91 (2) :208-228
[3] SW Mao, HL Huang, D Gan. Microstructure of aluminized stainless steel 310 after annealing[]J]. Surface and Coatings Technology, 2010, 205(2): 533-539
[4] WT Tsai, KE Huang.Microstructural aspect and oxidation resistance of an aluminide coating on 310 stainless steel[J]. Thin Solid Films, 2000, 366(1-2): 164-168
[5] NS Cheruvu, R Wei, DW Gandy.Influence of thermal exposure on the stability of metastable microstructures of sputter deposited nanocrystalline 304 and 310 stainless steel coatings[J]. Surface and Coatings Technology, 2010, 205(5): 1211-1219
[6] WANG En-ze, XU Yan-ping, BAO Chong-gao, et al. Fabrication of Al2O3/heat-resistant steel composite and its wear-resistance at high temperature and abrasive[J]. Acta Materials Composite Sinica, 2004, 21(1): 56-60
[7] L. Ajdelsztajn, J.A. Picas, G.E. Kim, F.L. Bastian, J. Schoenung, V. Provenzano, Mater.
Sci. Eng. A 2002, (338) 33–43.
[8] CAO Yu-xia, DU Ling-zhong, ZHANG Wei-gang, et al. Study on Preparation and Tribological Properties of Atmospheric Plasma-sprayed NiCoCrAlY/Al2O3 Wear-resistant Coatings [J]. Surface Technology, 2015, 44(5): 62-66
[9] CHANG Dai-zhan, YANG Wang-yue, ZHANG Lai-qi, et al. Microstructures and high temperature wear resistant properties of CoNiCrAIY-A12O3 Explosion Spraying Coatings[J]. Transactions of Materials and Heat treatment, 2006, 27(1): 75-78
[10] GUO Chun, CHEN Jian-min, YAO Run-gang, et al. Microstructure and High Temperature Wear Resistance of Laser Cladding NiCoCrAlY/ZrB2 Coating[J]. Rare Metal Materials and Engineering, 2013, 42(8): 1547-1551
[11] WANG Hong-yu, ZUO Dun-wen, WANG Ming-di, et al. High temperature frictional wear behaviors of nano-particle reinforced NiCoCrAlY cladded coatings[J]. Trans. Nonferrous Met. Soc. China, 2011, (21): 1322-1328
[12] ShibataM, Kuroda S,Murakami H, Ode M,Watanabe M, Sakamoto Y. Comparison of
microstructure and oxidation behavior of CoNiCrAlY bond coating prepared by different thermal spray process. Mater Trans, JIM 2006, 47(7):1638–42.
[13] J.A. Picas, M. Punset, S. Menargues, et al. Microstructural and tribological studies of as-sprayed and heat-treated HVOF Cr3C2–CoNiCrAlY coatings with a CoNiCrAlY bond coat. Surface & Coatings Technology, 2015, (268): 317–324
[14] A.S. Khanna, W.S. Rathod. Development of CoNiCrAlY oxidation resistant hard coatings using high velocity oxy fuel and cold spray techniques, Int. Journal of Refractory Metals and Hard Materials, 2015, (49)374–382

IOP Conf. Series: Materials Science and Engineering **250** (2017) 012004 doi:10.1088/1757-899X/250/1/012004

Quasi-Static and Dynamic Properties of Ti-4Al-3V-0.6Fe-0.2O Titanium Alloy Plates

Rui Liu[*], Songxiao Hui, Wenjun Ye, Yang Yu and Xiaoyun Song

State Key Lab of Nonferrous Metals & Processes, General Research Institute for Nonferrous Metals, Beijing 100088, China
email: h_o_n_o_r@163.com

Abstract. This paper aimed to study the quasi-static and dynamic properties of the titanium alloy with nominal composition Ti-4Al-3V-0.6Fe-0.2O. The ingots Ti-4Al-3V-0.6Fe-0.2O alloy were forged into slabs of 32 mm thickness by one heat above $T_{\beta\text{-transus}}$ temperature and one heat below $T_{\beta\text{-transus}}$ temperature, and then rolled into plates of 8 mm thickness by one heat below $T_{\beta\text{-transus}}$ temperature. The quasi-static tensile properties were tested by MTS™ testing system at strain rate of 10^{-3} s^{-1}. The dynamic compression properties were tested by Split Hopkinson Pressure Bar system at strain rate of 3000±200 s^{-1}. The results show that the quasi-static properties of the Ti-4Al-3V-0.6Fe-0.2O plate are comparable to ATI 425TM alloy plate and commercial Ti-6Al-4V plate. The average dynamic flow stress of Ti-4Al-3V-0.6Fe-0.2O plate is comparable to ATI 425™ alloy plate, over 100MPa higher than that of Ti-6Al-4V plate. The maximum strain during homogeneous dynamic plastic deformation of Ti-4Al-3V-0.6Fe-0.2O plate is approximately 15 ~ 20% higher than that of ATI 425™ alloy plate, and is approximately 80% of that of Ti-6Al-4V plate.

1. Introduction

Compared to steel or aluminum alloys, the annual consumption of titanium alloys is smaller due to its higher cost. Reducing the cost of titanium alloys has been urgently demanded in recent years. At present, the technology of low-cost process for titanium sponge has not been realized in industrial scale. Thus, sufficient use of recycled titanium scraps, which can be realized by the technology of electron beam cold hearth melting (EBCHM)[1-2], has been widely regarded as an efficient method of cost reduction. Titanium scraps are usually recycled from leftover bits and pieces of slabs or ingots. As Ti-6Al-4V alloy is the most common and widely used titanium alloy, it is easy to find abundant amount of Ti-6Al-4V scraps. Low-cost Ti-6Al-4V alloy products with recycled titanium scraps addition have already been used in combat vehicles and aircrafts [3]. In titanium scraps exist the impurities such as Fe, O. The content of iron or oxygen in scraps may increase during the process of recycling. Thus, the series of Ti-Al-V-Fe-O alloys, which has the potential for low-cost producing via using scraps with high impurity content, are designed on the base of Ti-6Al-4V alloy.

ATI 425™ titanium alloy, developed by Wah Chang (Oregon, USA), is one of Ti-Al-V-Fe-O alloys with comparable strength and improved formalilty to Ti-6Al-4V[4]. The nominal chemical composition of ATI 425™ alloy is as 3.5 wt%~4.5 wt% Al, 2.0 wt%~3.0 wt% V, 1.2 wt%~1.8 wt% Fe, 0.20 wt%~0.30 wt% O, with the balance of Ti. ATI 425™ alloy was initially designed for the purpose of ballistic protection, demanding the resistant to dynamic loading. The researches on Ti-6Al-4V showed that the dynamic load-bearing capacity is influenced by hot working and heat treatment[5-7], as well as the content of Al, V, Fe, O elements[8]. In our previous study[9], dynamic properties of ATI 425™ alloy were investigated by hot rolled plate with nominal composition Ti-3.5Al-2.5V-1.5Fe-0.25O, of which the aluminum content reached lower limit. In this paper, Ti-4Al-3V-0.6Fe-0.2O alloy plates were prepared for the purpose of investigating quasi-static and dynamic properties of ATI 425™ alloy with extensive composition.

2. Experimental

The ingots of Ti-4Al-3V-0.6Fe-0.2O titanium alloy were double VAR-melted with 20 kg in weight, which has a composition of 4.30±0.15 wt% Al, 2.90±0.09 wt% V, 0.58±0.02 wt% Fe, 0.20±0.01 wt% O, 0.012±0.002 wt% C, 0.012±0.002 wt% N, 0.0010 wt% H, with the balance of Ti. The b-transus temperature ($T_{\beta\text{-transus}}$) of the ingot was tested as 985±5 °C. The ingots were forged into slabs of 32 mm thickness by one heat at the temperature of 40~50 °C above $T_{b\text{-transus}}$ and one heat at the temperature of 40~50 °C below $T_{\beta\text{-transus}}$, and then hot rolled at the temperature of 930 °C into the plates of 8 mm thickness. The as-rolled Ti-4Al-3V-0.6Fe-0.2O plates were respectively heat treated as following: HT1, 750 °C/ 1 hour/ Air cooling; HT2, 930 °C/ 1 hour/ Air cooling; HT3, 940 °C/ 1 hour/ Air cooling; HT4, 960 °C/ 1 hour/ Air cooling.

The quasi-static tensile properties were performed on MTS™ testing system at strain rate of 10^{-3} s^{-1}. Standard specimens for tensile test were 5 mm in diameter and 25 mm in gauge length, with the axis perpendicular to rolling direction of the plate (T-direction). Dynamic compression properties were tested by a SHPB equipment at average strain rate of 2000~3000 s^{-1}. At the strain rate of 2000~3000 s^{-1}, visible damages can be observed on the specimens. Specimens for dynamic test were machined into cylinder shape, with 4 mm in diameter and 4 mm in length and the accuracy were controlled to 0.015 mm. The axis of the dynamic testing specimen was parallel to normal direction of the plates. Both top and bottom surface of the dynamic specimens were grounded to 600♯ sand paper to enhance a smooth contact with the compression platens during tests. The values of average flow stress σ, maximum strain ε and absorbed energy *E* during dynamic homogeneous plastic deformation were calculated via stress-strain curves obtained by SHPB compression testing.

The microstructures of the plates after various heat-treatment were examined by Axiovert 200 MAT optical microscopy (OM). The metallographic specimens were polished and etched by a solution with the volume fraction of 5% HF +10% HNO_3+85% H_2O at room temperature.

3. Results

The optical microstructures of the Ti-4Al-3V-0.6Fe-0.2O plates after heat treatment HT1~HT4 are shown in Fig. 1 (a)~(d). The microstructure of the plate after conventional annealing treatment HT1 is consisted of elongated α phase. Bimodal microstructure, consisting of primary α phase (α_p) and transformed β phase, can be obtained by heat treatment at elevated annealing temperature as HT2~HT4. The volume fractions of α_p for HT2, HT3, HT4 plates are 58±3%, 54±3%, 20±3% respectively. It is indicated that the microstructures of Ti-4Al-3V-0.6Fe-0.2O alloy can be controlled as similar types to that of commercial Ti-6Al-4V or ATI 425™ alloys after proper processing of rolling and heat treatment.

The quasi-static tensile properties of Ti-4Al-3V-0.6Fe-0.2O plates are shown in Table.1, which are comparable to commercial Ti-6Al-4V plates or ATI 425™ plates. For commercial Ti-6Al-4V plates in 8mm-thick after HT1 heat treatment, the typical quasi-static properties (in T-direction) are usually tested as 1040 MPa ~ 1070 MPa in ultimate tensile strength, 1015 MPa ~1040 MPa in yield strength, 11.0% ~ 13.0% in elongation and 28% ~ 40% in reduction of area. For ATI 425™ plates in our previous study, that were tested as 1085 MPa ~ 1100 MPa in ultimate tensile strength, 980 MPa ~1065MPa in yield strength, 18.5% ~ 21.0% in elongation and 43% ~ 48% in reduction of area.

Fig.2 shows typical true stress-strain curves of Ti-4Al-3V-0.6Fe-0.2O plates performed in SHPB compression experiment at the strain rate of 2000~3000 s^{-1}. The variation tendency of the curves is similar t (a) of ATI 425™ plates, but quite different from that of Ti-5Al-3V-0.025Fe-0.044O alloy [12] in our previous study, indicating that Fe element aggravates the effect of softening during dynamic deformation.

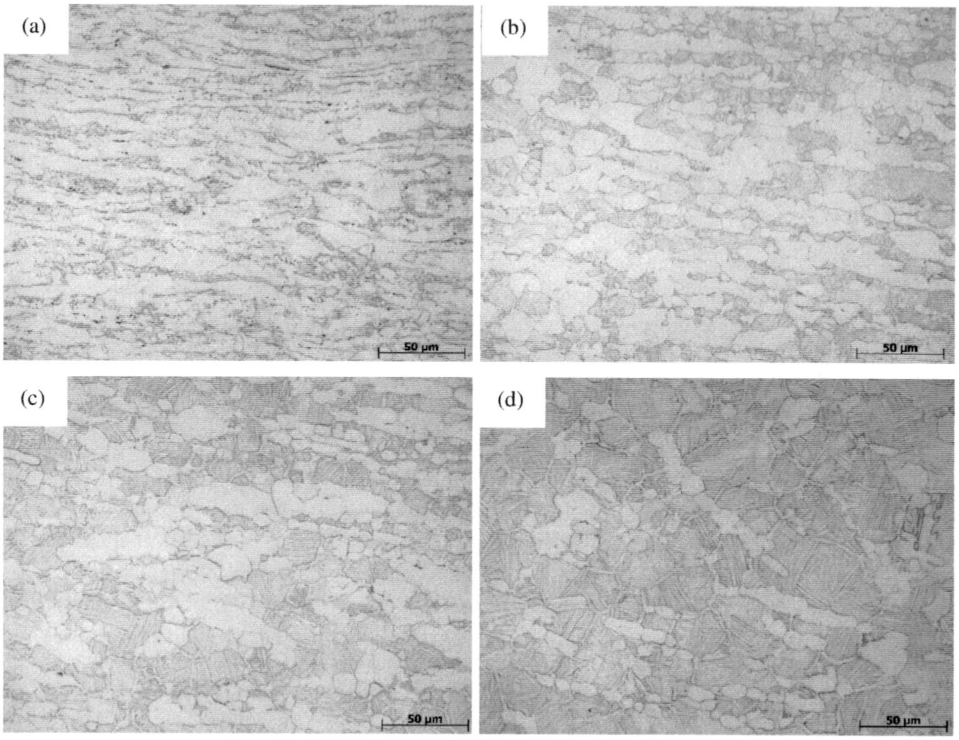

Figure 1. Microstructures of Ti-4Al-3V-0.6Fe-0.2O plate after various heat treatment.
(a) HT1; (b) HT2; (c) HT3 ; (d) HT4.

Table 1. quasi-static tensile properties of Ti-4Al-3V-0.6Fe-0.2O after various heat treatment.

Heat treatment	Ultimate tensile strength / (MPa)	Yield strength / (MPa)	Elongation /	Reduction of area /
HT1	1095±15	1040±10	16.0±1.0	42±3
HT2	1055±15	960±5	17.0±1.0	43±2
HT3	1040±10	950±10	15.0±1.0	40±1
HT4	1000±15	900±15	10±1.5	25±2

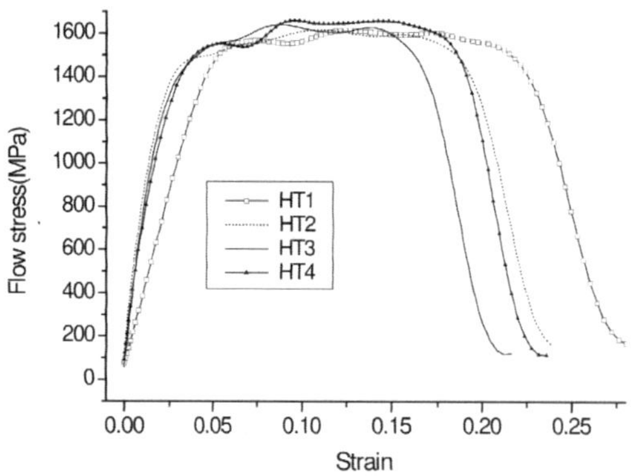

Figure 2. Typical true stress-strain curves of Ti-4Al-3V-0.6Fe-0.2O after various heat treatment.

Typical dynamic properties Ti-4Al-3V-0.6Fe-0.2O after various heat treatment are shown in Table.2. The average flow stress σ, maximum strain ε, and absorbed energy E during homogeneous plastic deformation, were calculated from true stress-strain curves for each specimens with various heat treatment. In our previous study[9,10], typical SHPB compression properties of 8 mm-thick commercial Ti-6Al-4V plates were as σ=1460±15 MPa, ε=0.17±0.02, E=255±20 J/cm^3, while that of ATI 425^{TM} plates were as σ=1550±30 ~1600±30 MPa, ε= 0.09±0.01 ~0.13±0.01, E=150±15 ~200±15 J/cm^3. As a comparison, Ti-4Al-3V-0.6Fe-0.2O plates exhibite comparable dynamic strength to ATI 425^{TM} alloy plate, over 100MPa higher than that of Ti-6Al-4V plate. The maximum strain during homogeneous dynamic plastic deformation of Ti-4Al-3V-0.6Fe-0.2O plate is approximately 80% of that of Ti-6Al-4V plate, and is approximately 15 ~ 20% higher than that of ATI 425^{TM} alloy plate.

Compared Ti-4Al-3V-0.6Fe-0.2O alloy with Ti-6Al-4V alloy, the increasement in iron content contributes to the improvement in both quasi-static strength and dynamic strength. Compared Ti-4Al-3V-0.6Fe-0.2O alloy with ATI 425^{TM} alloy, higher aluminum and vanadium content with lower iron content affects slightly on quasi-static strength and dynamic strength, and causes noticeable deterioration in quasi-static plasticity, but leads to enhancement in dynamic plasticity.

Table 2. Dynamic properties of Ti-4Al-3V-0.6Fe-0.2O after various heat treatment.

Heat treatment	Average flow stress σ / (MPa)	Maximum strain during homogeneous plastic deformation ε /	Absorbed energy during homogeneous plastic deformation E / (J/cm^3)
HT1	1570±35	0.16±0.01	240±20
HT2	1540±25	0.16±0.02	235±20
HT3	1570±30	0.14±0.01	215±25
HT4	1600±15	0.15±0.01	235±15

4. Conclusion

Manufactured in similar conditions of rolling and heat treatment, quasi-static properties of Ti-4Al-3V-0.6Fe-0.2O titanium alloy are comparable to the commercial Ti-6Al-4V alloy and ATI 425^{TM} alloy. The average dynamic flow stress of Ti-4Al-3V-0.6Fe-0.2O plate is comparable to ATI 425^{TM} alloy plate, over 100MPa higher than that of Ti-6Al-4V plate. The maximum strain during homogeneous dynamic plastic deformation of Ti-4Al-3V-0.6Fe-0.2O plate is approximately 80% of that of Ti-6Al-4V plate, and is approximately 15 ~ 20% higher than that of ATI 425^{TM} alloy plate.

5.Acknowledgement

The authors appreciate the financial support from the National Nature Science Foundation of China (NSFC, No. 51601016), and the International S&T Cooperation Program of China (ISTCP, No. 2015DFR50930).

6.References

[1] A. Mitchell 1999 The electron beam melting and refining of titanium alloys, *Mater. Sci. Eng. A* **263** 217-223

[2] A. N. Kalinyuk, N. P. Trigub, V. N. Zamkov, et. al. 2003 Microstructure, texture, and mechanical properties of electron-beam melted Ti–6Al–4V, *Mater. Sci. Eng. A* **346** 178-188

[3] J. S. Montgomery and M. G. H. Wells 2001 Titanium armor applications in combat vehicles, *JOM* **4** 29-32

[4] T. S. Srivatsan, M. Kuruvilla, L. Park 2010 A study at understanding the mechanisms governing the high cycle fatigue and final fracture behavior of the titanium alloy: Ti–4Al–2.5V, *Mater. Sci. Eng. A* **527** 435-448

[5] X. Q. Liu, C. W. Tan, J. Zhang, et al. 2008 Influence of stress-state on adiabatic shear

sensitivity of Ti-6Al-4V, *Rare Metal Mater. Eng.* **37** 1522-1525

[6] Y. M. Luo, J. X. Liu, X. W. Cheng, et al. 2015 Adiabatic shear banding of hot-rolling Ti-6Al-4V alloy subjected to dynamic shearing and uniaxial dynamic compression, *Rare Metals* **34** 632-637

[7] K. Sun, X. D. Yu, C. W. Tan, et al. 2014 Effect of microstructure on adiabatic shear band bifurcation in Ti–6Al–4V alloys under ballistic impact, *Mater. Sci. Eng. A* **595** 247-256

[8] R. Liu, S. X. Hui, W. J. Ye, et al. 2013 Dynamic stress-strain properties of Ti-Al-V titanium alloys with various element contents, *Rare Metals* **32** 555-559

[9] R. Liu, S. X. Hui, W. J. Ye, et al. 2017 Quasi-Static and Dynamic Properties of Ti-3.5Al-2.5V-1.5Fe-0.25O Titanium Alloy Plates, *Mater. Sci. Forum* **879** 1159-1163

[10] R. Liu, S. X. Hui, W. J. Ye, et al. 2014 Dynamic compression behavior of Ti-6Al-4V plates prepared by a single electron beam cold hearth melted ingot, *Adv. Mater. Res.* **1015** 328-331

Oxidation Behavior of NiTi-Al Based Alloy with Nb and Mo Additions

Xiaoyun Song [1*], Wenjun Ye [1], Songxiao Hui [1] and Yan Li [2]

[1] State Key Laboratory of Nonferrous Metals & Processes, General Research Institute for Nonferrous Metals, Beijing 100088, China.
[2] School of Materials Science and Engineering, Beihang University, Beijing 100191, China
Email: songxiaoyun@grinm.com

Abstract. Oxidation behavior of a Ni-45Ti-5Al-2Nb-1Mo (composition in at %) alloy was studied in comparison with the Ni-45Ti-5Al alloy by X-ray diffraction (XRD), scanning electron microscopy (SEM) and cyclic oxidation tests. The results show that the microstructure of Ni-45Ti-5Al-2Nb-1Mo alloy is composed of NiTi matrix, Ti_2Ni and (Nb, Ti) solid solution phases. By 2 at% Nb and 1 at% Mo addition, the Ni-45Ti-5Al-2Nb-1Mo alloy shows excellent oxidation resistance. The mass gain of Ni-45Ti-5Al-2Nb-1Mo alloy after 100 h exposure at 800 °C is 2.29 mg/cm^2, which is much lower than that of Ni-45Ti-5Al alloy. This is mainly resulted from the doping effect of Nb and Mo in TiO_2 and the formation of a continuous Al_2O_3 layer by Nb and Mo addition.The oxide scale of the present alloy is with multi-layer structure in order of TiO2 layer/ Al2O3 + NiAl2O4 mixture layer/ TiO2 layer/ Nb-rich TiO2 layer/ Ni3Ti layer from outside.

1. Introduction

Order intermetallic alloys based on aluminides have been developed as high temperature structural materials for decades [1]. Recent investigations indicate that NiTi-Al based intermetallic alloys have the potential to use as high temperature structural materials for aerospace applications because of their low density, high strength at high temperature and good corrosion resistance. The addition of small amount of Al (no more than 8 at%) into NiTi alloy can dramatically improve the strength especially at high temperature [2-3]. Alloying is an effective method to improve the mechanical properties of NiTi-Al based alloys. Many elements, such as Mo, Nb, Hf, Zr, B and RE, have been added to strengthen the NiTi-Al alloys [4-7]. It has been reported that a Nb- and Mo modified Ni-45Ti-5Al-2Nb-1Mo (at%) alloy is expected to become a potential high-temperature structural material with strength of 650 MPa at 650 °C and 350 MPa at 800 °C[8-9].

Actual applications in the aerospace industry require the alloys to be considered in a variety of high-temperature environments. The oxidation behaviors of ternary NiTi-Al alloys at temperatures between 600 °C and 800 °C have been systemically studied, which found that NiTi-Al alloys shows poor oxidation resistance with the mass gain more than 20 mg/cm^2 after exposure in air at 800 °C for 100h. By adding little quaternary element Nb (3 at%), the oxidation resistance of NiTi-Al alloys can be notably improved [10-11].

However, so for the studies on the oxidation behavior of NiTi-Al based alloys are limited, especially the knowledge of NiTi-Al based alloy simultaneously modified by two or more alloying elements are still lack. Therefore, in this paper, the isothermal oxidation behavior of Nb- and Mo-containing Ni-45Ti-5Al-2Nb-1Mo alloy is studied and discussed in detail.

IOP Conf. Series: Materials Science and Engineering **250** (2017) 012005 doi:10.1088/1757-899X/250/1/012005

2. Experimental Procedure

A 10 kg ingot with nominal composition of Ni-45Ti-5Al-2Nb-1Mo (at%) alloy was prepared by vacuum induction furnace using high purity Ni, Ti, Al, Nb and Mo as raw materials. To minimize the segregation of composition the ingot was repeatedly melted four times. The ingot was homogenized at 1000 °C for 24 h, and then isothermal forged at 900 °C into plates.

The specimens with dimensions of 10 mm × 10 mm × 8 mm for the oxidation tests were cut using an electro-discharge machine and then the surface was mechanically polished with 1000 grit SiC paper. The cyclic oxidation experiments were carried out at 800 °C in air for 100 h. Three samples were employed in order to obtain an average mass change. The mass gain was weighed by an electronic balance with the precision of 0.1 mg. The morphology on the surface and cross-sections of the oxide scales were observed by a JSM-5600 scanning electron microscopy (SEM) equipped with a Link ISIS energy-dispersive X-ray spectroscopy (EDS). The phase structure of the oxide scales was identified by a Regaku D/max 2200PC X-ray diffractometer (XRD) using Cu Kα radiation (λ=0.154 nm).

3. Results and Discussion

3.1. Microstructure

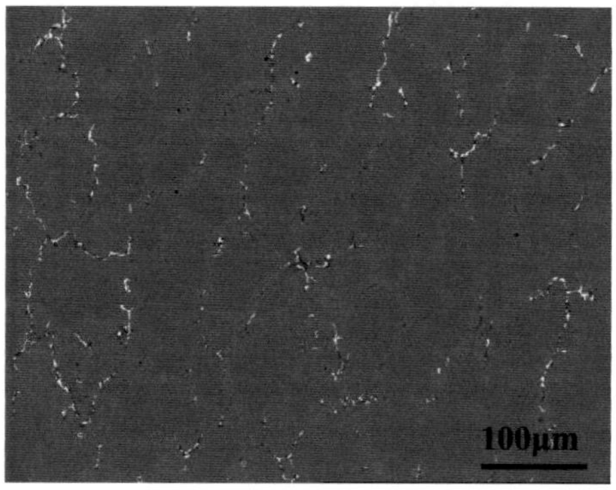

Figure 1. Microstructure for Ni-45Ti-5Al-2Nb-1Mo alloy before oxidation

Figure 1 shows the SEM micrograph of Ni-45Ti-5Al-2Nb-1Mo alloy before oxidation. According to the EDS and XRD results, the microstructure of Ni-45Ti-5Al-2Nb-1Mo alloy consists of NiTi matrix, black Ti_2Ni phase and white (Nb, Ti) solid solution phase. The latter two phases are precipitated along the grain boundaries. It is noted that Nb and Mo are dissolved within these three phases, and Nb is enriched in the grain boundaries.

3.2. Oxidation Kinetics

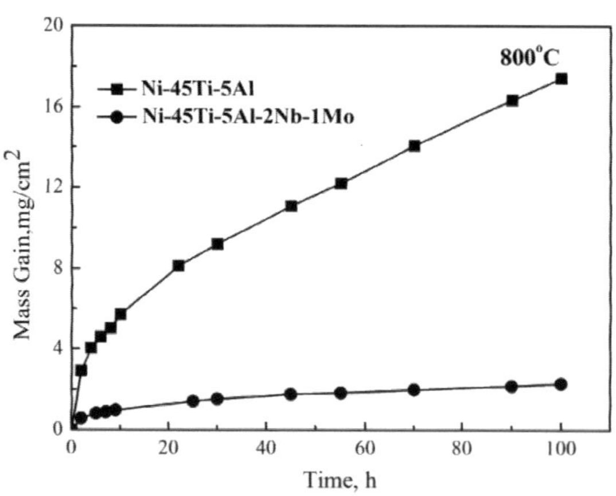

Figure 2. Isothermal oxidation kinetics of Ni-45Ti-5Al-2Nb-1Mo alloy exposed at 800 °C for 100 h

The oxidation kinetic of Ni-45Ti-5Al-2Nb-1Mo alloy at 800 ° C is shown in Figure 2. The result of Ni-45Ti-5Al alloy is also illustrated for comparison [11]. It can be found that the mass gain after 100 h exposure at 800 ° C is 2.29 mg/cm^2 for Ni-45Ti-5Al-2Nb-1Mo alloy, which is much lower than that for Ni-45Ti-5Al alloy of 17.42 mg/cm^2, indicating that the former alloy exhibits much better oxidation resistance than the latter. Thus, the oxidation resistance of NiTi-Al based alloy is remarkably improved by Nb and Mo addition.

3.3. Morphology and Microstructure of Oxide Scale

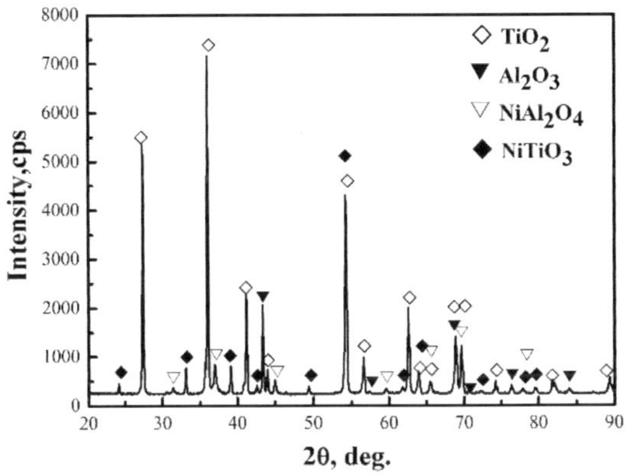

Figure 3. XRD spectra of the oxide scale on the Ni-45Ti-5Al-2Nb-1Mo alloy after 100 h isothermal oxidation at 800 ℃

Figure 4. SEM micrograph of the surface morphology of the oxide scale in Ni-45Ti-5Al-2Nb-1Mo alloy after oxidation at 800 ℃ for 100 h

Figure 3 shows the XRD spectra of the oxide scale on the sample of the Ni-45Ti-5Al-2Nb-1Mo alloy after oxidation for 100 h at 800 °C. It is indicated that the external oxide scale in the Ni-45Ti-5Al-2Nb-1Mo alloy mainly consists of TiO_2 and Al_2O_3 with a trace of $NiAl_2O_4$ and $NiTiO_3$. Figure 4 shows the surface morphology of the oxide scale in the Ni-45Ti-5Al-2Nb-1Mo alloy after oxidation for 100 h at 800 °C. Compared with ternary NiTi-Al alloy, the oxides on the sample are dense and fine, namely the addition of Nb and Mo improves the adherence of oxide scale to the substrate. It is notable that a rugged morphology is observed on the oxide surface, which is similar to the results reported in NiTiNb and NiTi-Al-Nb alloys [10, 12]. This is suggested to be associated with the inhomogeneous distribution of Nb elements, i.e. the NiTi matrix containing less Nb is much easier to be oxidized than the grain boundaries with more Nb content.

To clarify the mechanism for the formation of such oxide structure, cross-sectional morphology of the oxide scale of Ni-45Ti-5Al-2Nb-1Mo alloy is observed on SEM, as shown in Figure 5. It can be found that the oxide scale consists of a multi-layer with a total thickness of about 30 μm. The compositions of each layer detected by EDS are listed in Table 1. The outer oxide scale is composed of two fold TiO_2 layers (marked as A and C) with a thin dark Al-rich oxide layer (marked as B) embedded. The XRD result suggests that the Al-rich oxide layer is a mixture of Al_2O_3 and $NiAl_2O_4$ ($NiO\cdot Al_2O_3$). Beneath the sandwich oxide structure, a layer with discontinuous white particles dispersed (marked as D) is observed. The white particle is too small to analyze its composition, so the result shown in Table 1 is the average composition of layer D. It is supposed that layer D is also TiO_2 containing more Nb and Mo content than layer C, which is due to the outward diffusion of Nb and Mo in the substrate. And then a Ti-depleted area in light-grey colour (marked as E) next to the substrate can be observed due to the net Ti-consumption during the course of oxidation. The EDS analysis indicates that it is stoichiometric close to Ni_3Ti phase. Thus, the formed oxide scale of the Ni-45Ti-

5Al-2Nb-1Mo alloy is in order of TiO_2 layer/ Al_2O_3 + $NiAl_2O_4$ mixture layer/ TiO_2 layer/ Nb-rich TiO_2 layer/ Ni_3Ti layer from outside.

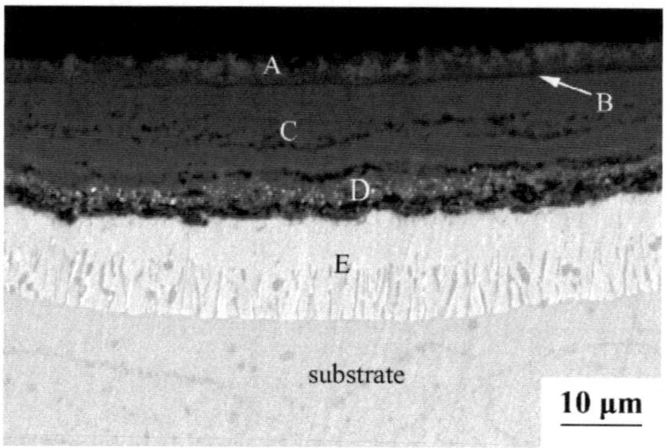

Figure 5. Cross-sectional SEM morphology of Ni-45Ti-5Al-2Nb-1Mo alloy afteroxidation at 800°C for 100 h

Table 1. Compositions of oxide layer for Ni-45Ti-5Al-2Nb-1Mo alloy after oxidation in Figure 5.

Layer	Composition (at %)					
	Ni	Ti	Al	Nb	Mo	O
A	0.18	26.97	2.3	0.24	-	70.31
C	3.08	21.01	4.2	0.87		70.84
D	4.09	29.23	-	3.07	1.40	62.21
E	68.63	25.36	4.43	0.82	0.76	-

In the present alloy, the oxidation resistance for Ni-45Ti-5Al-2Nb-1Mo alloy at 800 ° C is dramatically improved compared with ternary Ni-45Ti-5Al alloy. This is mainly attributed to the beneficial effect of 2 at% Nb and 1 at% Mo addition on the oxidation behavior. Firstly, the growth of TiO_2 is primary by oxygen diffusion through oxide scales via a vacancy mechanism [13]. In the studied alloy, 3.07 at% Nb and 1.4 at% Mo is detected in the TiO_2 layer by EDS. This will lead to a decrease of the oxygen vacancies, and thus slow the diffusion of oxygen. Secondly, the addition of Nb and Mo can increase the thermodynamic activity of Al, and accelerate the formation of a stable and continuous alumina scale, which can provide an oxidation barrier and prevent the diffusion of oxygen. Similar phenomenon has been found in Ni-50Ti-9Al-1Mo alloy, in which 1 at% Mo addition promotes to form a continuous dense Al-rich layer beneath the outmost TiO_2 layer.

4. Conclusions

In this study, the oxidation behavior of Nb- and Mo- modified NiTi-Al based alloy at 800 ° C was investigated. By 2 at% Nb and 1 at% Mo addition, the Ni-45Ti-5Al-2Nb-1Mo alloy shows an excellent oxidation resistance. Nb and Mo additions dramatically lower the oxidation rate and improve the adherence of oxide scale to the substrate. The mass gain is 2.29 mg/cm^2 after 100 h exposure at 800 ° C, which is much lower than that for ternary Ni-45Ti-5Al alloy. The oxide scale of the present alloy is about 30 μm in thickness with multi-layer structure in order of TiO_2 layer/ Al_2O_3 + $NiAl_2O_4$ mixture layer/ TiO_2 layer/ Nb-rich TiO_2 layer/ Ni_3Ti layer from outside. The improvement of oxidation resistance for Ni-45Ti-5Al-2Nb-1Mo alloy is mainly attributed to the doping effect of Nb and Mo in TiO_2 which can decrease the oxygen vacancy concentration and slow the diffusion of oxygen. Moreover, Nb and Mo elements enhance the formation of a dense and continuous Al_2O_3 layer to provide an oxidation barrier.

5.Acknowledgments
This work is supported by National Natural Science Foundation of China (NSFC, No. 51571036).

6.References

[1] Liu CT, Stringer J, Mundy JN, Horton LL and Angelini P 1997*Intermetallics***5**579-596
[2] Koizumi Y, Ro Y, Nakazawa S and harada H 1997 *Mater Sci Eng A***223**36-41.
[3] Meng LJ, Li Y, Zhao XQ, Xu J and Xu HB 2007 *Intermetallics***15**814-818.
[4] Song XY, Li Y, Zhang F and Li SS 2010 *Chinese Journal of Aeronautics***23**715-719.
[5] Xu HB, Meng LJ, Xu J, Li Y and Zhao XQ 2007*Intermetallics***15**778-782.
[6] Li Y, Yu KY, Song XY and Zhang F 2011 *Rare Met***30**522-526.
[7] Zheng L, Li T, Song XY and Li Y 2013 *Chinese Journal of Rare Metals***37**171-178
[8] Song XY, Li Y and Zhang F 2013 *Rare Met.* **32**475-479.
[9] Song XY, Li Y and Zhang F 2014 *Mater Sci Eng A***594**229-234
[10] Xu J, Zhao XQ, Gong SK and Xu HB 2006 *Acta Metall Sinica***42**820-826
[11] Xu J. Study on the oxidation behavior of TiNi-Al(Nb) intermetallic alloys [D]. Beijing: Beihang University, 2007.
[12] Xu J, Zhao XQ, Gong SK 2006 *Trans Nonferrous Met Soc China***16**s31-36
[13] Roy TK, Balasubramaniam R, Ghosh A 1996*Metall Mater Trans A***27**3993-4002

The Kinematic Analysis of Flat Leverage Mechanism of the Third Class

A Zhauyt[1*], G Mamatova[2], G Abdugaliyeva[3], K Alipov[1], A Sakenova[1] and A Alimbetov[1]

[1] Kazakh National Research Technical University named after K. I. Satpayev, Almaty 050013, Kazakhstan
[2]Academy of Civil Aviation, Almaty 050039, Kazakhstan
[3] Karaganda State Technical Universities, Karaganda 100000, Kazakhstan
E-mail: [1]ali84jauit@mail.ru

Abstract. It is necessary to make link mechanisms calculation to the strength at designing of flat link mechanisms of high class after definition of block diagrams and link linear sizes i.e. it is rationally to choose their forms and to determine the section sizes. The algorithm of the definition of dimension of link mechanism lengths of high classes (MHC) and their metric parameters at successive approach is offered in this work. It this paper educational and research software named GIM is presented. This software has been developed with the aim of approaching the difficulties students usually encounter when facing up to kinematic analysis of mechanisms. A deep understanding of the kinematic analysis is necessary to go a step further into design and synthesis of mechanisms. In order to support and complement the theoretical lectures, GIM software is used during the practical exercises, serving as an educational complementary tool reinforcing the knowledge acquired by the students.

1. Introduction

In the teaching of subjects related to Machine Theory, supporting and complementing theoretical lectures with a simulation and analysis software, helps the students to understand deeply and visually the theoretical bases of the Mechanisms Science [1]. The software has been developed focusing, not only on educational purposes but also on research in the field of computational kinematics and mechanism design applications [2-3]. The software presented in this article also has potential to be used by students of Bachelor or Master Degrees in Mechanical Engineering and other subjects related to Robotics, Mechanism Design, etc. GIM has been developed in a modular structure [4-5]. After defining the kinematic structure of a linkage in the Geometry module, the user can perform the motion simulation in the Motion module. GIM is mainly oriented to the field of kinematic analysis, motion simulation and dimensional synthesis of planar mechanisms. In any case, it also includes other modules for workspace and singularity evaluation [6-7] and static analysis of mechanical structures.

2. Materials and Methods

The mechanisms of the third class, thanks to their structure, can provide difficult laws of the movement of the points and the driven members. For example, by means of such mechanism it is possible to reproduce set movements of two driven members (output) at one driving (entrance) member [1]. The questions of synthesis and kinematic analysis of mechanisms of the third class are insufficiently studied therefore these mechanisms received limited application in the national economy. In this subsection the technique of the kinematic analysis of the transmission mechanism of the third class with usage of vector calculation on the example of the mechanism of the third class with a sliding bar 5 and driving member 1 is considered (see Fig. 1).

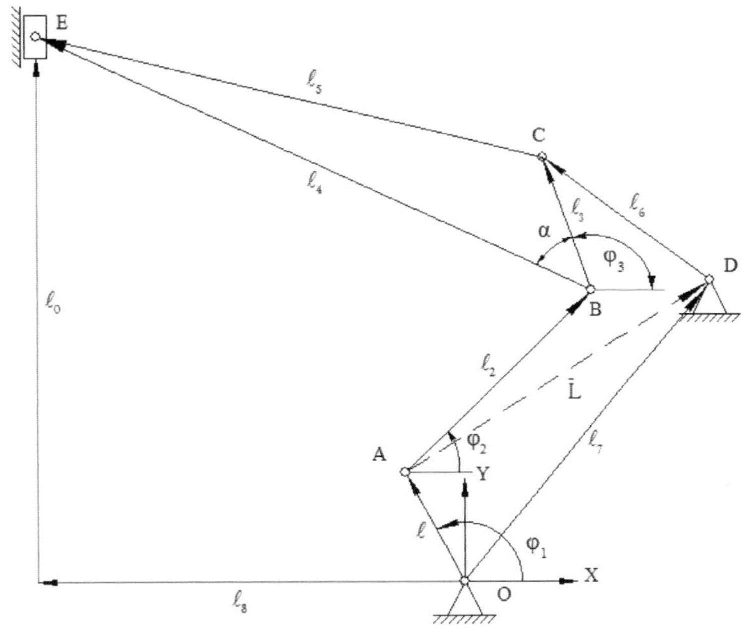

Figure 1. Kinematic scheme of flat lever mechanism of the third class.

For this mechanism it is possible to work out three following vector equations of isolation:

$$\left.\begin{aligned} \bar{l}_2 + \bar{l}_3 - \bar{l}_6 &= \bar{L} \\ \bar{l}_3 + \bar{l}_5 - \bar{l}_4 &= 0 \\ \bar{l}_2 + \bar{l}_4 - \bar{l}_0 &= \bar{l}_8 - \bar{l}_1 \end{aligned}\right\} \quad (1)$$

where $\bar{L} = \bar{l}_7 - \bar{l}_1$ the known vector-function of the generalized coordinate φ_1.
We choose any two from required vectors in the system Eq. (1) and we spread out them on orthogonal vectors $\bar{L}$ and $(\bar{k} \times \bar{L})$. Let it be a vector $\bar{l}_2$ and $\bar{l}_3$:

$$\left.\begin{aligned} \bar{l}_2 &= p_2 \bar{L} + q_2 (\bar{k} \times \bar{L}), \\ \bar{l}_3 &= p_3 \bar{L} + q_3 (\bar{k} \times \bar{L}), \end{aligned}\right\} \quad (2)$$

$\vec{i}$, $\vec{j}$, $\vec{k}$ cross heading of system of coordinates $Oxyz$; P_2, q_2 and P_3, q_3 are unknown decomposition coordinates [2].

The others vectors on the basis of equality Eq. (1) can be expressed through chosen. The vector $\bar{l}_4$ is defined by turning movement vector $\bar{l}_3$ on the corner α between them, and the other vector $\bar{l}_6$, $\bar{l}_5$ and $\bar{l}_0$ are defined from equalities Eq. (1). Substituting in the received expressions Eq. (2), we have:

$$\left.\begin{aligned}
\bar{l}_4 &= \frac{l_4}{l_3}\left[p\bar{L} + q\left(\bar{k}\times\bar{L}\right)\right]\\
\bar{l}_5 &= \left(\frac{l_4}{l_3}p - p_3\right)\bar{L} + \left(\frac{l_4}{l_3}q - q_3\right)\left(\bar{k}\times\bar{L}\right)\\
\bar{l}_6 &= \left(p_2 + p_3 - 1\right)\bar{L} + \left(q_2 + q_3\right)\left(\bar{k}\times\bar{L}\right)\\
\bar{l}_0 &= \bar{l}_1 - \bar{l}_8 + \left(p_2 + \frac{l_4}{l_3}p\right)\bar{L} + \left(q_2 + \frac{l_4}{l_3}q\right)\left(\bar{k}\times\bar{L}\right)
\end{aligned}\right\} \qquad (3)$$

where: $p = p_3\cos\alpha - q_3\sin\alpha$, $q = p_3\sin\alpha + q_3\cos\alpha$.

In system of the equations Eq. (3) parameters P_2, q_2, P_3, q_3 are unknown [3]. These parameters can be determined, having squared both parts of the equations Eq. (2) and the third equation of the system Eq. (2) and scalarly multiplying the last equation on $\bar{l}_8$:

$$\left.\begin{aligned}
&\left(p_2^2 + q_2^2\right)L^2 = l_2^2\\
&\left(p_3^2 + q_3^2\right)L^2 = l_3^2\\
&\left[\left(p_3 - p_2 - 1\right)^2 + \left(q_2 + q_3\right)^2\right]L^2 = l_6^2\\
&\left(p_2 + \frac{l_4}{l_3}\right)\left(l_8, \bar{L}\right) + \left(q_2 + \frac{l_4}{l_3 q}\right)\left[\bar{l}_8, \bar{k}, \bar{L}\right] = l_8^2 - \left(l_8, l_1\right)
\end{aligned}\right\} \qquad (4)$$

The nonlinear system of the equations generally has not the only solution and number of these solutions is equal to number of assemblies of the mechanism. The system Eq. (4) can be simplified, having reduced to two equations relatively q_2 and q_3. For this purpose from the second and the last system parameters Eq. (4) p_3 и p_2 we express through q_3 and q_2:

$$p_3 = \pm\left[\frac{l_3^2}{L^2} - q_3^2\right]^{\frac{1}{2}} \qquad (5)$$

$$p_2 = a_{11}p_3 + a_{12}q_3 + a_{13}q_2 + a_{14} \qquad (6)$$

where:

$$\left.\begin{aligned}
a_{11} &= -\frac{l_4}{l_3}\left(\cos\alpha - a_{13}\sin\alpha\right)\\
a_{12} &= \frac{l_4}{l_3}\left(\sin\alpha - a_{13}\cos\alpha\right)\\
a_{13} &= -\frac{\left[l_8, \bar{k}, \bar{L}\right]}{\left(\bar{l}_8, \bar{L}\right)}\\
a_{14} &= \frac{l_8^2 - \left(\bar{l}_8, \bar{l}_1\right)}{\left(\bar{l}_8, \bar{L}\right)}
\end{aligned}\right\} \qquad (7)$$

Substituting parameters p_2 and p_3 in the first and third equations of system Eq. (4), we have system of two equations relatively q_2 and q_3:

$$\left.\begin{array}{l}\left\{\pm\left[\dfrac{l_3^2}{L^2}-q_3^2\right]^{\frac{1}{2}}\cdot a_{11}+a_{12}q_3+a_{13}q_2+a_{14}\right\}^2+q_2^2=\dfrac{l_3^2}{L^2}\\ \left\{\pm\left[\dfrac{l_3^2}{L^2}-q_3^2\right]^{\frac{1}{2}}\cdot(1+a_{11})+a_{12}q_3+a_{13}q_2+a_{14}-1\right\}^2+(q_2+q_3)^2=\dfrac{l_6}{L^2}\end{array}\right\} \quad (8)$$

Here pluses and minuses before the radical correspond to different assemblies of the mechanism [4]. From solutions of the system Eq. (8) taking into account Eq. (6) and Eq. (7) we determine unknown parameters P_2, q_2, P_3, q_3. As initial approximations of projection of the corresponding vectors functions $\bar{L}$ and $\left(\bar{k}\times\bar{L}\right)$ for a certain position of the mechanism are set by synthesis conditions. Provisions of other links of the mechanism are from scalar products of cross heading of system of coordinates $Oxyz$ and the found vectors:

$$\left.\begin{array}{l} tg\varphi_2=\dfrac{\bar{l}_2\bar{j}}{\bar{l}_2\bar{i}}=\dfrac{p_2L_y+q_2L_x}{p_2L_x+q_2L_y}\\ tg\varphi_3=\dfrac{\bar{l}_3\bar{j}}{\bar{l}_3\bar{i}}=\dfrac{p_3L_y+q_3L_x}{p_3L_x-q_3L_y}\\ tg\varphi_4=\dfrac{\bar{l}_4\bar{j}}{\bar{l}_4\bar{i}}=\dfrac{pL_y+qL_x}{pL_x+q_3L_y}\\ tg\varphi_6=\dfrac{\bar{l}_6\bar{j}}{\bar{l}_6\bar{i}}=\dfrac{(p_2+p_3-1)L_y+(q_2+q_3)L_x}{(p_2+p_3-1)L_x-(q_2+q_3)L_y}\\ l_0=l_{1y}-l_{8y}+\left(p_2+\dfrac{l_4}{l_3}p\right)\cdot L_x+\left(q_2+\dfrac{l_4}{l_3}q\right)\cdot L_y \end{array}\right\} \quad (9)$$

where $L_x=l_{7x}-l_1\cos\varphi_1$, $L_y=l_{7y}-l_1\sin\varphi_1$.

We pass to definition of analogs of speeds and speedups. Differentiating Eq. (1) on φ_1 taking into account $\varphi_3'=\varphi_4'=\varphi_5'$, we receive:

$$\left.\begin{array}{l}\bar{l}_2\varphi_2'+\bar{l}_3\varphi_3'-\bar{l}_6\varphi_6'=-\bar{l}_1\\ \bar{l}_2\varphi_2'+\bar{l}_4\varphi_3'-\bar{i}\,l_0'=-\bar{l}_1\end{array}\right\} \quad (10)$$

Analogs of speeds of links can be defined from the solution of linear system Eq. (10):

$$\left.\begin{array}{l}\varphi_2'=\dfrac{1}{\Delta}\{(\bar{l}_1,\bar{j})[\bar{l}_3,\bar{k},\bar{l}_6]-(\bar{l}_4,\bar{j})[\bar{l}_1,\bar{k},\bar{l}_6]\}\\ \varphi_3'=-\dfrac{1}{\Delta}\{(\bar{l}_1,\bar{j})[\bar{l}_2,\bar{k},\bar{l}_6]-(\bar{l}_2,\bar{j})[\bar{l}_1,\bar{k},\bar{l}_6]\}\\ \varphi_6'=\dfrac{1}{l_6^2}[(\bar{l}_2,\bar{l}_6)\varphi_2'+(\bar{l}_3,\bar{l}_6)\varphi_3'+(\bar{l}_1,\bar{l}_6)]\\ l_0'=(l_1',\bar{i})+(\bar{l}_2,\bar{i})\varphi_2'+(\bar{l}_4,\bar{i})\varphi_3'\end{array}\right\} \quad (11)$$

where $\Delta=\left(\bar{l}_4\,\bar{j}\right)\left[\bar{l}_2\bar{k}\,\bar{l}_6\right]-\left(\bar{l}_2\,\bar{j}\right)\left[\bar{l}_3\bar{k}\,\bar{l}_6\right]$.

For definition of analogs of speedups of links we differentiate Eq. (10) on φ_1:

$$\left.\begin{aligned} \bar{\ell}_2\varphi_2'' + \bar{\ell}_3\varphi'' - \bar{\ell}_6\varphi_6'' &= -\bar{L}_1 \\ \bar{\ell}_2\varphi_2'' + \bar{\ell}_4\varphi_3'' - \bar{i}\ell_0'' &= -\bar{L}_2 \end{aligned}\right\} \qquad (12)$$

where:

$$\begin{aligned} \bar{L}_1 &= \bar{k}\times\bar{l}_1 + \bar{k}\times\bar{l}_2\varphi_2'^2 + \bar{k}\times\bar{l}_3\varphi_3'^2 - \bar{k}\times\bar{l}_6\varphi_6'^2 \\ \bar{L}_2 &= \bar{k}\times\bar{l}_1 + \bar{k}\times\bar{l}_2\varphi_2'^2 + \bar{k}\times\bar{l}_4\varphi_3'^2 \end{aligned} \qquad (13)$$

The solution of linear system Eq. (12) can be written down similarly in the following type:

$$\left.\begin{aligned} \varphi_2'' &= \frac{1}{\Delta}\{(\bar{L}_2,\bar{j})[\bar{l}_3,\bar{k},\bar{l}_6] - (\bar{l}_4,\bar{j})[\bar{L}_1,\bar{k},\bar{l}_6]\} \\ \varphi_3'' &= -\frac{1}{\Delta}\{(\bar{L}_2,\bar{j})[\bar{l}_2,\bar{k},\bar{l}_6] - (\bar{l}_2,\bar{j})[\bar{L}_1,\bar{k},\bar{l}_6]\} \\ \varphi_6'' &= \frac{1}{l_6^2}[(\bar{l}_2,\bar{l}_6)\varphi_2'' + (\bar{l}_3,\bar{l}_6)\varphi_3'' + (\bar{L}_1,\bar{l}_6)] \\ l_0'' &= (\bar{L}_2,\bar{i}) + (\bar{l}_2,\bar{i})\varphi_2'' + (\bar{l}_4,\bar{i})\varphi_3'' \end{aligned}\right\} \qquad (14)$$

The usage of formulas Eq. (9), Eq. (11) and Eq. (14) quite just allows making an algorithm of the solution tasks of the kinematic analysis of the mechanism of the third class on the computer [5].

3. Results and Discussion

In this paper, a kinematic model of the third class Assur group is presented. The method has the property of visibility, which makes it suitable for use in engineering practice [7]. Simulation of the kinematical analysis of third class plain lever mechanism with output dwell is done using Model GIM software application that gives opportunity to develop discrete, continuous and hybrid models of complicated technical systems (see Fig. 2-5).

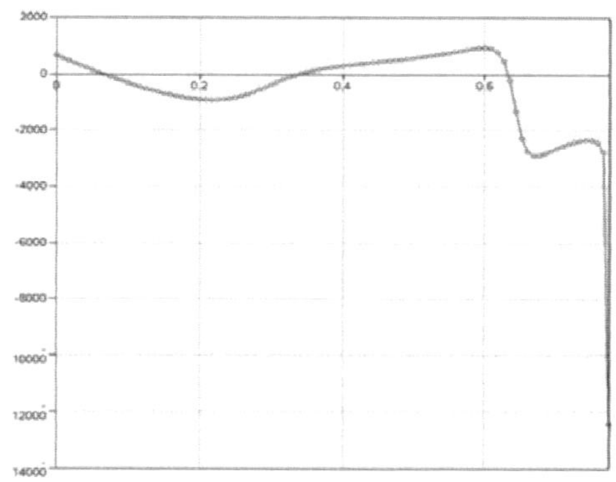

Figure 2. Coputed plot of the E point velocity module.

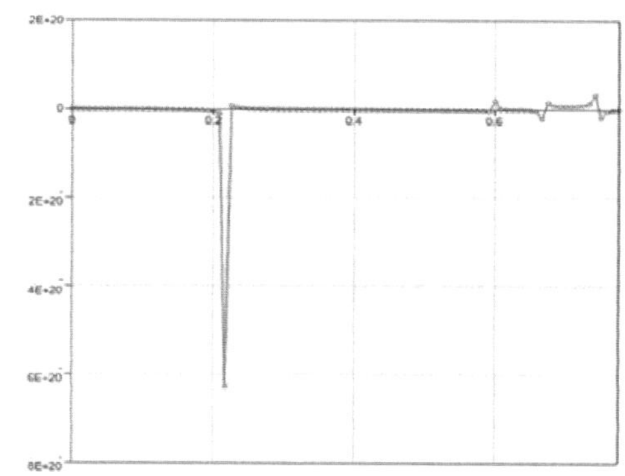

Figure 3. Coputed plot of the E point curvature center module.

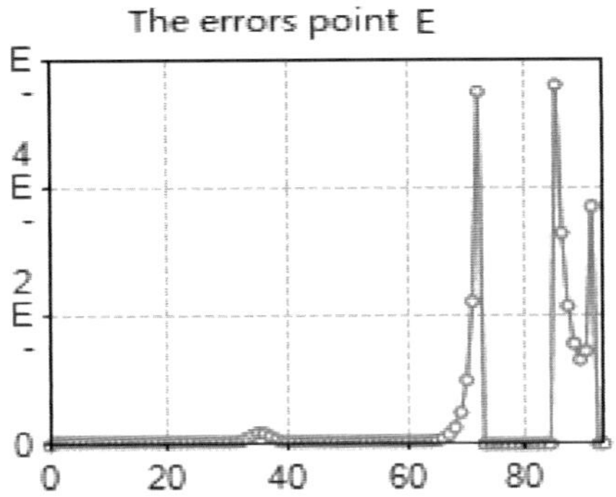

Figure 4. Computed plot of the errors point E.

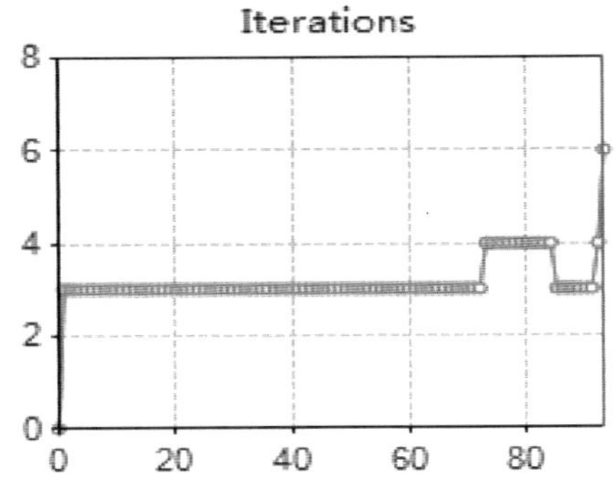

Figure 5. Computed plot of the iterations point E.

4. Conclusions

The combined analytic and graphic-analytical kinematic analysis method of the sixth class third order mechanism can be performed by accepting the inverse motion, considering the input element of the original mechanism as frame and one of the adjacent ternary element as drive element. The analysis of the mechanism is reduced to the study of third and second class Assur kinematic groups. The positional kinematic analysis shows that the most difficult problem is the solving of a trigonometric equations system by means of numerical methods. The kinematic analysis of a particular case indicated by Artobolevsky highlights the difficulties in computing the velocities in particular positions. However, some properties of the mechanism recommend it to be used in applications which need the preservation of parallel positions of the higher order couplers in self-locked extreme position and its compact size in folded position.

5. References

[1] Zhauyt A 2015 The substantiating of the dynamic parameters of the shaking conveyor mechanism *Vibroengineering Procedia* **5** 15-20

[2] Ren-Chung Soong 2008 An adjustable six-bar mechanism with variable input speed for mechanical forming presses *Transactions of the CSME /de la SCGM* **32** 453-466

[3] Hsieh W H and Tsai C H 2011 On a novel press system with six links for precision deep drawing *Mech Mach Theory* **46** 230-239

[4] Mitsi S, Tsiafis I and Bouzakis K D 2015 Force analysis of six-bar linkage for mechanical presses considering joint frictions *J Balk. Tribol. Assoc.* **21** 269-281

[5] Tso P L and Liang K C 2002 A nine-bar linkage for mechanical forming presses *Int J Mach Tools Manuf* **42** 126-139

[6] Zhauyt A, Zhahanova I, Smailova G, Murzakhmetova U, Kurmangalieva L, Kurenbay A 2017 The kinematic analysis of the third class mechanism *Vibroengineering PROCEDIA* **12** 208-212

[7] S Mitsi, I Tsiafis, K D Bouzakis 2017 Dynamic analysis of six-bar mechanical press for deep drawing *IOP Conf. Ser.: Mater. Sci. Eng.* **174** 1-8

Utilization of Aluminum Waste with Hydrogen and Heat Generation

O A Buryakovskaya[1,2], E A Meshkov[1], M S Vlaskin[1], E I Shkolnokov[1] and A Z Zhuk[1]

[1]Joint Institute for High Temperatures of the Russian Academy of Sciences, Izhorskaya str., 13, build. 2, 125412 Moscow, Russian Federation
[2]National Research Nuclear University MEPhI, Kashirskoe shosse 31, 115409 Moscow, Russian Federation
E-mail (O. A. Buryakovskaya): osminojishe@yandex.ru

Abstract.A concept of energy generation via hydrogen and heat production from aluminum containing wastes is proposed. The hydrogen obtained by oxidation reaction between aluminum waste and aqueous solutions can be supplied to fuel cells and/or infrared heaters for electricity or heat generation in the region of waste recycling. The heat released during the reaction also can be effectively used. The proposed method of aluminum waste recycling may represent a promising and cost-effective solution in cases when waste transportation to recycling plants involves significant financial losses (e.g. remote areas). Experiments with mechanically dispersed aluminum cans demonstrated that the reaction rate in alkaline solution is high enough for practical use of the oxidation process. In theexperiments aluminum oxidation proceeds without any additional aluminum activation.

1 Introduction

The total amount of waste (municipal and industrial) generated annually worldwide is about 4 billion tons [1]. Municipal solid waste contributes about 2.0 billion tons and the rate of waste generation is constantly growing. Municipal solid waste contains several percents of non-ferrous metals [2] and aluminum represented mostly by packaging waste accounts for about 60 % of the non-ferrous metals [3]. Another considerable source of aluminum waste is processing waste (generally scrap and chips). Aluminum waste also includes aluminum flakes, electrotechnical scrap, parts of vehicles, radiators, foil etc. In other words, a large amount of products made of aluminum is transformed into waste and the problem of its effective utilization becomes very actual.

Today the most widely used method of aluminum waste utilization is aluminum recycling from municipal or industrial waste withfurther production of secondary aluminum. Aluminum already achieves almost 60% recycling for all aluminum packaging and about 68% for beverage cans[4].

Main consumers of aluminum are transport industry, construction industry, packaging industry and mechanical engineering. To produce 1 kg of primary aluminum 4-5 kg of bauxite and 15-17 kWh of electricity is required.

To produce secondary aluminum from aluminum waste a number of technical procedures have to be carried out including collection, sorting and melting. Metal waste sorting is a labor-intensivemulti-staged process. Toremove organic coatings and paints from aluminum cans surface, two techniques can be applied: heating to a specific temperature and long-term thermal treatmentat this temperature or a short cyclical heating almost to the melting point of aluminum. Both methods are not environmentally friendly because they can lead to the formation of large quantities of environmentally harmful substances, for example, dioxins discharged into the atmosphere. After sorting and organic impurities removing the aluminum is melted. Due to inhomogeneous chemical composition a significant amount of slagis formedduring melting process. The weightofslagcan reach 15 % of the original metal weight. In some countries this slag is classified as hazardous waste [5].

Although the production of secondary aluminum reduces the consumption of primary aluminum the quality (chemical purity) and the cost of secondary aluminum in some cases don't meet the requirements ofprimary aluminum consumers. Due to the growth of aluminum waste generation a problem of aluminum waste utilization enhances.

It is an interesting was of aluminum waste utilization by corroding it in water with hydrogen production in according with the following equation:

$$Al + 2H_2O = AlOOH + 3/2H_2 \quad (1)$$

Thin passive layer of aluminum oxide produced onaluminumand aluminum alloy surfaces protects them against furtheroxidation and corrosion.

Themethods of intensification of the reaction between aluminum and water and development of approachesto efficient conversion of chemical and thermal energy into useful thermal and electrical energy were studied in[6-8].Chemical energy of aluminum can be converted into electricity with an efficiency of 40 % asreported in [9, 10].By-product of aluminum-fueled power plant – aluminum oxide can be used as separate markete product in different areas including catalysts and sorbent production or construction material [11, 12].

The aim of this study is to investigate oxidation of aluminum-containing waste in different forms in alkali aqueous solutionsfor hydrogen generation and analyze the structure of solid oxidation products.

2 Materials and Procedures

In this research waste aluminum cans mechanically dispersed into small pieces and aluminum wire in the form of cylinders with the length of 3 mm and diameter of 2 mm were used. Specific surface of aluminum cans was approximately 75.6 cm^2/g. In the experiments with alkali aqueous solutions preparation of waste aluminum cans included mechanical dispersion only without removingvarnish and paint. Only about 5 % of the aluminum canreaction surface was free of varnish and paint. Aluminum content in aluminum wire was 99.5 wt.%. Specific surface area of aluminum cylinders was approximately 11 cm^2/g. Experiments on oxidation of aluminum in forms of can particles and cylinders were carried out to establish the influence of paint and varnish (on the surface of waste aluminum cans) on oxidation process.

The experimental plant included a 500 cm^3reactor made of Simax glass with a thermostatic jacket which input and output were connected to the thermostat Huber CC-308. Reactor was located on a magnetic stirrer intended for adjustmentrotation speed of the stirring barat the bottom of the reactor. During the experiments the temperature of reaction medium was measured by a resistance thermometer with the accuracy of 0.01 °C. The scheme of experimental plant is shown in [13].

Hydrogen produced within the reactor from aluminum oxidation passed through a gas bubbler filled with distilled water forcooling down to room temperature. Then hydrogen was supplied into a collecting flask filled with water and the hydrogen pushed water out from the flask. The volume of water pushed out from the flask was equal to the volume of gas collected within the flask. Water replaced by hydrogen was collected in another flask placed on the Sartorius balance withthe accuracy of0.01 g. The balance was connected with a computer and transferred the mass data every 0.5 sec.

3 Results and Discussion

The results of the experiments on oxidation of waste aluminum cans and aluminum wire are presented in Table 1. Fig. 1 shows hydrogen generation curves for two experiments. As it can be seenin Fig. 4,regardless of the fact that the mass of aluminum can pieceswas twice more than that of aluminum wire pieces, the peak of hydrogen generation rate for aluminum wire is sharper and shorter than that of waste aluminum cans. For aluminum wire the maximum value of hydrogen generation rate was 61.5 cm^3/min per 1 g or 5.6 cm^3/min per 1 cm^2 of the initial reaction surface area; this value was achieved after 100 sec of the experiment, the induction period was 30 sec. For waste aluminum cans the maximum value of hydrogen generation rate was 27.1 cm^3/min per 1 g or 0.4 cm^3/min per 1 cm^2 of the initial interaction surface area; this value was achieved after 640 sec of the experiment, the induction time was 40 sec. Experimental results show that waste aluminum can piecesreactslower than the pieces of aluminum wire with a close value of specific surface area. The induction time for waste aluminum cans was longer than that for the aluminum wire due to the inhibiting effect of the paint. The specific surface area of the aluminum can pieces may reach its maximum value after total coating removal. The experiment with waste aluminum cans was stopped after about 5 hours when no hydrogen gas bubbling was observed. The conversion degree for aluminum cans was 73%.

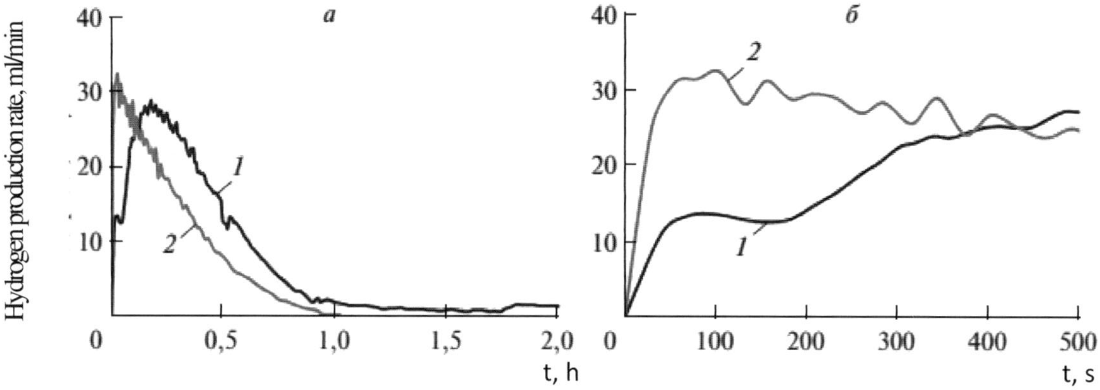

Figure1. Hydrogen production rate vstime: a) kinetic curves and b) initial sections of the kinetic curves for aluminum can pieces oxidation in KOH aqueous solution (1) and aluminum wire pieces in NaOH aqueous solution (2)

Table 1. Initial parameters and results of experiments on oxidation of waste aluminum cans and aluminum wire.

Parameter	Waste aluminum cans	Aluminum wire
Alkali	KOH	KOH
Initial pH of aquous solution	13.0	13.0
Temperature, °C	90	90
Stirring speed, rpm	300	300
The volume of alkali solution, cm^3	300	340
The mass of alkali, g	1.7	1.7
The mass of aluminum, g	1.07	0.52
Experiment duration, h	4.72	1.33
Aluminum conversion degree in the sample (without correction for the weight of the paint),% wt.	73	94

The solid oxidation product produced from aluminum cans was filtered, washed with distilled water, dried ina vacuum oven and then it was calcinated in air at a temperature of 1200 °C in order to obtain alumina.

According to XRD analysis the solid product producedfrom waste aluminum cans contains $Al(OH)_3$ in two crystallographic forms (gibbsite and bayerite). Gibbsite and bayeritecontribute 86 % and 14 % to the aluminum hydroxide volume respectively. Non-oxidized aluminum, as well as α- and γ-Al2O3 (< 5%) werealso registered in the product. The sample contained inconsequential amounts of potassium aluminate that may indicate low stability of this compound under the experimental conditions.

Surface morphology of solid products was studied using scanning electron microscopy (SEM). Fig. 2 shows microphotographs of the oxidation product obtained after the calcination at 1200 °C. As it can be seen the oxidation product obtained from waste aluminum cans represents a cluster of local build-up crystals, the size of a single crystal varies from several microns to several scores of microns. After heat treatment the crystal volume became highly porous with a typical pore size of 100 nm.

Figure2 – The texture of the solid oxidation products produced from aluminum cans

4. Conclusion

An alternative method for aluminum-containing waste utilization was investigated and compared with the conventional method of melting.

The solid oxidation products from aluminum cans were analyzed and the results proved formation of porous materials with complex structure.

The concept of using the method of aluminum-containing waste utilization in the remote areas was proposed.

5. Acknowledgment

The study was supported through a grant of the Russian Science Foundation (project No. 14-50-00124).

6.References

[1] 2012 *Globalization & waste management. Phase 1: Concepts and facts*: ISWA)

[2] 2014 Municipal Solid Waste Generation, Recycling, and Disposal in the United States. Tables and Figures for 2012. U.S. Environmental Protection Agency. Office of Resource Conservation and Recovery)

[3] Biganzoli L, Ilyas A, Praagh M v, Persson K M and Grosso M 2013 Aluminium recovery vs. hydrogen production as resource recovery options for fine MSWI bottom ash fraction *Waste Management* **33** 1174-81

[4] Bloxsome N September 15, 2014 Commitment to sustainable packaging in Europe *Aluminum International Today*

[5] Nappi C 2013 The global aluminium industry 40 years from 1972 *London: International Aluminium Institute*

[6] Larichev M N, Laricheva O O, Shaitura N S and Shkolnikov E I 2012 Possibilities of practical usage of dispersed aluminim oxidation by liquid water *Thermal Engineering* **59** 1000–9

[7] Dudoladov A O, Buryakovskaya O A, Vlaskin M S, Shkolnikov E I and Zhuk A Z 2015 Generation of hydrogen by aluminium oxidation in aqueous solutions at low temperatures *International Journal of Hydrogen Energy* **41**

[8] Vlaskin M S, Shkolnikov E I, Lisitsyn A V and Bersh A V 2010 Thermodynamic calculation of the parameters of a reactor for oxidizing aluminum in wet saturated steam *Thermal Engineering* **57** 794-801

[9] Franzoni F, et. al. 2011 Operating maps of a combined hydrogen production and power generation system based on aluminum combustion with water *Intern. J. Hydrogen Energy.* **36** 2803–16

[10] Franzoni F, et. al. 2010 Combined hydrogen production and power generation from aluminum combustion with water: analysis of the concept *Intern. J.Hydrogen Energy* **35** 1548–59

[11] Shkolnikov E I, Shaitura N S and Vlaskin M S 2013 Structural properties of boehmite produced by hydrothermal oxidation of aluminum *The Journal of Supercritical Fluids* **73** 10-7

[12] Lisitsyn A V, Dombrovsky L A, Mendeleyev V Y, Grigorenko A V, Vlaskin M S and Zhuk A Z 2016 Near-infrared optical properties of a porous alumina ceramics produced by hydrothermal oxidation of aluminum *Infrared Physics & Technology* **77** 162-70

[13] Ambaryan G N, Vlaskin M S, Dudoladov A O, Meshkov E A, Zhuk A Z and Shkolnikov E I 2016 Hydrogen generation by oxidation of coarse aluminum in low content alkali aqueous solution under intensive mixing *International Journal of Hydrogen Energy* **41** 17216-24

IOP Conf. Series: Materials Science and Engineering **250** (2017) 012008 doi:10.1088/1757-899X/250/1/012008

Analysis of the Source System of Nantun Group in Huhehu Depression of Hailar Basin

Yue Li [1,*], Junhui Li [2], Qi Wang [2], Bingyang Lv [2] and Guannan Zhang [3]

[1] Northeast Petroleum University, Earth Sciences Institute, Daqing 163318, China
[2] Exploration and Development Research Institute, Daqing Oilfield Company Limited, Daqing 163712, China
[3] Operation team of The NO.2 Oil Production Plant, Daqing Oilfield Company Limited, Daqing 163000, China
*Corresponding author e-mail: 418828568@qq.com

Abstract. Huhehu Depression will be the new battlefield in Hailar Basin in the future, while at present it's in a low exploration level. The study about the source system of Nantun group is little, so fine depiction of the source system would be significant to sedimentary system reconstruction, the reservoir distribution and prediction of favorable area. In this paper, it comprehensive uses of many methods such as ancient landform, light and heavy mineral combination, seismic reflection characteristics, to do detailed study about the source system of Nantun group in different views and different levels. The results show that the source system in Huhehu Depression is from the east of Xilinbeir bulge and the west of Bayan Moutain uplift, which is surrounded by basin. The slope belt is the main source, and the southern bulge is the secondary source. The distribution of source system determines the distribution of sedimentary system and the regularity of the distribution of sand body.

1. Introduction

The analysis of the sediment source is the very important content in analysis of sediment basin, which goes by reproduction of sedimentary basin evolution and recovering palaeoenvironment, its main methods is the composition of the terrestrial fragmental components and combination of heavy minerals. They are under the control of clastic rock handling distance and the types of rocks in the parent rock area. Sandstone is the main rock types of terrestrial clastic rock. The clastic material mainly contains the mechanical crushing of the mother rock, which reflects the important signs of the sediment source. The analysis of source system is significant to determine the location of sediment source, character, sediment transport path, and the sedimentary pattern of the whole basin, and prediction of sand body in basin.

This paper is in use of ancient landform, light and heavy minerals, seismic reflection characteristics, thickness of sandstone and percent content of sandstone, to do detailed study about ancient source system of Nantun group in Huhehu Depression in different angle and different levels. It has theoretical and practical significance to the destruction of sedimentary system and the further exploration of oil and gas in this area.

2. Geology

Huhehu Depression is in the southeast of Hailar basin, which is the secondary structure unit in Hailar basin. It spread from north to east, which is next to east of Selim Bell bulge, the west of Bayan Moutain uplift, the north is connected to Emmini Depressin, the south border is extended outside the

IOP Conf. Series: Materials Science and Engineering **250** (2017) 012008 doi:10.1088/1757-899X/250/1/012008

country and connected to Mongolia. The depression represents dustpan-like sag which is faulting in east and overlapping in west. Its square is 2500km^2.The structure can be divided into three tectonic units: actic region, gentle slope belt, excavated zone (fig.1). Sedimentary formation mainly composed of cretaceous, which divides from top to bottom is: lower Cretaceous Tongbomiao Formation, Nantun Formation, Damoguai River Formation, Yimin Formation and upper Cretaceous Qingyuangang Formation. Nantun Formation is the mainly oil layer system, its rock association from top to bottom is: Nantun Formation section one and Nantun Formation section two. Nantun Formation section one is mainly composed of dark grey mudstone, grey siltstone and gray, mixed sand conglomerate interbeded and there are grey coarse sandstone. Nantun Formation section two mainly contains a set of grey siltstone and thick black mudstone, grey conglomerate, grey-white packsand, black coal seam in some regions [1-3].

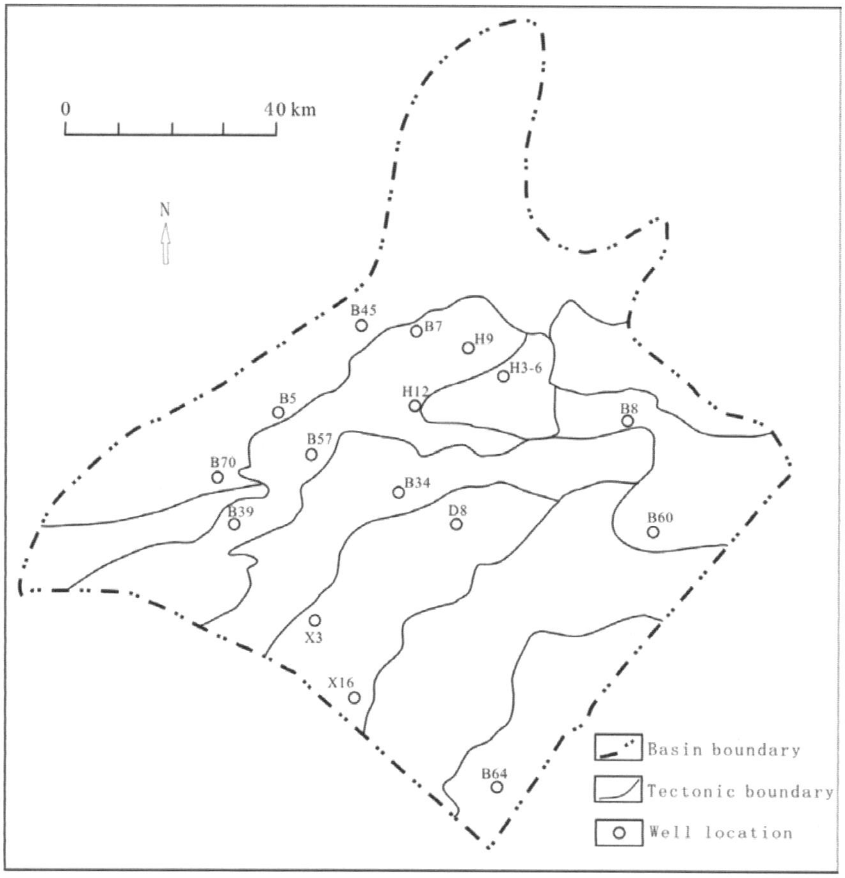

Figure 1. The location of the study area and the division of structural elements

3. Analysis of Combination of Syndepositional Fault

Different combination of syndepositional faults have different structure paleogeomorphology thus determine migration and distribution of source system, the space fabric of the depositional system. Therefore, to analysis combination of syndepositional fault in basin can help us to analysis the main sedimentation center, drawing out the path of source in basin, in order to provide macro direction of source analysis [4-5].

The parallel fault-order and sedimentary fracture combination system developes north-east direction which is in the northwestern gentle slope. It forms a multistage fault slope, comes to being the path of the multistage flow of sediment to the depression center, which can indicate the direction of the source. The development of a main fracture in the northwest slope and a parallel-associated secondary fracture, which formed the brush fracture system, the direction of source advanced from uplift of the main fracture to the lake basin. The area where the brush diverges can form a lower location of the structure easily, and they usually associate with thick "amputation of horn" sand body, therefore to control the local secondary depositional center; two groups of the same direction of the steep slope provide the main path to the migration of sediment source, they are low structure when they are in the fracture zone of synsedimentary faults, and they are usually in the injection position of the largest water system. The inside angle in the south forked fracture system corresponding to low structure location. The thick broken angle sand body controls the direction of source and distribution of sand body obviously (fig.2).

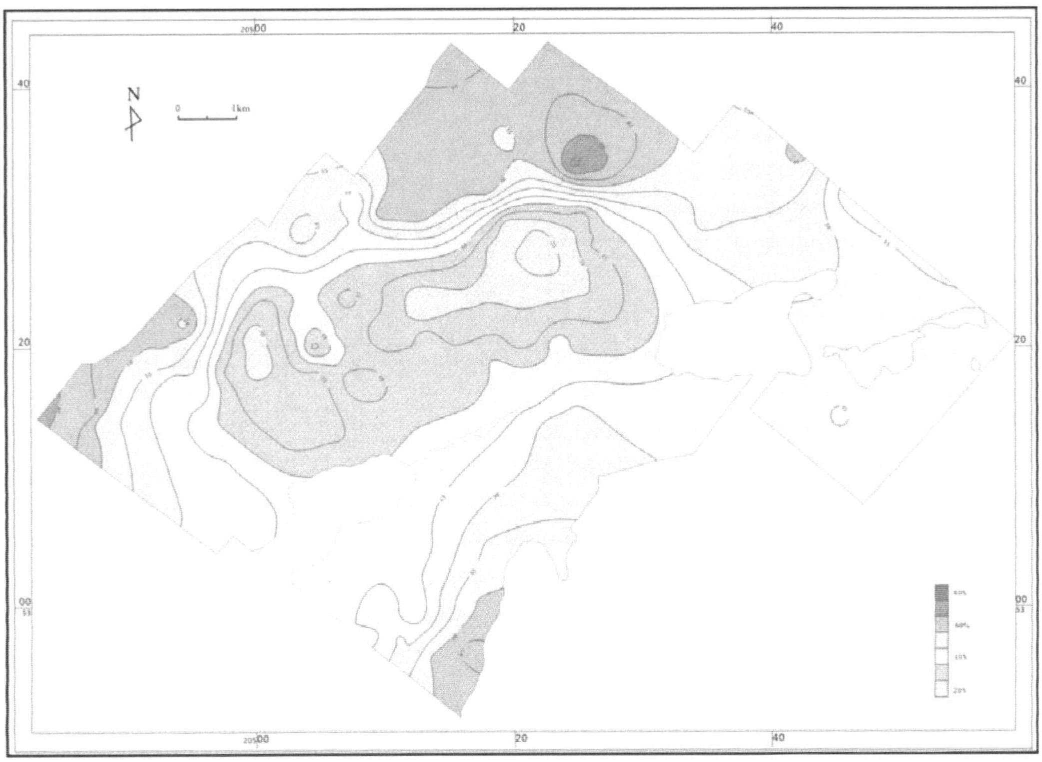

Figure 2. The characteristics of the pattern of fracture zone and the distribution of sand body

4. Analysis of Paleogeomorphologic Features

The paleogeomorphology has important control effect on the spread of sedimentary system. Through the analysis of paleogeomorphology, it can be determined that the spread pattern of upheaval and chase. By this way, we can confirm the source direction of the basin and the location of the source area. On the basis of construction grid interpretation and sequence stratigraphic structure in Huhehu sag, our research has been applied the settlement and stripping analysis technique and finally restore the early palaeogeomorphology of Nantun formation in Huhehu sag. We can see the changing process between the hollow and upheaval during this time visually. This study lays a solid foundation for the analysis of the source [6-9].

As a whole, the paleogeomorphologic features of the Nantun sedimentary period in Huhehu sag shows the tectonic characteristics of three hollows and two upheavals. The eastern Xlinbell bulge and the western Bayanshan uplift distribute surround the Huhehu sag. These two uplifts are the important source area of Huhehu sag during the Nantun sedimentary period. The distribution pattern of the hollow and upheaval in this depression is very evident. Sediment has been carried from the provenance to the low-lying district through valleys. By the recovery of Nantun group paleogeomorphology characteristic analysis, we think that the main source direction is the gentle slope and the steep slope. The southern uplift is the secondary source direction (fig.3).

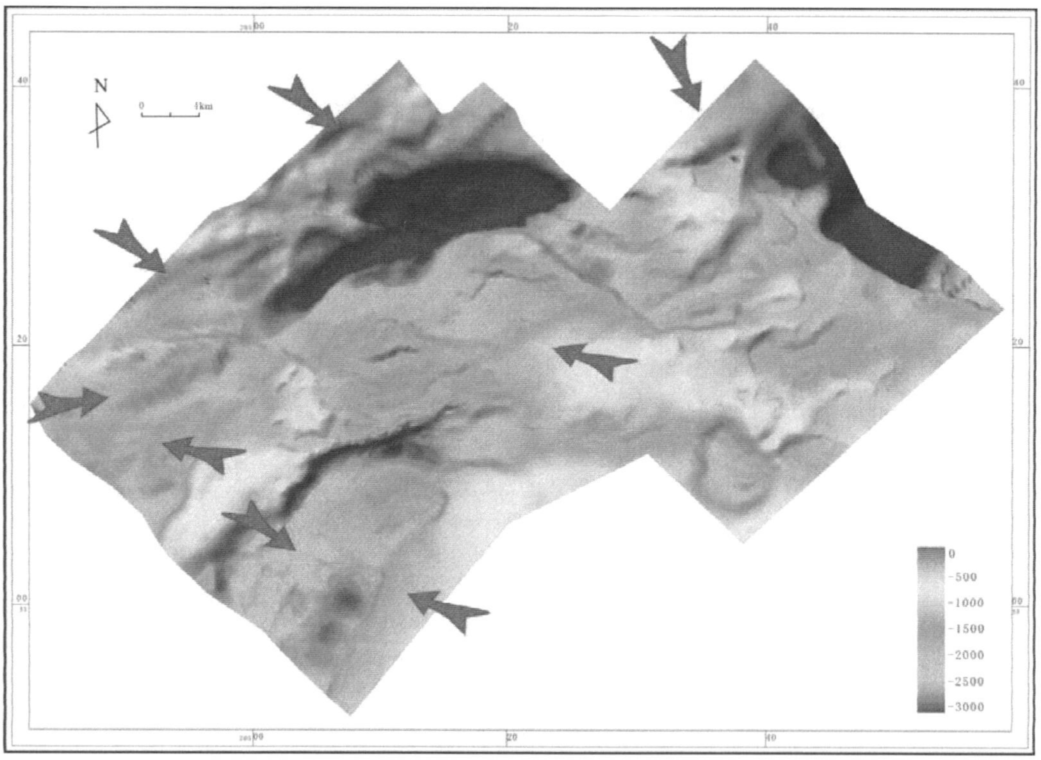

Figure 3. The palaeogeomorphology and source system analysis of Huhehu sag during Nantun period

5. The Distribution Pattern of Sedimentary System

Our research mainly used the date of core, logging and seismic progradation reflection characteristics. It confirms that Huhehu sag develop four kinds of sedimentary facies during Nantun period. They are fan dalta, braided river delta, sublacustrine fan and lake[4~5].

During the first half of the Nantun sedimentary period, the crust was sinking accompanied by a strong pull. Lakes became big and deep. Huhehu sag mainly developed the fan dalta, braided river delta and sublacustrine fan sedimentary facies. The fan dalta sedimentary system was developed in the southern region of sag. The northern region of sag most developed the braided river delta sedimentary system whose fan boby was very large. The subenvironment of lake mainly developed in the settlement center of sag. At the same time, because of gravity flow, the sublacustrine fan always developed in the shallow lake area.

The source direction of the second half of Nantun sedimentary period was same as the first period. Because of fractures continuous activity, the lake area expanded further. The sediment size was relatively smaller than before. Around the lake it mainly braided river delta sedimentary system and deep lake environment developed in subsags belt, near the center of sag. Because of gravity flow, the sublacustrine fan developed in the local area of deep lake and half - deep lake facies. Meanwhile, due to the characteristics of multi-cycle and volatile sedimentation, the lake were surrounded by swamp facies.

6. Results

To study ancient source system in Nantun formation of Huhehu Depression with using ancient landform, light-heavy minerals, seismic reflection characteristics, sandstone thickness and the percentage of sandstone in different levels and different angles. The results show that the source in Nantun Formation of Huhehu Depression has inheritance. The source comes from east of Xilinbeier bulge and west of Bayanshan Mountain uplift, and it gathers around the center. The gentle slope belt and the steep slope is the main source, the south of uplift is the secondary source. Nantun Formation develops four sedimentary facies including fan delta, braided river delta, sublacustrine fan and lake, which is according to core, logging, well-logging mainly and seismic reflection features, also combining with the regionally structural evolution. The north of Nantun 1 braided river delta, and south of it mainly contains of fan delta. Deeplake-half deep lake develops in some location. In Nantun 3 sedimentary period, developing braided river delta mainly, and sublacustrine fan facies in some location of deep lake area, and the marsh facies around the lake basin relatively developed at the same time.

7. References

[1] Yang Ren-chao, Li Jin-bu, Fan Ai-ping, et al. The progress and development trend of Terrigenous sedimentary rocks and provenance analysis[J]. Acta Sedimentologica Sinica, 2013, (01):99-107.

[2] Wei Ran, Li Hong-yang, Yu Bin, et al. Research Progress on provenance analysis of sedimentary basin[J]. Lithologic Reservoirs,2013,(03):53-57.

[3] Wu Lei,Xu Huai-min,Ji Han-cheng. Analysis of Paleogene sedimentary system evolution and source Bozhong sag in Bohai Bay Basin[J]. Quaternary Geology, 2006, (01):81-88.

[4] Huang Chuan-yan, Wang Hua, Xiao Dunqing, et al. Sequence Patterns, Characteristics of Depositional Systems and Model of Reservoirs of Fault Steep Slope Belt of the First M

ember of Shahejie Formation in Banqiao Sag[J].Acta Sedimentologica Sinica, 2007, 25(3):386-391.

[5] Feng You-liang, Xu Xiu-sheng. Syndepositional Structural Slope-Break Zone Controls on Lithologic Reservoirs-A Case from Paleogene BohaiBay Basin[J]. Petroleum Exploration and Development, 2006, 33(1): 22-31.

[6] Li Lin-lin, Guo Zhao-jie, Guan Shu-wei, et al. Characteristics of Cenozoic detrital heavy mineral assemblages and their palaeogeographic evolution in the southwestern margin of the Qaidam Basin[J]. Chinese Science: Earth Science, 2015, (06):780-798+1-6.

[7] Lv Lin, Jiao Yang-quan, Wu Li-qun, et al. Reconstruction of provenance sedimentary system of the Paleogene Shahejie Formation in Qikou Sag, Bohai Bay Basin [J].Acta Sedimentologica Sinica,2012,(04):629-638.

[8] Wang Hua, Bai Yun-feng, Huang Chuan-yan, et al. The Paleogene provenance system of the Dongying reconstruction and application of Qikou Sag[J]. Earth Science, 2009, (03):448-456.

[9] Chen Jun-liang, Wu He-yong, Zhu De-feng, et al. Tectonic Evolution of the Hailar Basin and Its Potentials of Oil-Gas Exploration[J]. Chinese Jour Geol, 2007, 42(1):147-159.

IOP Conf. Series: Materials Science and Engineering **250** (2017) 012009 doi:10.1088/1757-899X/250/1/012009

Synthesis and Thermal Stability of Novel Poly(*M*-Carborane-Siloxanes) with Various Pendant Groups

Xiaoxue Yang, Yanyan Zhao, Cuicui Wang, Xiaoa Zhang*, Shengling Jiang and YafeiLyu

Key Laboratory of Carbon Fiber and Functional Polymers of Ministry of Education, College of Materials Science and Engineering, Beijing University of Chemical Technology, Beijing 100029, China
*Email: zhangxiaoa1982@163.com

Abstract. Poly(*m*-carborane-siloxanes) with various pendant groups (**P15**-**P46**) were synthesized via polycondensation of *m*-carborane-containing disilanols (**1**-**4**) and highly active bisureidosilanes (**5** and **6**). The obtained polymers exhibit controlled molecular weight by carefully adjusting the monomer ratio. Standard spectroscopic techniques including FTIR and NMR were utilized to characterize these polymers and satisfactory results were obtained. TGA analysis indicated that the thermal cyclization of polysiloxanes under nitrogen was greatly postponed by the incorporated *m*-carborane cage, since the siloxane bonds within the main chain were strengthened by the inductive effect of the latter. DSC and FTIR results confirmed that both siloxane unit and carborane cage were oxidized at elevated temperature under air, which contributed to the transformation of the polymers into the mixture of SiO_2 and B_2O_3. Therefore, high char yield was obtained. Besides, the electronic effect of pendant groups greatly influenced the degradation behavior of *m*-carborane-containing polysiloxanes, having nothing to do with their position. The initial degradation temperature (T_{d5}) increases with varying substituent in the order: $CH_2CH_2CF_3 < CH_3 \approx Ph < CH{=}CH_2$.

1. Introduction

Polysiloxane materials have been widely used in various industries including aerospace, construction, electrical and electronics, motor industry, and medicals owing to their unique properties such as high thermal stability, weather resistance, low surface energy, excellent dielectric properties, and biocompatibility[1]. Polysiloxanes are generally composed of inorganic main chain linked by Si-O bonds and organic pendant groups such as methyl, phenyl, and vinyl, therefore their modification approaches can be divided into two types: (1) the attachment of functional pendant groups, and (2) the introduction of special building blocks into the polymer main chain. Systems of the first type can be prepared by direct (co)polymerization of functional organosilicon monomers or postpolymerization of reactive pendant groups[2]. Those of the second type are generally designed to improve the thermal stability of polysiloxanes. For instance, various metallic atoms such as aluminum, titanium, and iron were incorporated into the main chain of polysiloxanes, and the obtained polymers exhibited ultrahigh melting point (T_m) and glass transition temperature (T_g)[3]. Polysiloxanes with arylene in the main chain were also designed and synthesized so that polymers with high degradation temperature and char yield were obtained[4].

Poly(*m*-carborane-siloxanes) with exactly alternating structural units (scheme 1) exhibit excellent thermal stability and elastomeric properties simultaneously, and thus have been intensively studied[5]. It is believed that the carborane cage can inhibit the cyclization of linear polysiloxane at elevated temperature under nitrogen, so the depolymerization process is postponed. The polymers were first synthesized by ferric chloride-catalyzedpolycondensation of dichlorosilane with *m*-carborane-containing dimethoxysilane, nevertheless the obtained polymers showed crosslinked rather than linear structures due to the catalyst-induced crosslinking[6-9]. In order to synthesize linear polymers with satisfactory molecular weight, *m*-carborane-containing dihydroxysilanes were reacted with various reactive monomers such as bisaminosilane, bisureidosilane, and dichlorosilane[10-13]. Recently, the polycondensation of dihydroxysilanes and cyclosilazane was conducted by Chinese scientists, which gave polymers with high thermal stability as well[14].

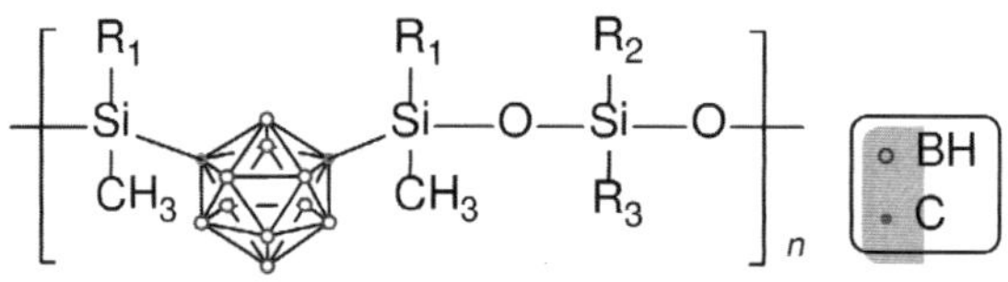

R_1=CH_3 or $CH_2CH_2CF_3$
R_2=CH_3, Ph, or $CH_2CH_2CF_3$
R_2=CH_3, Ph, or $CH_2CH_2CF_3$

Figure 1.Poly(*m*-carborane-siloxanes) with exactly alternating structural units.

It has been more than fifty years since the first poly(m-carborane-siloxane) was developed, however the synergistic effect of carborane moiety and pendant groups on the thermal stability has not been fully studied yet. In this work, four m-carborane-containing disilanols (1-4) with various substituents were designed and synthesized, which further reacted with bisureidosilanes (5 and 6) to give poly(m-carborane-siloxanes) (P15-P46) with various pendant groups and almost the same molecular weight (scheme 2). The obtained polymers were carefully characterized, and the effect of carborane moiety, atmosphere, and pendant groups on their thermal stability was studied.

2. Experimental

2.1. Materials

m-Carborane, tetrahydropyrrole, and *n*-butyllithium in *n*-hexane solution (2.4 M)were purchased from J&K Chemical. Phenyl isocyanate was obtained from Sigma Aldrich. Dimethyldichlorosilane, methylphenyldichlorosilane,methylvinyldichlorosilane, and methyltrifluoropropyldichlorosilane were obtained from TCI Chemical. Anhydrous diethyl ether, dry chlorobenzene, *n*-hexane, *n*-heptane, and acetone were from Sinopharm Chemical Reagent Beijing Co. Ltd. All materials were used without further purification.

2.2. Instrumentation

FTIR spectra were recorded on a Nexus 670 infrared spectrometer. Molecular weight (M_w) and polydispersity indices (M_w/M_n) of the polymers were tested in THF with a Waters 515 gel permeation chromatography system. Tetrahydrofuran (THF) was used as eluent at 30 °C at a flow rate of 1.0 ml/min. ^{1}H and ^{11}B NMR spectra were acquired in deuterated chloroform on a Bruker AV400 at a proton frequency of 400 MHz and the corresponding boron frequency. Tetramethylsilane was used as an internal standard. The melting temperature (T_m) of the *m*-carborane-containing monomers was determined by differential scanning calorimetric (DSC) method, which was conducted using a TA Q1000 DSC operating under nitrogen at a heating rate of 10 °C/min from 30 to 150 °C.Thermogravimetric analysis (TGA) and simultaneous differential scanning calorimetry (DSC) analysis were performed on a TA Instrument SDT-Q600 at a heating rate of 20 °C/min from 50 to 900 °C under nitrogen and air respectively.

*2.3. Synthesis of 1,7-Bis(Hydroxylmethylorganosilyl)-M-Carborane Monomers(**1-4**)*

1,7-Bis(hydroxylmethylorganosilyl)-*m*-carborane (**1-4**) were synthesized according to scheme 2. The detailed procedures for the synthesis of 1,7-bis(hydroxyldimethylsilyl)-*m*-carborane (**1**) were given below as an example. Into a 50 mL baked Schlenk flask with a side arm was added *m*-carborane (**7**, 2.58 g, 17.9 mmol). The flask was evacuated under vacuum and flushed with dry nitrogen three times through the side arm. Then 10 mL of anhydrous ether was injected into the flask to dissolve **7**. After the flask was cooled below 0 °C in an ice-salt bath, *n*-butyllithium in *n*-hexane solution (2.4 M, 16.4 mL, 39.4 mmol) was added dropwise, and then the system was intensely stirred for 3 h. After stirring was stopped, the reaction mixture separated into two layers. The upper layer is *n*-hexane, while the lower layer is 1,7-dilithio-*m*-carborane ether solution. After the upper layer was removed by double-ended needle under reduced pressure, the lower layer was washed with 20 mL of *n*-hexane on stirring. *n*-Hexane was removed by the above method. The flask was further cooled in an ice-salt bath, and 10 mL of anhydrous diethyl ether was injected to dilute the solution. Into the flask was added dropwisedimethyldichlorosilane (4.8 mL, 39.4 mmol), and then the system was stirred below 0 °C for 1 h. Then the reaction proceeded at room temperature for 20 h. The formed solid byproduct lithium chloride (LiCl) was removed by filtration under reduced pressure under nitrogen. The collected filtrate was condensed to remove solvent and excess silane. 1,7-Bis(chlorodimethylsilyl)-*m*-carborane was obtained as a light brown clear liquid. This liquid was transferred to reaction flask, into which 20 mL of anhydrous diethyl ether, 20 mL of acetone, and 5 mL of deionized water were also added. The system was stirred at room temperature for 5 h. After stirring was stopped, the reaction mixture separated into two layers. The upper layer is 1,7-bis(hydroxyldimethylsilyl)-*m*-carborane (**1**) ether solution, while the lower layer is mixed solvent of acetone and water. The upper layer was condensed under reduced pressure, and crude product was obtained. Then the obtained **1** was purified by recrystallization in *n*-hexane three times, and white solid was obtained in a yield of 52.8%. T_m = 101.1 °C (DSC method). FTIR, ν (cm^{-1}, in KBr pellet): 3292 (-OH), 2963 ($-CH_3-$), 2592 (B-H), 1403, 1260, 1103 (Si-O), 994, 866, 811, 781, 659. ^{1}H NMR (400 MHz, $CDCl_3$), δ (TMS, ppm): 3.50-1.50 (m, 10H), 1.94 (s, 2H), 0.26 (s, 12H).

1,7-Bis(hydroxylmethylphenylsilyl)-m-carborane(2)was synthesized from methylphenyldichlorosilane in a yield of 46.4% by similar method as a white solid. Tm = 87.6 oC (DSC method). FTIR, ν (cm-1, in KBr pellet): 3336 (-OH), 3077 (CH on Ph), 2973 (-CH3-), 2594 (B-H), 1590 (C=C on Ph), 1428, 1260, 1100 (Si-O), 993, 853, 773, 730, 694, 472, 428. 1H NMR (400 MHz, CDC13), δ (TMS, ppm): 7.46 and 7.37 (m, 10H), 3.10-1.40 (m, 10H), 2.21 (s, 2H), 0.49 (s, 6H).

1,7-Bis(hydroxylmethylvinylsilyl)-*m*-carborane (**3**) was synthesized from methylvinyldichlorosilane in a yield of 58.1% by similar method as a white solid. T_m = 113.2 °C (DSC method). FTIR, ν (cm^{-1}, in KBr pellet): 3299 (-OH), 3060 and 3020 ($=CH_2$), 2958 ($-CH_3-$), 2609 (B-H), 1634 and 1595 (C=C), 1406, 1260, 1103 (Si-O), 968, 858, 802. ^{1}H NMR (400 MHz, $CDCl_3$), δ (TMS, ppm): 6.11 (m, 2H), 5.89 (m, 4H), 3.30-1.40 (m, 10H), 2.01 (s, 2H), 0.33 (s, 6H).

1,7-Bis(hydroxylmethyltrifluoropropylsilyl)-m-carborane (4) was synthesized from methyltrifluoropropyldichlorosilane in a yield of 28.7% by similar method as a white solid. Tm = 82.0 oC (DSC method). FTIR, ν (cm-1, in KBr pellet): 3299 (-OH), 2973 (-CH3-), 2911 (-CH2-), 2609 (B-H), 1421 (Si-CH3), 1266, 1212, 1135, 1069, 1029, 898 (Si-CH2CH2CF3), 1103 (Si-O), 860, 807. 1H NMR (400 MHz, CDC13), δ (TMS, ppm): 3.30-1.40 (m, 10H), 2.20-2.00 (s, 4H, Si-CH2CH2CF3), 1.95 (hidden by peak at 2.20-2.00, Si-OH), 1.10-0.80 (m, 4H, Si-CH2CH2CF3), 0.30 (s, 6H, Si-CH3).

*2.4. Synthesis of bis(1,1-tetramethylene-3-phenylureido)methylorganosilane monomers(**5** and **6**)*

Bis(1,1-tetramethylene-3-phenylureido)methylorganosilane monomers (5 and 6) were synthesized according to scheme 2. The detailed procedures for the synthesis of bis(1,1-tetramethylene-3-phenylureido)dimethylsilane (5) were given below as an example. A 50 mL Schlenk flask with a side arm was evacuated under vacuum and flushed with dry nitrogen three times through the side arm. Then into the flask were added n-heptane (20 mL) and dimethyldichlorosilane (6.10 mL, 50 mmol). The flask was cooled to -10 oC in an ice-salt bath, and then tetrahydropyrrole (17.7 mL, 200 mmol) was added dropwise. The system was intensely stirred at -10 oC for 3 h, and white solid byproduct was generated. The reaction proceeded at room temperature for another 15 h. Then the solid byproducttetrahydropyrrole hydrochloride was removed by filtration under reduced pressure under nitrogen. The collected filtrate was evaporated under reduced pressure to remove solvent. The obtained crude product was further dissolved in 8 mL of anhydrous diethyl ether, and then the solvent was once again removed under reduced pressure. The dissolution and removal processes were conducted three times, and bis(tetramethyleneamino)dimethylsilane was obtained as a light brown liquid. The flask was once again cooled to -10 oC, and 25 mL of anhydrous diethyl ether was added. The system was stirred and phenyl isocyanate (10.8 mL, 98.3 mmol) was added dropwise. The reaction was conducted at -10 oC for 3 h, and plenty of white solid was generated. The reaction was further conducted at room temperature for 2 h. After stirring was stopped, the upper liquid layer was removed by double-ended needle under reduced pressure. The lower solid was washed with 10 mL of anhydrous diethyl ether, and the solvent was removed by the above method. After being washed three times, white solid 5 was obtaind in a yield of 86.4%. FTIR, ν (cm-1, in KBr pellet): 3300, 3056, 2970 (-CH3-), 2877 (-CH2-), 1665, 1645 (C=O), 1597 (Ph), 1531, 1488, 1445, 1395, 1296, 1238, 1181, 1074, 1024, 958 (Si-N), 909, 845, 802 (Si-C), 752, 695, 659, 550, 501. 1H NMR (400 MHz, CDCl3), δ (TMS, ppm): 7.41 (d, 4H, aromatic protons ortho to N), 7.27 (d, 4H, aromatic protons meta to N), 7.01 (m, 2H, aromatic protons para to N), 3.46 (s, 4H), 1.97 (s, 4H), 0.09 (s, 6H).

Bis(1,1-tetramethylene-3-phenylureido)methyltrifluoropropylsilane (**6**) was synthesized from methyltrifluoropropyldichlorosilane in a yield of 98.3% by similar method as a white solid. FTIR, ν (cm^{-1}, in KBr pellet): 3300, 3064, 2970 ($-CH_3-$), 2876 ($-CH_2-$), 1667 (C=O), 1595 (Ph), 1538, 1482, 1418, 1307, 1267, 1201, 1123, 1074, 1024, 966, 902, 836, 758, 701, 552, 509. ^1H NMR (400 MHz, acetone-d_6), δ (TMS, ppm): 7.56 (d, 4H, aromatic protons *ortho* to N), 7.28 (d, 4H, aromatic protons *meta* to N), 7.00 (m, 2H, aromatic protons *para* to N), 3.43 (s, 4H), 2.39 (m, 2H), 1.91 (s, 4H), 1.40-1.20 (m, 2H), 0.27 (s, 3H).

*2.5. Synthetic Routes to Poly(M-Carborane-Siloxanes) (P**15**-P**46**)*

Poly(*m*-carborane-siloxanes) with various pendant groups (**P15**-P**46**) were synthesized according to scheme 2. The detailed procedures for the synthesis of poly(*m*-carborane-dimethylsiloxane) (P**15**) were given below as an example. Into a 50 mL baked Schlenk flask with a side arm was added 1,7-bis(hydroxyldimethylsilyl)-*m*-carborane (**1**, 731.3 mg, 2.5 mmol). The flask was evacuated under vacuum and flushed with dry nitrogen three times through the side arm. Then 5 mL of dry chlorobenzene was injected into the flask to dissolve **1**. In another dry schlenk flask bis(1,1-tetramethylene-3-phenylureido)dimethylsilane (**5**, 1.146 g, 2.625mmol) was dissolved in 10 mL of dry chlorobenzene.After the first flask containing **1** was cooled to -10 °C in an ice-salt bath, the solution of **5** was added dropwise to the flask on stirring. The system was intensely stirred at -10 °C for 3 h, and then at room temperature for another 10 h. The solid byproduct was removed by filtration under reduced pressure under nitrogen, and then the filtrate was transferred to reaction flask. **1** (51.5 mg, 0.176mmol) was dissolved in 2 mL of dry chlorobenzene, and then this solution was added dropwise to the reaction flask at -10 °C under nitrogen. The reaction was conducted at -10 °C for 2 h and at room temperature for 3 h. The solid was removed by filtration, and the collected filtrate was condensed and added to a large amount of methanol on stirring. A white sticky product of polymer **P15** was precipitated in a yield of 68.2%. M_w: 11200; M_w/M_n: 2.05 (GPC, polystyrene calibration). FTIR, ν (cm^{-1}, in KBr pellet): 2963, 2597, 1410, 1262, 1100, 1046, 858, 802. ^1H NMR (400 MHz, $CDCl_3$), δ (TMS, ppm): 3.10-1.50 (B-*H*), 0.19, 0.08.

P**25** was synthesized from **2** and **5** by a similar method in a yield of 71.9%. M_w: 12900; M_w/M_n: 2.12 (GPC, polystyrene calibration). FTIR, ν (cm^{-1}, in KBr pellet): 3317, 3133, 3071, 2965, 2873, 2596, 1957, 1886, 1650, 1595, 1534, 1500, 1444, 1393, 1305, 1262, 1177, 1100-1000, 845, 803, 733, 695, 474, 429. ^1H NMR (400 MHz, $CDCl_3$), δ (TMS, ppm): 7.50-7.10 (aromatic protons), 3.30-1.50, 0.29, 0.06.

P**35** was synthesized from **3** and **5** by a similar method in a yield of 50.9%. M_w: 8800; M_w/M_n: 2.22 (GPC, polystyrene calibration). FTIR, ν (cm^{-1}, in KBr pellet): 3060, 3017, 2962, 2596, 1595, 1444, 1405, 1262, 1100-1000, 963, 848, 803, 770. ^1H NMR (400 MHz, $CDCl_3$), δ (TMS, ppm): 6.06 ($\boldsymbol{CH_2}$=CH-), 5.77(CH_2=C***H***-), 3.30-1.50, 0.26, 0.11.

P**45** was synthesized from **4** and **5** by a similar method in a yield of 44.1%. M_w: 9500; M_w/M_n: 1.63 (GPC, polystyrene calibration). FTIR, ν (cm^{-1}, in KBr pellet): 2964, 2910, 2600, 1444, 1419, 1368,

1316, 1266, 1211, 1130, 1100-1000, 1070, 1025, 900, 802, 768. ^{1}H NMR (400 MHz, $CDCl_3$), δ (TMS, ppm): 3.10-1.50, 2.10-2.00 (CH_2CF_3), 1.00-0.80 ($CH_2CH_2CF_3$), 0.26, 0.15.

P16 was synthesized from **1** and **6** by a similar method in a yield of 61.1%. M_w: 9200; M_w/M_n: 1.78 (GPC, polystyrene calibration). FTIR, ν (cm^{-1}, in KBr pellet): 2964, 2597, 1444, 1410, 1368, 1314, 1264, 1210, 1100-1000, 900, 858, 812. ^{1}H NMR (400 MHz, $CDCl_3$), δ (TMS, ppm): 3.20-1.50, 2.10-2.00, 0.90-0.70, 0.23, 0.13.

P36 was synthesized from **3** and **6** by a similar method in a yield of 65.6%. M_w: 16000; M_w/M_n: 2.05 (GPC, polystyrene calibration). FTIR, ν (cm^{-1}, in KBr pellet): 3060, 3010, 2961, 1596, 1444, 1406, 1368, 1315, 1264, 1211, 1130, 1100-1000, 1068, 966, 900, 850, 806, 767. ^{1}H NMR (400 MHz, $CDCl_3$), δ (TMS, ppm): 6.20-5.70, 3.30-1.50, 2.10-1.90, 0.90-0.70, 0.28, 0.13.

P46 was synthesized from **4** and **6** by a similar method in a yield of 52.4%. M_w: 12000; M_w/M_n: 2.11 (GPC, polystyrene calibration). FTIR, ν (cm^{-1}, in KBr pellet): 2964, 2910, 2600, 1444, 1419, 1368, 1316, 1266, 1211, 1130, 1100-1000, 1070, 1025, 900, 802, 768. ^{1}H NMR (400 MHz, $CDCl_3$), δ (TMS, ppm): 3.20-1.50, 2.20-2.00, 1.26, 0.79, 0.28, 0.19.

3. Results and Discussion

*3.1. Synthesis of Poly(M-Carborane-Siloxanes) (**P15-P46**)*

Poly(*m*-carborane-siloxanes) with various pendant groups (**P15-P46**) were synthesized via polycondensation of *m*-carborane-containing disilanols (**1-4**) and bisureidosilanes (**5** and **6**) according to scheme 2. The reactions and polymerizations proceeded smoothly, and desired polysiloxanes were obtained in acceptable yields (~ 41-71 %). It is worth noting that the bisureidosilane monomers are very sensitive to the moisture in the air, so they must be kept and dealt with under dry nitrogen in the glove box. The polycondensation is a nucleophilic substitution on bisureidosilane by disilanol as the nucleophile. The substitution can proceed rapidly below 0 °C, at which temperature the possible self-condensation side reaction of silanol can be greatly restrained. Chlorobenzene was selected as the polymerization solvent, for it is a good solvent for the polymer but cannot dissolve the urea byproduct, providing a driving force for the complete polycondensation and enabling an easy purification method of polymer from the byproduct.

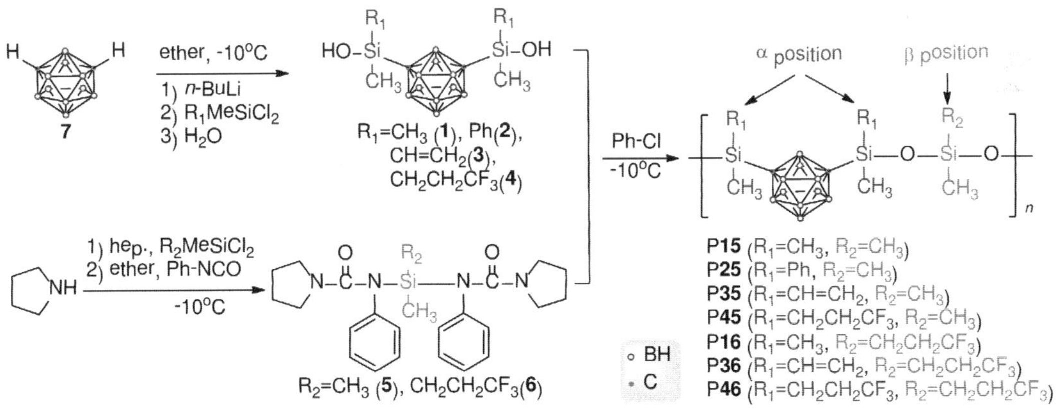

Figure 2. Synthetic routes to poly(*m*-carborane-siloxanes) with various pendant groups (**P15-P46**).

Considering the step-growth character of the polycondensation, the monomer ratio should be carefully adjusted so as to acquire polymers with high molecular weight. The synthetic process of poly(*m*-carborane-dimethylsiloxane) (**P15**) is taken as an example, and the results are shown in Table 1. The two monomers, 1,7-bis(hydroxyldimethylsilyl)-*m*-carborane (**1**) and bis(1,1-tetramethylene-3-phenylureido)dimethylsilane (**5**), were introduced to the reaction system by a semicontinuous charging, and thus the monomer ratio (r = [**5**] / [**1**]) was well regulated. Stoichiometric **1** was first placed in the flask, and then the chlorobenzene solution of **5** was added dropwise at -10 °C. During the addition, r increased from 0 to 1.05, while polymers with low molecular weight were obtained (Table 1, nos 1-3).

In order to increase the molecular weight of the product, additional **1** was introduced slowly to adjust the molar ratio, and polymer with the highest M_w was obtained when r is 0.98 (Table 1, no 6).

Table 1.The synthesis of poly(*m*-carborane-dimethylsiloxane) (**P15**) [a]

no.	t, h [b]	r [c]	yield, %	M_w [d]	M_w/M_n [d]
1	0	0	0	-	-
2	3	1.05	20.4	2940	1.58
3	13	1.05	39.6	3200	1.62
4	14	1.02	32.7	5030	1.60
5	15	1.00	41.1	7400	1.72
6	16	0.98	68.2	11200	2.05
7	17	0.96	66.5	8600	2.01

[a] Carried out in chlorobenzene at -10 °C under nitrogen; [b] t is the reaction time; [c] r is the molar ratio of **5** to **1**, r = [**5**] / [**1**]; [d] Estimated by gel permeation chromatograph (GPC) in THF on the basis of polystyrene calibration.

Similarly, the above semicontinuous charging was also utilized in order to synthesize poly(*m*-carborane-siloxane)s with various pendant groups. The optimal molar ratio of bisureidosilane to disilanol (r) was obtainted and the results are shown in Table 2. It is seen that polymers with relatively consistent M_w were obtaind by carefully regulating r. These polymers will be chosen to study the effect of pendant group on their thermal stability, without considering the effect of molecular weight.

Table 2.The synthesis of poly(*m*-carborane-siloxane)s (**P25**-**P36**) [a]

no.	polymer	t, h [b]	r [c]	yield, %	M_w [d]	M_w/M_n [d]
1	**P25**	16	0.98	71.9	12900	2.12
2	**P25**	17	0.96	68.3	29400	2.50
3	**P35**	16	0.98	50.9	8800	2.22
4	**P45**	17	0.96	44.1	9500	1.63
5	**P16**	16	0.98	61.1	9200	1.78
6	**P36**	17	0.96	60.8	30400	2.55
7	**P36**	18	0.94	65.6	16000	2.05
8	**P46**	17	0.96	52.4	12000	2.11

[a] Carried out in chlorobenzene at -10 °C under nitrogen; [b] t is the reaction time; [c] r is the molar ratio of bisureidosilane to disilanol; [d] Estimated by gel permeation chromatograph (GPC) in THF on the basis of polystyrene calibration.

3.2. Structure Characterization

The structures of the obtained poly(*m*-carborane-siloxane)s (**P15**-**P46**) were characterized by standard spectroscopic methods such as FTIR and NMR. The analysis data were quite satisfactory corresponding to the expected molecular structures (see Experimental section for details). An example of the FTIR spectrum of poly(*m*-carborane-dimethylsiloxane) (**P15**) is shown in Fig. 1; the spectra of its corresponding monomers, 1,7-bis(hydroxyldimethylsilyl)-*m*-carborane (**1**) and bis(1,1-tetramethylene-3-phenylureido)dimethylsilane (**5**), are also given in the same figure for comparison. It is seen from Figure 1 that monomer **1** shows strong absorption band at 2592 cm^{-1}, which is assigned to the stretching vibrations of B-H bonds. This band is also present in the spectrum of P**15**, indicating that the carborane cage is successfully incorporated into the polymer structure. Besides, both monomers show medium absorption peaks at 2963 cm^{-1}, ascribing to the asymmetry stretching vibrations of methyl attached to silicon atoms. This peak is also seen in the spectrum of P**15**, which may be attributed to the polymerization process that dimethyl silyl from the two monomers will constitute the polymer main chain. Furthermore, the broad double-peak appears in the range of 1100-1048 cm^{-1} attributed to the characteristic absorption of Si-O-Si structure, which indicates the formation of polydimethylsiloxane main chain.

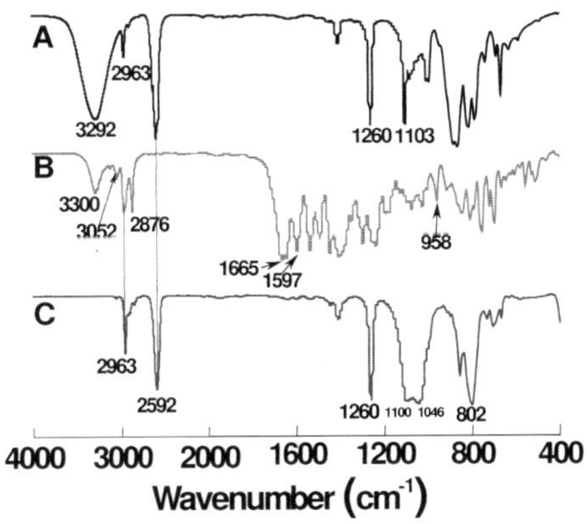

Figure3. FTIR spectra of (A) 1,7-bis(hydroxyldimethylsilyl)-*m*-carborane (**1**), (B) bis(1,1-tetramethylene -3-phenylureido)dimethylsilane (**5**), and (C) poly(*m*-carborane-dimethylsiloxane) (**P15**, Table 1 no. 6).

Figure 2 shows the ^{1}H NMR spectra of P**15** and their starting monomers (**1** and **5**). It is seen that monomer **1** shows three kinds of protons, which is well consistent with its molecular structure. Hydroxyl protons show their resonance peak at δ 1.94, while methyl protons resonate at δ 0.26. Baseline prominence and unrecognized peaks at δ 3.50-1.50 are ascribed to protons on the carborane cage. Monomer **5** also shows resonance peaks well associated with its expected structures. The aromatic protons resonate at δ 7.41, 7.27, and 7.01, and methyl protons resonate at δ 0.09, while the peaks associated with the resonance of methylene protons appear at δ 3.46 and 1.97. During the polymerization, dimethylsilane structures from both monomers are connected together to constitute the polymer chain, therefore **P15** shows two resonance peaks of methyl protons at δ 0.19 and 0.08. Besides, similar to monomer **1**, **P15** also shows baseline prominence at δ 3.50-1.50 ascribed to the resonance of protons on carborane cage.

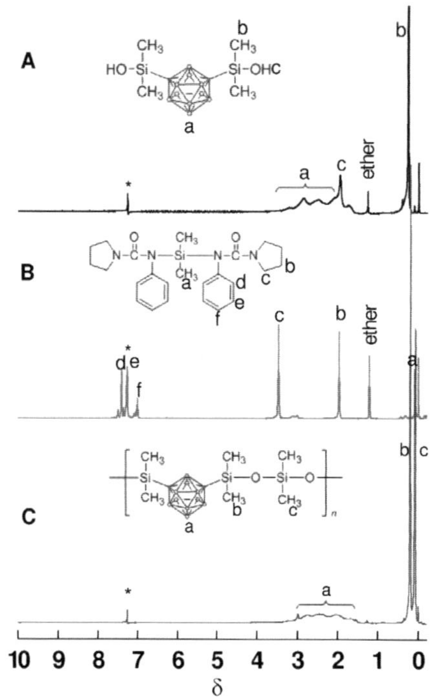

Figure4.^{1}H NMR spectra of (A) 1,7-bis(hydroxyldimethylsilyl)-*m*-carborane (**1**), (B) bis(1,1-tetramethylene -3-phenylureido)dimethylsilane (**5**), and (C) poly(*m*-carborane-dimethylsiloxane) (**P15**, Table 1 no. 6) in $CDCl_3$. The solvent peaks are marked with asterisks.

The most attractive point of this work lies in the introduction of carborane cage into the main chain of polysiloxanes with various pendant groups. Considering the existence of boron, ^{11}B NMR spectroscopy was also utilized to characterize the obtained polymers. The ^{11}B NMR spectrum of **P15** is shown in Figure 3. It is clear that theborons on carborane cage resonate in the range of δ -2 to -18, which is well consistent with the NMR data of *m*-carborane.

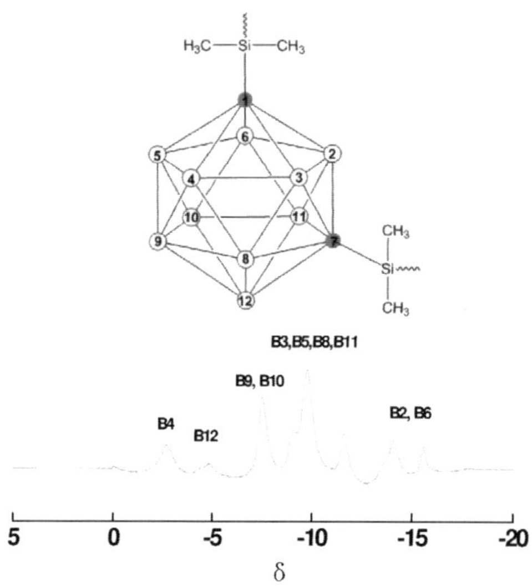

Figure5.^{11}B NMR spectrum of poly(*m*-carborane-dimethylsiloxane) (**P15**, Table 1 no. 6) in $CDCl_3$.

3.3. Thermal Stability

Carborane cage can serve as a heat resistant structural unit within the inorganic main chain of polysiloxane, therefore it is of great importance to carefully study the thermal stability of poly(*m*-carborane-siloxanes). Firstly, thermal gravimetric analysis (TGA) technique was utilized to evaluate the thermal stability of poly(*m*-carborane-dimethylsiloxane) (**P15**) under nitrogen and air, and the recorded TGA and DTG thermograms are shown in Figure6. The thermograms of a commercially available polydimethylsiloxane (PDMS, $M_w \approx 10000$) is also given for comparison. The initial decomposition temperature (T_{d5}) is defined as the temperature at which the polymers lose 5% of its original weight. T_{d5} values and char yield of the polymers are picked up from the TGA curves and shown in Table 3.

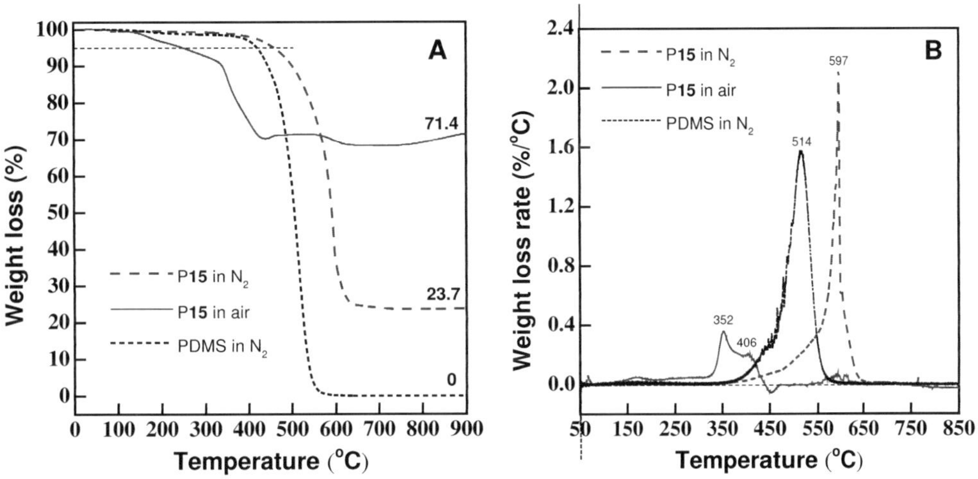

Figure6. TGA (A) and DTG (B) curves of poly(*m*-carborane- dimethylsiloxane) (**P15**, Table 1 no. 6) under nitrogen and air; TGA and DTG curves of carborane-free polydimethylsiloxane (PDMS, $M_w \approx 10000$) are also given for comparison.

Table 3.Effect of pendant groups on the thermal stability of poly(*m*-carborane-siloxane)s

no.	polymer	R_1	R_2	T_{d5} [a], °C	char yield at 900 °C, %
1	PDMS	-	-	421	0
2	P15	CH_3	CH_3	460	23.7
3 [b]	P15	CH_3	CH_3	250	71.4
4	P25 [c]	Ph	CH_3	476	81.6
5	P35	$CH{=}CH_2$	CH_3	555	87.6
6	P45	$CH_2CH_2CF_3$	CH_3	402	12.6
7 [b]	P45	$CH_2CH_2CF_3$	CH_3	371	44.7
8	P16	CH_3	$CH_2CH_2CF_3$	381	12.5
9	P36 [d]	$CH{=}CH_2$	$CH_2CH_2CF_3$	507	61.4
10 [b]	P46	$CH_2CH_2CF_3$	$CH_2CH_2CF_3$	326	42.6

[a] T_{d5} is defined as the temperature at which the polymers lose 5% of its original weight; [b] TGA analysis was conducted under air; [c] Sample from Table 2 no. 1; [d] Sample from Table 2 no. 7.

It is seen that, heated under nitrogen, **P15** shows much higher T_{d5} value (460 °C) than that of carborane-free PDMS (421 °C). We are also surprised to find that **P15** owns char yield of 23.7% at 900 °C under nitrogen, which is hard to explain by the usual degradation mechanism of linear polysiloxane. It has been well established that linear polysiloxane undergoes depolymerization via a cyclization mechanism under inert atmosphere, which gives rise to low molecular weight volatile cyclosiloxane, without forming any residue[15]. However, the introducted*m*-carborane cage within the main chain can serve to prevent the above extensive cyclization reaction, therefore the depolymerization process is postponed to higher temperature. DTG thermograms (Fig. 6B) provide more information on the depolymerization process. It is obvious that PDMS shows maximum weight loss rate (1.57%/°C) at 514 °C, indicating that at this temperature the highest cyclization rate is obtained. As for **P15**, this temperature transferred to 597 °C and the corresponding weight loss rate is 2.12%/°C. It is postulated that the siloxane bond can be strengthened by the inductive effect of electron-deficient *m*-carborane, thus the activation energy of the cyclization increases accordingly[16,17]. Additionally, the high char yield of **P15** can be attributed to the formation of boron oxide (B_2O_3), which is well confirmed by recent studies on carborane-containing polymers[18-20].

P15 owns different degradation behavior when it is heated under air. As is well-known, during the thermo-oxidative degradation of polysiloxane, the pendant group undergoes oxidation at lower temperature than depolymerization and gives rise to volatile materials such as carbon dioxide and water, and the polymer finally transforms into pure silicon dioxide (SiO_2) at elevated temperature [21]. Therefore, it is reasonable that **P15** owns lower T_{d5} (250 °C) and higher char yield (71.4%) under air. Moreover, it is believed that the polymer transforms into the mixture of SiO_2 and B_2O_3 at elevated temperature under air, since silicon and boron in the polymer structure were oxidized simultaneously. The formation of these oxides is further confirmed by FTIR analysis (Fig. 8) which will be discussed below.

Figure 5 shows the simultaneous differential scanning calorimetry (DSC) curve of **P15** under air, which gives two exothermic peaks at 406 and 579 °C. The first peak is ascribed to the oxidation of pendant methyl, indicating relatively high weight loss rate (0.21%/°C, Fig. 6B). The peak at 579 °C is associated with the cleavage and oxidation of *m*-carborane cage, which is consistent with significant energy release and tiny weight loss (Fig. 6B). Indeed, the oxidation of carborane cage absorbs oxygen from the environment, retarding the weight loss rate to some extent.

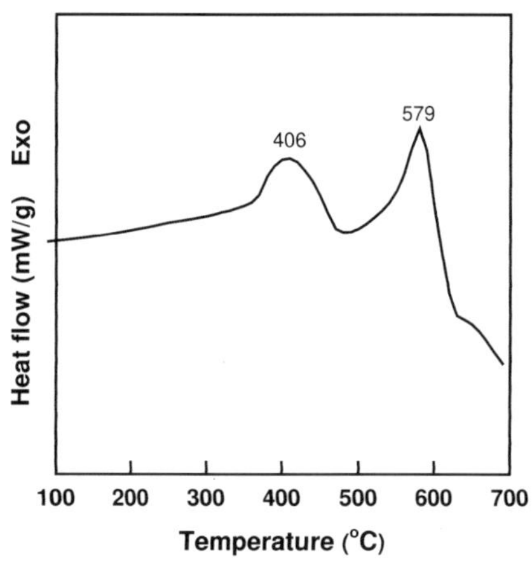

Figure7.Simultaneous differential scanning calorimetry (DSC) curve of **P15** (Table 1 no. 6) under air.

The oxidation behavior of poly(*m*-carborane-siloxanes) is further studied by FTIR technique. **P15** was first thermally treated under air at 450 and 650 °C for 20 min, and then the samples were analyzed by FTIR. It can be seen from Figure 6A that, after thermal treatment at 450 °C, **P15** owns similar characteristic absorption peaks as the original polymer (Fig. 1C), which indicates that both the siloxane chain and the carborane cage remain intact and only the pendant methyl is partly oxidized. Thus, it is again confirmed that the weight loss of **P15** below 450 °C under air (Fig. 4B) is caused by the oxidation of pendant methyl. However, both polymer main chain and carborane cage were damaged when **P15** was thermally treated at 650 °C. It is seen from Fig. 6B that the characteristic absorption band around 1100cm^{-1} disappears, indicating the breakdown of polymer main chain linked by Si-O-Si. Besides, the intensity of band at 2592 cm^{-1} attributed to B-H sharply decreases due to the cleavage and oxidation of carborane cage. A broad characteristic absorption band around 1440 cm^{-1} appears in the spectrum, corresponding to B-O-B asymmetric stretching vibration, indicating the formation of boron oxide (B_2O_3). New absorption peaks appear at 1055 and 800 cm^{-1}, corresponding to asymmetrical and symmetrical stretching vibrations of silicon dioxide (SiO_2) respectively.

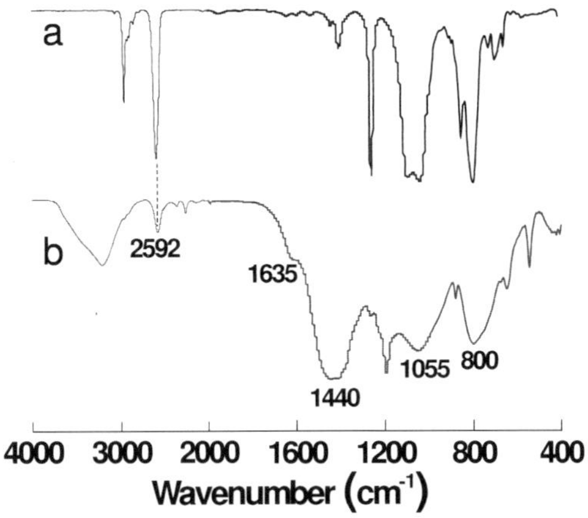

Figure8.FTIR spectra of P**15** after thermal treatment at different temperatures under air for 20 min: (A) 450 °C, and (B) 650°C.

TGA analysis was also utilized to study other poly(*m*-carborane-siloxanes) under nitrogen, and the results are given in Figure 7. The initial decomposition temperature (T_{d5}) and char yield are picked up and shown in Table 3. On basis of their distance from *m*-carborane cage, pendant groups are divided into two kinds of substituents, i.e., R_1 and R_2, which are located on α- and β-positions respectively (scheme 2). When β-substituent (R_2) is selected as methyl, the effect of α-substituent (R_1) on the thermal stability of the polymers is studied in detail. It is seen that the T_{d5} values increase with varying α-substituent in the order: $CH_2CH_2CF_3 < CH_3 \approx Ph < CH=CH_2$. This result is well consistent with the electronic effect of the substituents. Researches show that electron-donating substituent can increase the electron availability at oxygen, thereby facilitating the formation of cyclosiloxane and reducing the initial degradation temperature[22].Just like methyl, phenyl exhibits slight electron-donating rather than electron-withdrawing ability, on account of silicon *d*-orbital participation in the delocalized π-electronic system of phenyl. **P45** with pendant trifluoropropyl owns comparatively low T_{d5} value, probably because γ-fluoro elimination reaction occurs at a lower temperature, in which a fluorine atom migrates to the silicon atom in the main chain, generating 1,1-difluoropropene[23]. As an electron-withdrawing group, vinyl reduces electronic density of oxygen in the main chain, and thus increases the T_{d5} value of P**35**. Further analyses of TGA data give more useful information. It is seen that **P45** and P**16** exhibit quite close T_{d5} value and char yield, which indicates that the effect of pendant substituent on thermal stability mainly depends on its electronic effect, having nothing to do with its position. Besides, it is obvious from TGA data of these polymers that the effect of substituents on α- and β-positions satisfies superposition principle.

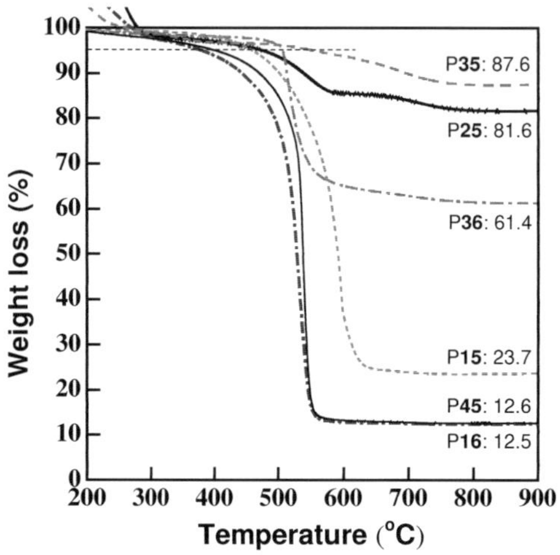

Figure9. TGA curves of poly(*m*-carborane-dimethylsiloxanes) under nitrogen.

4. Conclusion

In summary, *m*-carborane-containing disilanols (**1**-**4**) were reacted and condensed with highly active bisureidosilanes (**5** and **6**) to give a series of poly(*m*-carborane-siloxanes) with various pendant groups such as methyl, phenyl, vinyl, and trifluoropropyl (P**15**-P**46**). The monomer ratio was carefully adjusted by semicontinuous charging in order to obtain polymers with controlled molecular weight, which were selected to study their thermal stability. The obtained polymers were characterized with spectroscopic techniques including FTIR and NMR, which give satisfactory results. TGA analysis indicated that the thermal cyclization of polysiloxanes under nitrogen was greatly postponed by the incorporated *m*-carborane cage, since the siloxane bonds within the main chain were strengthened by the inductive effect of the latter. DSC and FTIR results confirmed that both siloxane unit and carborane cage were oxidized at elevated temperature under air, which contributed to the transformation of the polymers into the mixture of SiO_2 and B_2O_3. Therefore, high char yield was obtained. Besides, the electronic effect of pendant groups greatly influenced the degradation behavior of *m*-carborane-containing polysiloxanes, having nothing to do with their position. The initial degradation temperature (T_{d5}) increases with varying substituent in the order: $CH_2CH_2CF_3 < CH_3 \approx Ph < CH=CH_2$.

5.Acknowledgments

This work was financially supported by National Key R&D Program of China (No. 2016YFB0302105).

6.References

[1] LaiGQ and Xing SM 2010*Silicone products: synthesis process and application(in Chinese),second edition, Chemical Industry Press, Beijing* p.312
[2] Jones R G, Ando W and Chojnowski J2000 *"Silicon-containing polymers: the science and technology of their synthesis and application", Kluwer academic publishers, London* p.79
[3] Andrianov K A 1965 *"Metalorganic polymers", John Wiley & Sons, Inc., New York*p.276
[4] Dvornic, PR and Lenz, RW 1983 *Polymer***24** 763
[5] PetersEN1984*IndEng Chem. Prod. Res. Dev***23** 28
[6] Papetti S, Schaeffer BB, Gray AP andHeying TL1966*J PolymSci***Part A-14** 1623
[7] Dietrich HJ, Alexander RP, Heying TL, Kwasnlk H, Obenland CO,Schroeder HA 1974*MakromolChem***175**425
[8] Patel M, Swain AC, Skinner AR, Mallinson LG and Hayes GF 2003 *MacromolSymp***202** 47
[9] Patel M., Swain AC, Cunningham JL, Maxwell RS and Chinn SC 2006*Polym. Degrad. Stab***91**548

[10] Peters EN, Hedaya E, Kawakami JH, Kopf PW, Kwiatkowski GT, Mcneil DW and Owen DA 1977 *J Polym. Sci.: Polym. Chem. Ed***15** 2229
[11] Peters EN, Kawakami JH, Kwiatkowski GT, Hedaya E, Joesten BL, Mcneil DW and Owen DA 1977*J Polym. Sci.: Polym. Phys. Ed.***15** 723
[12] Stewart DD, Peters EN, Beard CD, Dunks GB, hedaya E, Kwiatkowski GT, Moffitt RB and Bohan JJ 1979*Macromolecules***12** 373
[13] Lai YC, Dvornic PR and Lenz RW 1982*J Polym. Sci.: Polym. Chem. Ed.***20** 2277
[14] ZhangX, Kong L, Dai L, Zhang X, Wang Q, Tan Y and Zhang Z 2011*Polymer***52** 4777
[15] Kendrick TC, Parbhoo B and White JW 1991*"The Silicon-heteroatom bond", John Wiley & Sons Ltd., London* p.97
[16] kesting RE, Jackson KF and Klusmann EB 1970*J Appl. Polym. Sci.***14** 2525
[17] Patel M and Swain AC2004 *Polym. Degrad. Stab.* **83** 539
[18] Bekasova NI1984 *Russ. Chem. Rev.***53** 61
[19] Qi S, Han G, Wang H, Li N, Zhang X, Jiang S and Lu Y 2015 *J Polym. Sci.***33** 1606
[20] Qi S, Han G, Wang H, Li N, Zhang X, Jiang S and Lu Y 2016*J Appl. Polym. Sci.*DOI: 10.1002/App.43488
[21] Jones RG, Ando W and Chojnowski J 2000 *'Silicon-containing polymers: the science and technology of their synthesis and application", Kluwer academic publishers, London*p.201
[22] Thomas TH and Kendrick TC 1970 *J Polym. Sci.***Part A-2, 8** 1823
[23] Knight GJ and Wright WW 1982*Polym. Degrad. Stab.***4** 465

IOP Conf. Series: Materials Science and Engineering **250** (2017) 012010 doi:10.1088/1757-899X/250/1/012010

Synthesis, Amphiphilic Property and Thermal Stability of Novel Main-chain Poly(*o*-carborane-benzoxazines)

Xiaoxue Yang, Guo Han, Zhen Yang, Xiaoa Zhang*, Shengling Jiang and Yafei Lyu

Key Laboratory of Carbon Fiber and Functional Polymers of Ministry of Education, College of Materials Science and Engineering, Beijing University of Chemical Technology, Beijing 100029, China

* Email: zhangxiaoa1982@163.com

Abstract. Five poly(*o*-carborane-benzoxazines) were synthesized via Mannich reaction of *o*-carborane bisphenol, paraformaldehyde, and diamine, and their structures were well characterized. Light transmission and ^{1}H NMR in D_2O confirmed that poly(*o*-carborane-benzoxazines) with PEG segments showed excellent water solubility and amphiphilic property. TGA analyses were conducted under nitrogen and air, and the results showed that the polymers own high initial decomposition temperatures owing to the shielding effect of carborane moiety on its adjacent aromatic structures. Besides, poly(*o*-carborane-benzoxazines) own high char yield at elevated temperatures, for the boron atom could combine with oxygen from the polymer structure or/and the air and be oxidized to form boron oxide, and thus the polymer weight is retained to a large extent. PEG segments had an adverse effect on the initial decomposition and char yield, and thus their concentration should be adjusted to control the polymer's thermal stability.

1. Introduction

As a new type of addition-curable phenolic polymer, polybenzoxazines have exhibited many superior performances such as high thermal stability, no evolution of volatiles and near-zero shrinkage during curing, flexibity of molecular design, and tailored mechanical properties [1]. However, polybenzoxazines possess low density and are rather brittle, which limits their application [2]. Furthermore, benzoxazine resins are increasingly required to own higher properties such as thermal stability in high-tech fields. On one hand, to obtain higher thermal stability, benzoxazine resins were incorporated with addition reaction groups such as acetenyl, allyl, cyano, maleimide, and furan which could participate in the curing reaction and thus improved the degree of curing [3-6]. On the other hand, bulky moieties such as fluorenyl, adamantine, and naphthalene were also introduced to the molecular structure of benzoxazine resins so as to enhance their heat resistance [7,8]. To date the search for new high-temperature units is still attracting much attention.

Closo-carboranes ($C_2B_{10}H_{12}$), with an icosahedral geometry, are excellent building blocks for thermally stable polymers [9]. Carboranes could absorb energy like a kind of "power gauge" at high temperature owing to its "superaromatic" electronic structure. Besides, the large volume of three-dimensional cage endows carboranes with strong shielding effect on adjacent segments. Therefore, carboranes were incorporated into many traditional polymers to improve their thermal stability including epoxy resins, phenolic resins, and aromatic polymers [10-12].

We have successfully synthesized carborane bisphenols and corresponding carborane-based epoxy resins, phenolic, and benzoxazine resins [10, 11, 13].. In this work, we report our studies on polybenzoxazines with *o*-carborane and PEG segments from *o*-carborane bisphenol and various functional diamines (Chart 1). Five Poly(*o*-carborane-benzoxazines) were synthesized by Mannich

IOP Conf. Series: Materials Science and Engineering **250** (2017) 012010 doi:10.1088/1757-899X/250/1/012010

reaction of *o*-carborane bisphenol, paraformaldehyde, and diamine according to Scheme 1 and Chart 2. Their chemical structures have been well characterized, and their amphiphilic property and thermal stability were studied.

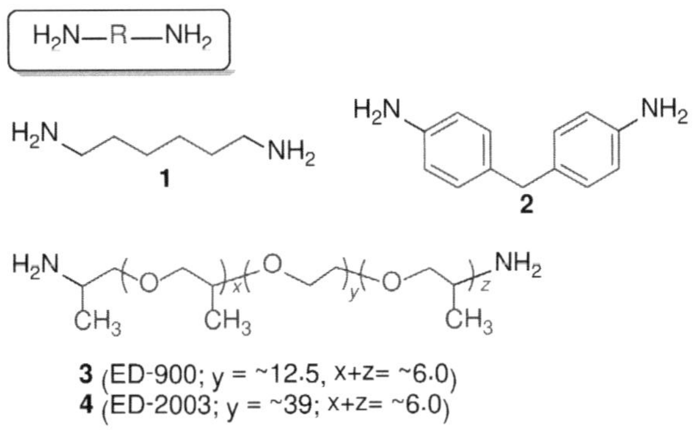

Figure 1. Functional diamines used in this work

2. Experimental

2.1 Materials

1,2-bis(4-hydroxyphenyl)-*o*-carborane (*o*-BC) was kindly supplied by Beihang University. Hexamethylenediamine (**1**) and 4,4-diaminodiphenyl methane (DDM, **2**) were purchased from Aladdin. Polyetherdiamines (**3** and **4**) were obtained from Huntsman, whose structures are shown in Chart 1. The molecular weight of **3** is 900 g/mol, lower than that of **4** (2000 g/mol). Paraformaldehyde, toluene, ethanol, methanol, and *n*-hexane were obtained from Sinopharm Chemical Reagent Beijing Co. Ltd. All materials were used without further purification.

2.2 Instrumentation

Molecular weight (M_w) and polydispersity indices (M_w/M_n) of the polymers were tested in THF with a Waters 515 gel permeation chromatography system. Tetrahydrofuran (THF) was used as eluent at 30 ^oC at a flow rate of 1.0 ml/min. FTIR spectra were recorded on a Nexus 670 infrared spectrometer. ^1H, ^{13}C, and ^{11}B NMR spectra were acquired in deuterated chloroform and acetone on a Bruker AV400 at a proton frequency of 400 MHz and the corresponding carbon and boron frequencies. Tetramethylsilane was used as an internal standard. Ultraviolet spectrum(UV) were tested by UV-3600 Under the Ultraviolet light at 250 cm^{-1}. Thermogravimetric analysis (TGA) was performed by a TA Instrument SDT-Q600 at a heating rate of 20 $^oC/min$ from 50 to 900 ^oC under nitrogen and air respectively.

*2.3 Synthesis of Poly(O-Carborane-Benzoxazines) (Poly(BC-**1**)-Poly(BC-**4**))*

Poly(*o*-carborane-benzoxazine)s (poly(BC-**1**)-poly(BC-**4**)) were synthesized according to Scheme 1. The detailed procedures for the synthesis of poly(BC-**1**) were given below as an example. Into a 100 mL three-neck round-bottom flask equipped with a magnetic stirrer and a reflux condenser were added 1,2-bis(4-hydroxyphenyl)-*o*-carborane (*o*-BC, 1.0 g, 3.03 mmol), **1** (0.352 g, 3.03 mmol), paraformaldehyde (0.39 g, 12.12 mmol), and 45 mL of toluene-ethanol mixed solvent (2:1 volume ratio). The system was stirred under reflux for 10 hours. The reaction mixture was allowed to cool to room temperature, followed by evaporation of toluene by stream of air. The resultant product was dissolved in 40 mL of chloroform, and then washed three times with 1 mol/L of sodium bicarbonate aqueous solution. The chloroform was removed by rotary evaporator, and the obtained product was placed under vacuum at 40 ^oC till constant weight was obtained. Besides, in order to increase the ring content, the chloroform solution could also be further reprecipitated in 80 mL of methanol, and the purified product was collected by suction filtration.

*2.3.1 Characterization Data for Poly(BC-**1**).*A yellow solid was obtained in a yield of 59.9 %. M_w: 8500; M_w/M_n: 6.55 (GPC, polystyrene calibration; Table 1, No. 9). FTIR, ν (cm^{-1}, in KBr pellet): 2933 (unsaturated C-H), 2856 (saturatedC-H), 2587 (B-H), 1612, 1580, 1500, 1462 (Ph), 1329, 1125 (C-N-C), 1238, 1024 (C-O-C), 936. ^{1}H NMR (400 MHz, Chloroform-d), δ (TMS, ppm): 7.20-6.48 (m, 6H), 4.78 (s, 2H), 3.83 (s, 2H), 2.56-0.9 (m, 10H), 2.61, 1.59, 1.31 (t, 2H). ^{13}C NMR (400 MHz, Chloroform-d), δ (TMS, ppm): 155.95, 130.49, 123.08, 119.85, 116.16, 86.25, 83.04, 50.17.

*2.3.2 Characterization Data for Poly(BC-**2**).*A yellow solid was obtained in a yield of 47.6 %. M_w: 9500; M_w/M_n: 2.75 (GPC, polystyrene calibration). FTIR, ν (cm^{-1}, in KBr pellet): 3026 (unsaturated C-H), 2898 (saturated C-H), 2588 (B-H), 1612, 1582, 1505, 1422 (Ph), 1371, 1183 (C-N-C), 1247, 1074 (C-O-C), 936. ^{1}H NMR (400 MHz, Chloroform-d), δ (TMS, ppm): 7.12-6.52 (m, 6H), 5.16 (s, 2H), 4.34 (s, 2H), 3.78, 3.74, 3.25-1.28 (m, 10H), 3.74 (m, 2H). ^{13}C NMR (400 MHz, Chloroform-d), δ (TMS, ppm): 155.95, 146.06, 134.94, 129.75, 123.32, 120.61, 118.88, 116.90, 114.93, 85.51, 80.32, 50.17.

*2.3.3 Characterization Data for Poly(BC-**3**).*The resultant product was precipitated in *n*-hexane, and a yellow liquid was obtained in a yield of 59.9 %. M_w: 8500; M_w/M_n: 6.55 (GPC, polystyrene calibration). FTIR, ν (cm^{-1}, in KBr pellet): 3475 (-OH), 2872 ($-CH_2-$), 2589 (B-H), 1611, 1582, 1501, 1454 (Ph), 1351 (C-N-C), 1248, 1108 (C-O-C), 936. ^{1}H NMR (400 MHz, Chloroform-d), δ (TMS, ppm): 7.62-6.48 (m, 6H), 4.89 (s, 2H), 3.94 (s, 2H), 3.65 (d, 2H), 3.51-3.04 (s, 2H), 2.60-1.25 (m, 10H), 1.20-1.00 (m, 3H). ^{13}C NMR (400 MHz, Chloroform-d), δ (TMS, ppm): 158.85, 132.26, 129.57, 125.52, 121.85, 115.18, 86.18, 81.50, 62.61, 55.58. ^{11}B NMR (400 MHz, Chloroform-d), δ (TMS, ppm): 30.67, -3.99, -12.36.

*2.3.4 Characterization Data for Poly(BC-**4**).*The resultant product was precipitated in *n*-hexane, and a yellow liquid was obtained in a yield of 59.9 %. M_w: 8500; M_w/M_n: 6.55 (GPC, polystyrene calibration). FTIR, ν (cm^{-1}, in KBr pellet): 3509 (-OH), 2871 ($-CH_2-$), 2589 (B-H), 1610, 1582, 1501, 1455 (Ph), 1106, 1285 (C-O-C), 1350 (C-N-C), 936. ^{1}H NMR (400 MHz, Chloroform-d), δ (TMS, ppm): 7.40-6.45 (m, 6H), 4.90 (s, 2H), 3.96 (s, 2H), 3.65 (d, 2H), 2.60-1.35 (m, 10H), 1.33-0.93 (m, 3H). ^{13}C NMR (400 MHz, Chloroform-d), δ (TMS, ppm): 156.47, 132.22, 129.56, 125.49, 122.63, 115.18, 86.15, 81.41, 62.61, 55.56. ^{11}B NMR (400 MHz, Chloroform-d), δ (TMS, ppm): 31.05, -4.43, -12.20.

*2.4 Synthesis of Poly(O-Carborane-Benzoxazines) (Poly(BC-**2-3**))*

Poly(BC-**2**-**3**) was synthesized by the reaction of 1,2-bis(4-hydroxyphenyl)-*o*-carborane (*o*-BC) with the mixed amines of **2** and **3**, and the polymer structure is shown in Chart 2. The resultant product was precipitated in *n*-hexane, and a yellow liquid was obtained in a yield of 62.1%. M_w: 8500; M_w/M_n: 2.67 (GPC, polystyrene calibration). FTIR, ν (cm^{-1}, in KBr pellet): 3157 (-OH), 2874 ($-CH_2-$), 2590 (B-H), 1610, 1590, 1515, 1452 (Ph), 1108, 1289 (C-O-C), 1351 (C-N-C), 936. ^{1}H NMR (400 MHz, Chloroform-*d*), δ (TMS, ppm): 7.36-6.47 (m, 6H), 4.89 (s, 2H), 3.85 (s, 2H), 3.59 (d, 2H), 2.60-1.42 (m, 10H), 1.32-0.90 (m, 3H). ^{13}C NMR (400 MHz, Chloroform-*d*), δ (TMS, ppm): 156.48, 131.32, 128.57, 122.41, 115.19, 88.15, 82.41, 62.69, 55.58. ^{11}B NMR (400 MHz, Chloroform-d), δ (TMS, ppm): 30.25, -4.30, -12.45.

3. Results and Discussion

*3.1 Synthesis of Poly(O-Carborane-Benzoxazines) (Poly(Bc-**1**)-Poly(Bc-**4**))*

Poly(*o*-carborane-benzoxazines) (poly(BC-**1**)-poly(BC-**4**)) were synthesized from 1,2-bis(4-hydroxyphenyl)-*o*-carborane (*o*-BC), diamines (**1-4**), and paraformaldehyde in toluene-ethanol solvent according to Scheme 1. The polymerization reactions proceeded smoothly, and desired polybenzoxazines were obtained in satisfactory yields (~ 67.1 %). The obtained polymers were characterized by standard spectroscopic methods, from which satisfactory analysis data were obtained (see "Experimental" for details).

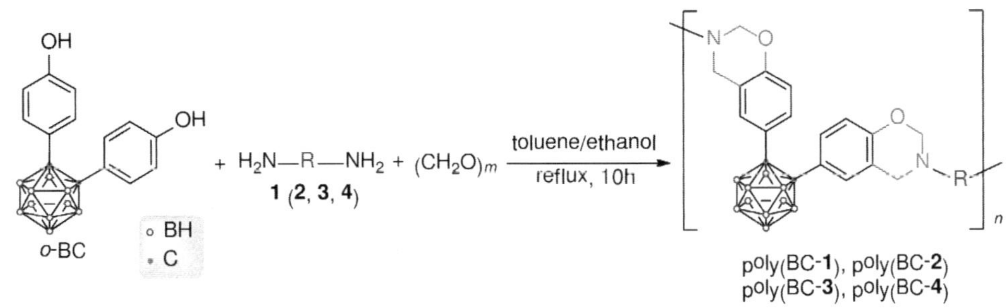

Figure 2. Synthetic routes to poly(*o*-carborane-benzoxazines) (poly(BC-**1**)- poly(BC-**4**))

To synthesize poly(BC-**1**) from *o*-carborane bisphenol (*o*-BC) and hexamethylenediamine (**1**), 1,4-dioxane was first selected as polymerization solvent, but the product was completely insoluble in most organic solvents such as THF, DMF, DMSO, and chloroform (Table 1, no 3). Then the polymerization was conducted in chloroform and xylene respectively, whereas the obtained products were partially soluble in organic solvents and the soluble fraction owned relatively low molecular weight (Table 1, nos 1 and 2). After trial and error, it was found that the polymerization solution became homogeneous in toluene-ethanol mixed solvent (2:1 volume ratio). The resultant product was well dissolved in the mixed solvent. High yield (87%) and acceptable ring content (64.5%) were obtained when the polymerization lasted for 10 h (Table 1, no 6). It is noted that the ring content could be greatly increased by further purifying the obtained polymers via precipitation in methanol, while the yield decreased accordingly (Table 1, nos 8-10). The obtained poly(BC-**1**) can be dissolved in common organic solvents such as chloroform, THF, and dichloromethane.

Table 1 The synthesis of Poly(*o*-carborane-benzoxazine) (poly(BC-**1**))from hexamethylenediamine (**1**)

no.	T, °C	t, h	solvent	M_w [a]	M_w/M_n [a]	yield,%	ring content,% [b]
1	60	5	chloroform	1400 [c]	1.81	13.5	41.1
2	150	10	xylene	1500 [c]	1.96	8.30	34.5
3	100	10	1,4-dioxane	- [d]	-	50.6	-
4	80	2	toluene/EtOH	5800	7.97	7.90	67.8
5	80	5	toluene/EtOH	5900	6.28	14.5	53.6
6	80	10	toluene/EtOH	5900	4.52	87.0	64.5
7	80	15	toluene/EtOH	4400	5.90	68.0	66.5
8 [e]	80	5	toluene/EtOH	14000	7.36	47.3	86.4
9 [e]	80	10	toluene/EtOH	8500	6.55	59.9	89.9
10 [e]	80	15	toluene/EtOH	6100	5.79	57.7	89.6

[a] Estimated by gel permeation chromatograph (GPC) in THF on the basis of polystyrene calibration. [b] Calculated from NMR data, see "Structure Chacterization" for detail. [c] Data for the soluble fraction; for the polymerization product was only partially soluble in THF. [d] The product was completely insoluble in common organic solvents such as THF, DMF, DMSO, and chloroform. [e] The product was further purified by precipitation in *n*-hexane.

Similarly, other diamines (**2**-**4**) were utilized to synthesize poly(*o*-carborane-benzoxazine)s and the results were shown in Table 2. Poly(BC-**2**) was synthesized from **2** by the same procedures, and the obtained polymer exhibits high molecular weight (M_w = 9500) and high ring content (92.3%). During the synthesis of poly(BC-**3**) and poly(BC-**4**), the crude products were reprecipitated in *n*-hexane rather than methanol, since the target products can be dissolved in methanol. Furthermore, poly(BC-**2**-**3**) (chart 2) was also designed and synthesized in order to carefully study the effect of hydrophilic chain on the water solubility of the obtained polymers.

Table 2 The synthesis of poly(*o*-carborane-benzoxazine)s from other diamines (**2**-**4**) [a]

no.	diamine	poly(*o*-carborane-benzoxazine)	*t* ,h	M_w [b]	M_w/M_n [b]	yield,%	ring content,% [c]
1	**2**	poly(BC-**2**)	10	9500	2.75	47.6	92.3
2	**3**	poly(BC-**3**)	10	7700	2.37	67.1	82.0
3	**4**	poly(BC-**4**)	10	15000	2.70	63.5	95.0
4	**2** and **3**	poly(BC-**2**-**3**)	10	8500	2.67	62.1%	85.7

[a] Carried out at refluxing temperature in toluene/EtOH. [b] Estimated by gel permeation chromatograph (GPC) in THF on the basis of polystyrene calibration. [c] Calculated from NMR data, see "Structure Chacterization" for detail.

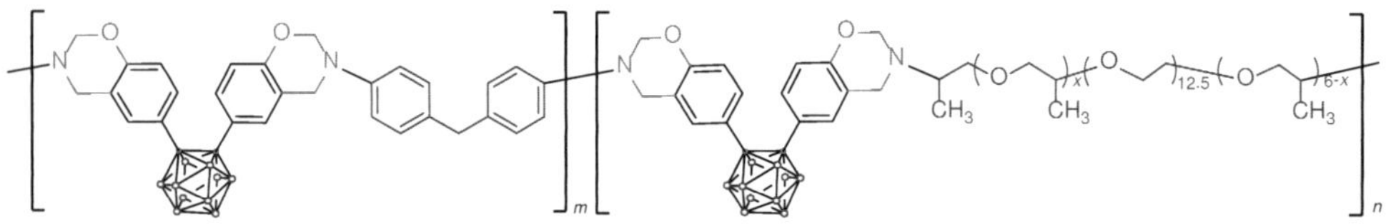

Figure 3. Poly(BC-**2**-**3**)

3.2 Structure Characterization

The structures of the obtained poly(*o*-carborane-benzoxazine)s were characterized by standard spectroscopic methods such as FTIR and NMR. The analysis data were quite satisfactory corresponding to the expected molecular structures (see Experimental section for details). An example of the FTIR spectrum of poly(*o*-carborane-benzoxazine) (poly(BC-**2**)) is shown in Figure 1; the spectrum of starting bisphenol (*o*-BC) is also given in the same figure for comparison. It is seen that *o*-BC exhibits absorption peak at 2588 cm^{-1}, which is ascribed to the stretching vibration of B-H on carborane cage. This band is still present in the spectrum of poly(BC-**2**), indicating that the carborane cage remains intact during the polymerization. Besides, new peaks appear at 1074 and 1247 cm^{-1} associated with the symmetric and asymmetric vibrations of C-O-C, while absorption peaks at 1183 and 1371 cm^{-1} are related with the symmetric and asymmetric vibrations of C-N-C, which confirms the formation of cyclic ether of benzoxazine structure. Furthermore, wide absorption band around 3332 cm^{-1} ascribe to phenolic hydroxyl group almost disappears, which is further evidence that carborane bisphenol has been transformed into polybenzoxazine.

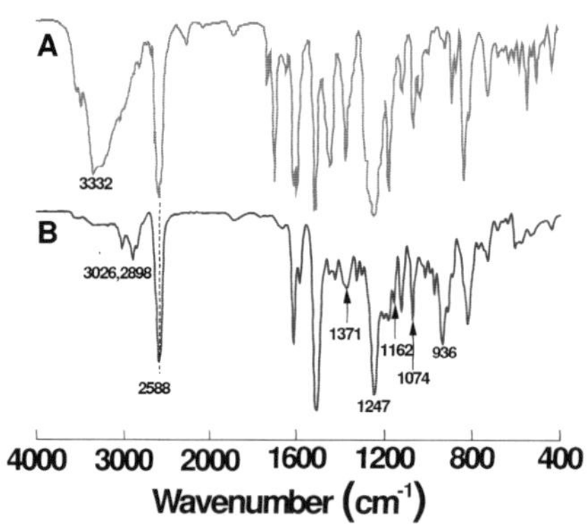

Figure 4. FTIR spectra of (A) *o*-carborane bisphenol (*o*-BC) and (B) poly(*o*-carborane-benzoxazine) (poly(BC-**2**), sample from Table 2 No. 1).

Figure 2 shows the ^{1}H NMR and ^{13}C NMR spectra of poly(BC-**2**) and its starting bisphenol (*o*-BC). It is seen that *o*-BC shows resonance peak at δ ~ 8.85 associated with phenolic hydroxyl group (Figure 2A), which disappears in the spectrum of poly(BC-**2**) (Figure 2B), indicating the complete transformation of bisphenol to benzoxazine structure. New resonance peaks at δ ~ 5.16 and 4.34 appear which are attributed to methylene protons on the oxazine ring, while a resonance peak at δ ~ 3.78 is related with methylene protons of the opened structure (Figure 2B). Since the synthesized poly(*o*-carborane-benzoxazines) own both oxazine ring and opened structure, the integrated intensities of the above resonance peaks were utilized to calculate the percentage of benzoxazine ring, i.e. ring content, of the polymer and the results were shown in tables 1 and 2. It is seen from Figure 2D that new peaks appear at δ ~ 80.1 and 50.1 which are attributed to methylene carbons on the oxazine ring, further verifying the formation of benzoxazine structures. Since the obtained polymers contain *o*-carborane in their structures, ^{11}B NMR data were also given in Experimental section, which are well consistent with their precursory carborane bisphenol[13].

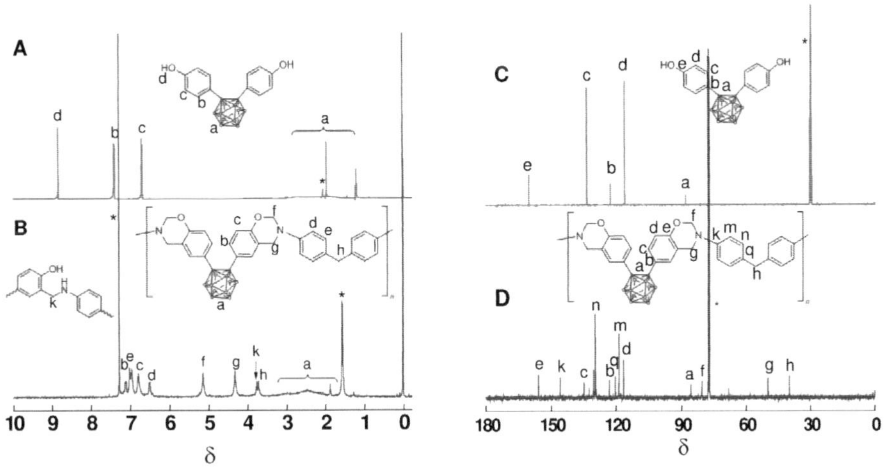

Figure 5. ^{1}H NMR and ^{13}C NMR spectra of *o*-carborane bisphenol (*o*-BC) in $(CD_3)CO$ (A, C) and poly(*o*-carborane-benzoxazine) (poly(BC-**2**), sample from Table 2 No. 1) in $CDCl_3$ (B and D). The solvent peaks are marked with asterisks.

3.3 Water Solubility and Amphiphilic Property

PEG chains have been widely incorporated into polymer molecules to prepare materials with functions such as water solubility, amphiphilicity, biocompatibility, and hydrogel formability. Therefore, PEG-containing poly(*o*-carborane-benzoxazines) (poly(BC-**3**) and poly(BC-**4**)) were designed and synthesized by using polyetherdiamines with different PEG chain length. To demonstrate their water solubility, different concentrations of polymer aqueous solutions were prepared and their light transmission spectra were recorded at room temperature.

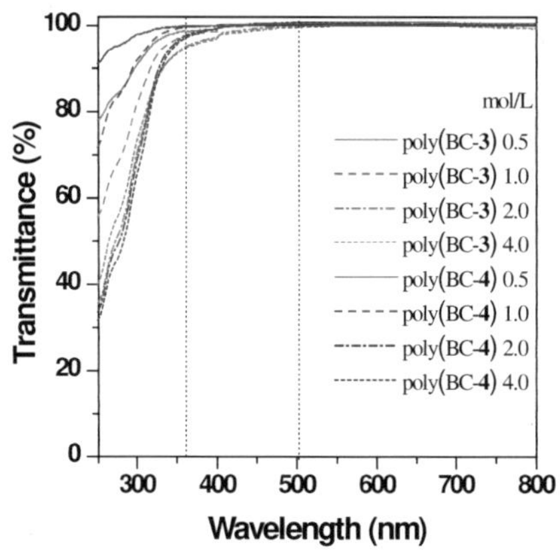

Figure 6. Light transmission spectra of the aqueous solution of poly(BC-**3**) and poly(BC-**4**) at polymer concentration of 0.5-4.0 mol/L.

As depicted in Figure 3, it is obvious that the transmittances of polymer aqueous solutions are 100% at wavelength longer than 510 nm, even when the polymer concentration is as high as 4.0 mol/L. In the region where the wavelength is shorter than 520 nm, the transmittances of polymer aqueous solutions decrease, and the declination is associated with the electronic absorption of the carborane bisphenol. The above phenomena suggest that PEG-containing poly(*o*-carborane-benzoxazines) can be dissolved in water to form clear aqueous solution.

The incorporation of PEG chains into polybenzoxazine backbone bestowed the resultant polymers with a segmented structure, which consists of hydrophobic segments and hydrophilic chains with different length. In aqueous solution, the hydrophobic part is encapsulated by the hydrophilic PEG segments. As a result, a single polymer chain takes a micelle-like configuration, and carborane bisphenol units of the polymers are insoluble in water, although clear aqueous solutions have been observed. This deduction was validated by the measurements of the ^{1}H NMR spectra of the polymers in D_2O at room temperature. The ^{1}H NMR spectra of poly(BC-**3**) in D_2O and $CDCl_3$ are shown in Figure 4 as an example. In $CDCl_3$, the resonance peaks for the protons on phenyl groups and PPG chains can be found, and the resonance peaks for the protons on PEG segments are also clearly recognized (Figure 4A). This indicates the good solubilities of both the hydrophobic (carborane bisphenol unit and PPG chain) and hydrophilic (PEG segment) parts are soluble. While in D_2O, there exist only one peak at $\delta \sim 3.65$, which can be assigned to the resonance of protons of water soluble PEG chain (Figure 4B); but no other peaks are observed, indicating that hydrophobic carborane bisphenol units and PPG chains cannot be dissolved in water.

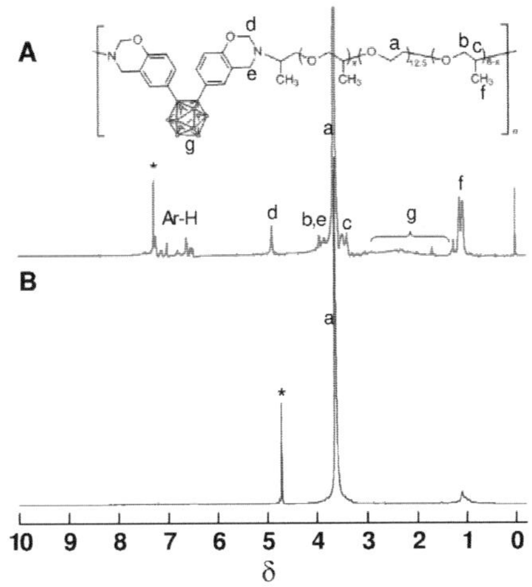

Figure 7. ^{1}H NMR spectra of poly(*o*-carborane-benzoxazine) (poly(BC-3), sample from Table 2 No. 2) in (A) $CDCl_3$ and (B) D_2O. The solvent peaks are marked with asterisks.

3.4 Thermal Stability

Poly(*o*-carborane-benzoxazines) were investigated by thermal gravimetric analysis (TGA) under nitrogen and air, and the results are shown in Figure 5. It is seen that poly(BC-**1**) and poly(BC-**2**) lose their 5% weight under nitrogen at 420 and 543 °C (T_5), respectively, much higher than that of carborane-free polybenzoxazine. The initial decomposition temperatures of polybenzoxazine were highly improved by introduction of carborane cage into their backbones. The incorporated carborane moiety shows excellent shielding effect on its adjacent aromatic structures, and thus the thermal stability of polybenzoxazine is improved. It is interesting to note that poly(BC-**1**) and poly(BC-**2**) own much higher T_5 under air (553 and 708 °C) than under nitrogen. It is known that the carborane cage within polymer backbone can be oxidized to form boron oxide by absorbing oxygen from the polymer structure and/or the environment, so the weight loss is largely postponed. The oxygen in air can promote the oxidation reaction, and thus the polymer may absorb so much oxygen that the weight gain occurs over 500 °C for poly(BC-**2**). Besides, poly(BC-**1**) and poly(BC-**2**) own high char yield under nitrogen and air at elevated temperatures, which is consistent with other carborane-containing thermosets[10, 11, 13, 14]. The oxidation-caused weight gain also contributes to higher char yield under air. Poly(BC-**3**) and poly(BC-**4**) are polybenzoxazines with PEG segments. It is seen that these two polymers exhibit much lower T_5 and char yield, indicating that the thermal degradation occurs easily. By selecting two kinds of diamines, poly(BC-**2**-**3**) was designed and synthesized with adjusted PEG concentration and controlled thermal stability.

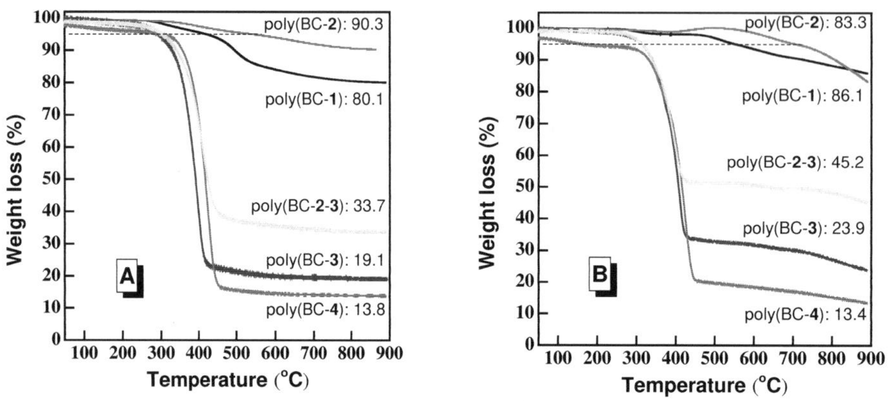

Figure 8. TGA curves of poly(*o*-carborane-benzoxazines) under nitrogen (A) and air (B). (**P1** sample from Table 1 No. 9).

4. Conclusion

In summary, five poly(*o*-carborane-benzoxazines) were synthesized via Mannich reaction of *o*-carborane bisphenol, paraformaldehyde, and diamine. The synthetic conditions were well optimized to give acceptable molecular weights. The obtained polymers were characterized with spectroscopic techniques including FTIR, ^{1}H NMR, and ^{13}C NMR, which gave satisfactory results. Light transmission and ^{1}H NMR in D_2O confirmed that poly(*o*-carborane-benzoxazines) with PEG segments showed amphiphilic property. TGA analyses were conducted under nitrogen and air, and the results showed that the polymers own high initial decomposition temperatures owing to the shielding effect of carborane moiety on its adjacent aromatic structures. Besides, poly(*o*-carborane-benzoxazines) own high char yield at elevated temperatures, for the boron atom could combine with oxygen from the polymer structure or/and the air and be oxidized to form boron oxide, and thus the polymer weight is retained to a large extent. PEG segments had an adverse effect on the initial decomposition and char yield, and thus their concentration should be adjusted to control the polymer's thermal stability.

5.Acknowledgments

This work was financially supported by National Key R&D Program of China (No. 2016YFB0302105).

6.References

[1] Sawaryn C, Kreiling S, Schonfeld R, Landfester K and Taden A. In Handbook of Benzoxazine Resins, Ishida H, Hatsuo I, Tarek A, Eds, Elsevier: Amsterdam 2011 Chapter 35 605-620.
[2] Nair C P R 2004 Prog Polym Sci 29 401.
[3] Chernykh A, Agag T, Ishida H 2009 Polymer 50 3153
[4] Kumar K S S, Nair C P R, Radhakrishnan T S, Ninan K N 2007 Eur. Polym J 43 2504
[5] Cao G P, Chen W J, Liu X B 2008 Polym Degrad Stab 93 739
[6] Liu Y, Zhao S, Zhang H, Wang M, Run M 2012 Thermochim Acta 549 42-48
[7] Chang H C, Lin H, Tian Y W, Feng Y R, Chan L H 2012 J Polym. Sci. Part A: Polym. Chem 50 2201
[8] Lu Y, Li M, Zhang Y, Hu D, Ke L, Xu W 2011 Thermochim Acta 515 32
[9] Grimes R N, Carboranes Second Edition Elsevier: New York 2011 Chapter 9 301-540 Chapter 10 541-674
[10] Qi S, Wang Y, Han G, Yang Z, Zhang X, Jiang S, Lv Y 2015 Acta Polymerica Sinica (8) 921
[11] Qi S, Han G, Wang H, Li N, Zhang X A, Jiang S, Lu Y 2015 Chinese J Polym. Sci. 33 (11) 1606
[12] Vinogradova S V, Valetskii P M, Kabachii Y A 1995 Russ. Chem. Rev. 64 365

[13] Qi S, Wang H, Han G, Yang Z, Zhang X A, Jiang S L, Lu Y. 2016 J Appl Polym Sci. 133 43488
[14] Han G, Yang Z, Yang X, Zhang X, Jiang S and Lu Y 2016 Chin.J Polym. Sci. 34(9) 1103

Electric Power Engineering Cost Predicting Model Based on the PCA-GA-BP

Lei Wen[1], Jiake Yu[1] and Xin Zhao[2]

[1]Department of Economics and Management, North China Electric Power University, Baoding, Hebei, 071003, China
[2]Economic and Technology Research Institute of xinjiang electric power company, Wulumiqi, xinjiang, 840000, China
Email:wenlei0312@163.com

Abstract. In this paper a hybrid prediction algorithm: PCA-GA-BP model is proposed. PCA algorithm is established to reduce the correlation between indicators of original data and decrease difficulty of BP neural network in complex dimensional calculation. The BP neural network is established to estimate the cost of power transmission project. The results show that PCA-GA-BP algorithm can improve result of prediction of electric power engineering cost.

1.Introduction

Since the 21st century, China's economic level has been rapid development, power construction project investment put forward higher requirements for power engineering cost management. It is the significant work that the prediction of power engineering cost to improve the management of electric power engineering project. The accurate engineering cost prediction can reduce the waste of money and is the embodiment of scientific development.

In this paper, an hybrid artificial neural network algorithm: PCA-GA-BP model are proposed by combining Principal Component Analysis,Genetic Algorithm and the BP (Back Propagation) neural networks.

In this algorithm, the PCA algorithm is established to reduce the correlation between data indicators and decrease the difficulty of foresting model in the high dimensional calculation. The Back Propagation neural network is used to predict electric power engineering cost, and Genetic model are used to optimize the initial parameters of Back Propagation neural network to prevent it into over-fit.

2. Material and Methods

2.1 Principal Component Analysis

PCA is a statistical analysis technology proposed by Scientist Pearson in 1901 and developed by Scientist Hotelling later. PCA utilizes an orthogonal transformation to convert multiple indicators into a few comprehensive indicators called principal components. these principal components can reflect the most information of original indicators . they are usually expressed as a linear combination of the original indicators.

In the theory of PCA, original data is transformed to a new coordinate system. Consider a process with input data set matrix $X \in \mathfrak{H}^{n \times p}$,where each of the n rows represents observations and each of p columns represents process indicators. The role of PCA is to reduce the difficulty in the high dimensional calculation and eliminate the multicollinearity between the variables.

The standard PCA algorithm can be explained as follows:

Step 1: Standardize the process indicators and indicators of each column becomes equal to 0 and 1respectively.

Step2: Find out the correlation between standardized indicators. The covariance matrix Λ by following expression.

$$\Lambda = \frac{1}{n-1} X^T X = V \Lambda V^T \quad (1)$$

where $\Lambda = diag(\lambda_1 \geq \lambda_2 \ldots \geq \lambda_m)$.

Step3: Determine the number of principal component by parallel analysis method.

Step4: Determine the expression of principal component

$$F_i = \lambda_i Z_i \quad (2)$$

where Z is Standardized value of process indicator.

2.2 Standard BP Neural Network

The BP neural network is one of the most widely used neural network models, which is put forward by Rumelhart and Mcelland in 1986.The back propagation neural network s a multilayer feed-forward model trained by the principle of error inverse propagation algorithm. The standard topology of BP neural network includes input layer, hidden layer and output layer.

three-layers BP neural network is established in this paper. The main parameters of standard BP model can be seen as follows:

(a) Network Training Algorithm: gradient descent algorithm
(b)The Number of Neurons in the Input Layer :10
(c)The Number of Neurons in the Hidden layer:15
(d)The Number of Neurons in the Outputlayer:1
(e)Maximum Number of iterations:50

2.3 BP Model Optimized by GA

The GA is a randomized search method that evolves from the evolution of the biological world which is proposed by Holland in 1962. GA is used to generate useful solutions for optimization and search problems. Genetic algorithms provide a general framework for solving complex system problems. In this paper, the Genetic algorithms is established to optimized the initial parameters of BP model in order to improve the accuracy of the forecast results.

The GA-BP algorithm can be described as follows:

Step1:Use GA to find the optimal initial weights and threshold value of BP neural network。

The following equation is utilized to decide the length of a chromosome :

$$N_c = N_k + N_i \times N_j + N_j \times N_k + N_j \quad (3)$$

where N_c is the length of a chromosome, N_k is the number of neurons in output layer of BP model, N_i is the number of neurons in input layer of BP model, N_j is the number of neurons in hidden layer of BP model.

The fitness function is calculated as:

$$\mathrm{F(i)} = \mathrm{k}\left(\sum_{i=1}^{n}\left(\mathrm{abs}(y_i - o_i)\right)\right) \quad (4)$$

where i is the number of the whole chromsomes, n is the number of neurons in output layer of BP model, y_i is the forecasting electric power engineering cost signals,o_i is the guiding electric power engineering signals. k is the coefficient of fitness function.

In this paper, the main parameters of GA are selected as:

(a)Size of chromosomes:50
(b) Probability of Crossover :0.2
(c)Probability of Mutation:0.15
(d)Algorithm of Selection Operation: Roulette Wheel Algorithm
(e)Number of Iterations:500

Step2: Complete the electric power engineering cost predicting by utilizing the BP neural network whose initial weights and thresholds have been optimized by GA.

3. The Proposed Method

Use the PCA algorithm to convert 30 original indicators into 10 principal components. The purpose of doing so is to reduce the difficulty of BP model in the high dimensional calculation and eliminate the multicollinearity between the variables.

Establish the BP neural network optimized by the GA and MEA respectively. MEA and GA are used to optimized the initial parameters of BP model. The purposed of this step is to make BP model predict more accurately.

Calculate the predictions by using the BP model, PCA-GA-BP model and PCA-MEA-BP model. Making comparisons between predicted and actual values. three error criterions and three percentage error criterions were used in this paper.

Make comparisons between different prediction models of the forecasting accuracy. The involved models include the PCA-GA-BP model, the PCA-MEA-BP model and the single PCA-BP model.

4.Result

The Root Mean Square Error(RMSE), the Mean Absolute Percentage Error(MAPE) , the Mean Absolute Error(MAE) of these three indexes are selected in this paper to test the prediction accuracy of the proposed PCA-GA-BP model and PCA-MEA-BP model. The equation of three indexes are adopted as:

$$MAPE = \frac{1}{M}\sum_{t=1}^{N}\left|\frac{X(t)-\hat{X}(t)}{\hat{X}(t)}\right| \quad (5)$$

where $X(t)$is the raw data of electric power engineering cost, $\hat{X}(t)$ is the forecasted data series of electric power engineering cost and N is the number of the sample in the data series.

$$MAE = \frac{1}{N}\sum_{t=1}^{N}\left|X(t)-\hat{X}(t)\right| \quad (6)$$

$$RMSE = \sqrt{\frac{1}{N-1}\sum_{t=1}^{N}[X(t)-\hat{X}(t)]^2} \quad (7)$$

Table 1. Result

Indexes	PCA-GA-BP	PCA-BP
MAE(%)	3.858	5.975
MAPE(%)	7.721	11.906
RMSE(%)	3.668	5.66

5.Discussion

From the result, We can draw the following conclusions:

When comparing the hybrid GA-BP neural networks with the standard BP neural networks, the hybrid GA-BP neural networks performs better than the latter in all indexes obviously.

Since the PCA algorithm is established to reduce the correlation between data indicators and the MEA/GA algorithms select the best parameters for standard BP neural networks, the optimized BP neural networks can achieve high accuracy prediction results.

6.Conclusion

Above all, the hybrid artificial neural network algorithm PCA-GA-BP model a are proposed for the electric power engineering cost by combining Principal Component Analysis, Genetic Algorithm and the Back Propagation model. The contribution rate of the Principal Component Analysis and Genetic model to promote the standard BP model are investigated based on electric power engineering cost forecasting experiment. The results in the experiment show the GA algorithm also can improve the predictive performance of BP neural network and are suitable for the electric power engineering cost predictions for the cost management.

7.References

[1] H Liu, H Tian, Y Li, Comparison of new hybrid FEEMD-MLP, FEEMD-ANFIS, wavelet packet-MLP and wavelet packet-ANFIS for wind speed predictions 2015 *Energy Convers* **11** 21

[2] O B Shukur, M H Lee, Daily wind speed forecasting through hybrid KF-ANN model based on ARIMA 2015 *Renew. Energy* **33** 637-647

[3] A U Haque, P Mandal, J Meng, M Negnevitsky, Wind speed forecast model for wind farm based on a hybrid machine learning algorithm 2015 *Int. J. Sustain. Energy* **34 (1)** 38-51

[4] H Liu, C Chen, H Tian, Y Li, A hybrid model for wind speed prediction using empirical mode decomposition and artificial neural networks 2012 *Renew. Energy* **48** 545-556

[5] D Petkovi c, S Shamshirband, NB Anuar, H Saboohi, AW Abdul Wahab, M Proti c, E Zalnezhad, SMA Mirhashemi, An appraisal of wind speed distribution prediction by soft computing methodologies: A comparative study 2014 *Energy Convers* **21** 133-139

[6] H Liu, H Tian, C Chen, Y Li, An experimental investigation of two wavelet- MLP hybrid frameworks for wind speed prediction using GA and PSO optimization 2013 *J. Electr. Power Energy Syst* **2** 161-173

[7] A Troncoso, S Salcedo-Sanz, C Casanova-Mateo, JC Riquelme, L Prieto, Local models-based regression trees for very short-term wind speed prediction 2015 *Renew. Energy* **81** 589-598

[8] JC Palomares-Salas, A Agüera-Pe rez, JJG. Rosa, JM. Sierra-Ferna ndez, A Moreno-Mun~oz, Exogenous measurements from basic meteorological Stations for wind speed forecasting 2013 *Energies* **6 (11)** 5807-5825

[9] Z Su, J Wang, H Lu, G Zhao, A new hybrid model optimized by an intelligent optimization algorithm for wind speed forecasting 2014 Energy Convers **12** 443-452

[10] H Liu, HTian, Y Li, Comparison of two new ARIMA-ANN and ARIMA-Kalman hybrid methods for wind speed prediction 2012 *Appl. Energy* **16** 415-424.

Microstructure Characteristics of Fe-Matrix Composites Reinforced by In-Situ Carbide Particulates

Xiaodong Huang[1] **and Yanpei Song**[2*]

College of Materials, Henan University of Science and Technology, Luo yang 471023, China

*E-mail: sypei@haust.edu.cn

Abstract. Carbide particulates reinforced iron-matrix composites were prepared by in-situ synthesis reaction between Ti, V and C on liquid alloys surface. The microstructure of the composite was characterized by SEM, TEM and OM. The results showed that the main phases were α-Fe, carbide particulate; besides, there were small amounts of γ-Fe and graphite (G) in the composite. The carbides were $TiVC_2$ and VC in the shape of short bar and graininess. The matrix consisted of martensite and small amounts of retained austenite.

1. Introduction

Particulates reinforced metal matrix composites (MMCs) have developed rapidly due to its promising advanced properties [1-2]. Mainly matrix materials in the composites were aluminum, magnesium and titanium; their applications were restricted due to high cost. Though Fe matrix composite developed latest, it has been attracted much attention owning to its excellent wear-resistance property. The most common reinforcement in Fe-matrix composites were VC, WC and NbC because of high hardness and thermal stability [3-5]. In the metal matrix composites, in-situ particulates reinforced composites had advantages over external additive ones [6]. Those particulates synthesized by in-situ reaction were fine in size, thermodynamically stable and evenly distributed [7]. In addition, in-situ techniques involved a chemical reaction on surface of the melt matrix alloy and formed clean and uncontaminated reinforcement interface with a stronger bonding interface [8].

As TiC and VC have the same lattice type and similar lattice constants (lattice constant of TiC is 4.329 Å, and that of VC is 4.16 Å) and also have proximate melting point, they can form a solid solution. Fe-matrix composites reinforced by TiC or VC particulates prepared by powder metallurgy and in-situ reaction casting method have been reported, but there were few special reports of hybrid-reinforced Fe-matrix composites synthesized by in-situ reaction casting method. In this paper, a hybrid iron matrix composite reinforced with $TiVC_2$ and VC particulates was prepared by in-situ reaction casting method, and then the microstructure of the composite was researched.

[1] Author: Xiao-Dong HUANG, male, master, research on particulate reinforced steel matrix composites, E-mail: huangxiaodong1772@163.com.
[2] Corresponding author: Yan-Pei SONG, male, PhD, Professor, research on new wear-resistant materials and metal matrix composites; E-mail: sypei@haust.edu.cn

2. Experiment Procedure

2.1. Preparation

The main raw materials used for the synthesis of the composite were Ti–Fe, V-Fe, C and low carbon steel (Steel 20). The composition of the low carbon steel, Ti–Fe and V-Fe were shown in Table 1. The mixed powders of Ti–Fe, V-Fe and C between 50 μm and 100 μm were taken which Ti, V and C was kept in an atomic ratio of 1:1:2. The investigated alloy was melt in a 50 kg medium-frequency induction furnace. When the melt temperature increased to 1650° C, put 11.80 kg mixed powder on the surface of molten liquid-metal, the heat of liquid-metal impelled the reactions of Ti, V and C in the mixed powders. After all mixed powders' reaction completed, the melt should be degassed before it was poured in the casting ladle. Then the melt was poured into the moulds that were prepared with a mixture of sodium silicate and sand [9]. After cooled down to room temperature naturally, a standard Y-shape test block was obtained [10]. The specimens were heated at 950°C for 1 h and then oil quenching, subsequently tempered at 200°C for 3 h.

Table 1. The chemical composition of raw materials (wt. %)

Material	C	Si	Mn	S	P	Al	V	Ti	Fe
Ti-Fe	0.1	4.0	2.5	≤0.03	≤0.05	8	—	29.87	Balance
V-Fe	0.4	≤2	0.5	0.04	0.07	≤0.5	50	—	Balance
Steel 20	0.19	0.21	0.40	≤0.035	≤0.035	—	—	—	Balance

Table 2. The chemical composition of the composite (wt. %)

Elements	C	Si	Mn	S	P	V	Ti	Fe
Content (wt. %)	1.61	0..63	0.77	0.017	0.027	3.36	0.87	Balance

2.2. Characterization

The specimens were etched with a solution of 4% nitric acid and alcohol after polished by diamond paste, and they were examined by an optical microscope (OM) and a JSM-5800 scanning electron microscope (SEM) equipped with an energy dispersive spectrometer (EDS). Meanwhile the microstructure morphology and crystal structure of the composites were determined by a JEM-2100 transmission electron microscope (TEM). The chemical composition of the composite was tested by Optical Emission Spectrometry (showed in Table 2). The volume fraction of carbides in the composite was determined for 10 fields by line analysis method, in which the total lengths of carbides in specimen crossed by ruler (10 mm) were measured under 100 times microscope. Then calculate the ratio of total lengths of carbides to lengths of ruler. The formula is as follows:

$$vol.\% = \frac{\sum_{1}^{n} a_i}{\mathrm{L}} \times 100\% \tag{1}$$

Where: *vol. %* is the volume percentage, a_i is the length of carbide particle crossed by ruler, and L is the length of ruler.

3. Results and Discussion

3.1. Thermodynamic Analysis

In this work, the feasibility of the reactions related to the synthesis carbide particulates in iron-based composites was considered. The thermodynamic of in situ chemical reactions was researched [11-13]. The potential reactions' equations between the raw materials are as follows:

$$Ti + C = TiC \quad \Delta G_1 = -181.251 + 0.013T\ (kJ \cdot mol^{-1}) \quad (25\text{-}1670°C) \tag{2}$$

$$Fe + Ti = FeTi \quad \Delta G_2 = -41.261 + 0.012T\ (kJ \cdot mol^{-1}) \quad (727\text{-}1727°C) \tag{3}$$

$$2Fe + Ti = Fe_2Ti \quad \Delta G_3 = -90.885 + 0.027T\ (kJ \cdot mol^{-1}) \quad (727\text{-}1727°C) \tag{4}$$

$$2V + C = V_2C \quad \Delta G_4 = -146.4 - 0.00634T\ (kJ \cdot mol^{-1}) \quad (273\text{-}1727°C) \tag{5}$$

$$V + C = VC \quad \Delta G_5 = -117.04 + 0.0067T\ (kJ \cdot mol^{-1}) \quad (273\text{-}1727°C) \tag{6}$$

$$8V + 7C = V_8C_7 \quad \Delta G_6 = -98.7 + 1.26\times10^{-3}T\ln T + 0.835\times10^{-6}T^2 + 0.83\times10^{2}T^{-1} - 1.08T\times10^{-3}T\ (kJ \cdot mol^{-1}) \quad (273\text{-}1727°C) \tag{7}$$

The change of Gibbs free energies (ΔG) of TiC, FeTi, Fe_2Ti, V_2C, VC and V_8C_7 were calculated and plotted in Fig. 1 as a function of temperature. It was known that if the more negative of the ΔG in the interest temperature range, the higher tendency to form the substance. From Fig.1, it can be seen that TiC and V_2C had the most negative ΔG values, which indicated that TiC and V_2C phase were the most stable products under 1650°C. It was also found that the ΔG values were more and more negative from FeTi, Fe_2Ti, and V_8C_7 to VC.

3.2. Microstructure Analysis

Fig. 2 showed optical microscope images of the Fe-based composites reinforced by in-situ carbide particulates. It was evident from Fig.2 that there were three different forms of carbide in the composites, which included bar-shaped particles (marked as 1), square-shaped particles (marked as 2) and irregular-shaped particles (marked as 3). In addition, there was a small quantity of chrysanthemum graphite in the matrix microstructure (marked as 4). After etched by 4% Nital solution, the matrix microstructure was plate martensite (seen Fig. 2c, d). The hardness value of the composite achieved HRC54.3 and impact toughness reached to 9.0 $J \cdot cm^{-2}$. Through statistic analysis and calculations for 10 field, the volume fraction of carbide in the composite was about 15% .

In order to confirm the type of carbide particles, the line scanning analysis of the different type carbides was performed, and the analysis results were shown in Fig. 3. It was seen from Fig.3 that there were V and C elements in the bar-shaped carbides (see Fig. 3a), and V, Ti, and C element existed in the grainy carbide (see Fig. 3b). Therefore, it was inferred that the grainy carbides were multicarbide and the bar-shaped carbides were vanadium carbide. The reason for forming the different type of carbides was that the composition of the alloy melts surface area was hypereutectic after adding the mixed powders in liquid alloy, leading to multicarbide formation in the melts by in-situ reaction directly, and grown up to grainy carbides [14-15]. By electromagnetic stirring of the melts, the composition of the alloy melts transformed to uniform hypoeutectic state. Austenite precipitated from the liquid alloy and grown into dendrite during the freezing of cast, which led to melts achieved eutectic composition, and then the bar-shape carbide was formed during eutectic transformation.

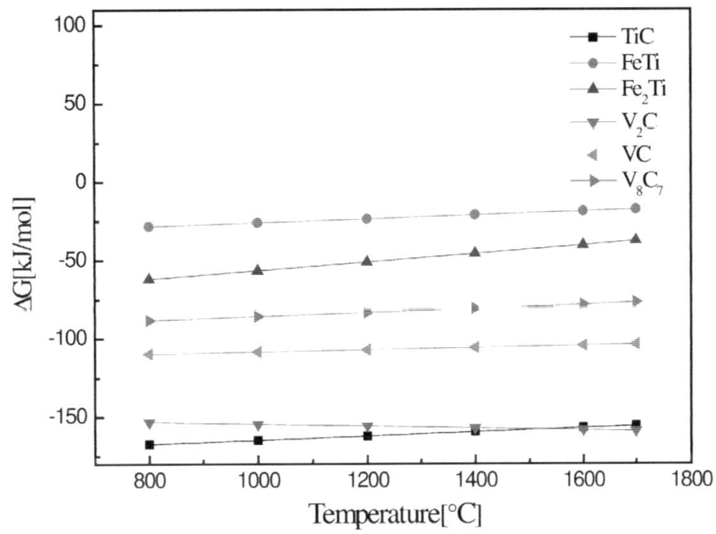

Figure 1. Change of Gibbs free energies (ΔG) as a function of temperature for TiC, FeTi, Fe_2Ti, V_2C, VC and V_8C_7.

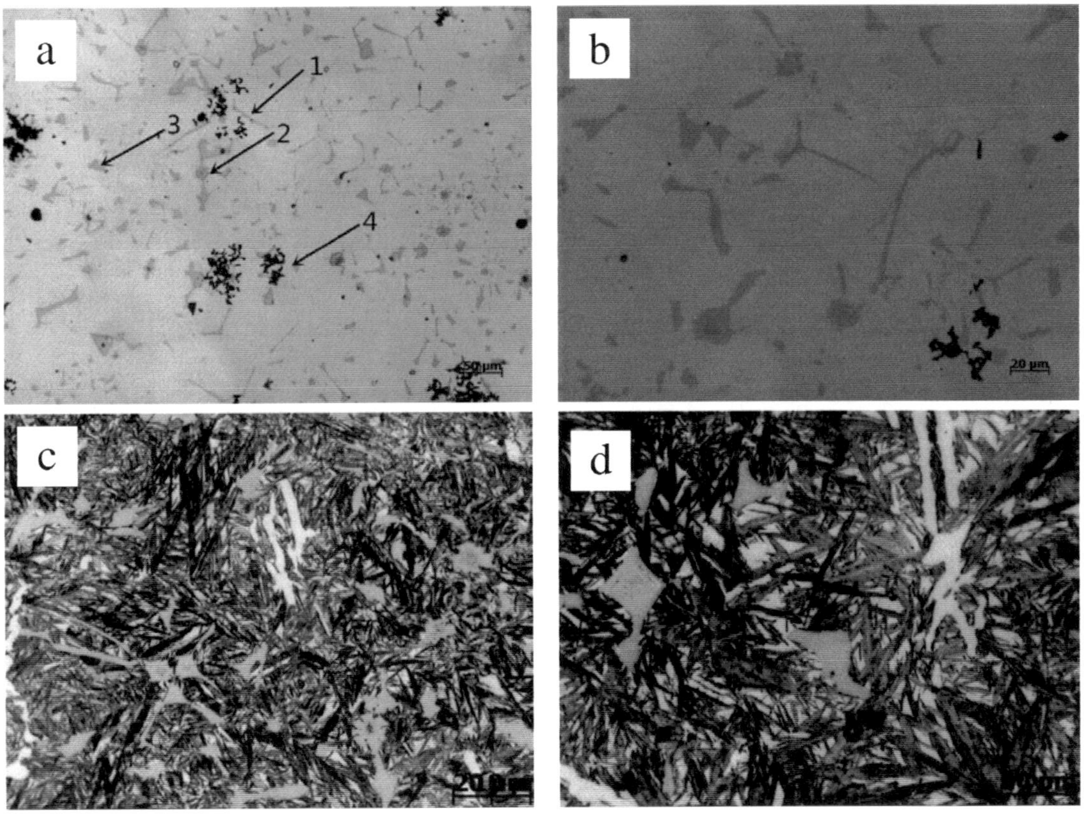

Figure 2. Optical microscope images: (a) without etched under 200 times; (b) without etched under 500 times; (c) etched by 4% Nital solution under 500 times; (d) etched by 4% Nital solution under 1000 times.

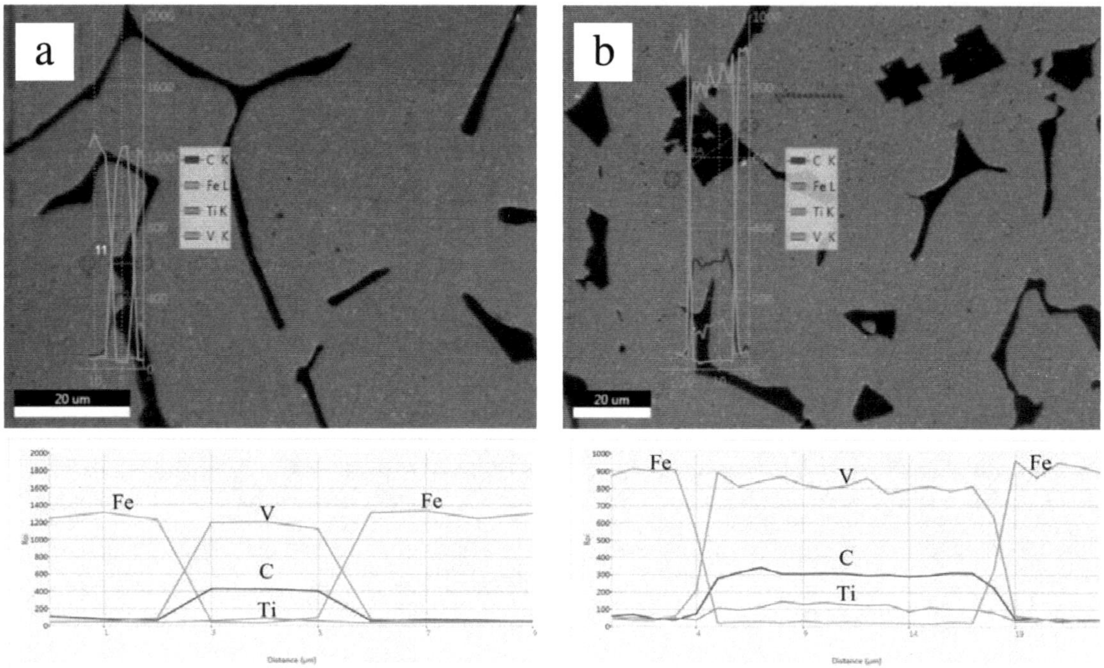

Figure 3. Line scanning of Fe, C, Ti and V in carbide particles.

The in-situ carbide morphology and micro-crystal structure of the composites were determined by transmission electron microscopy. The results were shown in Fig. 4. Plate martensites and carbide can be seen in Fig. 4a. Through selected area electron diffraction (SAED) of the carbide (see Fig. 4b), the type of the carbide was determined as $TiVC_2$ crystal with a face-centered cubic (fcc) structure and a lattice parameter of a=4.255Å. There was twin structure in area A (see Fig. 4c). Fig. 4(d) was the SAED pattern of reinforced particulates and matrix microstructure in Fig. 4(c). It can also be found that there were α-Fe, γ-Fe and VC in the composites. Therefore, the composite was consisted of martensite, small amounts of retained austenite, $TiVC_2$ and VC.

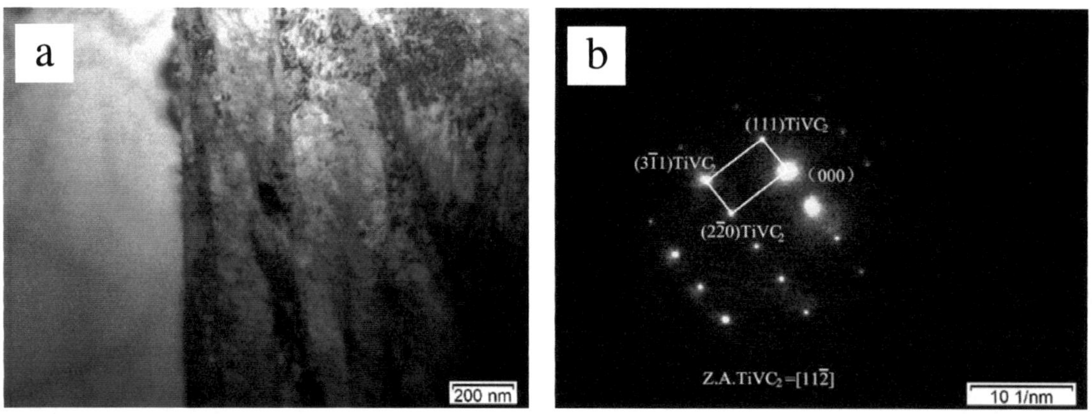

IOP Conf. Series: Materials Science and Engineering **250** (2017) 012012 doi:10.1088/1757-899X/250/1/012012

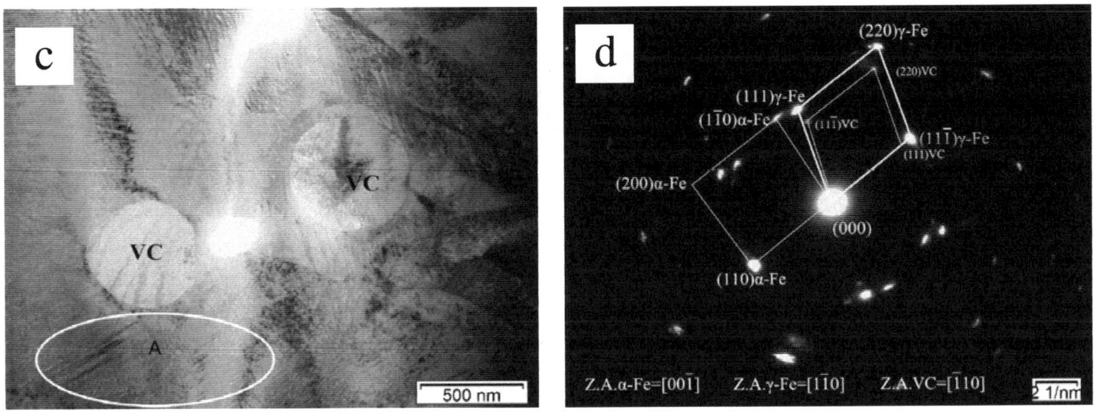

Figure 4. TEM characterization of (Ti,V)C/Fe composites: (a) TEM image of selected area; (b) selected area electron diffraction of reinforcing particulate in (a); (c) TEM image of selected area; (d) selected area electron diffraction of reinforcing particulates and matrix in (c).

4. Conclusions

An iron-based hybrid composite reinforced by carbide particles was prepared through in-situ reactions between Ti, V and C. There were grainy shape and bar shape carbides in the composite. Grainy carbides were $TiVC_2$ multicarbide formed from liquid alloy by in-situ reaction directly. Bar-shaped carbides were VC formed during eutectic solidification.

The matrix of the composite was consisted of plate martensite, small amounts of retained austenite and graphite.

5. References

[1] Kim, C. S., Cho, K., Manjili, M. H., Nezafati, M. (2017). Mechanical performance of particulate-reinforced al metal-matrix composites (MMCs) and al metal-matrix nano-composites (MMNCs). Journal of Materials Science (1), 1-31.

[2] Pramanik, S., Cherusseri, J., Baban, N. S., Sowntharya, L., Kar, K. K. (2017). Metal Matrix Composites: Theory, Techniques, and Applications. Composite Materials. Springer Berlin Heidelberg.

[3] Zhang, Z., Kovacevic, R. (2016). Multiresponse optimization of laser cladding steel + VC using grey relational analysis in the taguchi method. JOM, 68(7), 1762-1773.

[4] He, S., Fan, X., Chang, Q., Xiao, L. (2017). TiC-Fe-based composite coating prepared by self-propagating high-temperature synthesis. Metallurgical & Materials Transactions B, 1-6.

[5] Zhang, Z., Chen, Y., Zuo, L., Zhang, Y., Qi, Y., Gao, K. (2017). The effect of volume fraction of WC particles on wear behavior of in-situ WC/Fe composites by spark plasma sintering. International Journal of Refractory Metals and Hard Materials.

[6] M. J. Hamedi, M. J. Torkamany, J. Sabbaghzadeh. (2011). Effect of pulsed laser parameters on in-situ TiC synthesis in laser surface treatment. Optics and Lasers in Engineering 49: 557–563.

[7] Wang, J. J., Hao, J. J., Guo, Z. M., Wang, S. (2015). Plasma preparation and low-temperature sintering of spherical TiC –Fe composite powder. International Journal of Minerals Metallurgy and Materials 22(12), 1328-1333.

[8] Wang, J., Ding, Y. C., Cheng, H., Wang, Y. S. (2011). In situ production of Fe-TiC surface composite. Advanced Materials Research, 335-336, 80-85.

[9] Xian, Y., Wang, Y., Ding, Y., Wang, J. (2012). Research on in- situ synthesis of (V,Nb)C reinforced iron matrix composite and its microstructure. Iron Steel Vanadium Titanium. 33(6):22-25.

[10] Zhong, L., Xu, Y., Hojamberdiev, M., Wang, J., Wang, J. (2011). In situ fabrication of titanium carbide particulates-reinforced iron matrix composites. Materials & Design, 32(7), 3790-3795.

[11] Zhong, L., Hojamberdiev, M., Ye, F., Wu, H., Xu, Y. (2013). Fabrication and microstructure of

in situ vanadium carbide ceramic particulates-reinforced iron matrix composites. Ceramics International, 39(1), 731-736.

[12] Ye, F., Hojamberdiev, M., Xu, Y., Zhong, L., Zhao, N., Li, Y. (2013). Microstructure, microhardness and wear resistance of VCp/Fe surface composites fabricated in situ. Applied Surface Science, 280(8), 297-303.

[13] Liang, Y. J., Che, Y. C., Liu, X. X. (1993). Thermodynamics Data Handbook of Inorganic Substances. Northeastern University Press, Shenyang, China.

[14] Xu, L., Wei, S., Xing, J., Wang, Q., Zhang, Y., Long, R. (2007). Microstructure and wear properties of Fe–V–Cr–Mo alloy rolls. Rare Metal Materials & Engineering, 36, 423-426.

[15] Liujie, X., Wei, S., Long, R., Feng, N. I., Liu, Y., Wang, S. (2003). Research on morphology and distribution of vanadium carbide in high vanadium high speed steel. Foundry. 52(11): 1069-1073.

Serrated Flow Behavior of Aisi 316l Austenitic Stainless Steel for Nuclear Reactors

Qingshan Li[1], Yinzhong Shen[1,*] and Pengcheng Han[1]

[1]School of Mechanical Engineering, Shanghai Jiao Tong University, No.800 Dongchuan Road, Shanghai 200240, China

Email: shenyz@sjtu.edu.cn

Abstract. AISI 316L austenitic stainless steel is a candidate material for Generation IV reactors. In order to investigate the influence of temperature on serrated flow behavior, tensile tests were performed at temperatures ranging from 300 to 700 °C at an initial strain rate of 2×10^{-4} s^{-1}. Another group of tensile tests were carried out at strain rates ranging from 1×10^{-4} to 1×10^{-2} s^{-1} at 600 °C to examine the influence of strain rates on serrated flow behavior. The steel exhibited serrated flow, suggesting the occurrence of dynamic strain ageing at 450-650°C. No plateau of yield stresses of the steel was observed at an initial strain rate of 2×10^{-4} s^{-1}. The effective activation energy for serrated flow occurrence was calculated to be about 254.72 kJ/mol^{-1}. Cr, Mn, Ni and Mo solute atoms are expected to be responsible for dynamic strain ageing at high temperatures of 450-650 °C in the steel.

1. Introduction

AISI 316L austenitic stainless steel has been extensively applied in reactor vessel and piping systems. This steel is a candidate material for Generation IV nuclear reactors due to its excellent corrosion resistance, tensile and creep strength at elevated temperature, and enhanced resistance to sensitization and intergranular cracking [1]. Serrated flow, negative strain rate sensitivity, ductility minima and peak/plateau in yield strength are often observed in the austenitic stainless steels, such as types 304L and 316L stainless steels, which are characteristics of dynamic strain ageing (DSA). The phenomenon of DSA is also known as Portevin Le-Chatelier (PLC) effect, which is significant to austenitic stainless steels for the service temperature and DSA temperature regime overlap [2].The occurrence of DSA has significant influence on mechanical properties, for example, increasing strength and decreasing ductility [3]. As a consequence, it is of great significance to research the characteristics and mechanism of DSA of AISI 316L austenitic stainless steel at elevated temperatures.

The test parameters such as applied strain rate, temperature and several material variables have a great influence on the behavior of serrated flow. In this paper, we focus on the serrated flow behavior of AISI 316L austenitic stainless steel through tensile tests at temperatures of 300-700 °C at a strain rate of 2×10^{-4} s^{-1} and at various strain rates at 600°C. Additionally, the mechanism of serrated flow in AISI 316L austenitic stainless steel at elevated temperatures is discussed.

2. Experiment Procedures

The steel employed in this study is commercial AISI 316L austenitic stainless steel. The steel was solution treated at 1100 °C for 40 min, then cooled in water to room temperature. The composition (in wt pct) of the steel is 0.033 C, 0.8937 Mn, 0.271 Si, 16.56 Cr, 9.985 Ni, 2.094 Mo, 0.028 P, 0.0088 S. Specimens originated from a forged bar with a diameter of 18 mm. Tensile specimens with a gauge diameter of 10 mm and a gauge length of 25 mm were machined by wire-electrode cutting. Before tensile test, specimens were carefully polished using metallographic sandpaper.

Tensile tests were carried out on Shimadzu AG-100KNA machine within ±0.3% of indicated test force. For elevated temperature tests, temperature can be controlled within ±10 °C. Data was recorded using a data acquisition system attached to the tensile test system. Tensile tests were carried out at temperatures ranging from 300 to 700 °C at an initial strain rate of 2×10^{-4} s^{-1} to investigate the influence of temperature on mechanical properties of the steel. Another group of tensile test was performed at strain rates ranging from 1×10^{-4} to 1×10^{-2} s^{-1} at 600 °C to examine the influence of strain rate on serrated flow behavior of the steel.

3. Results and Discussion

The figure 1(a) illustrates the variation of yielding and ultimate tensile strengths with temperature at a strain rate of 2×10^{-4} s^{-1}. Normally, ultimate tensile strength drops significantly as temperature rise in elevated temperature tensile test. However, ultimate tensile strength decreases slightly, while yielding strength changes inconspicuously at intermediate temperatures of 450-650 °C. Figure 1(b) shows the variation of elongation with temperature. With increasing tensile temperature, elongation first increases slightly, and then decrease occurred at 400-500 °C and 500-600 °C. DSA is the phenomenon that strain and ageing occur simultaneously during deformation process. Since ageing has significant influence on strength and ductility, not only increase in ultimate and yielding strengths but also decrease in elongation are attributed to the occurrence of DSA, which can be explained by the mechanism of ageing strengthening. With the increase of interactions of solute atoms with dislocations during DSA, the dislocation density is enhanced so that ductility is decreased [4].

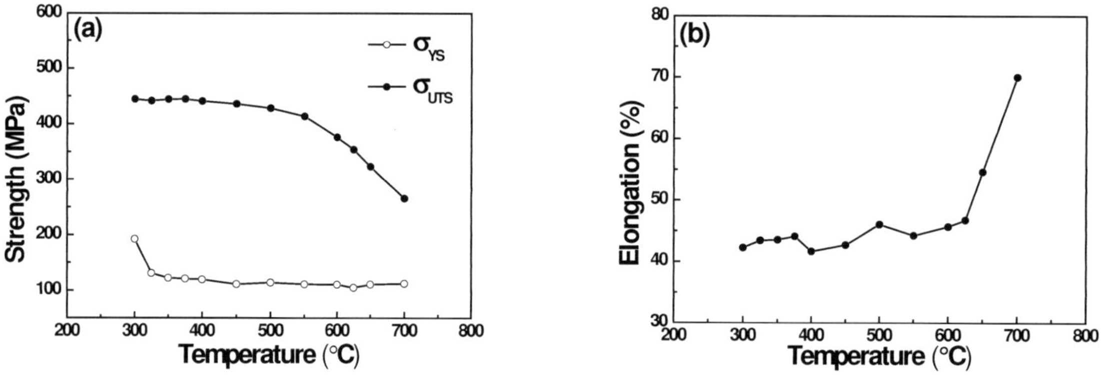

Figure 1. The variation of tensile properties with temperature at an initial strain rate of 2×10^{-4} s^{-1} (a) ultimate tensile strength and yielding strength, (b) elongation.

In general, stress-strain curves are smooth during tensile plastic deformation, in the case of present tensile tests, stress-strain curves remain smooth at low temperatures below 400°C, while serrations occurred on stress-strain curve when tensile temperature reached 450°C. Figure 2 shows segments of tensile curves at various temperatures at a strain rate of 2×10^{-4} s^{-1}. Serrated flow can be observed at intermediate temperatures of 450-650 °C. However, serrated flow disappeared at 700 °C and tensile curve became smooth again. Other researchers studied the DSA of AISI 316L austenitic stainless steel. Kim [5] found serrations in the temperature of 400-600 °C at a strain rate of 2×10^{-4} s^{-1}. Samuel's [6] research suggested that serrations were observed in the temperature of 450-650 °C at a strain rate of 3×10^{-4} s^{-1}. Choudhary's [2] tests were carried out at a strain rate of 3.16×10^{-4} s^{-1}. They found that serrations were first found at 300 °C, mild serrations appeared at 327 °C, stress-strain curves were smooth at 350°C-375 °C, mild serrations appeared again at 400-425 °Cand serrations reappeared in the temperature range of 450-650 °C.

Different types of serrations appearing on the stress-strain curves at various temperatures have been classified into types A, B and C serrations. Type A serrations exhibit an abrupt rise in the loads followed by discontinuous drops to or below the general level of tensile curves due to periodic locking of dislocations. They usually occur in the condition of low temperature or high strain rate. The type B serrations are discontinuous and fine scale oscillations about the general level of the stress-strain curve and Type B serrations often develop following the occurrence of type A serrations. Type C serrations

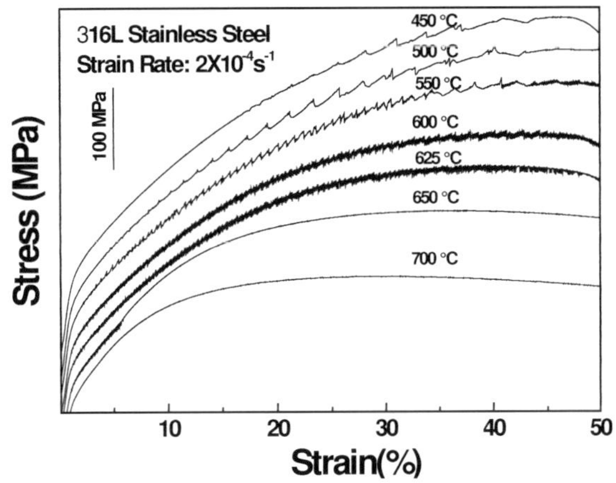

Figure 2. Tensile curves at various temperatures at an initial strain rate of 2×10^{-4} s^{-1}

are always below the general level of stress-strain curve. The segments of various types of serrations at different temperatures are showed in figure 3(a) and figure 3(b) Type A serrations appeared at 450 and 500 °C, and mixed type A+B serrations appeared at 550-625 °C. However, mixed type A+B serrations converted to Type C serrations at 625 °C. At 650 °C, type C serrations occurred during the early stage of tensile deformation, and serrations disappeared as deformation proceeds. In Choudhary's test, type C serrations were observed to intersperse between type A serrations at 650 °C [2].

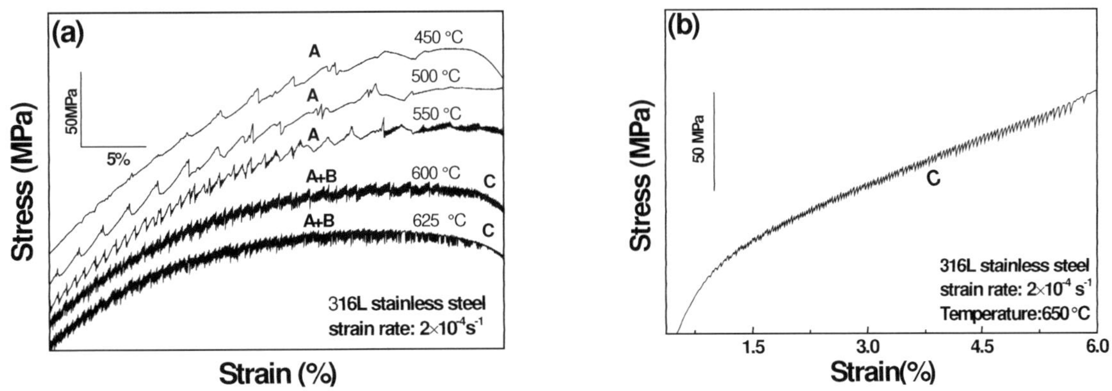

Figure 3. Segment of tensile curves (a) at various temperatures at an initial strain rate 2×10^{-4} s^{-1} and (b) at 650 °C at an initial strain rate of 2×10^{-4} s^{-1}

The segment of stress-strain curves are shown in figure 4. It was found that mixed type A+B serrations appeared at 600°C at strain rates of $1\times10^{-4}$and 2×10^{-4} s^{-1}, while type A serrations appeared at strain rates of 2×10^{-3}, $6\times10^{-3}$and 10^{-2} s^{-1}. It was suggested that type B serrations occurred at a lower strain rate. Summarily, in DSA regime, type A serrations often occurred at low temperature, type B serrations developed with type A serrations as strain increasing, and type B serrations generally occurred at higher temperature and lower strain rate. Type C serrations often took place at a relatively high temperature compared with the temperature for the occurrence of type A and B serrations. Serrated flow always appears after critical plastic strain (ε_c) for the onset of the first serration, which is vital parameter being dependent on both strain rate and temperature. Figure 5 (a) illustrates the change of critical strains for serration occurrence with different temperatures at a strain rate of 2×10^{-4} s^{-1}. The value of ε_c decreases as temperature increasing at 450-625°C, which verifies the occurrence of normal PLC effect. However, the value of ε_c increases with increasing temperature at 625-650 °C, indicating the occurrence of inverse PLC effect. Analogous result by Shankar [7], an inverse PLC was noticed in the temperature ($T>550$ °C) at 3×10^{-3} s^{-1}, 3×10^{-4} s^{-1} and 3×10^{-5} s^{-1}, where type C serrations occurred. Figure 5 (b) displays a change of critical strain for serrations with stain rates at 600 °C, it is shown that the value of ε_c increases when strain rate increased from 10^{-4} to 10^{-2} s^{-1}.

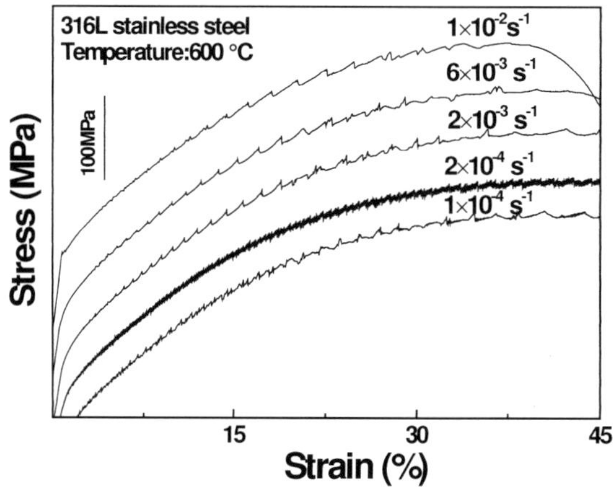

Figure 4. Segment of tensile curves at various Strain rates at 600 °C

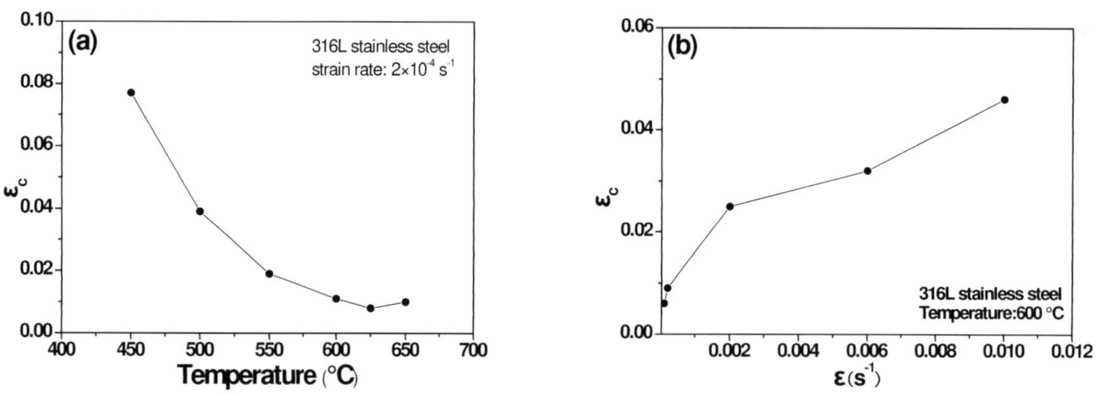

Figure 5. Plot of critical strains vs (a) temperatures at an initial strain rate of 2×10^{-4} s^{-1} and vs (b) stain rates at 600 °C

Due to strain can provide vacancy that contributes to the diffusion of solute atoms, a critical plastic strain is indispensable for the occurrence of DSA, which is known as vacancy-assisted solute diffusion. Diffusion velocity of solute atoms is affected significantly by temperature and strain rate. It is difficult to diffuse for solute atoms when temperature is low, then a greater stain is need at low temperature. For inverse PLC effect, some researcher guessed that solute atoms decrease at high temperature because of precipitation so that only a great strain can produce solute atoms atmosphere to pin dislocations. Likewise, the number of vacancy is small at high strain rate, therefore, there is no enough time for solute atoms to diffuse. So a greater critical plastic strain is necessary. Various models have been put forward to calculate activation energy of serrated flow causing by solute diffusion. Among these models, Mccormick model is widely accepted, so activation energy associated with the serrations can be calculated by acknowledged Mccormick model [8] described by the following equation

$$\varepsilon_c^{m+\beta} = K\dot{\varepsilon}T\exp[Q_m/(RT)] \quad (1)$$

where $\dot{\varepsilon}$ is strain rate, K is Boltzmann constant, R is ideal gas constant, T is the temperature, Q_m is the activation energy for solute migration. Taking the logarithm on both sides of the Eq. (1), we can obtain the relationship as follows

$$(m+\beta)\ln\varepsilon_c = \ln K + \ln\dot{\varepsilon} + \ln T + Q_m/(\mathrm{RT}) \quad (2)$$

If temperature is constant, the relationship can be described as

$$\ln\dot{\varepsilon} = (m+\beta)\ln\varepsilon_c + c \quad (3)$$

So the plot of $\ln\dot{\varepsilon}$ vs. ln ε_c is linear with a slope corresponding to the value of Q_m and $(m+\beta)$. The average $(m+\beta)$ value was primarily computed to be 2.17 from critical plastic strains at various strain rates at 600 °C, as showed in figure 6. The $(m+\beta)$ value for the AISI 316L stainless steel falls into the range between 2 and 3, indicating a substitution diffusion controlled mechanism of DSA.

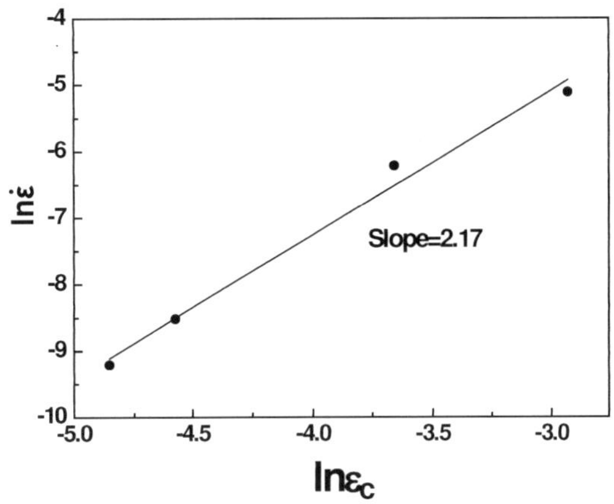

Figure 6. Relationship between critical and strain rate at 600°C

Analogously, critical plastic strain at 450, 500, 550, 600 and 625 °C is respectively 0.213, 0.082, 0.022, 0.012 and 0.009, when $\dot{\varepsilon}$ is constant. Then we can obtain the relationship as following

$$\ln(\varepsilon_c^{m+\beta}/T) = Q_m/RT + c \quad (4)$$

Hence, Q_m can be calculated according the value of $(m+\beta)$, an activation energy of 254.72 kJ/mol was obtained for types A and A+B serrations based on Eq. (4), as shown in figure 7, in agreement with previous studies. Kim et al. [5] and Samuel et al. [6] obtained an activation energy of 272 and 277-278 kJ/mol for AISI 316L stainless steel, respectively. Samuel [9] summarized activation energy data for serrated yielding in austenitic alloys, and different mechanisms were suggested. The diffusion of interstitial solute atoms (C, N and Cr for 316L) was responsible for the occurrence of DSA. Peng *et al.* [10] suggested that the mechanism of DSA in 18-8 austenitic stainless steel was attributed to solution atoms (C, Ni) atmosphere pinning the moving dislocation at a relatively low temperature range. The activation energy of these atoms for diffusion in γ-Fe is $Q_c \approx 131.2$ kJ/mol, $Q_{Mn} \approx 275.9$ kJ/mol, $Q_{Ni} \approx 282.2$ kJ/mol, $Q_{Mo} \approx 269.3$ kJ/mol and $Q_{Cr} \approx 334.4$ kJ/mol [11]. The diffusion of solute atoms in dislocation pipes need an activation energy ranging between 0.4 and 0.7 times the activation energy for bulk diffusion.

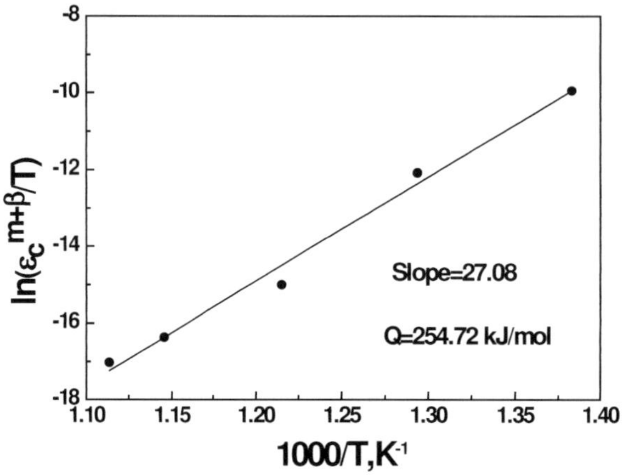

Figure 7. Variation of $\ln(\varepsilon_c^{m+\beta}/T)$ with 1000/T at an initial strain rate of 2×10^{-4} s^{-1}

Considering the value of $(m+\beta)$, we considered that the mechanism of serrated flow occurred at 450-625 °C in the present 316L steel is DSA arising from the interactions between diffusing solute Cr atoms (and/or some other substitution Mn, Mo, and Ni atoms) and mobile dislocations. The species of diffusing substitution atoms giving rise to the serrated flows requires further investigation to understand detailed mechanism of serrated flow in the steel.

4. Conclusions

The serrated flow behaviour of AISI 316L austenitic stainless steel has been investigated through tensile tests at various temperatures and strain rates in combination with the calculation of activation energy for the occurrence of serrated flow. AISI 316L austenitic steel exhibits serrated flow during tensile deformation at intermediate temperatures ranging from 450 to 650 °C at an initial strain rate of 2×10^{-4} s^{-1}. Normal PLC effect is observed at 450-625 °C, while inverse PLC effect occurs at 625-650 °C. In DSA regime, type A serrations often occur at low temperature, type B serrations develop with type A serrations as strain increasing, and type B serrations generally occur at a relatively high temperature and low strain rate. Type C serrations often take place at a higher temperature. DSA in AISI 316L austenitic stainless steel does not accompany with a plateau of yield stress, DSA leads to an increase in strength and a decrease in plasticity. Activation energy for the occurrence of serrated flows at 450-650 °C is obtained to be about 254.72 kJ/mol, suggesting that serrated flows are attributed to DSA caused by interactions between substitution Cr atoms (and or some other substitution Mn, Mo and Ni atoms) and mobile dislocations.

5. Acknowledgments

This investigation was supported by the Key Program of National Natural Science Foundation of China (Joint Funds of the National Natural Science Foundation of China and Shanghai Baosteel Group Company) (51034011), National Science and Technology Major Project (2011ZX06004-009), and ITER-National Magnetic Confinement Fusion Program (2011GB113001).

6. References

[1] Hong S.G. and Lee S.B. 2004 *J. Nucl. Mater.* **328** 232
[2] Choudhary B.K. 2014 *Metall. Mater. Trans. A* **45** 302
[3] Dehghani K. and Jones J.J. 2000 *Metall. Mater. Trans. A* **31A** 1375
[4] Tsuchida Y., Okamoto K. and Tokunaga Y. 1996 *Weld Int.* **10** 27
[5] Kim D.W., Ryu W.S., Hong J.H. and Choi S.K. 1998 *J Mater. Sci.***33** 675
[6] Samuel K.G., Mannan S.L. and Rodriguez P. 1988 *Acta Metall.* **36** 2323
[7] Shankar V., Valsan M., Rao K.B.S. and Mannan S.L. 2004 *Metall. Mater. Trans. A* **35** 3129
[8] Mccormick P.G. 1973 *Acta Metall.* **21** 873

IOP Conf. Series: Materials Science and Engineering **250** (2017) 012013 doi:10.1088/1757-899X/250/1/012013

[9] Samuel K.G., Ray S.K. and Sasikala G. 2006 *J. Nucl. Mater.* **355** 30
[10] Peng K.P., Qian K.W. and Chen W.Z. 2004 Mater. Sci. Eng. A **379** 372
[11] Balluffi R.W. 1970 Phys. Status Solidi B **42** 11

IOP Conf. Series: Materials Science and Engineering **250** (2017) 012014 doi:10.1088/1757-899X/250/1/012014

Experimental Study on Basic Mechanical Properties of BFRP Bars

Xiaochun Fan[1], Ting Xu[1,2], Zhengrong Zhou[1] and Xun Zhou[1]

[1]School of Civil Engineering and Architecture, Wuhan University of Technology,Wuhan 430070, China

[2]Corresponding author. Tel.:13027183668,E-mail:2267296315@qq.com

Abstract. Basalt Fiber Reinforced Polymer (BFRP) bars have the advantages of corrosion resistance, high strength, light weight, good dielectric properties, and they are new type of green reinforced alternative material. In order to determine the mechanical properties of BFRP bars, the tensile strength of basalt fiber bars was necessary to be studied. The diameters of the basalt fiber bars were compared by means of uniaxial tensile test in this article. Then the stress-strain curve can be drawn out. The results show that the stress - strain curve of BFRP bars present straight line relation, and there is no sign before failure; there is no yield platform on the stress-strain curve of BFRP bars, which are typical brittle material;the tensile strength of BFRP bars is about 3 times higher than that of ordinary steel bars. and the elastic modulus is about 1/5 of that of ordinary steel; the ultimate tensile strength of BFRP bars varies little with the increase of diameter, but there exist some differences in modulus values.

1. Introduction

Basalt continuous fiber is the fiber that used basalt ore as raw material, and melt the basalt stone under the temperature of 1450 ~ 1500°C, then formed through the platinum and rhodium alloy wire drawing slab at the high-speed. Basalt Fiber Reinforced Polymer (BFRP) bar is a new kind of non metal composite reinforced material, which is composed of the basalt fiber as reinforced material, synthetic resin as base material and mixed with appropriate auxiliary agent, formed by pultruded process and special surface treatment. This kind of new composite material has the advantages of corrosion resistance, high strength, light weight, anti-fatigue and insulation, which can replace reinforced steel for concrete structure. It can fundamentally solve the problem of steel corrosion, and has wide application prospect in civil engineering[1-2].

However, there are few researches on the basic mechanical properties and failure modes of BFRP bars[3-6]. In this paper, the tensile tests of BFRP bars with diameters of 10mm, 12mm and 16mm were tested. The failure modes of BFRP bars and the phenomena during the test were observed. The mechanical properties such as ultimate tensile strength and elastic modulus were measured, and the stress-strain curves of BFRP bars were drawn according to the test results.

2. Test Design

2.1. Test Material

This experiment uses the deep thread BFRP bars provided by Jiangsu Green Materials Vally New Material T&D Co., Ltd (GMV), and the diameter is 10mm, 12mm and 16mm. In order to ensure the reliability of the test results, each kind of bars is made of 5 specimens, and the parameters of the specimens are shown in table 1:

Table 1. Parameters of tensile specimens

Surface form	Specimen number	Diameter/mm	Single length/mm	Number of test pieces
	BFRP10	10	1300	5
Deep thread	BFRP12	12	1300	5
	BFRP16	16	1300	5

[a] BFRP: Basalt fiber reinforced polymer;
[b] BFRP**: ** indicates the diameter of the bar.
[c] Refer to ACI440.3R-04 "Guide Test Methods for Fiber-Reinforced Polymers (FRPs) for Reinforcing or Strengthening Concrete Structures": The length of the test section should not be less than 100mm, nor should it be less than 40 times the diameter of the FRP bar.

2.2. Production of the Specimens

The shear strength of BFRP bars is low, in order to avoid the shear failure at both ends of the tensile specimen due to the clamp action, the seamless steel sleeve with length 400mm is used for anchoring at both ends of the BFRP bars in the tensile test.Seamless steel sleeve needs to have a certain wall thickness to withstand compressive stress produced by the instrument at the end of the test piece, The wall thickness chosen in this experiment is 4mm. The anchoring procedure is as follows:

(1)The pressure sensitive adhesive tape is wrapped from the bottom of 400mm so that it can block one end of the steel sleeve to prevent leakage of resin glue, it can also ensure that BFRP bars can be centered in the seamless steel sleeve;

(2)The tensile specimen at both ends of 400mm anchorage length need to be handled by sticky sand to increase its adhesion with the resin glue;

(3)The inner wall of the steel sleeve need to be derusted, and use hand to block one end of seamless steel sleeve, pour resin glue into the other side of the sleeve, then make the BFRP bars rotated into the steel sleeve so that the bubbles in the sleeve can be avoided and the effective bond between them can be ensured;

(4)The resin glue on the sleeve surface need to be wiped off after inserting, and then reverse the material to ensure the full bonding between the resin glue and the reinforcing bar;

(5)The resin glue in the sleeve need to be replenished in time to ensure that it is fully bonded, and place it for about 5 days until the resin glue is fully hardened, the other end of the anchor is the same as above.

Enlargement of BFRP bars' ribs shown in Figure 1, the preparation of BFRP bars before tensile test shown from Figure 2 to Figure 7.

Figure 1. Enlargement of ribs

Figure 2. Wrapping tape

Figure 3. Enlargement of the sticky sand

Figure 4. Sand sticking at both ends

Figure 5. Filling with resin glue

Figure 6. Reverse anchored specimen

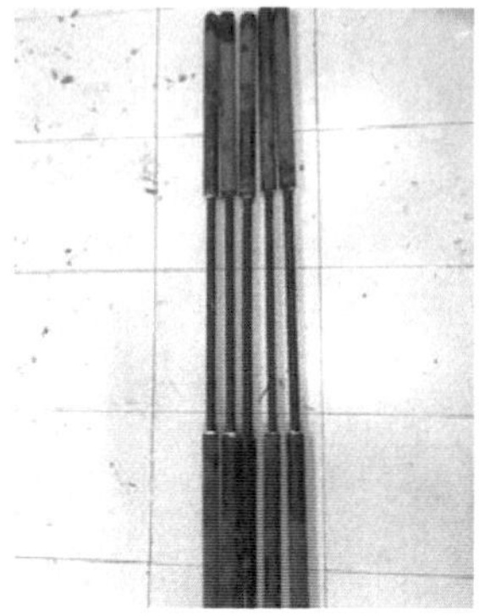

Figure 7. Specimens with anchors at both ends

The tensile test was carried out on the Servo hydraulic Dynamicigue Testing System Series (LFV-1000kN) in the Civil Engineering Structural Laboratory of Southeast University (see Figure 8). The strain in the testing process is measured by the extensometer. Specific steps are as follows:

Figure 8. Servo hydraulic Dynamic Fatigue Testing System Series(LFV-1000kN)

(1)Place the tensile specimen on the test machine, adjust the extensometer's position, set the parameters of the test machine, record and take photos in real time;

(2)Using displacement control to load, set the maximum displacement value of 50mm, loading rate of 0.12mm / s;

(3)Observe the test phenomena during loading, loading to the specimen damage, record the maximum load value and the mode of damage;

3. Test Results and Analysis

3.1. Typical Test Phenomenon of BFRP Bars

In the process of stretching, the surface appears adhesive peeling firstly, issued the sound of fiber stripping resin, followed by part of the fiber breaks and produces burrs. As the loading continues, the noise of fiber peeling becomes larger and larger, and white spot cracks appear on the surface of BFRP bars. As the load increases further, the fiber was constantly being pulled off and issued the "crackling" sound, the sound is getting more and more frequent, accompanied by a loud sound, the fiber broke suddenly, The middle of the BFRP bar is pulled off and the appearance is like explosive. The details can be seen from Figure 9 to Figure 11. (The number in the figure is the tensile force and fatigue displacement when the first wire broken.)

Figure 9. 10mmBFRP bar damage process

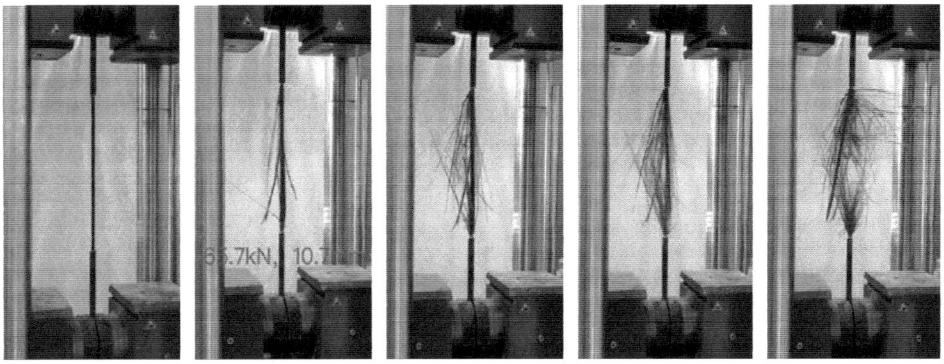

Figure 10. 12mmBFRP bar damage process

Figure 11. 16mmBFRP bar damage process

3.2. Mechanical Properties

The tensile strength and elastic modulus of the BFRP bars were calculated by the following methods.

The tensile strength is calculated according to formula (1):

$$f_u = \frac{F_u}{A} \tag{1}$$

Where f_u - the tensile strength (MPa); F_u - the maximum tensile strength (N) of the BFRP bars tested; A -BFRP tendons cross-sectional area (mm^2).

The elastic modulus is calculated according to formula (2):

$$E = \frac{F_1 - F_2}{(\varepsilon_1 - \varepsilon_2) * A} \quad (2)$$

Where E - tensile elastic modulus (MPa); F_1 - 50% maximum load (N); F_2 -20% maximum load (N) ε_1 -the strain corresponding to 50% of the maximum load; ε_2 -the strain corresponding to 20% of the maximum load

According to the data measured in the tensile test, the tensile strength and modulus of elasticity are calculated according to formula (1) and formula (2). The test results are summarized in table 2:

Table 2. Tensile test results

Specimen number	Bottom diamete r /mm	Maximum pulling force /kN	Ultimate strength /Mpa	Average ultimate strength /Mpa	Elastic Modulus /Gpa	Average elastic modulus /Gpa
BFRP10-1		79.16	1302		44.00	
BFRP10-2		77.99	1283		43.16	
BFRP10-3	8.8	75.16	1236	1275	43.41	43.37
BFRP10-4		77.34	1272		42.51	
BFRP10-5		77.98	1283		43.75	
BFRP12-1		109.09	1246		43.62	
BFRP12-2		105.83	1209		47.11	
BFRP12-3	10.56	106.88	1221	1230	50.32	46.11
BFRP12-4		109.13	1247		44.76	
BFRP12-5		107.24	1225		44.73	
BFRP16-1		196.11	1260		41.10	
BFRP16-3	14.08	188.94	1214	1212	42.95	44.17
BFRP16-4		191.26	1229		49.45	
BFRP16-5		178.26	1145		43.19	

[a] The second specimen in the 16mmBFRP bar has a sleeve slippage after the tensile force up to 140kN, so its sample data is invalid.

It can be seen from Table 2 that the tensile strength of BFRP bars is more than 1100MPa, about 3 times of HRB400 ordinary steel (tensile strength design value = 360MPa), the elastic modulus is about 1/5 of that of HRB400 ordinary steel (elastic modulus = 200GPa). The tensile strength of the materials with different diameters is similar, but there is a big difference in the elastic modulus, the larger the diameter, the more discrete the data measured.

3.3. Stress - Strain Curve

According to the tensile load and strain values collected in the experiment, the stress-strain curves of the BFRP bars are plotted (Figure. 12 to Figure. 14). From the figure we can see that the stress-strain curves are straight line, that is to say, the stress and strain of the BFRP bars are in good linear relationship. BFRP bars are linear elastic materials, there is no sign before failure and there is no plastic yield platform, which is completely different from ordinary steel bars.

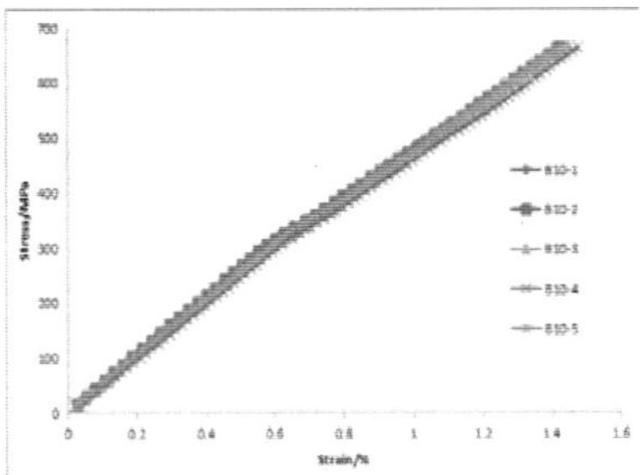

Figure 12. Stress-strain curve of BFRP bars with diameter of 10mm

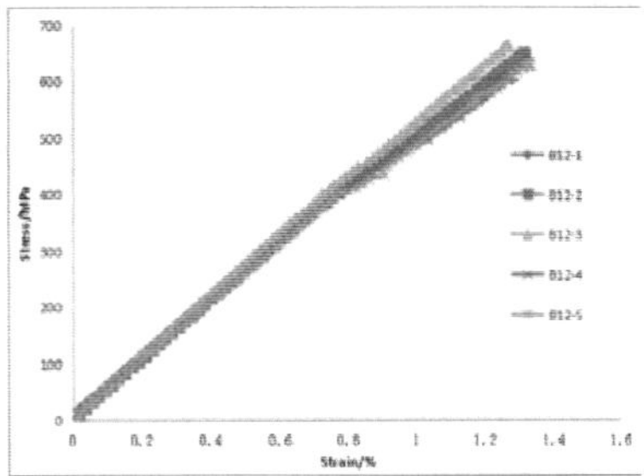

Figure 13. Stress-strain curve of BFRP bars with diameter of 12mm

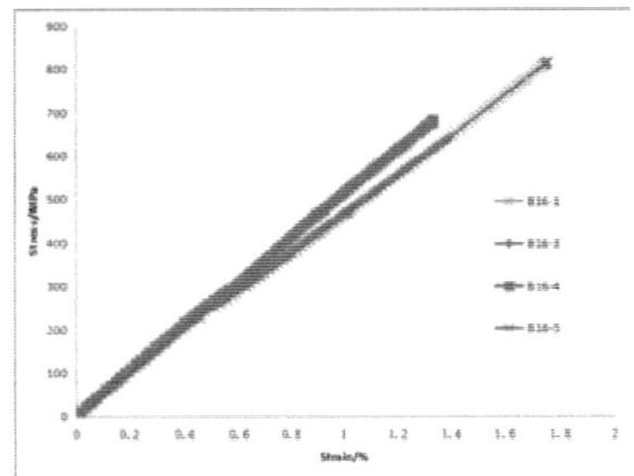

Figure 14. Stress-strain curve of BFRP bars with diameter of 16mm

4. Conclusion

(1)BFRP bar is a typical brittle material, it has no sign when it breaks and the fracture is abrupt.

(2)In this experiment, the tensile strength of BFRP bars is about 1212 ~ 1275MPa, the elastic modulus is about 43.37 ~ 46.11GPa, the tensile strength is about 3 times higher than that of ordinary steel bars, and the elastic modulus is about 1/5 of the ordinary steel bars or so.

(3)The BFRP bars are basically linear elastic deformed before they are destroyed, and their stress-strain relationship is approximately a straight line and there is no yield platform in the curve.

(4)The tensile strength of the fiber bars produced by the same raw material type, the same processing conditions and methods depends mainly on the performance of fiber itself, so BFRP bars with different diameters have similar ultimate tensile strength.

(5)Because with the diameter increases, the unit volume of fiber content is reduced, and the defects inside specimen increased, resulting in a large deviation of data for a set of 5 specimens in elastic modulus

5.Acknowledgements

This paper is supported by the National Science and Technology Support Program (2014BAB15B01).

6.References

[1] Hu Xianqi. 2008. The overview of the present development of basalt fiber in china[J]. *Hi-Tech Fiber & Application*. 33(6):12-19.

[2] Huo Baorong and Zhang Xiangdong. 2010 Consolidation test of RC columns bound with two kinds of fiber BFRP, CFRP [R] // *International Conference on Civil Engineering in China-Current Practice and Research Report*(Bejiing:Aussino Academic Publishing House) pp836-840.

[3] Huo Baorong and Zhang Xiangdong. 2012. Experiment study of BFRP bars' mechanical properties. *Applied Mechanics & Materials*, 174-177, 830-833.

[4] Cao Xiaofeng, Zhao Wen, Xie Qiang and Yang Guoliang. 2016. Experimental study on basic mechanical properties of BFRP bars[J]. *Highway Engineering*, 41(5):215-217.

[5] Liu Xiao and Liu Shunfeng. 2014 Experimental study on tensile properties of basalt fiber reinforced plastic [J]. *Shanxi Architecture*, 40 (16): 32-33.

[6] Li Yun. 2016 Experimental study on mechanical properties of basalt fiber reinforced plastic [J]. *Traffic Science and Technology*, (01): 140-142.

IOP Conf. Series: Materials Science and Engineering **250** (2017) 012015 doi:10.1088/1757-899X/250/1/012015

Comparison of different etching methods on the morphology and semiconductor characters of black silicon

Wenlin Meng [1], Hao Zhong [2], Wei Hou [1], Yanhao Gou [1], Yuhao Song [1] and Wei Li [*2]

[1] School of Optoelectronic Information, University of Electronic Science and Technology of China, Chengdu, 610054 China
[2] State Key Laboratory of Electronic Thin Films and Integrated Devices, University of Electronic Science and Technology of China, Chengdu, 610054 China
[*] Email: wli@uestc.edu.cn

Abstract. Femtosecond laser etching, deep reactive ion etching and metal-catalyzed chemical etching are used to fabricate black silicon. It has been found that the light absorption is significantly enhanced in the wavelength of 400~2200nm, in which the absorption in near infrared band of black silicon etched by femtosecond laser with SF_6 has the highest value. It is observed that, however, the minority carrier lifetime of crystalline silicon is shortened to some extent, which can be adjusted and controlled effectively by depositing SiN_x film to passivate the surface of black silicon. Finally, a PIN photodetector is manufactured based on black silicon and a higher responsibility of 0.57 A/W at 1060 nm is obtained compared to the PIN silicon photodetector without etching process.

1. Introduction

Silicon is the most commonly used semiconductor in microelectronics and optoelectronic devices, but the high reflectivity and low absorption of crystalline silicon (c-Si) restricts the sensitivity and efficiency of silicon-based devices. Moreover, the band gap of c-Si renders it transparent to wavelengths longer than 1100 nm, and makes it useless for near-infrared detecting. Any development that can reduce the reflection and enhance the absorption of silicon from ultraviolet to near-infrared wavelengths will assist in making silicon-based wideband detectors a reality [1-4]. Nano-structured and micro-structured c-Si (with cones, wires, or holes on the wafer surface) [5-8] have received steadily increasing interest as possible architectures for high efficiency and low cost material, because they exhibit very low reflection especially in the visible region. Many approaches such as femtosecond laser etching (FLE), deep reactive ion etching (DRIE) and metal-catalyzed chemical etching (MCE), have been developed to prepare these structures, by witch one can increase the light absorption of c-Si effectively. The lifetime of minority carrier in c-Si, however, could be shortened due to the increased recombination loss and contact resistance arising from the nanostructures and micro-structures of black silicon (BS) [9]. Over the past decades, minority carrier lifetime in silicon-based materials have been intensely investigated because it significantly affects the optical and electrical properties of the materials. Wang [10] believes that the defects in the silicon nanowire are related to the dangling bonds of the amorphous silicon itself. Kajjam [11] shows that oxygen increases deep defect densities measured between 0.35 eV and 0.55 eV below the conduction band. It is found that oxygen-induced defects can be reduced and that lifetime can be increased both by compensating with boron. Therefore, how to increase light absorption and adjust minority carrier life are the main thoughts of our research proposal. A passivation layer and a subsequent annealing process should be considered as effective methods to control the minority carrier lifetime of BS.

In this paper, three methods named as FLE, DRIE and MCE to fabricate BS and several anti-reflection/light-trapping structures are simultaneously obtained directly on the c-Si substrates have been carried out. Meanwhile the comparison of morphology, light absorption, minority carrier lifetime with/without SiN_x passivation have been made among three different BS materials. And finally a PIN photodetector is manufactured based on BS and the responsibility is compared to the PIN silicon photodetector without etching process.

2. Experimental Details

N-type silicon wafers with a thickness of 500 μm and a resistivity of 2500 to 3000 Ω·cm were used. MCE was carried out silver-plated nanoparticles and catalytic corrosion on the silicon surface. During the silver-plated nanoparticles process, the cleaned wafers were placed in a mixture solution of HF:$AgNO_3$ (0.009mol/ L) =1:4. Two groups were made for comparison by varying the silvering duration to be 30 s and 60 s. During catalytic corrosion process, silicon wafer coated with silver nanoparticles were placed into HF:H_2O_2=1:2 for 10 min. When the etching processes were over, the silicon pieces were dipped into an aqueous solution of HNO_3 and then rinsed with deionized water to remove any residual Ag.

The same N-type silicon substrates were irradiated by a regeneratively amplified Ti: sapphire femtosecond laser at 800 nm with a laser pulse width less than 100 fs, which was supplied by Spectra-Physics (Solstice). During scanning, the light power was set at 200 mW with 1000 Hz, the scanning speed was 1 mm/s and the scanning interval of each line was 0.1 mm. Three kinds of medium called SF_6, N_2 and air were chosen as the atmosphere during femtosecond laser irradiation.

In process of DRIE, photoresist was applied as the mask and the same N-type silicon substrates were etched by using SF_6 and C_4F_8. During the process, SF_6 was used as etching gas and C_4F_8 as preventing gas, respectively, and the flow ratio of them was 4:3. The upper electrode power was set at 1000 W for SF_6 etching or 1300 W for C_4F_8 preventing, and the working time was 3 min for etching or 2 min for preventing, separately. When the processes were over, the silicon pieces were etched with O_2 to remove any residual photoresist.

The morphologies of BS materials were characterized by a field emission scanning electron microscope (SEM, JSM-7500F). The light absorption was obtained at room temperature using a fiber optic spectrometer (NIR2500) equipped with an integrating sphere (Idea Optics, IS-20-5). The minority carrier lifetime of BS was measured by LT-100C. The responsivity of experimental PIN detector was calculated by using an optical power meter (OPHIR, Vega), an optical chopper (Scitec Instruments, Model-300CD) and a Keithley 2400 apparatus under dark room environment. In order to ensure the accuracy of the measurement, we carried out calibration before test and each of measurements was performed on a few samples (usually 4 to 6).

3. Results and Discussion

3.1. Morphologies of BS Materials

Figure 1 presents typical morphologies of BS etched by FLE in SF_6 and air atmosphere. It can be seen from the figure that under the same laser parameters, the morphologies of BS etched in two kinds of atmosphere are different obviously. In figure 1(a), some cones are larger but the others are smaller, presenting flat blanket-like structure. In fact, the same morphologies can also be obtained when etched in N_2 atmosphere. In figure 1 (b), the radius of each cone is about 40 μm, the cross-sectional area of the substrate is about 60 μm^2 and the diameter of the tapered cone is close to 800 nm. It is observed that the BS prepared in SF_6 have sharp pointed cones on the surface with nearly no difference between pointed cones, and the direction of the cones is consistent with the incident light. In addition, R. Younkin [12] explained that in the process of FLE, the halogen element and Si can produce volatile compounds, whereas air and N_2 can't. Therefore, the spike surface of BS etched in SF_6 are much sharper, while the surface etched in air are more blunt.

IOP Conf. Series: Materials Science and Engineering **250** (2017) 012015 doi:10.1088/1757-899X/250/1/012015

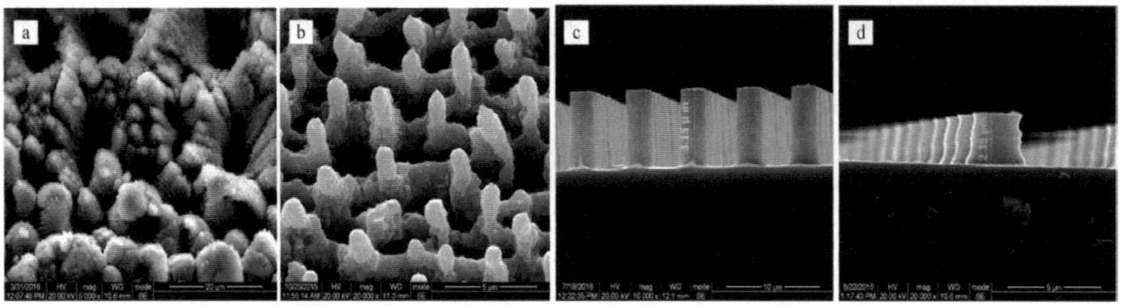

Figure 1. SEM images of BS produced by FLE in air (a), in SF_6 (b), and by DRIE (c, d).

Figure 2. SEM images of BS produced by MCE with silver plated 30s (a) or 60s (b).

Figure 1(c,d) show the BS etched by DRIE. It can be seen from figure 1(c,d) that there are two kinds of cycle as 6 μm or 10 μm, but one same diameter as 4 μm. It can be understood that the BS etched by DRIE is of a cylinder or a round table shape, the diameter is related to the size of the mask plate in lithography step, while the height is depended on repeating times in etching process. The final height is about 3.15 μm after 100 times or 2.35 μm after 70 times.

Figure 2 are SEM graphs of BS etched by MCE, in which silver was plated with 30s(a) or 60s(b), respectively. The diameter of the hole in (a) is obviously larger than that of in (b). It is considered that in the prosesse of our MCE, with the increase of plating time, much more areas are attached on the silicon substrate with silver nanoparticles, resulting in bigger etched areas.

3.2. *Absorption spectra and minority carrier lifetime of BS materials*

Figure 3(a) is the absorption spectra of BS etched by FLE, DRIE and MCE, respectively. It can be seen that the absorption of BS prepared by FLE and MCE both are 70% in the near-infrared, while the absorption of BS could reach as high as 90% by FLE in the SF_6 atmosphere.

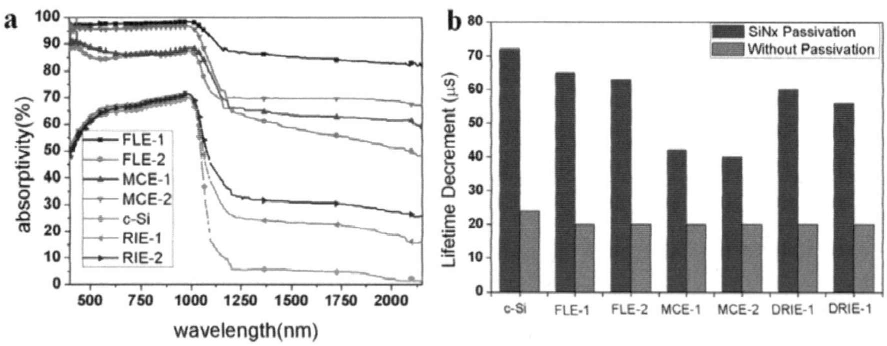

Figure 3. Absorption spectra of different BS (a) and minority carrier lifetime of BS with/without SiN_x passivation (b).

IOP Conf. Series: Materials Science and Engineering **250** (2017) 012015 doi:10.1088/1757-899X/250/1/012015

It is well known that the c-Si has a lower light absorption in the wavelength of 190~1200nm. There are two reasons that can explain above phenomenon: one is that the c-Si surface has a high reflectivity of ultraviolet-infrared light (about 40%), and the other is that the band structure of c-Si limits the photon energy below its band width of the absorption in near infrared, and so it can only absorb incident light with the photon energy larger than the band gap width of 1.12eV. In this experiment, however, the BS materials have excellent optical properties with high spectral absorption, which can be attributed to the following factors. Firstly, the nano- and micro-structures produce multiple reflection to the surface incident light and form the trapping effectiveness. However, no matter what the structure of BS surface is, the absorption of BS in the near infrared is not obviously improved, such as black silicon perpared by DIRE showed in figure 4. It is observed that the BS etched by FLE in the SF_6 atmosphere can still maintain a high absorption of more than 90% in the wavelengths greater than 1100 nm, which means that there need other explanations. Mazur and his co-workers [13] had used femtosecond laser to fabricate BS with H_2S, SF_6 and Ar+SF_6. The surface composition of the sample was analyzed by Rutherford Back Dispersion method, and it was found that BS contains more than 0.6% (atomic percent) of sulfur element. As early as in 1984, Janzen [14] et al. reported that eight kinds of dominate levels could be formed when doping sulfur in silicon. The sulfur element could be doped in two forms of 0.318eV and 0.614eV energy levels. Therefore, the doping of the sulfur group elements can change the silicon band structure, increasing the absorption of the infrared light. In our experiment, the energy of femtosecond laser could exceed the melting piont of c-Si, and the decomposed sulfur elements from SF_6 could be doped to the surface layer of the c-Si sample by means of liquid to solid transformation. Therefore, BS etched by FLE in SF_6 could still maintain a higher light absorption more than 1100nm.

Figure 3(b) is the minority carrier lifetime of different BS with/without SiN_x passivation. As shown in the figure, the minority carrier lifetime of BS prepared by different methods has no obviously change without passivation. It is known that LT100-C is a device-scale instrument based on high frequency photoconductive decay, it is commonly used for testing bulk and stick silicon materials, so it can only give a relatively minority carrier lifetime, not the body lifetime of the carriers, though the surface recombination of carriers is far larger than that of bulk recombination without passivation. Nevertheless, the lifetime of BS can be adjusted by film passivation like SiN_x, which has been demonstrated early by Yamazaki et al [15]. Here, by comparing three different etching methods after the same SiN_x passivation, we found that the minority carrier lifetime of BS etched by FLE and DRIE are higher than that by MCE, resulting from the fact that the morphologies of BS prepared by FLE and DRIE are rougher than those by MCE. It is also found that passivation with special film like SiN_x could decrease the surface recombination and enlarge the difference of minority carriers lifetime among above mentioned three etching methods.

3.3. Device Application of Black Silicon

Figure 4 gives the skeleton of a PIN photodetector (a) and the responsibilities of two silicon photodetectors (b). For the photodetector that the surface is etched by FLE with SF_6, it can be found that the peak responsivity reaches at 1000nm, and the responsivity at 1060nm can maintain as high as 0.57A/W. For the photodetector that the surface is not etched anymore, however, the peak responsivity reaches at 950nm but the responsivity at 1060nm is only 0.26A/W. It can be concluded that the responsivity of Si-PIN photodetector with BS is higher than that of Si-PIN photondetector without BS in the near infrared band.

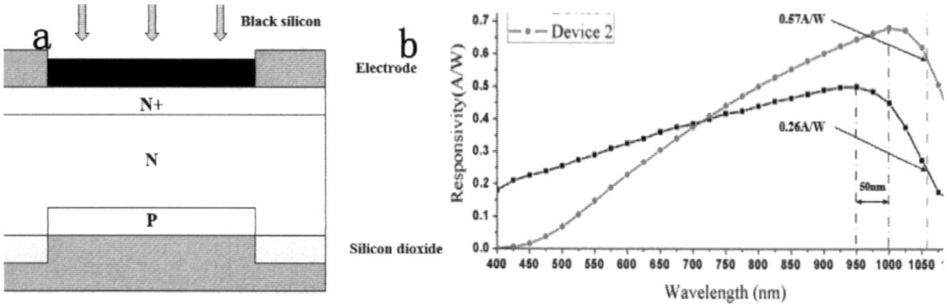

Figure 4. The skeleton of a PIN detector (a) and responsivities of two different devices (b).

4. Conclusions

In summary, black silicon (BS) materials are fabricated by FLE, DIRE and MCE respectively, and the nanocone arrays or holes on the surface of silicon substrate have diameter and length of 100~400nm and 1.5~3μm, differently. Greatly enhanced light absorptance of black silicon has been observed in a wide wavelength range from 400 to 2200 nm, and the maximum absorptance reaches 90%. This enhancement is explained by reduced reflectance, light-trapping effect, and scattering effect caused by the specific nanostructures and/or microstructures on the surface of silicon substrate. However, the minority carrier lifetime of c-Si is shortened due to the increased recombination loss and contact resistance arising from above BS structures. Based on three different etching methods, we have found that the minority carrier lifetime can effectively be adjusted by SiN_x passivation. A novel Si-PIN photodetector with BS formed on the front surface has been fabricated and the comparison of device responsivities has been made. It is found that the Si-PIN photodetector with BS formed on the front surface has an apparently increase in responsivity particularly in the near-infrared wavelengths.

5.Acknowledgements

This work was partially supported by the National Science Foundation of China (Grant No. 61235006, 61421002).

6.References

[1] Zhong H, Guo A, Guo G, Li W and Jiang Y 2016 *Nanoscale research letters* **11** 322
[2] Su Y, Li S, Zhao G, Wu Z, Yang Y, Li W and Jiang Y 2012 *Journal of Materials Science: Materials in Electronics* **23** 1558-61
[3] Jiang J, Li S, Jiang Y, Wu Z, Xiao Z and Su Y 2012 *Journal of Materials Science: Materials in Electronics* **24** 463-6
[4] Aida M S, Attaf A and Benkhedir M L 2008 *Philosophical Magazine Part B* **73** 339
[5] Chen Y, Xu Z, Gartia M R, Whitlock D, Lian Y and Liu G L 2011 *ACS NANO* **5** 8002-12
[6] Koynov S, Brandt M S and Stutzmann M 2006 *Applied Physics Letters* **88** 203107
[7] Branz H M, Yost V E, Ward S, Jones K M, To B and Stradins P 2009 *Applied Physics Letters* **94** 231121
[8] Oh J, Yuan H C and Branz H M 2012 *Nature nanotechnology* **7** 743-8
[9] Garnett E and Yang P 2010 *Nano letters* **10** 1082-7
[10] Wang R P 2006 *Applied Physics Letters* **88** 142104
[11] Kajjam S, Konduri S and Dalal V L 2013 *Applied Physics Letters* **103** 093506
[12] Younkin R, Carey J E, Mazur E, Levinson J A and Friend C M 2003 *Journal of Applied Physics* **93** 2626-9
[13] Sheehy M A, Winston L, Carey J E, Friend C M and Mazur E 2005 *6* **17** 3582-6
[14] Janzén E, Stedman R, Grossmann G and Grimmeiss H G 1984 *Physical Review B* **29** 1907-18
[15] Sulima O V, Conibeer G, Yamazaki T, Kato S, Miyajima S and Konagai M 2014 *Proceedings of SPIE* **9178** 91780L

IOP Conf. Series: Materials Science and Engineering **250** (2017) 012016 doi:10.1088/1757-899X/250/1/012016

Simulation Research on Micro Contact Based on Force in Silicon Wafer Rotation Grinding

Qinglei Ren, Xin Wei, Xiaozhu Xie and Wei Hu

Faculty of Elcctromcchanical Engineering, Guangdong University of Technology, Guangzhou 510006, China

Email: 15920179558@139.com

Abstract. Silicon wafer rotation grinding with cup type diamond wheel is a typical ultra precision grinding process. In this paper, a simulation model based on force for micro contact between wheel micro unit and silicon wafer is established from the stable ductile grinding process. Micro contact process in grinding is simulated using the nonlinear explicit finite element analysis software LS-DYNA. The stress-strain results on silicon wafer and wheel micro unit are analyzed by finite element method. The results show that the critical displacement and load corresponding elastic to plastic - plastic to brittle exist on silicon wafer. In silicon plastic zone tangential sliding can produce plastic groove and uplift. Wear of wheel micro unit can be based on the simulation data to judge. The research provides support for wafer grinding and wheel wear mechanism.

1. Introduction

Semiconductor devices are the foundation of electronics industry-the largest industry in the world [1]. Silicon wafers are used as the substrates to build the vast majority of semiconductor devices [2]. Manufacturing of high-quality silicon wafers is the key to the initial fabrication of semiconductor devices [3]. Especially for large diameter silicon wafers ($\phi \geqslant$300mm) flattening and back thinning processing, using ultra precision grinding than the traditional lapping process is more effective, and the application of cup type diamond grinding wheel to silicon wafer rotation grinding is a typical form of silicon wafer ultra precision grinding [4].

Silicon wafer rotation grinding method was first proposed by Matsui [5] in 1988, and its principle is shown in Figure 1. The diameter of the rotary table is slightly larger than that of the silicon wafer, and the center line of the rotary table is coincident with that of the silicon wafer. The silicon wafer is adsorbed on the working table through a vacuum. The working surface center line of the grinding wheel is adjusted to align with the center of the silicon wafer. During grinding, the grinding wheel and the silicon wafer are rotated around the respective axis, and the grinding wheel is only axial feed. The grinding depth can be controlled by adjusting the axial feed rate of the grinding wheel and the rotation speed of the silicon wafer. Ductile grinding with plastic removal mode can be realized.

Research on silicon wafer rotation grinding has attracted much attention, and its grinding mechanism is the key point of the research [6]. Representative studies have been reported. Pei Z J et al. [7] have made a series of grinding experiments, and systematically studied the influence of grinding process on the surface quality of silicon wafers. These studies focus on the process of experimental and macroscopic in order to obtain the high quality silicon wafer. However, it is difficult to analyze the micro dynamic mechanism during the grinding process, and the experimental results have high cost, but the research on the finite element simulation analysis method is more time saving and labor saving.

At present, there is no report on the micro finite element simulation of silicon wafer rotation grinding. LS-DYNA is a general-purpose finite element program capable of simulating complex real world problems [8]. It is used by the automobile, aerospace, construction, military, manufacturing, and bioengineering industries. It is especially suitable for solving nonlinear dynamic impact problems for a variety of 2D and 3D nonlinear structures.

This paper is organized into four sections. Following this introduction section, Section 2 describes that the micro contact model of grinding wheel micro unit and silicon wafer is established, including the mechanical model and the simulation model. In Section 3, simulation results are presented and analyzed. Finally, conclusions are drawn in section 4.

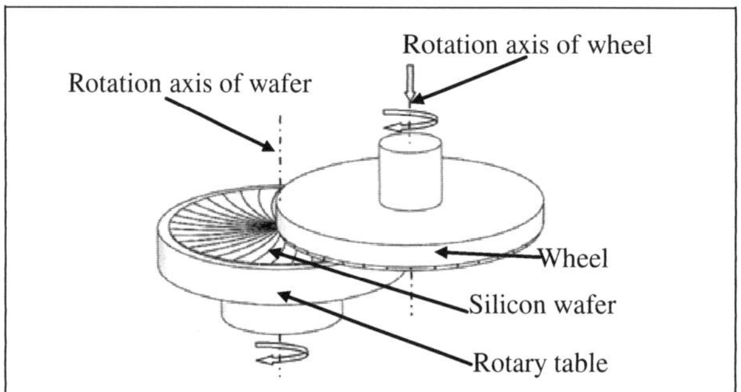

Figure 1. Illustration of silicon wafer rotation grinding.

2. Establishment of Micro Contact Model Of Grinding Wheel Micro Unit and Silicon Wafer

2.1. Micro Contact Mechanics Model

The ideal mechanical model of the micro contact between the grinding wheel micro unit and the silicon wafer is shown in Figure 2. The grinding wheel micro unit is extracted from the macro grinding wheel, which is in contact with the silicon wafer, and is composed of a unit volume of the bond holding a single diamond abrasive grain.

The actual grinding is a complex process, here to make simplifying assumptions: During grinding, the silicon wafer relative to the grinding wheel is stationary and homogeneous isotropic; bond surface of the grinding wheel end face is smooth; the diamond abrasive particle size distribution is uniform in the grinding wheel, and the height of the cutting edge of the abrasive grain is 1/3 of the average diameter of the spherical particle. The unit volume of the bond is determined by the concentration of the grinding wheel, the radius of the abrasive grain $\boldsymbol{R}$ and the density of the diamond.

In the mechanical model, the grinding wheel micro unit is the contact action on the silicon wafer with the synthesis speed $\boldsymbol{V}$ and the cutting depth $\boldsymbol{t}_w$. The decomposition analysis of the contact process is carried out. Normal z, the wheel micro unit is pressed into the silicon wafer, and the distance to be pressed is the cutting depth $\boldsymbol{t}_w$. Normal interaction force can be formed between the wheel micro unit and the silicon wafer. Stress field can be generated in the silicon wafer by normal load $\boldsymbol{F}_n$. With the increase of cutting depth, the stress field will be expanded, and the silicon wafer will be depressed. When the deformation exceeds the elastic yield limit and the fracture strength, the plastic and brittle deformation can occur respectively. In the same way, the normal reaction load F_n' is also applied to the wheel micro unit, and the stress field is generated. The wheel micro unit can be compressed to lead to deformation. Tangential x, the wheel micro unit slides on the silicon wafer with the speed V, along with the grinding motion trace. With the normal load, the sliding friction force f and f'are formed between the wheel micro unit and the silicon wafer, and the stress field is formed on them. In this way, the silicon wafer will be crushed and produce plastic flow along the moving direction of the grinding wheel micro unit, and wear will be produced on the wheel micro unit.

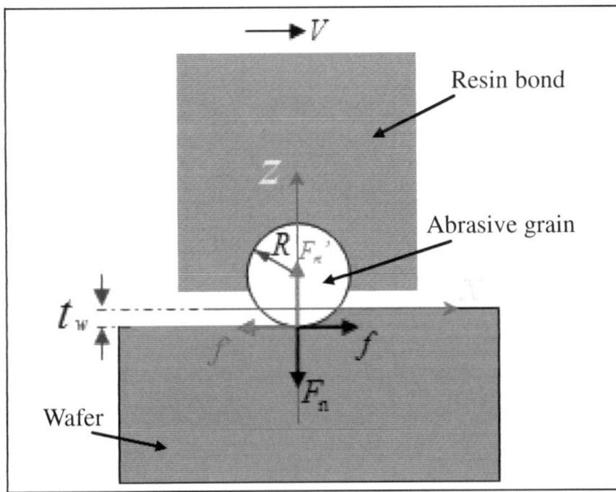

Figure 2. Micro contact mechanics model.

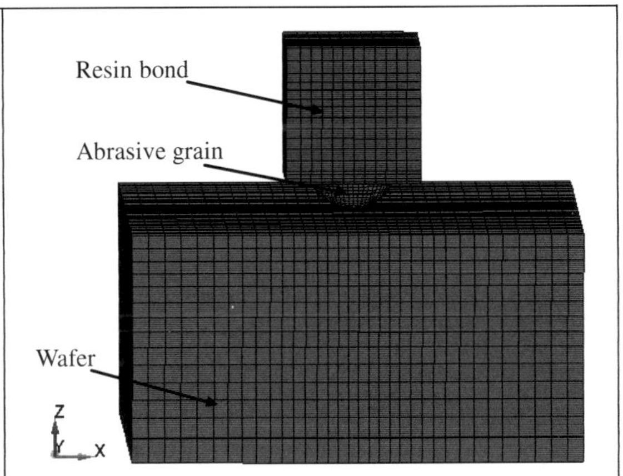

Figure 3. Micro contact simulation model.

2.2. Micro Contact Simulation Model

According to the analysis of the mechanical model, the three-dimensional simulation model of the micro contact between the grinding wheel micro unit and the silicon wafer is established, as shown in Figure 3. The geometric parameters of the simulation model are as follows. The size of resin bonded cube is 6μm. The diameter of the ball diamond abrasive grain is 4μm. The height of the resin embedded in the resin is 2.6μm. The size of the silicon wafer is 20*20*10μm.

In order to facilitate the calculation, the simulation model is simplified and assumed. As the elastic modulus of the diamond abrasive grain is much larger than that of silicon wafer, the material model of the diamond abrasive grain is considered as a rigid body. For the resin bond, only the factor of the force is considered, and other factors such as the thermal effect are not considered. The material model of the resin bond is considered as elastic plastic material. The silicon wafer is defined as *MAT_JOHNSON_HOLMQUIST_CERAMICS ceramic material model. The element types are all used in 3D explicit structure solid element Solid164. At the bottom of the silicon wafer, the full constraint is applied. The normal z negative and tangential x positive displacement loads are respectively applied to the bond, to realize the simulation of the grinding process of normal pressing and tangential sliding. In the simulation test, the mesh division and calibration of the model is based on the experimental data of micro scratch in silicon wafer [9].

3. Simulation Results and Analysis

3.1. Simulation Results and Analysis of Normal Pressing

In the process of simulating the grinding wheel micro unit pressing into the silicon wafer, the deformation strain of the silicon wafer is shown in Figure 4. Figure 4(a) shows the silicon wafer is starting to produce plastic strain when indentation displacement is about 45 nm, which indicates that the silicon wafer is in elastic phase before pressing to 45 nm. Figure 4(b) shows the plastic strain on the silicon wafer when pressed into about 330 nm, and the plastic strain has reached its limit. Figure 4(c) shows the strain on the silicon wafer when pressed into about 350 nm, and some elements have failure removal in the silicon wafer, which indicates that the silicon wafer begins to enter the stage of brittle fracture and deformation after pressing to 330 nm.

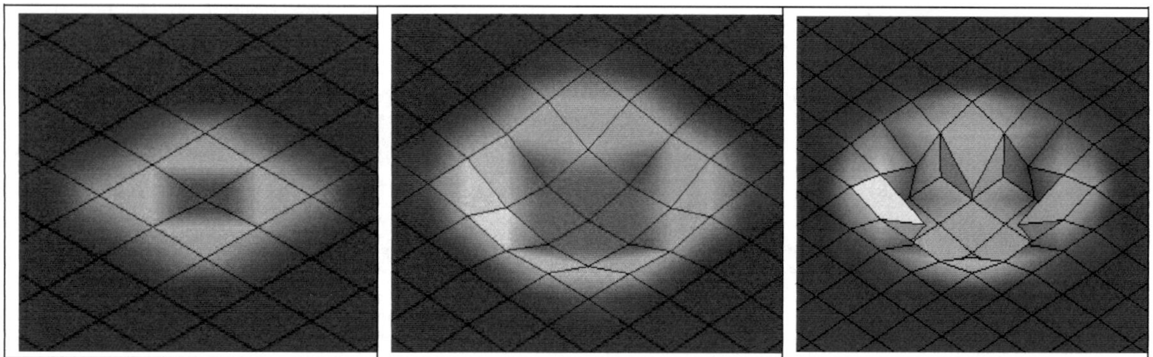

(a) Indentation displacement about 45 nm | (b) Indentation displacement about 330 nm | (c) Indentation displacement about 350 nm

Figure 4. Deformation strain of silicon wafer in normal pressing simulation.

During the simulation, the load curve on the silicon wafer with indentation displacement is shown in Figure 5. The figure shows that the contact force on the silicon wafer is about 2 mN when indentation displacement is about 45 nm, and the contact force on the silicon wafer is about 32 mN when indentation displacement is about 330 nm.

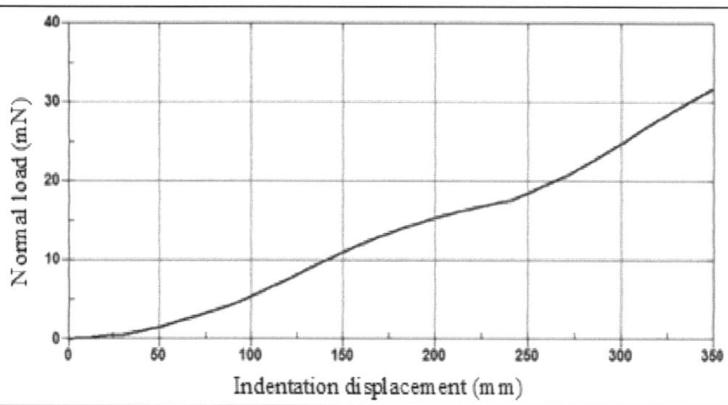

Figure 5. Load curve on silicon wafer with indentation displacement.

Simulation results of normal pressing show that under the external force, the critical values of load (2 mN) and displacement (45 nm) from elastic to plastic strain exist on the silicon wafer and the critical values of load (32 mN) and displacement (330 nm) from plastic to brittle deformation also exist on the silicon wafer.

3.2. Simulation Results and Analysis of Tangential Sliding

In the precision grinding of silicon wafer, the surface quality of the silicon wafer is high, so the grinding should be controlled in the ductile region, that is, the plastic stage of the silicon wafer. For the simulation of tangential sliding, the indentation displacement (i.e. the cutting depth of the abrasive grain) should be controlled in the plastic stage of the silicon wafer. The range of plastic stage has been obtained through the simulation of normal pressing, that is, the indentation displacement is between 45 nm and 330 nm.

The stress and strain simulation diagrams of the silicon wafer in tangential sliding are shown in Figure 6 (corresponding to the indentation displacement of 100 nm). The figures show that in the plastic deformation stage of the silicon wafer, the sliding stress is produced under the action of tangential sliding friction force. At the same time, the plastic flow deformation is formed under the stress in the silicon wafer, which creates the surface of a silicon wafer with plastic sliding groove and uplift.

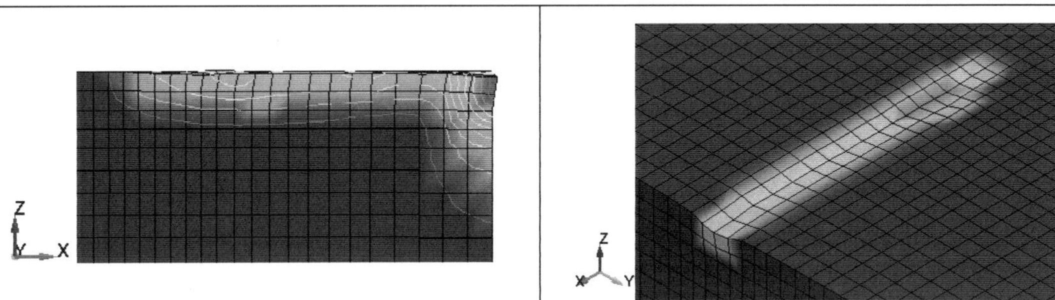

(a) Stress diagram of silicon wafer | (b) Strain diagram of silicon wafer

Figure 6. Stress and strain simulation diagrams of the silicon wafer in tangential sliding.

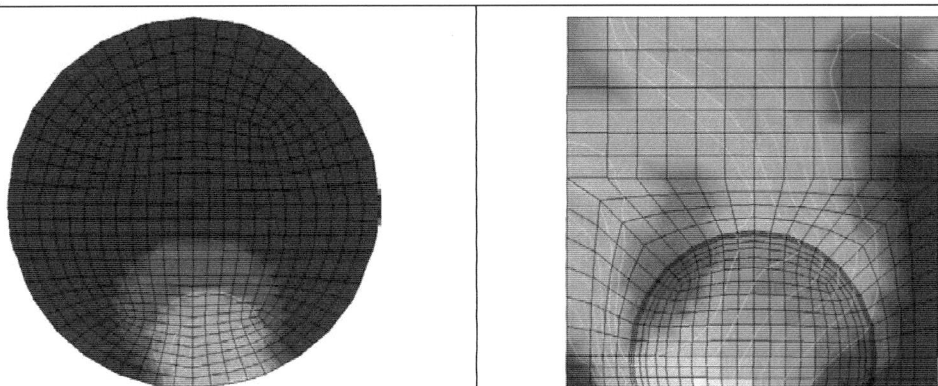

(a) Stress diagram of abrasive grain | (b) Stress diagram of resin bond

Figure 7. Stress diagrams on the grinding wheel micro unit during the tangential sliding simulation.

The stress diagrams on the grinding wheel micro unit during the tangential sliding simulation are shown in Figure 7. The stress distribution on the grinding wheel micro unit in the ductile region grinding process can be intuitively understood. Figure 7(a) shows the stress diagram of the abrasive grain in the grinding wheel micro unit. It is easy to find the location of the stress concentration point from the figure. The diamond abrasive grain has high fracture strength, which makes it difficult to produce fracture at the stress concentration point. Therefore, only attrition wear occurs on the grinding wheel abrasive grain. Figure 7(b) shows the stress diagram of the resin bond in the grinding wheel micro unit. If the stress value of the stress concentration point reaches the fracture strength of the grinding wheel bond, the bond fracture wear could occur, which could lead to pull-out of the abrasive grain.

4. Conclusions

Three dimensional three body simulation model based on force is established for micro contact between wheel and silicon wafer from grinding process. The following conclusions can be drawn from simulation analysis of the model:

(1) The critical load and displacement from elastic to plastic and plastic to brittle exist on the silicon wafer under the external force, and the critical values are obtained by normal pressing simulation.

(2) The stress and strain of the silicon wafer and the grinding wheel micro unit are obtained by tangential sliding simulation. The tangential sliding force can produce plastic groove and uplift in silicon plastic zone. Wear of wheel micro unit can be based on the simulation data to judge. The results of this paper can provide support for the research of silicon precision grinding and wheel wear mechanism.

5.Acknowledgements

This study was supported by the National Natural Science Foundation of China (Grant No.U0734008), the Natural Science Foundation of Guangdong Province of China (Grant No.8151009001000048) and the Youth Foundation of Guangdong University of Technology (Grant No. 082042).

6. References

[1] May G S and Sze S M 2004 *Fundamentals of Semiconductor Fabrication* (New York: Wiley) p 1
[2] Van Zant P 2000 *Microchip Fabrication* (New York: McGraw-Hill) p 37
[3] Pei Z J, Fisher G R and Liu J 2008 *Int. J. of Machine Tools and Manufacture* **48** 1297-1307
[4] Liu J H, Pei Z J and isher G R 2007 *Int. J. of Machine Tools and Manufacture* **47** 1-13
[5] Matsui S 1988 *Bull Japan. Soc. Prec. Eng.* **22** 295-300
[6] Zhou L, Tian Y B and Huang H 2012 *Proceedings of the Institution of Mechanical Engineers, Part B: Journal of Engineering Manufacture* **226** 66-75.
[7] Pei Z J, Fisher G R and Bhagavat Milind 2005 *Int. J. of Machine Tools and Manufacture* **45** 1140-1151
[8] LSTC, LS-DYNA, retrieved from: < http://www.lstc.com/products/ls-dyna>
[9] Gassilloud R 2005 *physica status solidi (a)* **202** 2858-2869

Effect of Yttrium on the Microstructure and Properties of Pt-Ir Electrical Contact Materials

Saibei Wang[1,2,3], Yong Sun[3], Song Wang[1,2], Mingjun Peng[3], Manmen Liu[1,2], Yonghua Duan[3], Yongtai Chen[1,2], Youcai Yang[1,2], Song Chen[1,2], Aikun Li[1,2]and Ming Xie[1,2,4]

[1]State Key Laboratory of Advanced Technologies for Comprehensive Utilization of Platinum Metals, Sino-platinum Metals Co. Ltd., Kunming, Keji Road 988, China
[2]Kunming Institute of Precious Metals, Kunming, Keji Road 988, China
[3]School of Material Science and Engineering, Kunming University of Science and Technology, Kunming, Wenchang road 68, China
Email:77481311@163.com

Abstract. The Pt-10Ir and Pt-10Ir-1Y were prepared by high frequency induction melting, then the samples were obtained by powder metallurgy, hot extrusion and drawing. The influence of Y addition on microstructure and electrical contact properties of Pt-10Ir alloy has been investigated by using optical microscopy, SEM, electronic balance and the contact material test system. The results show that the addition of Y leads to the micro-structural refinement and directional change of material transfer, but has almost no influence on erosion morphology.

1.Introduction

As compared with other electrical contact materials, Pt-Ir alloy features high hardness, high melting point, high corrosion resistance, low and stable contact resistance, making it a excellent electrical contact material. It is widely used in aircraft engine ignition contact, electrical contacts of high sensitive relay, micro-motor relay, and ship gravity pendulum relay, potentiometer and conducting ring of the heading device, the horizon and the gyroscope in the planes, the missiles and the ships and so on[1,2].
But because the preciousmetalisascarceresource,itshouldtryto use other alloying elements substitution of Pt-Ir alloy. Doping rare earth element such as Y is an important approach to improve properties and reduce cost at the same time.

Studies have shown that the grain of Pt dopanted with Y is obviously refined [3,4]. Chen et al. [3,5] showed that Y concentrations of 0.23wt% and 0.5wt%remarkably increase Vickers hardness and tensile strength at room temperature, but reduce elongation rate. The Y doped in Pt can enhance high temperature strength[3,6]. Moreover, Xie et al.[7] investigated the electrical contact performance of Pt-25Ir alloy and found that the welding resistance, arc erosion resistance was obviously increased due

[4]corresponding author, Email:powder@ipm.com.cn.
The present work was supported by the National Natural Science Foundation of China (No.51507075, U1302272,51461023,U1602271,U1602275), Yunnan Applied Basic Research Projects (2014FB164,2015FA042, 2016FB092,2014XT02)and the fund of the State Key Laboratory of Advanced Technologies for Comprehensive Utilization of Platinum Metals(No. SKL-SPM-201526)

IOP Conf. Series: Materials Science and Engineering **250** (2017) 012017 doi:10.1088/1757-899X/250/1/012017

to the adding of 0.5wt% Y. To the best of our knowledge, there is no report on the microstructure and the electrical contact performance of the Pt-10Ir alloy doped trace Y.

In this study, therefore, the Y was selected as the additional element in Pt-10Ir binary alloy forrefined microstructure and reduced costs. The influence of Y on as-cast microstructure,arc erosion morphology and material transferof Pt-10Ir was researched.

2.Experiments

The Pt-10Ir and Pt-10Ir-1Y alloys were prepared by high frequency induction melting with platinum (99.95%), iridium(99.95%)and yttrium(99.90%)under vacuumed condition. Cast samples, which were fabricated into a cylindrical rod-shape zirconia crucible, were solidified by air cooling. Microstructure analysis of the polished as-cast samples (finished with 0.5μmdiamond paste) was observed byoptical microscopy (LEICA DM4000M, German).The electrical contact samples were obtained by annealing and cold drawn or rolling. Afterwards, they were assembled on the contact material test system(JINFENG JF04C, China) undergoing test (breaking DC arcs) under different conditions as listed in Table 1.

Table 1. Test conditions.

Text voltage/ V	24
Load current/ A	15
Contact gap/ mm	1
Contact force/ N	0.2
Load type	resistive load

After finishing the predetermined number of operations, themorphology of arced spot with the tested contacts were observed by scanning electron microscopy(HITACHI S-3400N, Japan). The mass of materials was determined by electronic balance(METTLER TOLEDO AB135-S, Switzerland).

3.Result and Discussion

3.1 Microstructure

The microstructures of cast Pt-10Ir and Pt-10Ir-1Y contact materials were observed byoptical microscopy, and the images are shown in figure 1(a)~(b).

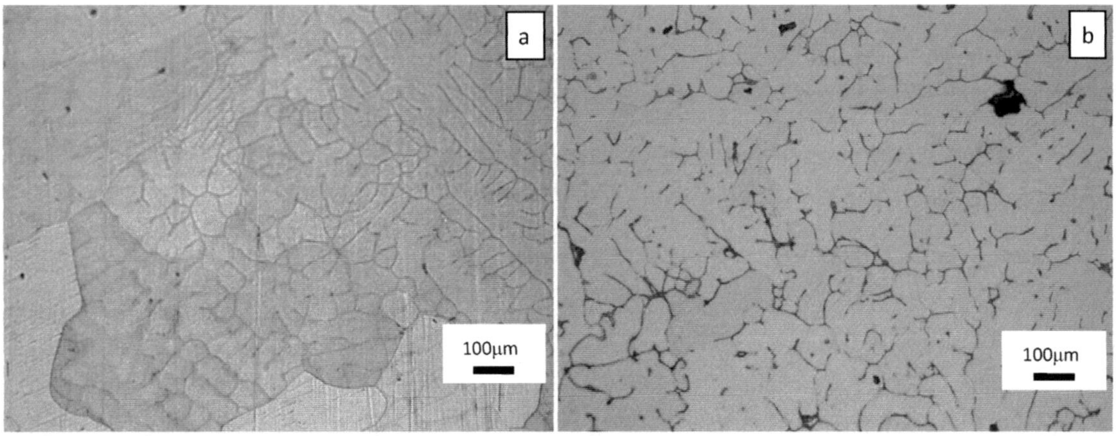

Figure 1. OM images of cast alloys: (a) Pt-10Ir alloy; (b) Pt-10Ir-1Y alloy.

In figure1, the microstructure of Pt-10Ircasthassignificantlydifferentgrainsize. There is seriously alloying elements segregation in some grain, which is characterized by cystiform-dendritic

structure.
The microstructure ofPt-10Ir-1Y isprincipally irregularcell crystal.By contrast, thecastmicrostructureb ecame more uniform and more fine by addition of Y.

3.2 Material Transfer

Usually, the erosion of anode or cathode contacts is different because of the two contactor material's different physical properties, or the different amount of the input arc energy and heat-flow density when the circuit turn on or turn off. For the dispersity of speed direction during material separating from contacts, one part of the erosion contact transfers to the surrounding space, one part of it deposits on the self surface, and another part of it transfers to the surface of the other contact, as shown in figure 2. If the looses and obtainment of the contact materials is off-balance, there would be increases or decreases of the contact weight.

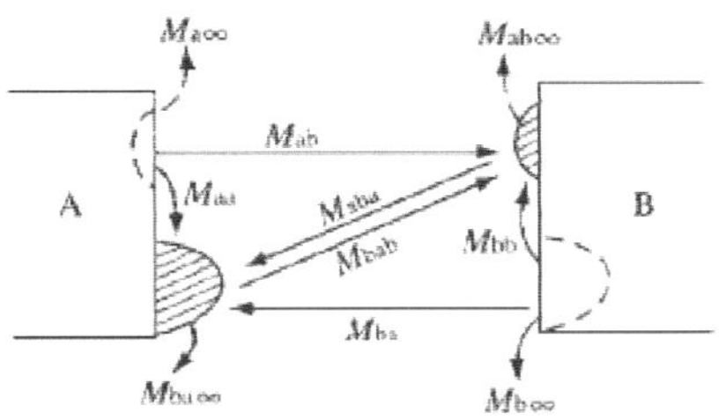

Figure 2. Material transfer schematic[8].

Due to the limitations of existing analysis methods, the degree of contact erosion is often characterized by the variation of contact weight amount Δm after multiple operations. Δm > 0means the net increase in contact material, otherwise weight reducing. The Δm calculated from measured results of Pt-10Ir and Pt-10Ir-1Y for cathode and anode are shown respectively in figure 3.

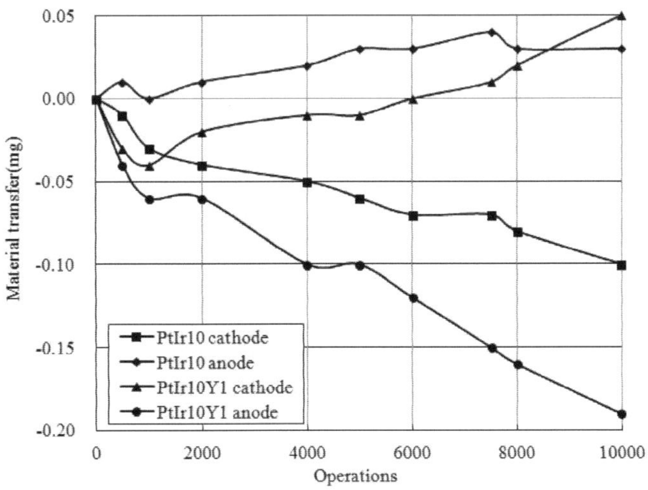

Figure 3. Contacts weight changes after 10000operations.

It can be observed from figure 3 that the material transfer of anode of Pt-10Ir is increases with increasing of the operations. The trend of cathode is the opposite of the anode, and the variation range of cathodeis obviously larger than that of the cathode. Researchers have proved that the material transfer direction is dependent on the arc duration in the metallic and gaseous phases[9].For Pt-10Ir, it means the gaseou sphase arc duration plays a main part under the current experimental condition when cathode loss and anode gain takeplace. For Pt-10Ir-1Y, it shows that the expected cathode loss and anode gain is reversed with cathode gain and anode loss. This phenomenon indicate that the addition of Y changes the arc behavior of Pt-10Ir alloy. That is, the metallic phase arcis dominant lead to the anode gain.

3.3 Erosion Morphology

Arc energy transfers heat to the contact pairs, thus resulting in the surface damage of the contacts. The surface profiles on each sample surface measured with a scanning electron microscopy are shown in figure 4 and 5.

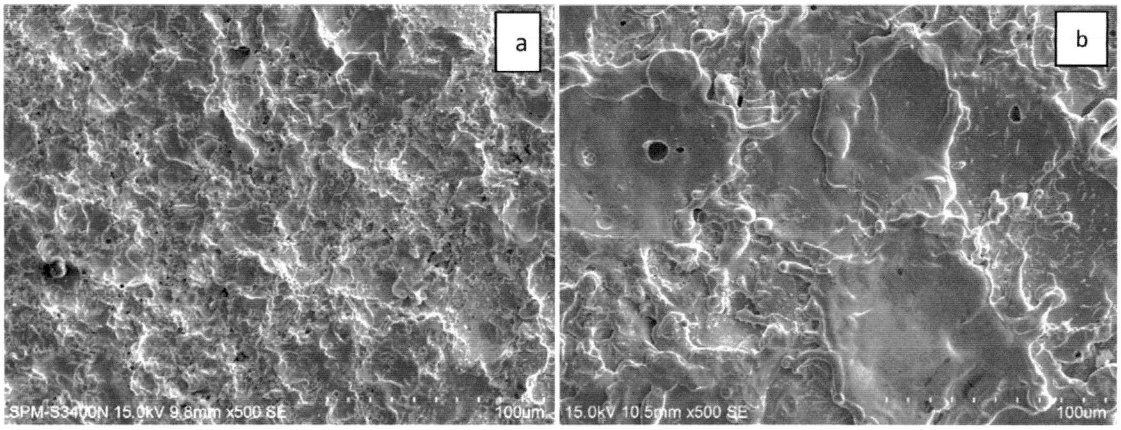

Figure 4. The morphology of Pt-10Ir contact surface after 10000 operations(a)cathode(b)anode.

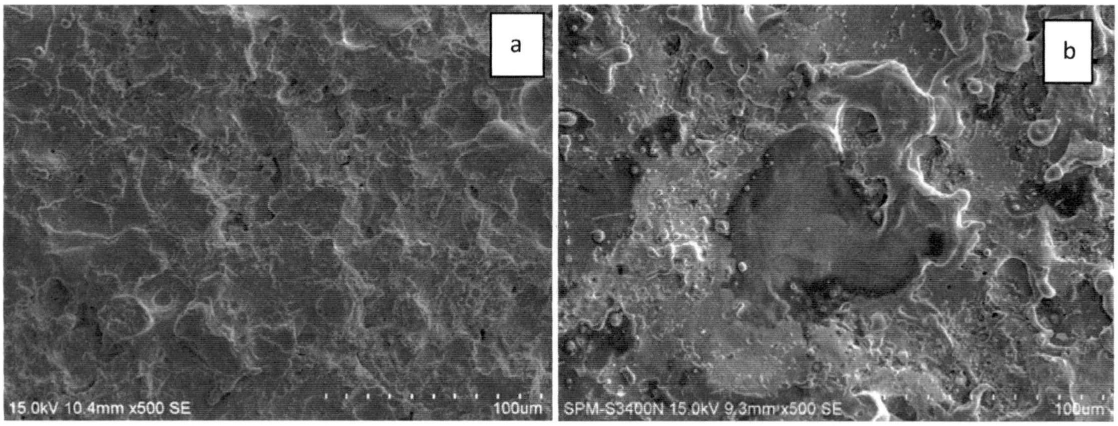

Figure 5. The morphology of Pt-10Ir-Y1 contact surface after 10000 operations(a)cathode(b)anode.

Figure 4 shows the typical SEM photomicrographs of theeroded surfaces of the Pt-10Ircontacts. It is seen from figure 4(a) that the interior of the eroded cathode surface presents a arcactionappearance where a large variety of small shallowspots are superimposed and interlaced one another. Only a very small area around the spot becomes molten due to concentrated Jouleheating.Figure 4(a) showsthe interior of the eroded anode surface which ischaracterizedbymolten pool and thesplashing droplets around it.Since the anode molten pool isinduced by electron bombardment due to the electric field,the material heat up and melt ,meanwhile, the molten droplets are splashed into strip-like metallicparticles on the anode surface. Their results indicate that the arc is short- arc. It is found from figure 5 that the eroded surface of Pt-10Ir-Y1is quite similar to figure 4. It means that arc of Pt-10Ir-Y1is also short-arc under aforementioned conditions. The addition of Y(1wt%) does not significantly affect the arc type of Pt-10Ir.

4.Conclusions

1) By comparisons in microscopic structure, it was shown that the shape and distribution of the crystals are refined due to the doping of Y.
2)The transfer direction of Pt-10Ir whose arc behavior is dominated by gaseous phase arc is from the cathode to the anode. On the contrary, material transfer of Pt-10Ir-Y1 is from the anode to the cathode. This is due to the Y effects the arc behavior.
3)The arc erosion observed in this study is obviously caused by anode arcs. Y is not without significant influence upon the erosion morphology of Pt-10Ir.

5.References

[1] Huang B Y, Li C G, Shi L K , Qiu G Z and Zuo T Y, 2006 *Chinese Materials Engineering Ceremony Volume 5 Nonferrous Metal Materials Engineering* (Beijing: Chemical Industry Press)
[2] Ning Y T, Yang Z F and Wen F, 2010 *Platinum* (Beijing: Metallurgical Industry Press)
[3] Xu G F. 1993 *Rare Metals and Cemented Carbides*, **4** 54
[4] Zhao W B, Tao Z Y, Ye Z Z and Song L, 1992 *Precious Metals*, **13** 17
[5] Chen F S, Hu C Y, Tong L Z and He H C, 1988 *Precious Metals*, **9** 15
[6] He H C, Hu C Y, Guo J X and Chen F S, 1991 *Precious Metals*, **12** 19
[7] Xie M, Chen J, Chen L, Fu S J, Ning D K, Zheng R C, Yang Y C, Li R M, Duan Y X and Zheng F Q, 2004 *Precious Metals*, **25** 25
[8] Rong M Z, Zhao Z Y and Yang W, 1998 *Low Voltage Apparatus*, **5** 16
[9] Chen Z K and Sawa Koichiro,1998 *IEEE transactions on components, packaging, and manufacturing technology*, 21 310

IOP Conf. Series: Materials Science and Engineering **250** (2017) 012018 doi:10.1088/1757-899X/250/1/012018

First-Principles Research on the Structural and Electric Properties of the Graphene-Like Alkali-Metal Absorbed InSe-M

Wei Kang, Tingyan Zhou and Jun Liu

School of Science, Chongqing University of Posts and Telecommunications, Chongqing 400065, China;

Corresponding author e-mail:liujun@cqupt.edu.cn

Abstract.The structural and electric properties of the alkali-metalabsorbedmonolayer InSe-M (M=Li, Na, K and Rb) are studied by the first-principles calculations. The absorption energies of the monolayer InSe-M are negative so that theyhave stableabsorption structures,indicating that it is possible to prepare them in experiments.The pristinemonolayer InSe is typical conductor, but the alkali-metalabsorbedmonolayer InSe-Mare electronically conductive. Therefore, the alkali-metal absorption causes important influence on the electric properties of the monolayer InSe. However, too little calculated magnetic moments show that the alkali-metal absorption causes little influence on the magnetic properties of the monolayer InSe.

1.Introduction

The discovery of the Graphenein 2004 has been viewed as thevery important finding in new materials in that the Graphenehave muchnewpotential application in photonics, nonlinear optics, flexible transparent electrodes, sensors, conductive composites, and gas separation membranes[1]. Since then thetwo-dimension (2D) materials which have unusual electric, mechanical and optical properties have drawn much attention, and then caused extensive interest on experimental, theoretical and application researches.Especially, the III-VI group van der Waals layered materials such as GaS[2] and InSe[3], have become the material focus recently. InSe is a typical semiconductor. For InSe layers, each layer has hexagonalstructure and the interaction between In and Se-ions is the covalent bond[4]. On the other hand, the interaction between layers is weak Van der Waals-type interaction. InSe is attracting much attention because of its much new possible application in photonics, nonlinear optics, spintronics, etc. More recently, few-layer InSe exhibits promising characteristics for optoelectronic application [4-6].Furthermore, a few layers of InSe flakes had been prepared by using the mechanical exfoliation technique established by Novoselov et al[7].The transport properties of InSe along the layers have been widely investigated over the past twenty-five years[8]. Moreover, InSe nanosheetsused to design field effect transistors (FET) show high electron mobility values such as 4.7, 27.6, 74 and 162 (cm^2 $V^{-1}s^{-1}$) for Al, Ti, Cr and In contacts, respectively[9]. Besides, the on-off ratios of the FET are up to 10^7–10^8[10].

Recently, there is increasing interest to study the influence of metal absorption or doping on the physical properties of Graphene-like materials. In order to realize the miniaturization of electronic devices, magnetic materials with single atom-layer thickness have caused extensive researches, such as the boron nitride single-layer[11,12]with the honeycomb structure. Nonmetal-doped metal oxide semiconductorshaveimportant effect on the application of magnetic devices in that they represent a new class of d^0-type magnetic materials. It was found that zinc oxide doped by the non-metallic elements such as carbon and nitrogen even can show ferromagnetismat room temperature [13]. The cavitation doping in the layered GaSe can induce the change of ferromagnetic phase. The average magnetic moments of the carrier increase with the increase of the hole density[14]. With the sharp decline of the magnetic moments, the system appears half-metallic. Besides, Ashwin Kishore studied the effect of Te on the monolayer InSe for the photocatalytic water splitting via the density functional calculations.They found that Te-doped InSe is more favorable for the hydrogen reduction reaction than the pristine monolayer InSe.Therefore, doped Te can improve the photocatalytic water ability in InSe monolayer[14, 15].However, there are few studies on alkali-metal absorbed InSe (InSe-M). Therefore, in this paper, based on the calculationof the absorption energies, we systematically investigate the stable crystal structures and electric properties of alkali-metal absorbed InSe monolayer.

2.Model and Method

The calculated supercell of the monolayer InSe-M is shown in Figure.1. The structure of the monolayer InSe consists of In-Se-Se-In layers as depicted in Figure.1a (side view) and Figure.1b (top view). 'M' represents the absorbed alkali-metals (M=Li, Na, K and Rb). Each layer has the hexagonal structure and the D3h symmetry from Figure.1b.The structural relaxing and electronic structure calculation of the pristine and adsorbed monolayer InSe are carried out by using Vienna Abinitio Simulation Package (VASP) based on the spin-polarized density functional theory (DFT)[16,17]. The projector augmented wave method (PAW) is used to describe the ion-electron potential[18]. The Perdew–Burke–Ernzerhof (PBE) parameterization of the generalized gradient approximation (GGA) is employed as the exchange–correlation function[19].In order to avoid the interlayer interaction, the Van der Waals force is considered.Here,a 3×3×1supercell is set with avacuum layer with the thickness 20Åalong the z direction to prevent interaction between slabs. Thecutoff energy is set as 400 eV and the Monkhorst-Pack k-grid meshis 7 × 7× 1 in slab (and monolayer) calculations. The lattice constant of the monolayer InSeare 12.225Å, 12.225Å and 20 Å,respectively.The valence electronic structures of isolated atoms are $5s^2 5p^1$,$4s^2 4p^4$,$(n\text{-}1)s^2(n\text{-}1)p^6 ns^1$ (n=1, 2, 3 and 4) for In, Se and alkali-metals, respectively.

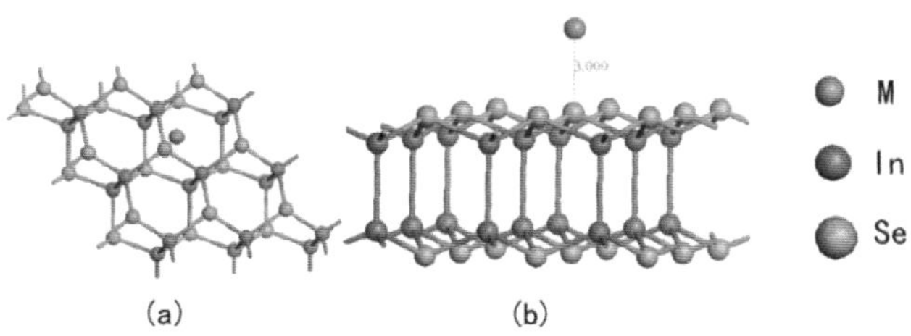

Figure.1.Ball-and-stick model for InSe-M monolayer, where subfigures (a)and (b)show side view and top view, respectively.

3 Results and Discussions

The band structure of the pristine monolayerInSeis showninFigure.2. From Figure.2, the pristinemonolayerInSeis a typical semiconductor with indirect band gap. Its valence band maximum (Ev=-0.0021eV) is between the K and M-points, but the conduction band minimum (Ec=1.4960 eV)is situatedat the Γ-point. Thus, there is an evident energy gap 1.4981eV, which is in excellent agreement with previous report[20].

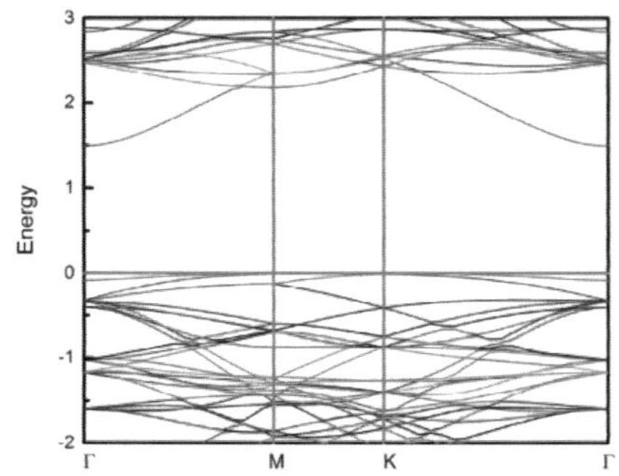

Figure.2The band structure of thepristine monolayerInSe.

From Figure.1, the calculatedsupercell of the alkali-absorbedInSe-Mmonolayer has 37 atoms including 18 In-atoms, 18 Se-atoms and one absorbed alkali-atom.In order to get the preferredabsorptionsite,the alkali-atom is placed above the Se and In-atom, respectively, and then corresponding absorption energy is calculated by the followed formula.

$$E_a = E_{total} - (E_{atom} + E_{InSe})$$

Here the E_{total}, E_{atom}, and $E_{InSe}$are the total energy of the alkali-absorbed monolayerInSe-M, the alkali-atom, and the pristine monolayerInSe, respectively. Based on our calculated results and the energy minimum principle, the alkali-atom prefers the top site above the Se-atom.According to the above formula, the absorption energies for Li, Na, K and Rb-absorbed monolayerInSe-Mare-1.38, -1.16, -1.58 and -1.63, respectively. These energies are negative, implying that the absorption process is exothermic, so that the above absorption process is possible in experiments.The absorption energy of the Rb-absorbed monolayer is lowest, implying that Rb absorption above the Se-ion is easier than other atoms. The main reason is possibly that the radius of Rb-5s orbital is larger than Li-2s, Na-3s and K-4s orbitalso that the Coulomb interaction between the Rb-5s electron and the nucleus is less than that between Li-2s, Na-3s and K-4s electrons and their nucleus.Less Coulomb interaction will result in that the 5s-electron is more easilytransferred to the Se-ion, and then form a more stable absorption structure.

Spin-polarized energy band structuresand state densities of themonolayerInSe-M (M= Li, Na, K and Rb) are shown in Figure. 3, where subfigure(1), (2), (3) and (4) are for Li, Na, K and Rb absorbed monolayer, respectively.There are spin-down(left) and spin-up (right) sub-band structures and spin-polarized densities ofstates (middle) in every subfigure, respectively. From Figure.3, both the spin-down sub-bands and the spin-up sub-bands of the monolayer InSe-M (M= Li, Na, K, Rb)go through the Fermi level, indicating that these materials are electrically conductive. However, from Figure.2, the pristine monolayer InSeis semi-conductor. Therefore, the absorbed alkali metals can improve the electrically conductive properties of the pristine monolayerInSe.

On the other hand, the magnetic properties are also calculated in this paper. The total magnetic moments of the monolayer InSe-M are 0.0001, 0.0006, 0.0008 and 0.0009 μ_B for M=Li, Na, K and Rb, respectively, indicating the absorbed alkali-metalscause little influence on the magnetic properties the monolayer InSe. In fact, from Figure.3 the spin-up sub-bands are very similar to spin-down sub-bands totally. Furthermore, the spin-up andthe spin-down TDOS are almost axially symmetric about theenergy axial. These resultsalso indicate that there is no spin-polarization in the monolayer InSe-M.

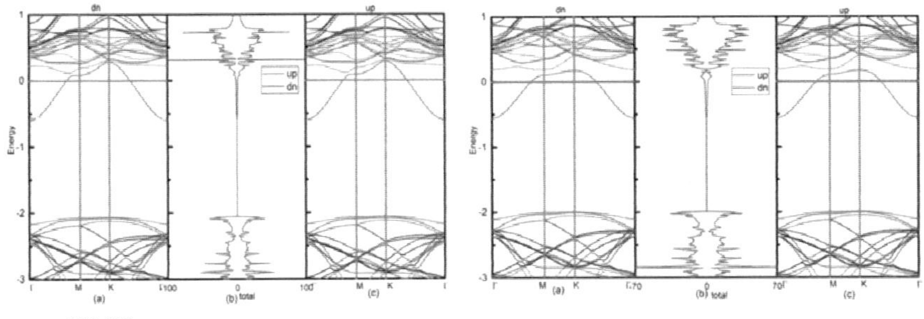

(1)(2)

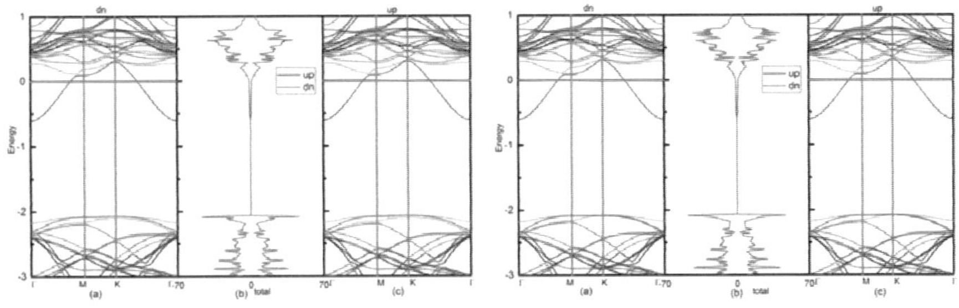

(3) (4)

Figure.3.The spin-polarized band structures and the total densities of states of the alkali-metal absorbedmonolayer InSe-M (M=Li, Na, K and Rb), where the spin-down (left) and spin-up (right) sub-band and spin-polarized TDOS (middle) are in every subfigures, respectively.

4.Conclusions

In this paper, the structural and electronic properties of the alkali-metal absorbedmonolayer InSe-M (M=Li, Na, K, Rb) are studied by the first-principles calculations. Results show that the monolayer InSe-M (M=Li, Na, K, Rb) have stable absorption structures so that it is possible to prepare them in experiments. The absorbed alkalimetal can cause important influence on the electric properties, but little influence on the magnetic properties of the monolayer InSe.

5.Acknowledgements

The Paper issupported by Chongqing InnovationResearch Project forGraduate Students (Grant No. CYS17228)

6.References

[1] Anasori B, Lukatskaya M R, Gogotsi Y. 2D metal carbides and nitrides (MXenes) for energy storage[J]. Nature Reviews Materials, 2017, 2: 16098.

[2] Hu P A, Wang L, Yoon M, et al. Highly responsive ultrathin GaS nanosheet photodetectors on rigid and flexible substrates[J]. Nano letters, 2013, 13(4): 1649-1654.

[3] Ateş A, Yıldırım M, Gürbulak B. Absorption measurement and Urbach's rule in InSe and InSe: Ho 0.0025, InSe: Ho 0.025 single crystals[J]. Optical Materials, 2006, 28(5): 488-493.

[4] Bandurin D A, Tyurnina A V, Yu G L, et al. High electron mobility, quantum Hall effect and anomalous optical response in atomically thin InSe[J]. arXiv preprint arXiv:1608.08950, 2016.

[5] Lei S, Ge L, Najmaei S, et al. Evolution of the electronic band structure and efficient photo-detection in atomic layers of InSe[J]. ACS nano, 2014, 8(2): 1263-1272.

[6] Feng W, Wu J B, Li X, et al. Ultrahigh photo-responsivity and detectivity in multilayer InSe nanosheets phototransistors with broadband response[J]. Journal of Materials Chemistry C, 2015, 3(27): 7022-7028.

[7] Novoselov K S, Geim A K, Morozov S V, et al. Electric field effect in atomically thin carbon films[J]. science, 2004, 306(5696): 666-669.
[8] Segura A, Mar B, Martinez-Pastor J, et al. Three-dimensional electrons and two-dimensional electric subbands in the transport properties of tin-doped n-type indium selenide: Polar and homopolar phonon scattering.[J]. Physical Review B Condensed Matter, 1991, 43(6):4953-4965.
[9] Garni S E A, Qasrawi A F. Absorption and optical conduction in InSe/ZnSe/InSe thin film transistors[J]. Functional Materials Letters, 2016, 09(02).
[10] Lu A J, Zhang R Q, Lee S T. Stress-induced band gap tuning in □ 112□ silicon nanowires[J]. Applied Physics Letters, 2007, 91(26):263107-263107-3.
[11] Zhang Z, Guo W. Energy-gap modulation of BN ribbons by transverse electric fields: first-principles calculations[J]. Physical Review B Condensed Matter, 2011,77(7):439-446.
[12] Guo H Y. Theoretical design and performance controlling of two dimensional nanomaterials[D]. University of Science and Technology China, 2014.
[13] Cao T, Li Z, Louie S G. Tunable magnetism and half-metallicity in hole-doped monolayer GaSe[J]. Physical Review Letters, 2015, 114(23).
[14] Kishore M R A, Ravindran P. Te doped indium (II) selenide photocatalyst for water splitting: A first principles study[C]//AIP Conference Proceedings. AIP Publishing, 2017, 1832(1): 090029.
[15] Blöchl PE. Projector augmented-wave method[J]. Physical Review B Condensed Matter, 1994, 50(24):17953.
[16] Kresse G, Furthmüller J. Efficiency of ab-initio total energy calculations for metals and semiconductors using a plane-wave basis set[J]. Computational Materials Science, 1996, 6(1):15-50.
[17] Kresse G, Furthmüller J. Efficient iterative schemes for ab initio total-energy calculations using a plane-wave basis set[J]. Physical review B, 1996, 54(16): 11169.
[18] Perdew J P, Burke K, Ernzerhof M. Generalized gradient approximation made simple[J]. Physical review letters, 1996, 77(18): 3865.
[19] Perdew J P, Burke K, Ernzerhof M. Generalized gradient approximation made simple[J]. Physical review letters, 1996, 77(18): 3865.
[20] Demirci S, Avazlı N, Durgun E, et al. Structural and electronic properties of monolayer group III monochalcogenides[J].Physical Review B, 2017, 95(11): 115409.

A Metamaterial with Dual-Band Perfect Terahertz Transmission

Guochao Wei[1,2], Yawei Zheng[1,2], Xiran Chen[1,2], Zhengpeng Qin[1,2], Luman Qin[1,2], Nangang Zhang[1,2,3],Kan Liu[1,2,3], Songzhan Li[1,2,3] and Shengxiang Wang[1,2,3*]

[1] School of Electronic and Electrical Engineering, Wuhan Textile University, Wuhan, 430020, P. R. China
[2] Advanced Micro-Nano Textile Innovation Research Center, Wuhan Textile University, Wuhan, 430020, P. R. China
[3] State Key Laboratory for Hubei New Textile Materials and Advanced Processing Technology, Wuhan Textile University, Wuhan, 430020, P. R. China
* Corresponding author: shxwang@wtu.edu.cn

ABSTRACT. In recent years, the electromagnetic metamaterial has been development rapidly. Meanwhile, its peculiar electromagnetic properties have been widely studied in the fields of electromagnetism and optics. Compared to conventional natural materials, metamaterials show some unusual phenomena, for example, negative refractive index. Because of their exotic properties, metamaterials have been found with many applications, such as perfect lens, cloaking, solar cells and so on. In this paper, we report a metamaterial with dual-band perfect transmission in terahertz range. By tailoring the periodicity of the unit cell, significantly high transmission can be obtained with the metamaterial. The simulation results show that significant transmission peaks at 3.56THz and 7.16THz with the magnitudes of 99.8% and 99.7%, respectively.

1. Introduction

Electromagnetic metamaterials have been researched for many years due to their peculiar optical and electromagnetism properties[1, 2]. Studies of the spectral response of metamaterials included various types of resonance structures. Because of the unusual properties, metamateials have been developed with many applications, such as perfect lenses, negative refractive index materials, invisible cloak[3-5]. Normally, metamaterials have been created with periodically unit cells consisting of metal structure and dielectric layers. The surface of metal structure resonates with the incident electromagnetic waves. Metamaterial composed of periodically splitting ring resonators (SRRs) have been widely studied[6]. The periodic SRR array has certainly electromagnetic response by changing its configurations. Metamatieral with high transmission properties are important in some potential applications[7].

Terahertz waves are electromagnetic waves with wavelength range of from 0.1 THz to 10THz. The terahertz wavelength varies from 0.03 mm to 3 mm, which falling in the regime between the microwaves and the infrared[8]. In 1896 and 1897, Rubens and Nichols were involved in this band, and over the next 100 years, far-infrared technology has achieved many achievements and has been industrialized[9]. But there are very few studies involving terahertz, the main reason is the limitation of effective terahertz sources and sensitive detectors. As in the 1980s, the development of a series of new technology, new material, making stable broadband pulsed THz sources became a kind of conventional technology, so the THz technology development rapidly[4, 10, 11].

IOP Conf. Series: Materials Science and Engineering **250** (2017) 012019 doi:10.1088/1757-899X/250/1/012019

In this paper, we designed a metamaterial with dual band high terahertz transmission. By tailoring the split ring resonator structure, we can achieve THz wave amplitude manipulation at the resonance frequency in the transmission spectra[12, 13]. This structure can select the electromagnetic waves of the characteristic frequencies through the surface of metamaterials. Based on the metamaterial with high terahertz transmission, some potentially various sensors could be developed.

2. Structure and Design

By changing the resonant frequency features of metamaterials, we build resonance units of different structures within the same plane in order to realize the structure of metamaterial with multiple transmission peaks. According to the SRR resonance principle[14], with a particular structure formed by the SRR array, there will be an equivalent LC circuit and plasma resonance frequency corresponding to it[15-17]. Based on this feature, we designed a metamaterial that presents two peaks with high transmission at terahertz frequencies.

The unit cell of the metamaterial is shown in figure 1 a, which consist of two metallic plate and dielectric layers. The metallic layers laid on the top and bottom of the dielectric layer. The periodicity is P=15um, the length of the arm is L1=10um, and the wide of the arm is L2=L3=1um. The thickness of the dielectric layers is t=2um. The top layer and bottom layer are illustrated in figure 1 b-c. The thickness of the metal (Cu) is 0.1um. The permittivity of the dielectric is $\varepsilon = 2.1 + i0.005$. The direction of the electric field of the incident wave is along the x-axis. Perfect matched layers (PML) in the z direction and the periodic conditions in the x and y directions have been applied during the simulation.

3. Simulation and Discussion

We simulated the metal-dielectric (MD) structure and metal-dielectric-metal (MDM) structure in range of 1~10THz. The results were shown in figure 2. For MD structure, there is only one transmission peak at 4.62THz and the transmission can reach 99.7%. For MDM structure we can find there are two high-frequency transmission peaks and the high transmission band is very narrow. Two transmission peaks are at 3.56THz and 7.16THz, and the transmission rates are 99.8% and 99.7% respertively. This indicates that the metamaterials only have high transmission rates within certain frequencies. Due to increase a metal layer, the relative dielectric and permeability of this metamaterial get changed. This led to variation the resonant frequency. So the frequency of transmission peak varied.

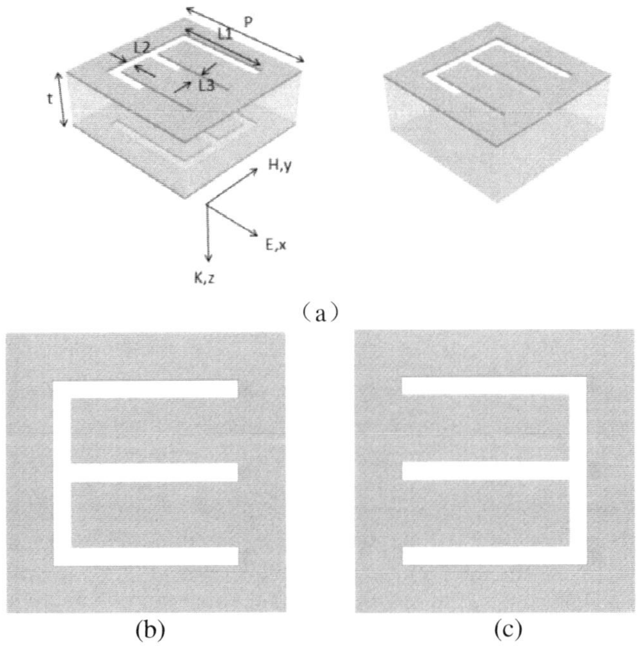

Figure 1 Schematics of the metamaterial. (a) unit cell; (b)the top layer; (c) the bottom layer.

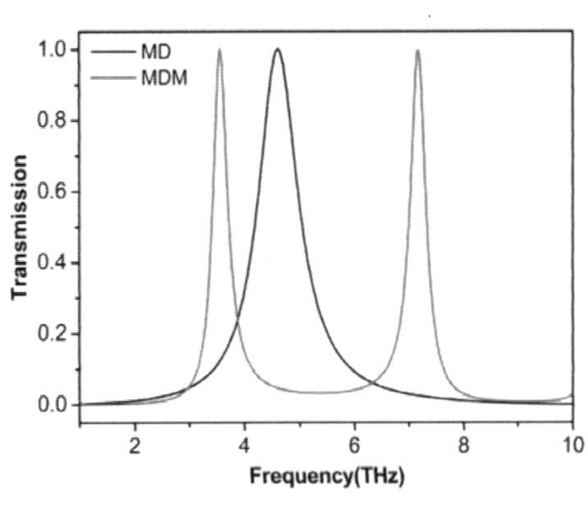

Figure 2 Simulated transmittance spectra of the MD and MDM

The response of the metamaterial transmission with obliquely incident radiation was examined. The transmissions with differently incident angles are shown in figure 3. When electromagnetic wave incident with different angles, there are still two high-transmission peaks near the frequency of 3.56THz and 7.16THz. Thus, the proposed metamaterial provides high transmission for wide incident angles.

IOP Conf. Series: Materials Science and Engineering **250** (2017) 012019 doi:10.1088/1757-899X/250/1/012019

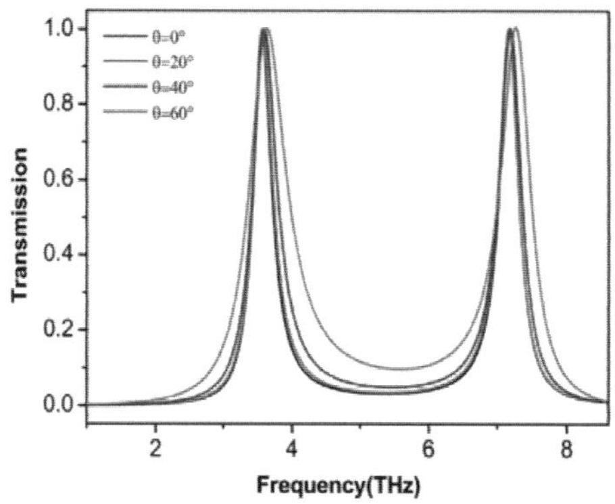

Figure 3 Transmission efficiencies as a function of incident angle

In order to further study the electromagnetic properties of the metamaterials. We change the structure of the metamaterial slightly. As shown in figure 4, by changing the distance of *a* and fixing other parameters, we can get the difference transmission amplitude and transmission peaks. We simulated differences distance of *a*, and the result is showen in figure 5. From 2 THz to 6 THz as the distance *a* increases, the first transmission peak moves to high frequency. According to this feature, we can change the distance of *a* to get the metamaterial which has specific frequency resonance.

Figure 4 The surface structure of the metamaterial.
(a) top layer.(b) bottom layer.

In order to study the influence from the thickness of the arm length, we keep other parameters constant and only change the value of L (L=L2=L3) as L=1 mm, 1.2 mm, 1.4 mm, 1.6 mm, 1.8 mm and 2 mm. Figure 6 plots the corresponding transmittance spectra. With the increase of the thickness L, the transmission peaks move to the high frequency.

4. Conclusion

In this paper, we proposed a metamaterial with dual band terahertz transmission that consists of two metal layers separated by a dielectric layer. We present simulation of this metamaterial, and get two maximum transmission peaks of 99.8% at 3.57 THz 99.7% at 7.16 THz, respectively. Among the narrow frequency band, the metamaterial shows high transmission at the characteristic frequency. Furthermore the structures have quite distinct advantages like multiband transmission peaks, flexibility and mechanical tunability. The transmission peaks can be adjusted by changing the structure of the metamaterial. Such kind of metamaterial could be used as candidate to achieve substantial electromagnetic response, various potential sensors and THz applications.

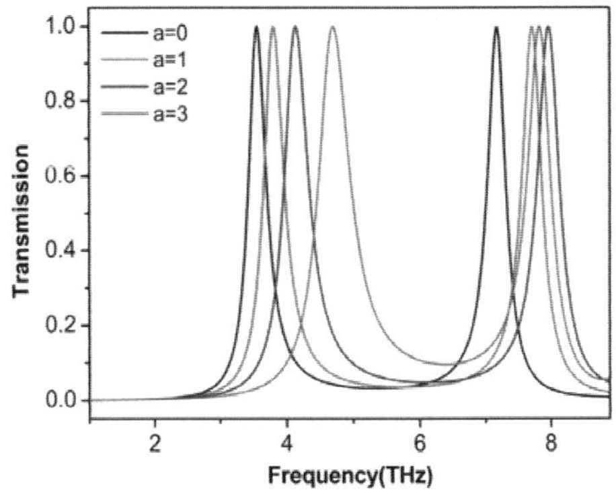

Figure 5 Simulated transmittance spectra of the metamaterial with different distance of a.

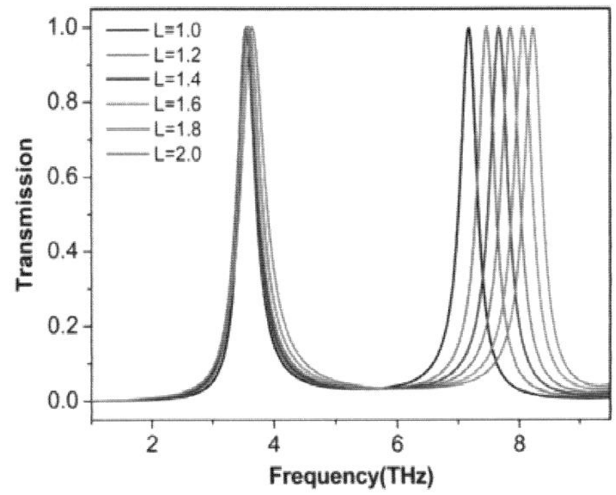

Figure 6 Simulated transmittance spectra of the metamaterial with different wall thickness of L.

5.Acknowledgements

This study was supported by National Natural Science Foundation of China (51302196) and State Key Laboratory for Hubei New Textile Materials and Advanced Processing Technology (ZDSYS201711).

6.References

[1] Liang L, Jin B, Wu J, Zhou G, Zhang Y, Tu X, Jia T, Jia X, Cao C, Kang L, et al. 2013 Terahertz narrow bandstop, broad bandpass filter using double-layer s-shaped metamaterials *Science China Information Sciences* **56** 1-7

[2] Park S J and Ahn Y H 2015 Substrate effects on terahertz metamaterial resonances for various metal thicknesses *Journal of the Korean Physical Society* **65** 1843-7

[3] Li X, Yang T, Zhu W and Li X 2012 Continuously tunable terahertz metamaterial employing a thermal actuator *Microsystem Technologies* **19** 1145-51

[4] Kim J H, Hokmabadi M P, Balci S, Rivera E, Wilbert D, Kung P and Kim S M 2016 Investigation of robust flexible conformal thz perfect metamaterial absorber *Appl. Phys. A* **122**

[5] Yao Z, Huang Y, Wang Q, Hu F, Quan B, Li J, Gu C and Xu X 2016 Tunable surface-plasmon-polariton-like modes based on graphene metamaterials in terahertz region *Computational Materials Science* **117** 544-8

[6] Zheng W, Li W and Chang S J 2015 A thermally tunable terahertz metamaterial absorber *Optoelectronics*

Letters **11** 18-21
[7] Wang B X, Wang L L, Wang G Z, Huang W Q, Li X F and Zhai X 2014 Frequency continuous tunable terahertz metamaterial absorber *J. Lightwave Technol.* **32** 1183-9
[8] Bilgin H, Zahertar S, Sadeghzadeh S, Yalcinkaya A D and Torun H 2016 A mems-based terahertz detector with metamaterial-based absorber and optical interferometric readout *Sensors and Actuators A: Physical* **244** 292-8
[9] Liu J, Zhou Q, Shi Y and Zhang C 2012 The rotation of polarization of a terahertz wave through subwavelength metallic structures *Science China Physics, Mechanics and Astronomy* **56** 514-8
[10] Seren H R, Keiser G R, Cao L, Zhang J, Strikwerda A C, Fan K, Metcalfe G D, Wraback M, Zhang X and Averitt R D 2014 Optically modulated multiband terahertz perfect absorber *Advanced Optical Materials* **2** 1221-6
[11] Grant J, Escorcia-Carranza I, Li C, Mccrindle I J H, Gough J and Cumming D R S 2013 A monolithic resonant terahertz sensor element comprising a metamaterial absorber and micro-bolometer *Laser & Photonics Reviews* **7** 1043-8
[12] Chen W, Zhou Q, Shi Y, Li C and Zhang C 2016 Trapped-mode resonances in asymmetric terahertz subwavelength structures *Appl. Phys. B* **122**
[13] Kozlov D S, Odit M A, Vendik I B, Roh Y G, Cheon S and Lee C W 2011 Tunable terahertz metamaterial based on resonant dielectric inclusions with disturbed mie resonance *Appl. Phys. A* **106** 465-70
[14] Lee D-H, Ling K, Lim S and Baek C W 2015 Fabrication of polarization-insensitive, multi-resonant metamaterial absorber using wafer bonding of glass dielectric substrate *Microelectron. Eng.* **136** 42-7
[15] Ning R, Liu S, Zhang H, Bian B and Kong X 2015 Tunable absorption in graphene-based hyperbolic metamaterials for mid-infrared range *Physica B: Condensed Matter* **457** 144-8
[16] Wang B X, Wang L L, Wang G Z, Huang W Q, Li X F and Zhai X 2014 A simple design of a broadband, polarization-insensitive, and low-conductivity alloy metamaterial absorber *Appl. Phys. Express* **7** 082601
[17] Wang G D, Liu M H, Hu X W, Kong L H, Cheng L L and Chen Z Q 2013 Broadband and ultra-thin terahertz metamaterial absorber based on multi-circular patches *The European Physical Journal B* **86**

IOP Conf. Series: Materials Science and Engineering **250** (2017) 012020 doi:10.1088/1757-899X/250/1/012020

Simulation of Piezoelectric Energy Harvester Based on the Vortex Flow

Meng Li, Haifeng Wang†, Yiliang Cui and Kaili Sun

School of Electromechanic Engineering, Qingdao University, Qingdao 266071, China
Email: hfwang@qdu.edu.cn

Abstract. In this article, numerical research on the fluid-structure interaction between the flexible piezoelectric energy harvester (FPEH) and the Von Karman vortex street forming behind a bluff body is carried out to optimize the oscillation of FPEH to obtain more electrical energy. Using ANSYS Workbench platform, the simulation is performed. The numerical results show that the maximal deformation of the PEH is 1.7428 mm, meanwhile the maximal voltage is 4.6144 V. Besides, these numerical results generated by the ANSYS simulation are in good agreement with the experimental results.

1. Introduction

With the increasing consumption of fossil energy in the industrial production, the energy crisis has been placed in front of the world. In recent years, people have developed a variety of renewable energy to replace the traditional fossil energy, such as solar energy, wind energy, wave energy and so on. Among them, ocean wave energy has been focused more attention because of its huge energy reserves and environmental pollution-free features. Piezoelectric materials are well known for their ability to generate electrical charge when they are deformed. With this feature, piezoelectric flags have been placed in the fluid flow using as interested candidates [1-2]. Periodic deformation of the piezoelectric flags leads a periodic charge transfer between the electrodes of piezoelectric patches positioned on the surface of the piezoelectric flags as they exhibit spontaneous self-sustained flapping when the surrounding flow exceed a critical velocity. There are several literatures about the response of flexible piezoelectric energy harvester (FPEH) placed in a cross-flow and obtained the maximal strain energy and output power. To maximize the amount of power generated by the eel, many factors have been considered in the investigation, such as the thickness and stiffness of the eel materials, the eel length, the bluff body width and the spacing between the body and eel head [3-4]. The feasibility and performance of harvesting the electrical charge by a resistive circuit was recently studied, and it was shown that both the stability of the system and the fluttering dynamics are influenced by electrical coupling [5-6]. The energy harvesting eel is a device that uses piezoelectric polymer to convert energy from fluid flow into electricity power, which can harvest energy in both ocean and river. In this research, interaction between FPEH and the vortex behind a bluff body is studied.

2. Basic Principle of Analysis Model

Under the condition that the fluid model is a continuous model, the law of conservation of mass in fluid dynamics refers to the continuity equation, and the velocity and density of matter in the model will be the function of time and space coordinates. The incompressible viscous fluid is considered in modeling. The governing equations, which are continuous equations, are [6]:

$$\frac{\partial \vec{u}}{\partial x}+\frac{\partial \vec{v}}{\partial y}+\frac{\partial \vec{w}}{\partial z}+\frac{\partial \rho}{\partial t}=0 \tag{1}$$

Where $\vec{u}$, $\vec{v}$, and $\vec{w}$ are speed vectors. ρ is a constant for the incompressible viscous fluids. The governing equations for the FSI model are [7]:

$$\rho\frac{\partial \vec{u}_{fluid}}{\partial t}+(\vec{u}_{fluid}\cdot\nabla)\vec{u}_{fluid}=\nabla\cdot\left[-pI+\mu\left(\nabla\vec{u}_{fluid}+(\nabla\vec{u}_{fluid})^{T}\right)\right]+\vec{F} \tag{2}$$

$$\frac{\partial^2 \vec{u}_{solid}}{\partial t^2}-\nabla\sigma=\rho\vec{F} \tag{3}$$

Where $\vec{u}_{fluid}$ and $\vec{u}_{solid}$ represent the velocity of the fluid flow and the solid mechanics respectively; $\vec{F}$ is the volume force field, σ is the Cauchy stress tensor which is related to the solid material being used for the FPEH.

In this paper, the k-ε turbulence model provided in ANSYS Fluent is used. In the standard k-ε model, ε is the turbulent dissipation factor. And the corresponding transport equation is:

$$\frac{\partial(\rho k)}{\partial t}+\frac{\partial(\rho k u_i)}{x_i}=\frac{\partial}{\partial x_j}\left[(\mu+\frac{\mu_i}{\sigma_k})\frac{\partial k}{\partial x_j}\right]+G_k+G_b-\rho\varepsilon-Y_M+S_k \tag{4}$$

Where G_k represents the turbulent kinetic energy caused by average velocity gradient; G_b represents the turbulent kinetic energy caused by buoyancy; Y_M represents effects of pulsating expansion on total turbulent kinetic energy in compressible turbulence.

3. Piezoelectric Equation

The piezoelectric equation is an expression that describes the relationship between the mechanical and electrical quantities of piezoelectric materials. This kind of piezoelectric equation is used in the ANSYS coupling solutions for the analysis about piezoelectric materials. In linear piezoelectricity, the equations of elasticity are coupled to the charge equation of electrostatics by means of piezoelectric constants [8]:

$$\{S\}=\left[s^{E}\right]\cdot\{T\}+[d]\cdot\{E\} \tag{5}$$

$$\{D\}=[d]^{t}\cdot\{T\}+\left[\varepsilon^{T}\right]\cdot\{E\} \tag{6}$$

Where $\{S\}$ is the elastic strain vector; $\{T\}$ is the stress vector (N/m^2); $\{D\}$ is the electric flux density vector (C/m^2); $\{E\}$ is the electric field intensity vector (V/m); $[s^E]$ is the compliance matrix evaluated at constant electric field, i.e. short circuit; $[d]$ and $[d]^t$ are the piezoelectric strain coefficient matrix (C/N), and the superscript$\{t\}$is the matrix transpose.

4. Case Simulation

When fluid flows from the left inlet side and passes through the bluff body, the Von Karman vortex street is generated, which takes on the repeating pattern of swirling vortices. The driving force coming from the shedding vortex makes the FPEH vibrate continuously. In this study, in order to perform the numerical simulation of fluid structure interaction, a three-dimensional model has been set up by ANSYS (shown in figure 1), which includes the fluid flow domain, a circular cylinder acting as a bluff body, and a FPEH. D is the cylinder diameter and used as the characteristic length, so the size of fluid domain can be described as $20D\times10D\times7.5D$ in length, width and height respectively. The FPEH is fixed on the outlet side of the cylinder, and its length, width and height are 200 mm, 0.1 mm and 20 mm respectively. Detailed geometry and material parameters for the model simulation is given in table

1. And the local meshing of both blunt body and FPEH is given in figure 2. The volume meshing method is adopted for the FPEH and the bluff body, and their minimum element sizes are 0.0005m and 0.01m respectively. Obviously, the grid density of the FPEH is more intensive than that of the blunt body.

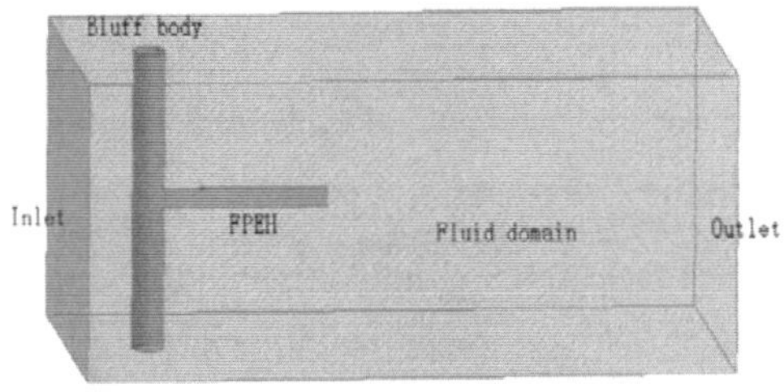

Figure 1. Computational domain.

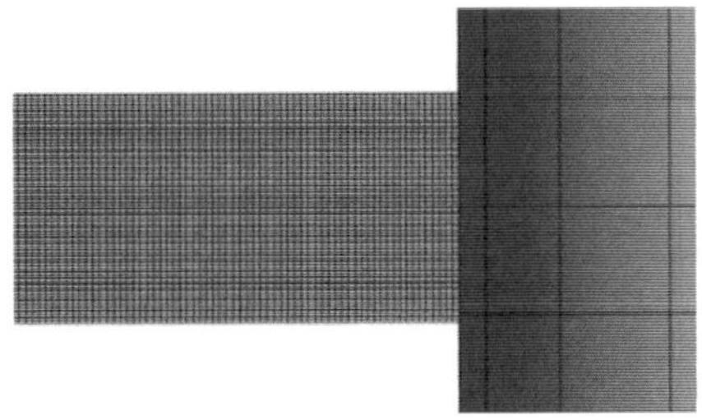

Figure 2. Mesh of partial enlargement.

Table 1 Parameters for the model simulation.

L [mm]	W [mm]	H [mm]	l [mm]	w [mm]	h [mm]	D [mm]	ρ_{solid} [kg/m^3]	μ_{fluid} [kg/ms]	E [GPa]	v
800	400	300	200	0.1	20	40	1770	1.420	2.5	0.32

There are many kinds of flexible piezoelectric materials. In this study, PVDF is used, and its detailed material parameters for the model simulations are given in table 2.

Table 2 Details of flexible piezoelectric materials.

Symbol	Parameter	PVDF	Units
d_{31}	Piezo Strain Constant	23	[(10^{-12})C/N]
d_{33}		-33	
g_{31}	Piezo Stress Constant	216	[(10^{-3})Vm/N]
g_{33}		-330	
k_{31}	Electromechanical Coupling Factor	12%	
k_t		14%	
C	Capacitance	380	[pF/cm^2,@1kHz]
V_0	Speed of Sound	2.2	[10^{3}m/s in thickness direction]

5. Results and Discussion

Before simulation, the voltage boundary on one side of the FPEH is set to 0 V. The uniform flow is used for the motion of the fluid, and the speed of fluid is set as 0.3 m/s. After calculation, the cloud maps of the generated voltage, the deformation and the pressure of the piezoelectric material are given in figure 3, figure 4 and figure 5 respectively. And the maximal deformation, maximal and minimal voltages of the FEPH at different times are given in table 3.Because of the influence of the shedding vortex, the voltage generated by the FPEH mainly concentrates in the fixed end of the piezoelectric sheet (shown in figure 3), and the maximal voltage is 4.6144V. The maximal deformation is

1.7428mm and occurs at the free end of the FPEH. For comparison, an actual FPEH with same dimensions was tested in a water tank. The maximal output voltage was measured as 4.78 volts under open-circuit condition with an oscilloscope. It can be found that the values of simulated voltage and experimental voltage are very close.

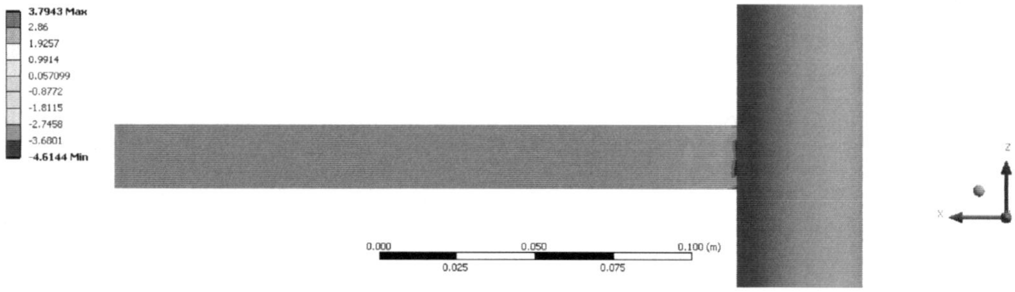

Figure 3. The output voltage of the FPEH.

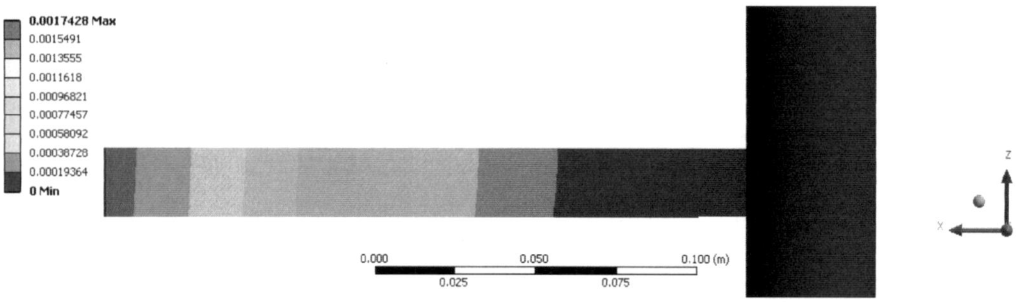

Figure 4. The deformation of the FPEH.

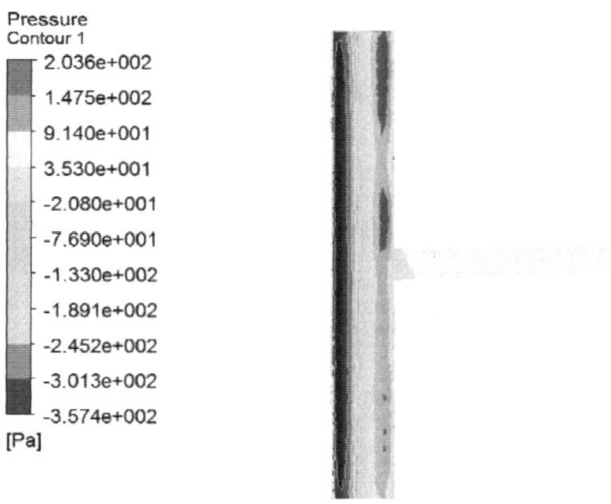

Figure 5. The pressure acting on the FPEH.

6. Conclusion

In this article, based on the phenomenon of Von Karman vortex street, the driving force originating from the shedding vortex acts on the FPEH and makes it vibrate continuously. Results indicate that the simulation result of power generation of FPEH is in good agreement with that of the actual measurement. The validity of the ANSYS software in the analysis of this kind of problem about fluid-structure interaction has been proved.

Table 3 Max deformation, max and min voltage at different time.

Time[s]	Max deformation [mm]	Max voltage [V]	Min voltage [V]
0.4	0.438	3.756	-2.044
0.8	1.682	3.785	-3.963
1.2	0.602	1.871	-1.605
1.6	0.174	3.441	-4.064
2.0	0.214	3.521	-4.526
2.4	0.238	3.584	-4.252
2.8	0.215	3.414	-4.177
3.2	0.162	3.542	-4.255
3.6	0.242	3.678	-4.433
4.0	0.249	3.461	-4.313
4.4	0.152	3.477	-4.153
4.8	0.237	3.708	-4.423
5.2	0.269	3.517	-4.404
5.6	0.179	3.532	-4.306
6.0	0.233	3.792	-4.614

7.Acknowledgments

The authors are grateful to the financial supports by the Natural Science Foundation of China (No. 51405255), the Science and Technology Project for Higher Education of Shandong Province (No. J12LN95), the Science and Technology Development Plan of Shandong Province [Policy Guidance Project] (No. 2012YD04038).

8.References

[1] Williamson C H K and Govardhan R A brief review of recent results in vortex-induced vibrations 2008 *J. Wind Eng. Ind. Aerod.* **96** 713

[2] Taylor G W, Burns J R, Kammann S A, Powers W B and Welsh T R The energy harvesting eel: a small subsurface ocean/river power generator 2002 *IEEE J. Oceanic Eng.* **26** 539

[3] Calio R, Rongala U B, Camboni D, Milazzo M, Stefanini C, Petris G and Oddo C M Piezoelectric energy harvesting solutions 2014 *Sensors* **14** 47

[4] Doare O and Michelin S Piezoelectric coupling in energy-harvesting fluttering flexible plates: linear stability analysis and conversion efficiency 2011 *J. Fluid. Struct.*, **27** 1357

[5] Allen J J and Smits A J Energy harvesting eel 2001 *J. Fluid. Struct.* **15** 629

[6] COMSOL A B 2010 *COMSOL Multiphysics User's Guide*

[7] Sin V K, Deng W Y and Xiao W H Study of motion of flexible eel from the wake of bluff body in a cross flow 2012 *Int. Conf. Numerical Analysis and Applied Mathematics* **1479** 161

[8] Zeng C X and Jiang L The finite element analysis of piezoelectric transducer based on ANSYS 2011 *Instrum. Technol.* **6** 31

IOP Conf. Series: Materials Science and Engineering **250** (2017) 012021 doi:10.1088/1757-899X/250/1/012021

Crystal Orientation and Electrical Properties of Tin Oxide Transparent Conducting Films Deposited on Rutile Surface

Y Sawada[1], Y Hashimoto[1], Y Hoshi[1], T Uchida[1], S Kobayashi[1], L Sun and B Yue[3]

[1]Center for Hyper Media Research, Graduate School of Engineering, Tokyo Polytechnic University, 1583 Iiyama, Atsugi, Kanagawa 243-0297, Japan.
[2]Guilin University of Electronics Technology, 1 Jinji Rd., Guilin, Guangxi, China.
[3]Rigaku Corporation, 3-9-12 Matsubara-cho, Akishima, Tokyo 196-8666, Japan.

Abstract. Thin films of tin oxide (SnO_2) without doping are attractive transparent conducting film since environmentally unfavorable elements of antimony or fluorine are eliminated. Tin oxide films without doping were fabricated very cheaply on (001) and (100) planes of single crystal of rutile (TiO_2) by spray chemical vapor deposition (mist CVD). The film deposited on rutile (001) surface was poorly epitaxial (double domain) but with higher mobility (24 $cm^2\ V^{-1}\ s^{-1}$) and lower resistivity ($1.6\times10^{-3}\ \Omega$ cm) than that deposited on glass substrate (16 $cm^2\ V^{-1}\ s^{-1}$ and $2.4\times10^{-3}\ \Omega$ cm) for reference. Deposition on rutile (100) surface resulted in better epitaxial growth (single domain). The mobility (39 $cm^2\ V^{-1}\ s^{-1}$) and the carrier electron density (2.7×10^{20} cm^{-3}) were much higher. The resistivity ($6.2\times10^{-4}\ \Omega$ cm) was compatible with those doped with antimony or fluorine and will be the lowest among tin oxide films without doping.

1. Introduction

Transparent conducting films are necessary for solar cells, displays, touch panels etc. A typical TCF material at present is tin-doped indium oxide (indium-tin-oxide, ITO) is often fabricated by physical vapor deposition such as magnetron sputtering. The authors reported elsewhere [1, 2] very cheap ITO films fabricated by original spray chemical vapor deposition (mist CVD). The lowest resistivity ($7.5\times10^{-5}\ \Omega$ cm) when deposited on a glass substrate and annealed in reducing atmosphere agreed with the value reported by Ohata et al. [3] who deposited on an yttrium-stabilized zirconia single crystal substrate by expensive pulsed laser deposition.

Indium is a rare element and fine powders of indium oxide damage human lungs. Aluminum or gallium-doped zinc oxide (AZO or GZO) has been developed as an alternative material. Low resistivity in the order of $10^{-4}\ \Omega$ cm has been possible by physical vapor deposition such as magnetron sputtering or ion plating while deposition of AZO or GZO films has been impossible or very difficult by chemical process including our spray CVD.

Another alternative material, tin oxide (SnO_2), needs doping of antimony or fluorine which are environmentally unfavorable. The authors reported elsewhere [4] tin oxide film without doping on glass substrate by spray CVD. The resistivity ($6.2\times10^{-3}\ \Omega$ cm) as-deposited at 255°C unfortunately did not seem sufficiently low; the mobility, 21 $cm^2\ V^{-1}\ s^{-1}$ and carrier electron density, $4.3\times10^{19}\ cm^{-3}$. Exceptionally low resistivity ($7.5\times10^{-3}\ \Omega$ cm) was reported by Banerjee and Das [5] for undoped tin oxide film deposited on glass by electron beam evaporation and annealing in air; mobility, 29 $cm^2\ V^{-1}$ s^{-1} and carrier electron density, $2\times10^{20}\ cm^{-3}$.

Crystal structure of SnO_2 (cassiterite) is rutile type, tetragonal system. Tin oxide films without doping were deposited in the present research on single crystal of rutile (TiO_2) substrates to improve crystallinity and mobility of the carrier electron to lower the resistivity. Post-deposition annealing in reducing atmosphere was also executed to increase the carrier electron density.

Undoped tin oxide films on rutile substrate were deposited by magnetron sputtering [6], metal-organic CVD (MOCVD) [7], atomic layer deposition (ALD) [8] and CVD [9]. However, none of them reported the resistivity. Undoped tin oxide films were deposited on sapphire substrate by pulsed-laser deposition (PLD) [10], CVD [11, 12] and molecular beam epitaxy (MBE) [13]. These achieved high mobility but failed to achieve sufficient carrier electron density and resulted in high resistivity. Exceptionally low mobility tin oxide film was deposited on sapphire by PLD [14]. These reported values were scattering. The electrical properties seem to highly depend on the deposition process.

2. Experimental

Mirror-polished rutile (single crystal of TiO_2) substrates (10 mm×10 mm×0.5 mm) with (001) and (100) surfaces were supplied from Crystal Base Co. and used without thermal treatment prior to the film deposition. Non-alkali glass substrates (Corning EAGLE XG, 10 mm×10 mm×0.7 mm) were used for reference. These substrates were ultrasonically cleaned in Semicoclean 56 (Furuuchi Chemical Co.), rinsed with pure water and boiled in acetone. Tin chloride, $SnCl_4 \cdot 5H_2O$, (Kanto Chemical) was dissolved in industrial alcohol (Imazu Chemical; ethanol 88.40%, isopropanol 10.49%, methylethylketone 1.11%) to prepare spray solution. Concentration of tin chloride was 0.1 M. A rutile substrate and a glass substrate were heated simultaneously at 400°C on a hotplate (Corning PC-400D). The surface temperatures of the substrates were measured by contact-type resistance thermometer (RKC Instrument, DP-300). The solution was sprayed manually 300 times at the interval of 5 s using an atomizer (SHO-BI, PV14033-00). Distance between the nozzle and the substrates was 15 cm. The films were annealed in the electric furnace at 600°C for 1 h in reducing atmosphere of N_2-0.1%H_2 (flow rate, 300 mL/min) to lower the resistivity. The film thickness was determined by X-ray fluorescence analysis (energy-dispersive type, JEOL JSX-3200) using fundamental parameter method assuming the film density as 6.45 g cm^{-3}. The film thicknesses were, 180, 145 and 123 nm, respectively, for on glass, (001) and (100) rutile surface. The thickness depended on the substrate. The crystal state of the films was evaluated with MiniFlex300 (θ-2θ mode), Rigaku and SmartLab3kW (in-plane and out-of-plane mode etc.), Rigaku.

3.Results and Discussion

3.1 Crystal Orientation

Figure 1 shows X-ray diffraction spectra for SnO_2 films after post-deposition annealing in reducing atmosphere.

The spectrum of the film deposited on a glass is amplified by 10 times since the peak intensity was weaker. Ten SnO_2 peaks were detected. The crystal orientation was approximately random. The average crystallite size determined from the peaks were approximately 41 nm. Unidentified weak peaks were detected at 63.92 and 77.14°.

The XRD spectrum for SnO_2 film deposited on rutile (001) surface showed weak and wider (002) and (301) peaks of SnO_2. The crystallite sizes were, respectively, 16 and 14 nm which were much smaller than that deposited on glass substrate. Results of in-plane and out-of-plane measurement are omitted. Epitaxial growth (double-domain) as TiO_2 001 // SnO_2 001 // SnO_2 301/031 (out-of-plane direction) and TiO_2 010 // SnO_2 010 // SnO_2 101/011 (in-plane direction). It should be kept in mind that the film was low crystallinity with small crystallite size although epitaxial. Unidentified weak peaks were detected at 31.52 and 56.14°.

The X-ray diffraction spectrum for SnO_2 film deposited on rutile (100) surface showed very sharp and strong (200) and (400) peaks of TiO_2 substrate and (200) and (400) peaks of SnO_2 film. The crystallite size (63 nm) was larger than those deposited on glass substrate (41 nm). Successful epitaxial growth (single-domain) of SnO_2 film was confirmed as TiO_2 100 // SnO_2 100 (out-of-plane direction) and TiO_2 010 // SnO_2 010 (in-plane-direction). Very weak (101) and (212) peaks of SnO_2 were detected. Unidentified weak peaks at 35.24 and 72.18° were detected.

IOP Conf. Series: Materials Science and Engineering **250** (2017) 012021 doi:10.1088/1757-899X/250/1/012021

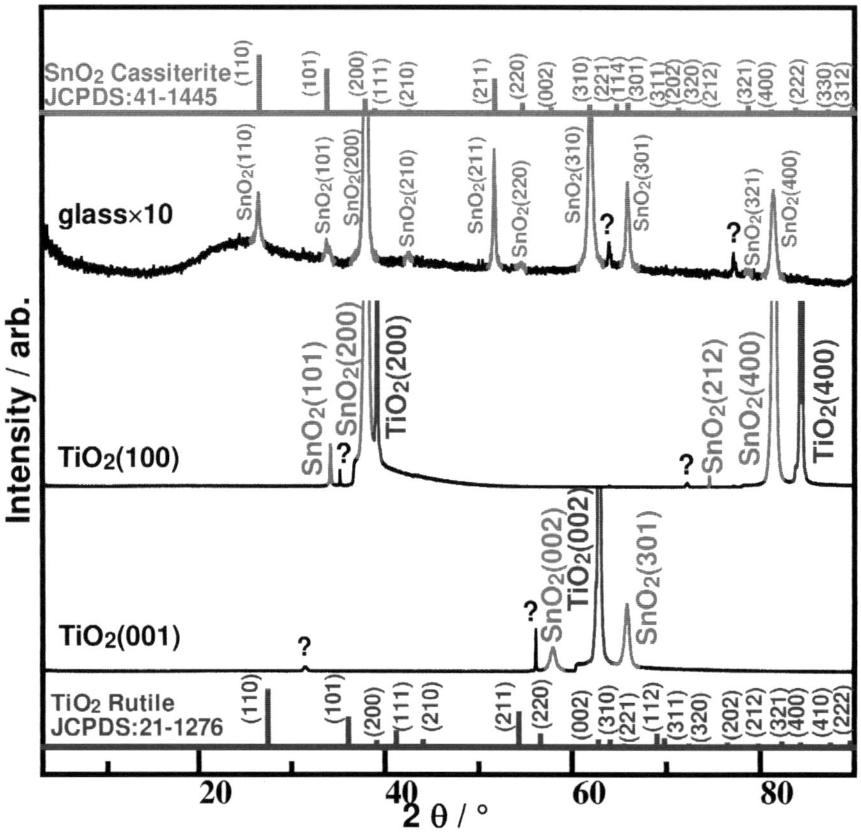

Figure 1. X-ray diffraction spectra for tin oxide films after post-deposition annealing in reducing atmosphere.

3.2 Electrical Properties

Figure 2 shows dependence of mobility on the carrier electron density. The dotted lines show the resistivity 1×10^{-3} and 1×10^{-2} Ω cm.

In case of SnO_2 film deposited on glass for reference, the mobility (16 cm^2 V^{-1} s^{-1}) was approximately same with our previous film [4]. Post-deposition annealing in reducing atmosphere increased remarkably (4 times) the carrier electron density (1.7×10^{20} cm^{-3}) than the previous as-deposited film. The resistivity (2.4×10^{-3} cm^{-3}) was much lower than the previous film although much higher than that that deposited by Banerjee and Das [5] by physical process.

Deposition on single crystal substrates increased the mobility and lowered the resistivity remarkably. Deposition on rutile (001) surface increased the mobility to 24 cm^2 V^{-1} s^{-1}, which is 1.5 times of that deposited on glass. Thus poorly-epitaxial growth (double domain) lowered the resistivity to 1.6×10^{-3} Ω cm although the carrier electron density (1.7×10^{20} cm^{-3}) unchanged from that deposited on glass.

Deposition on rutile (100) surface increased the mobility drastically to 39 cm^2 V^{-1} s^{-1}, which is 2.4 times of that deposited on glass. This should be attributed to the successful epitaxial growth (single domain) of the SnO_2 film. The carrier electron concentration also increased to 2.7×10^{20} cm^{-3}, which is 1.6 times of that deposited on glass. The reason of increasing carrier electron concentration is not clear but tentatively attributed to the crystal growth dependent on the substrate surface. Thus, the resistivity was lowered to 6.2×10^{-4} Ω cm. This value seems the lowest among the SnO_2 films without doping and compatible with antimony or fluorine-doped tin oxide (ATO or FTO) films.

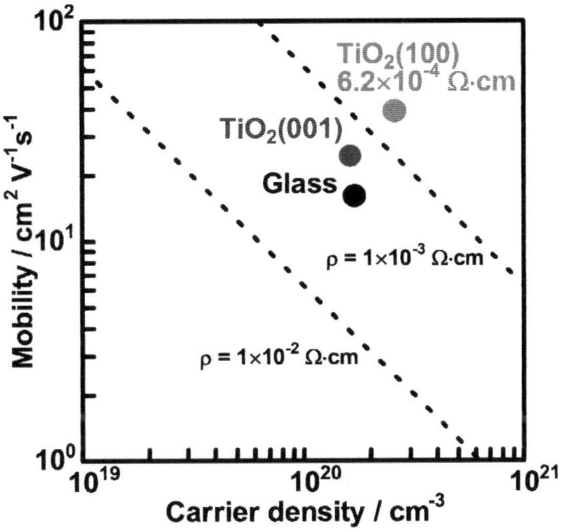

Figure 2. Dependence of mobility on carrier electron density for SnO_2 films after post-deposition annealing.

3.3 Optical Properties

Average transmittance in visible range was approximately 80% when deposited on a glass substrate. Reflection from the rutile substrate is high (approx. 35%) because of high refractive index of rutile (approximately 2.7). Transmittance improved slightly by depositing SnO_2 film. This should be attributed to lower refractive index of SnO_2 film. Low visible transmittance will be no problem in many cases since deposition of antireflective coating with much lower reflective index on SnO_2 film will increase the transmittance if necessary.

4. Conclusions

Thin films of tin oxide (SnO_2) without doping were fabricated on single crystal substrate of rutile (TiO_2) by spray CVD.

The film deposited on rutile (001) surface was poorly epitaxial (double domain). The mobility (24 cm^2 V^{-1} s^{-1}) was higher than that (16 cm^2 V^{-1} s^{-1}) deposited on glass. The carrier electron density (1.7×10^{20} cm^{-3}) was same as on glass. The resistivity (1.6×10^{-3} Ω cm) was lower than that deposited on glass substrate (2.4×10^{-3} Ω cm).

The film deposited on rutile (100) surface was epitaxial (single domain) with much higher mobility (39 cm^2 V^{-1} s^{-1}) and higher carrier electron density (2.7×10^{20} cm^{-3}). The resistivity (6.2×10^{-4} Ω cm) was compatible with those of tin oxide films doped with antimony or fluorine which are environmentally unfavorable elements.

5. References

[1] Sawada Y, Kobayashi K, Seki S and Funakubo H 2002 *Thin Solid Films* 46-50 409
[2] Sawada Y 2003 *Maters. Sci. Forum* 437/438 23
[3] Ohta H, Onita M, Hirano M, Tanji H, Kawazoe H and Hosono H 2000 *Appl. Phys. Lett.* 76 2740
[4] Kondo T, Funakubo H, Akiyama K, Wang M, Uchida T and Sawada Y 2008 *Trans. Maters. Res. Jpn.* 33 1363
[5] Babaerjee R and Das D 1987 *Thin Solid Films* 149 291
[6] Rachut K, Körber K, Brötz J and Klein A 2014 *Phys. Stat. Sol.* A211 1998

IOP Conf. Series: Materials Science and Engineering **250** (2017) 012021 doi:10.1088/1757-899X/250/1/012021

[7] Luan C, Ma J, Zhu Z, Kong L and Yu Q 2011 *J. Cryst. Growth* 318 599

[8] Kim D H, Kim W S, Kee S B and Hong S H 2010 *Sensors Actuators B: Chemical* 147 653

[9] Nagano M 1984 *J. Cryst. Growth* 67, 639

[10] Hyosik M, Hyeonseok Y, Jining C P, Canon J. and Kookrin C 2015 *Appl. Phys. Lett. Maters.* 3 076107

[11] Lu Y M, Jiang J, Xia C, Kramm B, Polity A, He Y B, Klar P J and Meyer B K 2014 *Thin Solid Films* 594 B2 270

[12] Okuno T, Oshima T, Lee S D and Fujita S 2011 *Phys. Stat. Sol.* 8 540

[13] White M E, Tsai M Y, Wu F and Spec J S 2008 *J. Vac. Sci. Technol.* A26 1300

[14] Tien L C, Norton D P and Budai J D 2009 *Maters. Res. Bull.* 44 10

IOP Conf. Series: Materials Science and Engineering **250** (2017) 012022 doi:10.1088/1757-899X/250/1/012022

Simulation of Wave Energy Harvesting by Piezoelectric Seaweed

Yiliang Cui, Haifeng Wang†, Meng Li and Kaili Sun

School of Electromechanic Engineering, Qingdao University, Qingdao 266071, China
Email: hfwang@qdu.edu.cn

Abstract. In the last decade, the research about energy harvesting for low-power electricity generation has received growing interest all around the world. In this study, a piezoelectric seaweed made by piezoelectric polymer (PVDF) has been used as the energy harvester. ANSYS Workbench is utilized to construct the flow domain and piezoelectric energy harvester, set up working conditions for the simulation, and calculate the power generating capacity of the piezoelectric seaweed actuated by the vibration. The results show that the maximal and minimal output voltages are 117.65V and -112.53V respectively, and it is possible to harvest energy from ocean wave using piezoelectric seaweed.

1.Introduction

In last several years, the research about energy harvesting for low-power electricity generation has received growing interest. Recent developments have renewed interest in the ocean wave energy conversion [1]. Since the frequency of ocean waves is very low, the actual operation is relatively difficult, and the current use of ocean wave energy is still lacking. With the development of science and technology, people have a good understanding of power generation with many functional materials. The capability of harvesting electrical energy from mechanical vibrations in a dynamic environment through piezoelectric transducers has been the topic of discussions for many years. Actuated by the ocean wave, the piezoelectric energy harvester can convert the ocean wave energy into electricity, while the generated energy mainly supports systems of low power consumption, such as portable electronic devices, autonomous sensors. The advantages of using a piezoelectric power supply are that it is ecological, embedded, and almost need no maintenance [2]. In this article, piezoelectric polymer (PVDF) is used as the energy harvester, and ANSYS Workbench is utilized to construct the model, set the relevant parameters, and calculate the amount of electricity generation.

2.Theory Foundation

2.1 PVDF

PVDF is a typical representative of piezoelectric polymers. Since PVDF has flexibility, good workability, and high piezoelectric constant, it can be made into sensors of different sizes [3]. The parameters of PVDF are shown in table 1.

2.2 ANSYS Workbench

ANSYS Workbench is a collaborative simulation environment combining many core simulation tools, which can carry out the structural analysis, fluid analysis, thermal analysis, coupling analysis, and so on. In ANSYS Workbench, the fluid-structure interaction can be simulated in one-way or two-way coupling methods by connecting the coupling participants to a component called system coupling [4].

Table 1 Parameters of PVDF

Symbol	Parameter	Value	Unit
ρ	Density	1.78	10^3kg/m^3
Y	Young's modulus	2	GPa
μ	Poisson's ratio	0.3	
d_{31}	Piezo strain constant	23	10^{-12}
g_{31}	Piezo stress constant	216	10^{-3}
k_{31}	Electromechanical coupling factor	12%	

2.3 Piezoelectric Equations

The piezoelectric equations are expressions that describe the relationship between the mechanical and electrical quantities of piezoelectric materials [5]. The piezoelectric effect is described by the coupling equations:

$$\{S\}=\left[s^E\right]\cdot\{T\}+[d]\cdot\{E\} \tag{1}$$

$$\{D\}=[d]^t\cdot\{T\}+\left[\varepsilon^T\right]\cdot\{E\} \tag{2}$$

Where $\{S\}$ is the elastic strain vector; $\{T\}$ is the stress vector (N/m^2); $\{D\}$ is the electric flux density vector (C/m^2); $\{E\}$ is the electric field intensity vector (V/m); $[s^E]$ is the compliance matrix evaluated at constant electric field, i.e. short circuit; $[d]$ and $[d]^t$ are the piezoelectric strain coefficient matrix (C/N), and the superscript$\{t\}$is the matrix transpose.

2.4 Airy Wave Theory

Open channel wave boundary conditions allow one to simulate the propagation of waves, which is useful in the marine industry. Wave steepness is generally defined as the ratio of wave height to wave length, and relative depth is defined as the ratio of wave height to the liquid depth. The wave profile for a linear wave is given as [6]:

$$\zeta(X,t)=A\cos\alpha \tag{3}$$

where α is defined as follows:

$$\alpha=k_x x+k_y y-\omega_e t+\varepsilon \tag{4}$$

x and y are the space coordinates in the $\hat{x}$ and $\hat{y}$ directions, respectively, ε is the phase difference, and t is the time.

The wave frequency ω is defined as follows for shallow/intermediate waves:

$$\omega=\sqrt{gk\tanh(kh)} \tag{5}$$

and as follows for short gravity waves:

$$\omega=\sqrt{gk} \tag{6}$$

where h is the liquid height, k is the wave number, and g is the gravity magnitude [7].

The velocity components for the incident wave boundary condition can be described in terms of shallow/intermediate waves and short gravity waves.

3.Numerical Analysis

3.1 Meshing

A 3D model is constructed by ANSYS DesignModeler, and meshed by ANSYS Meshing. As shown in figure 1, the piezoelectric seaweed is made of a sandwiched structure, including a flexible foam core of 1mm thickness and two PVDF films of 0.05mm. The length of this seaweed is 100mm. According to Ref. [8], a wave tank (figure 2) was modelled with sizes of about 1000mm long, 200mm wide, and 200mm high.

The computational domain is meshed by 7692 triangular elements for total degrees of freedom of 38462. To get better results, the element refinement has to be done at the fluid-solid interaction boundary to avoid mesh warnings, and increase the lattice density to guarantee accuracy in the coupling interface (figure 2).

3.2 Solution Method

The control equations are discretized to use the finite volume method based on the center of the cell. The viscous model is standard *k-ε* model with enhanced wall treatment. The second order upwind differential scheme is adopted in the convective term, and the central difference scheme is adopted in the diffusion term. The system coupling analysis is carried out between transient structure and fluid.

3.3 Initial and Boundary Conditions

It is important to define the appropriate boundary conditions and the initialization conditions for the whole computational domain, such as the flow velocity and the outlet pressure [9]. In this study, the rigid column wall boundary condition is used. In other words, there is no slip at the interface between fluid and solid. The inlet velocity is set as 1 m/s, and the outlet is the pressure outlet. Flow domain is divided into two phases of water and air, and the initial water height is 120 mm. A sinusoidal wave boundary is simulated, whose amplitude is 50 mm, and length is 500 mm (shown in figure 3).

Figure 1. Mesh of piezoelectric seaweed

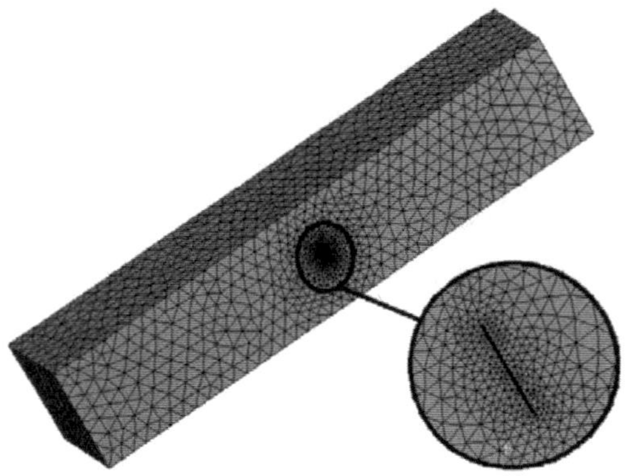

Figure 2. Mesh of computational domain

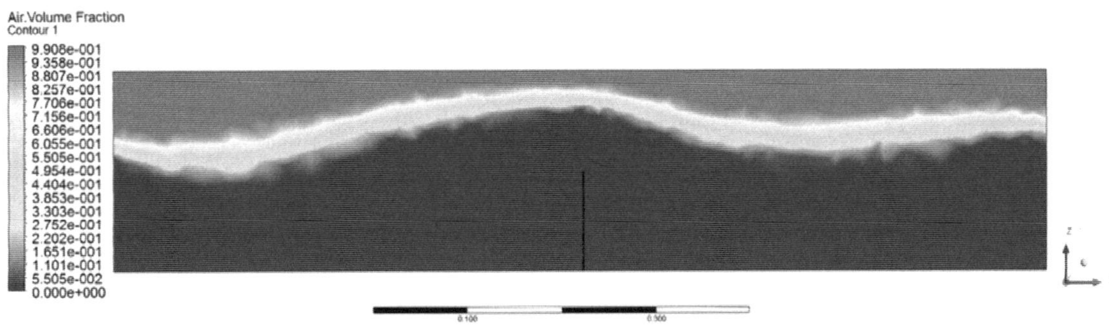

Figure 3. Wave cloud map

IOP Conf. Series: Materials Science and Engineering **250** (2017) 012022 doi:10.1088/1757-899X/250/1/012022

4.Results and Discussion

The piezoelectric seaweed is deformed by the impact of water flow, the maximum deformation is about 20.1 mm, as shown in figure 4 (a). During reciprocating oscillation, the piezoelectric seaweed recovers part of the deformation, and the corresponding maximum deformation is reduced to 18.9 mm, as shown in figure 4 (b).

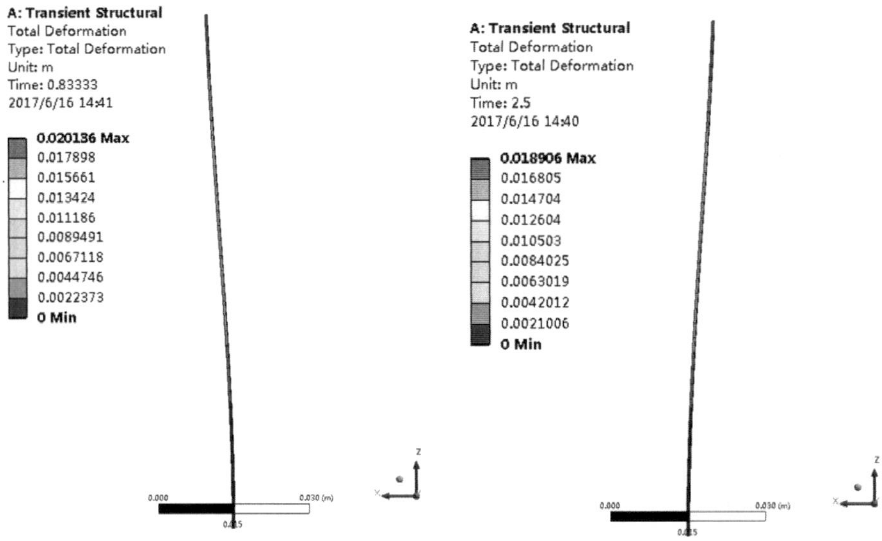

(a) (b)

Figure 4. Deformation of piezoelectric seaweed at two extreme positions

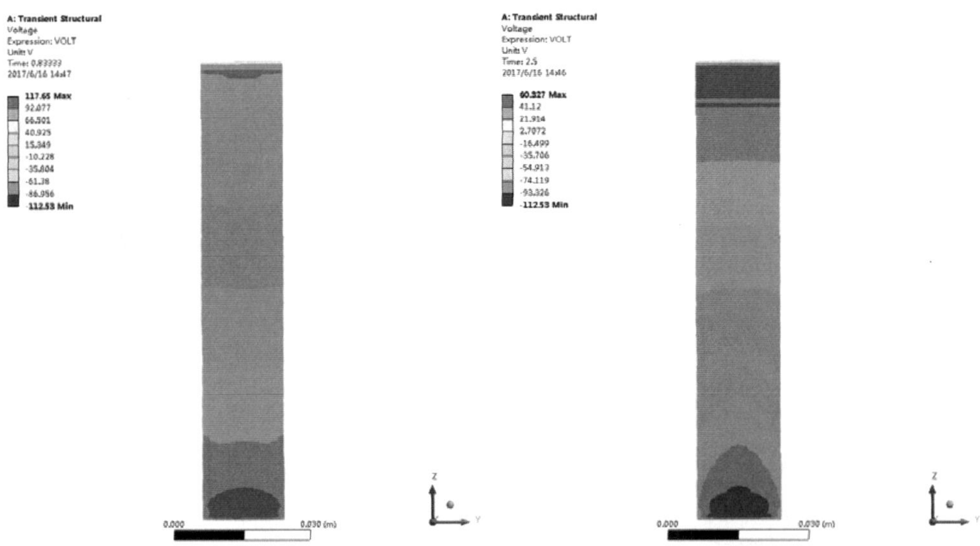

(a) (b)

Figure 5. Output voltage of piezoelectric seaweed at two extreme positions

As shown in figure 5, the generated voltage of the piezoelectric seaweed varies with the deformation directions. Since both ends of the piezoelectric seaweed may be deformed in different directions, the maximal and minimal voltages are 117.65V and -112.53V respectively in case that the deformation occurs along the direction of the water flow (figure 5 (a)). While in figure 5(b), the maximum and minimum are 60.33V and -112.53V in the opposite swing direction. According to figure 5(b), the detailed voltage generation values are given in table 2.

Table 2 Detailed voltage generation values of piezoelectric seaweed .

Time(s)	Minimum(V)	Maximum(V)	Time(s)	Minimum(V)	Maximum(V)
0.0	0	0	1.3	-52.372	86.188
0.1	-2.5200	14.946	1.4	-39.338	69.447
0.2	-11.916	32.797	1.5	-24.968	52.007
0.3	-24.883	52.045	1.6	-12.123	32.823
0.4	-39.262	69.637	1.7	-2.7261	15.331
0.5	-52.274	86.300	1.8	-0.2479	0.2792
0.6	-64.097	98.910	1.9	-14.823	1.7384
0.7	-72.758	108.79	2.0	-33.319	8.4949
0.8	-78.116	114.98	2.1	-53.154	18.284
0.9	-80.308	116.53	2.2	-71.496	30.513
1.0	-78.188	115.18	2.3	-88.823	41.929
1.1	-72.965	108.90	2.4	-102.00	52.504
1.2	-64.182	99.062	2.5	-112.53	60.327

5.Conclusions

Driven by the wave force, the piezoelectric seaweed can harvest energy from ocean or river wave. As a simple, lightweight electromechanical converter, PVDF can be widely used in the power generation from fluid wave. Although the generated voltage is not strong (-112.53V~117.65V), after signal conditioning it still can be used for electrical equipment with low power consumption, i.e. wireless sensors.

6.Acknowledgments

The authors are grateful to the financial supports by the Natural Science Foundation of China (No. 51405255), the Science and Technology Project for Higher Education of Shandong Province (No. J12LN95), the Science and Technology Development Plan of Shandong Province [Policy Guidance Project] (No. 2012YD04038).

7. References

[1] Wang C K 2009 *Analysis Method and Reserves Assessment of Ocean Energy Resource* (Beijing: China Ocean Press)
[2] Pankonien A M 2010 *Proc. ASME 2010 Conf. on Smart Materials, Adaptive Structures and Intelligent Systems* (Philadelphia) p 223
[3] Schmidt V H 1986 *Proc. 6th IEEE Int. Symp. on Applications of Ferroelectrics* p 538-42
[4] ANSYS Inc. 2016 *Workbench User's Guide* Release 17.2
[5] Zeng C X and Jiang L 2011 *Instrum. Technol.* **6** 31
[6] ANSYS Inc. 2016 *ANSYS Fluent Theory Guide* Release 17.2
[7] Nitulescu T, Talu S 2001 *Applications of descriptive geometry and computer aided design in engineering graphics* (Romania: Risoprint Publishing house)
[8] Zurkinden A S, Campanile F and Martinelli L 2007 *Proc. COMSOL Users Conference* (Grenoble)
[9] Hobeck J D, Inman D J 2012 *Smart Mater. Struct.* **21** 105024

IOP Conf. Series: Materials Science and Engineering **250** (2017) 012023 doi:10.1088/1757-899X/250/1/012023

Modeling Fluid and Heat Transport in a New Type Thermal Isomerization Fluidized Bed Reactor

Xiaoxiao Yang, Zewu Fu, Yuying Zhao, Liujun Liu and Rui Li*

College of materials science and technology, Beijing Forestry University, 100083, Beijing, China
Email: 13020021531@163.com

Abstract. In the current work,with a new concept of resident ratio which impacts the reaction time, a fluid flow and heat transfer model were employed for simulating pressure drop, temperature profile and fluid flow properties of new type thermal isomerization reactor. The thermal isomerization experiment of β-pinenewas performed using the reactor. Momentum equation, energy equation and kinetic equationswere used to describe the fluid flow and heat transfer. The experimental results were in good agreement with theoretical simulation which indicated that the temperature difference between boundary and initial can be decreased by using steel balls and this modified fluidized bed can improve the yield and selectivity of the products effectively.

1. Introduction

Myrcene iswidely used in the commercial production of various terpenic alcohols, chemical aromas,it can also be used as the starting point of vitamins A and E. Myrcene can be produced fromβ-pinene, which is obtained through the distillation of turpentine.The yield ofmyrcene fromβ-pinene could be increased if the reaction conditions were optimized.Varies type reactors have been used in relatively researches. He Jindong et al. compared gas-phase and liquid-phase thermal isomerization reaction of α-pinene, and obtained the difference of their selectivity and best temperature. Regularly, pipe type reactors were utilized for monoterpenes isomerization. Goldblatt et al. used a reactor of 20mm diameter, 1m long, and nitrogen as carrier gas at 403°C for β-pinene rearrangement to produce myrcene. A steel pipe of 1.4mm diameter, 30cm long was employed by Savich et al. for myrcene production at 580-750°C, whose resident time is about 0.01 to 0.1 second. Tanaka et al. employed a quartz tube of 20mm diameter, 300 mm long to rearranged gaseous pinene at 400-500° C. Zhang Zhongfu used the stainless steel tube coiled into three Φ1m circles which can effectively improve the fluid turbulence and increase the chance of molecular collision. But in these reactors, the inputted heat mainly transferred nearby the pipe wall through fluid convection and finite heat radiation what resulted in the total heat transfer coefficient is low, so the wall temperature was much higher than the interior which caused the tendency that carbon deposition appeared in the pipe wall, so it was necessary to use a slender pipe to ensure a high flow rate and avoid local overheating. In production, it showed that the coke cleaning cycle of this type reactor was short, along with side reactions.

We proposed a sump type fluidized bed reactor which combined the advantages of packed bed and fluidized bed which consisted of large steel balls and copper powder. The steel balls make the channel narrow and variable what resulted in the turbulence.The copper powder could distributed more uniform in the axial direction, its disordered collision make the reactor had the self-cleaning function. In addition, this reactor could be made as the common reactor for conveniently arrangement,

IOP Conf. Series: Materials Science and Engineering **250** (2017) 012023 doi:10.1088/1757-899X/250/1/012023

maintenance and reducing investment, it also has the potential for the use of flue gas heating and reducing operating costs.Based on the experimental data and mechanism of the reactor, the present work mainly focused on the modelling of the fluid flow and heat transfer for the pressure drop and the temperature profile in an isomerization fluidized bed reactor.

2. System Descriptions

2.1 Experimental Device

The diagrammatic sketch of reactor was shown in Figure1, the feedstock was preheated and transported to the bottom of the reactor, rapidly isomerized in 400~550 ℃. Then the target product was obtained after cooling by the condenser pipe, the whole system was carried out under auxiliary gas assisted streaming and different pressures.The reactor is made of stainless steel pipe with the specification of Φ45×3.5 and 600mm long, which filled with particles of Φ16mm and Φ150μm copper powder and automatically controlled by the resistant furnace.

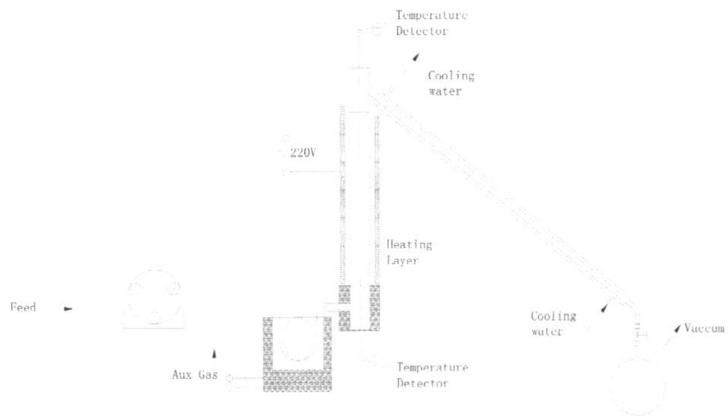

Figure1. Diagrammatic sketch of reactor for thermal isomerization.

2.2 Reactor Model and Mechanical Properties of Cold State Fluid

In Figure 2, when the fluid flowed across, the original flow region was distributed into each direction and converged again before the next ball, by repeating these units, the reaction time was prolonged and the temperature was well-distributed. Li Jian et al. simulated the effect of sphere size on flow properties by CFD, and turned out that larger sphere could minimize the boundary effect. In order to study the mechanical properties of cold fluid, the influence from the steel balls to fluid resistance and flow lengthwere considered.

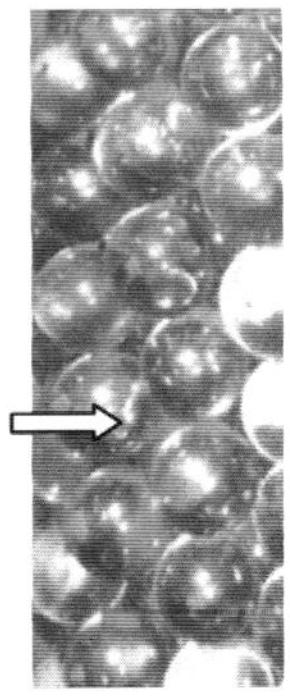

Figure 2. The fluid flow state in the reactor.

IOP Conf. Series: Materials Science and Engineering **250** (2017) 012023 doi:10.1088/1757-899X/250/1/012023

In the cold state experiment, by adding copper powder, we observed some obviously resident areas performing behind the steel balls. In this study, the fluid unit in each channel was divided into three zones which is shown in Figure 2:

Steady distribution zone A: The fluid is hindered by the downstream particles and thanks to the expansion of the flow channel, the main flow direction of the 3 dimensions began to diffuse, it can be simply expressed as the flow past body in the 2 dimensions;

Hedging mixture zone B: The fluid generates a periodic trailing vortex in this zone. Fluids from different directions come together and mix totally. In this zoon, some particles will stay at the vortex, so we put forward a new accept-resident ratio;

Steady reduced zone C: The blended fluid goes through the minimum channel with increasing speed and decreasing flow pressure, meanwhile under the limitation from particles, the fluid has a definite flow direction and the turbulence has been suppressed sharply.The volume of resident area is related to the particle scale, following the experience formula, the relationship between void content and resident content is proposed by Dai Gance et al.:

$$\varepsilon_b = 1.6(\varepsilon - 0.2)(1 - \varepsilon)(1)$$

Where ε is the void content of the entire bed and ε_b is resident content.The resident area is assumed to be a symmetrical triangle whose base is an arc in two dimensions which illustrated in Figure 2. This area can be calculated by following equation:

$$\frac{Rh}{2} - \left(\frac{1}{6}\pi R^2 - \frac{\sqrt{3}}{4}R^2\right) = A_b = A_{\varepsilon_b} \quad (2)$$

2.3 Kinetic model

As shown in Figure 3, during the isomerization of β-pinene (CAS: 127-91-3), side reactions lead to the formation of not only myrcene (CAS: 123-35-3), but also little limonene (CAS: 5138-86-3) and psi-limonene. Kolicheski et al. compared two situations for the absence and presence of myrcene decomposition reactions, suggested that the second model is better. However, the decomposition model appears not to be sufficient, since predicted yields are greater than experimental yields.

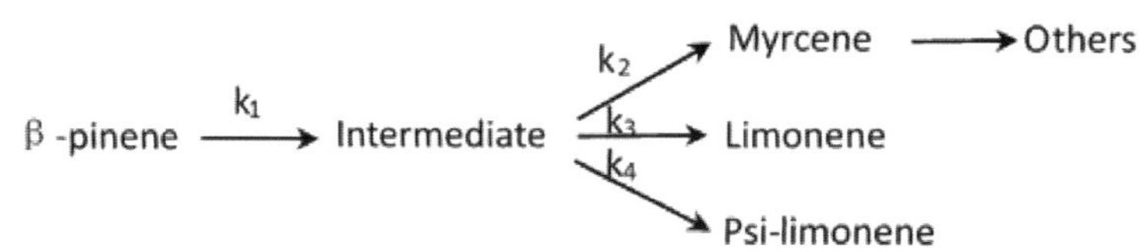

Figure 3. The molecular rearrangements during the isomerization of β-pinene.

Depending on Arrhenius equation, the apparent rate constant k_i can be calculated by following equation:

$$k_i = \frac{1}{T} ln \frac{[A_i]_0}{[A_i]} (i = 1,2,3) \quad (3)$$

Based on the theory of parallel reaction, the rate of β-pinene reduction is equal to the sum of product formation rate:

$$k_1 = k_2 + k_3 + k_4 \quad (4)$$

3. Numerical Analysis

3.1 Fluid Dynamics

In order to implement the flow models, tube-to-particle diameter ratios and drag from reactor wall were disposed like packed bed: Churchill et al. described the wall drag by taking the surface area of the tube into account in the definition of the hydraulic diameter. Winterberg et al. handled the radial porosity variation and wall drag in packed tubes. As the pressure drop consisted of pipe flow drop and resident drop owing to the resident effect and the resident drop, Cheng Yi et al. has concluded the pressure drop by using Ergun equation:

$$\Delta P = 150\frac{\mu u}{d_p^2} \times \frac{(1-\varepsilon)^2}{\varepsilon^3} + \frac{7}{4} \times \frac{\rho u^2}{d_p} \times \frac{1-\varepsilon}{\varepsilon^3} L_{eff} + (1-\sigma)(1-\varepsilon_b) L_{eff}\rho \quad (5)$$

Where d_p is particle diameter, uis flow rate, μ is viscosity and σ is resident ratio.

3.2 Heat Transfer

Quantities of works have been carried out on heat transfer. The approach using effective thermal conductivity of the packed bed is most acceptable. Slavin et al. used the measurable parameter, surface roughness of the particle to avoid the use of any adjustable parameters in the determination of effective thermal conductivity for uniform metallic spheroids filled with ideal gas with no compression of particles. Sharma et al. considered the effective thermal conductivity model for radiation transport by modifying shape factor using the extinction coefficient and also developed the heat transfer model using single control volume applied to the reactor. Guangzhan Xu et al. have reported that the temperature is much higher in the gap and dorsal to the ball, and the heat transfer coefficient increases with the decrease of the porosity.

The effective thermal conductive can be regarded as the sum of thermal conductivity and thermal emissivity:

$$\lambda_{eff} = \lambda_{bed}^{k} + \lambda_{bed}^{j} \quad (6)$$

Where λ_{bed}^{k} is radiation thermal conductivity, λ_{bed}^{j} is contact thermal conductivity.Kandula et al. had proposed Cell Model while investigating the effective thermal conductivity of particle bed; Breitbach et al. improved equation (7) for thermal radiation. Depending on the investigation of different rules of particle volume of the heat flow by Prasad et al., the contact thermal conductive could be obtained by equation (8).

Table 1. The equations of radiation thermal conductivity and contact thermal conductivity

$\lambda_{bed}^{k} = \left\{\left[1-(1-\varepsilon)^{1/2}\right]\varepsilon + \frac{(1-\varepsilon)^{1/2}}{\frac{2}{\varepsilon_r}-1} \cdot \frac{B+1}{B} \cdot \frac{1}{1+\frac{1}{\left(\frac{2}{\varepsilon_r}-1\right)\Lambda}}\right\} 4\sigma T^3 d \ (7)$	radiation thermal conductivity, $\varepsilon_r = 0.7$
Where $B = 1.25\left(\frac{1-\varepsilon}{\varepsilon}\right)^{\frac{10}{9}}$	directly heat radiation
$\Lambda = \frac{\lambda_s}{4\sigma T^3 d}$	conduction-radiation-conduction
$\frac{\Lambda_{bed}^{j}}{\lambda_s} = \left[\frac{3(1-\mu_p^2)}{4E_s} fR\right]^{\frac{1}{3}} \cdot \frac{1}{0.531S}\left(\frac{N_A}{N_L}\right) \ (8)$ Where $f = p\frac{S_F}{N_A}$, $N_A = \frac{1}{4R^2}$, $N_L = \frac{1}{2R}$ and $\mu_p = 0.25$	contact thermal conductivity

4. Results and Discussion

4.1 Resident Ratio

Resident ratio is a novel parameter in present study, there is the recirculation region at the downstream of each sphere, it provides a stagnant space for fluid, which results in the increase of reaction time. Although the resident area is constant, the resident ratio varies with fluid density which is influenced by flow rate.The influence from flow rate to resident ratio is shown Figure4, with the increasing of the flow rate and fluid mass, the resident ratio is decreasing apparently. The resident ratio can be regressed from the experimental data as:

$$\sigma = -0.51 lnu - 0.4\frac{m}{V} + 2.145 \quad (9)$$

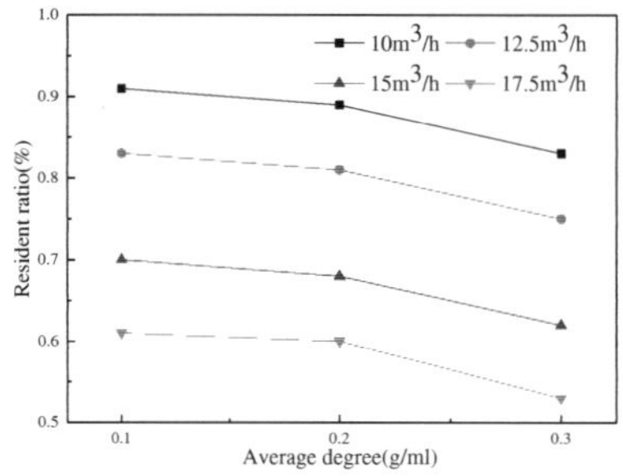

Figure 4. Resident ratios under different flow rates and average degree.

4.2 Fluid Flow Validation of Cold State Fluid

The experimental pressure drop with cooper powder was compared with calculation data in Figure5. As expected, the pressure drop increased with the increase of flow rate. Meanwhile, by increasing fluid mass, marked pressure drop embodied that higher flow mass much more fit the model.

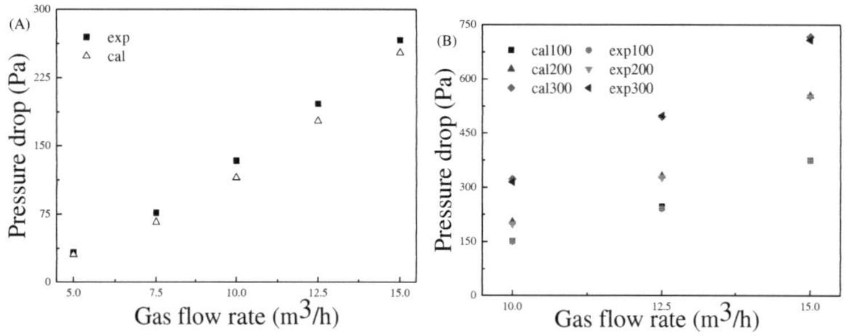

Figure 5. Pressure drop of calculation and experimental, (A) without copper powder, (B) with copper powder, 'exp' is experimental data, 'cal' is calculation data,100, 200 and 300 are weight of copper powder.

4.3 Temperature and Reaction Simulation

Figure6 shows the CFD results of temperature distribution and concentrition of β-pinene and myrcene in blank pipe and filled pipe that outlet temperature were 850K and 765K, respectively. In the blank pipe, temperature is higher near the wall, β-pinene has higher concentration at bottom and central region, myrcene is less wherever in the wall or central region, percent conversion of the reaction is low. The situation improves obviously in the filled pipe, the temperature can distrubute uniform in radial direction and reach the highest temperature quickly, β-pinene still concentrats at bottom but sharp falling to a low level, meanwhile myrcene reaches a high level as expected. This enhencement proves that this new type reactor can prolong the reaction time so that β-pinene can almost entirely convert to myrcene at a lower temperature.

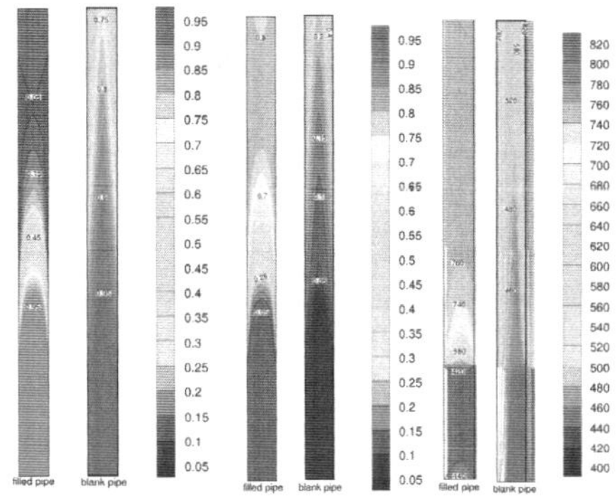

(A) (B) (C)

Figure 6. Simulation results of β-pinene, myrcene and temperature distribution

Table.2 Experimental results of productions through rearrangement of β-pinene

Temperature /°C	β-pinene content /%	β-pinene selectivity /%	myrcene content /%	myrcene selectivity /%	psi-limonene content /%	psi-limonene selectivity /%	limonene content /%	limonene selectivity /%
720	54.7	43.4	36.4	86.6	1.6	3.8	4.1	9.8
730	36.1	62.7	53.0	87.3	2.4	3.9	5.8	9.6
740	21.2	78.1	65.2	86.2	3.0	4.0	7.0	9.3
750	8.3	91.4	75.5	85.4	3.6	4.1	8.0	9.1
760	2.5	97.5	78.8	83.6	4.0	4.2	8.5	9.0
770	0.8	99.2	77.8	81.1	4.2	4.4	8.7	9.1

5. Conclusion

Fluid flow and heat transfer model for new type fluidized bed thermal isomerization reactor was presented for simulating the concentration, temperature profile and fluid flow properties. Depending on the present theory of particle bed fluidity mechanics, a flow physical model was proposed, and resident ratio was firstly putted forward the pressure drop correlation in reactor.The selectivity and yield of monoterpenes was influenced by temperature and resident time. Typical fluidized bed reactor could not control the resident time effectively, which could bring byproducts. The filling of particles gave uniformity channels on section area and better distribution. The new type thermal isomerization fluid bed reactor combines the advantages of the fluid bed and fix bed reactor.The experiments proved that the flow rate of 15m^3/h at 760K for myrcene was the best conditions.

6. Acknowledgements

The authors gratefully thank for special fund for Beijing common construction project.

7. References

[1] He Jindong, Gong Y, Zhao W. A,*comparative study on the gas-phase and liquid-phase thermal isomerization reaction of α-pinene,* Journal of Physical Organic Chemistry, 2013
[2] Peng YuXing, Liu JL, Cun LF, Dai HS,*Vapor phase thermal isomerization of α- and β-pinene*,* Journal of the American Chemical Society, 1941
[3] Tanaka, J., Katagiri, T., Ozawa, K.,*The thermal isomerization of pinane and its products,* Bulletin of the Chemical Society of Japan, 1971

[4] Zhang Zhongfu, *On the production process of beta pinene pyrolysis of myrcene,* Chemical engineering and equipment, 2013
[5] Li Jian, Song XM, Lu JC, *CFD simulation study of the influence of the sphere bed scale on the pore flow characteristics,* Nuclear power engineering, 2013
[6] Dai Gance (2008). Momentum transfer. In: Transfer phenomenon introduction, vol 3. Chemical Industry Press.
[7] Kolicheski, M. B., Cocco, L. C., Mitchell, D. A., Kaminski, M,*Synthesis of myrcene bypyrolysis of β-pinene: analysis of decomposition reactions,* Journal of Analytical and Applied Pyrolysis, 2007
[8] Churchill, S. W,*Viscous flows,* 1988
[9] Winterberg, M., E. Tsotsas,*Impact of tube-to-particle-diameter ratio on pressure drop in packed beds,* Aiche Journal, 2000
[10] Cheng Y, Liu LJ, Li R,*Fluid characteristics and cold state experiment validation of novel sump type fluidized bed,* Chemistry and industry of forest products, 2015
[11] Xu GZ, Sun ZN, Meng XK,*Numerical simulation of fluid flow and heat transfer in pebble bed reactor,* Nuclear Power Engineering, 2013
[12] Kandula, M,*On the effective thermal conductivity of porous packed beds with uniform spherical particles,* Journal of Porous Media, 2011
[13] Breitbach, G., Barthels, H,*Radiant heat transfer in the htr core after failure of the afterheat removal systems,* Trans. Am. Nucl. Soc.1978

IOP Conf. Series: Materials Science and Engineering **250** (2017) 012024 doi:10.1088/1757-899X/250/1/012024

The Nonlinear Changes of Temperature Response Birefringence of Side-Hole Fiber

Ning Wang, Jianbo Fu, Dailin Li, Huafeng Zhu and Tao Wang

College of Science, China University of Petroleum (huadong), Qingdao, Shandong, 266580, People's Republic of China

E-mail: qfwangning@163.com

Abstracts: For deeply investigate the birefringence responding of side-hole fiber caused by temperature, the response performance are analyzed by finite element method and elastic-optic theory, especially the nonlinear changes. The simulation results showed that the response birefringence is linear with temperature no matter how the geometry changed, but the response sensitivity is nonlinear with geometry parameters. The hole radius and hole-core distance both have their limit value to get obvious temperature sensitivity. Large hole radius and small hole-core distance is good to get high response sensitivity. Furthermore some interesting results are discussed about the response sensitivity changed with the square value of hole-core separation angle. The sensitivity is nonlinear with the square of hole-core separation angle for temperature, which is different to pressure response sensitivity.

1. Introduction

The side-hole optic fiber was proposed as pressure sensor because of its special side-hole structure[1]. Two air tunnels are symmetrically surrounded with fiber core. The internal stress is anisotropic when the pressure is loaded on the fiber. That is favorable to pressure sensing because of large birefringence caused by stress anisotropy. In recent several decades[2-7], the side-hole optical fiber sensor is attractive, such as writing fiber gratings onto side-hole fiber[8-9] or filling some materials into side-hole[10-11]. Simultaneously, this birefringence optical fiber has good thermal stability and pressure-temperature crossing response. Although its pressure response sensitivity is higher than temperature response sensitivity, the temperature sensing performance also gave us great interest [12-13].

The birefringence of side-hole fiber can be attributed to intrinsic birefringence and response birefringence. The intrinsic birefringence is got in optical fiber fabrication. The response birefringence is caused by temperature, pressure or some other parameters. Some analysis [12] has demonstrated that the response birefringence is linear with temperature. By the analysis in this paper, the temperature response performance is not linear with its geometry parameters. Some linear and nonlinear response birefringence performance curves are discussed in this paper, especially its nonlinear performance. The research results give some supplement to the temperature response birefringence theory of side-hole fiber. That is helpful to deeply understand the sensing performance of side-hole fiber.

2. Simulation and Analysis

The structure of side-hole fiber is showed in Figure 1. This kind of anisotropy structure caused the obvious birefringence. The radius of side hole is showed as r. And the hole-core distance is d. The hole-core separation angle, the angle between x axis and the tangent line of fiber core center point to side-hole outer boundary, is showed as ψ. Because of this structure, the response birefringence is got by stress-optic effect caused some loaded stress. So the anisotropy refractive index distribution of side-hole fiber is got when the outer temperature changed, which caused the temperature response birefringence. According to the photoelastic theory, the birefringence value is got as followed formula showed.

$$B = c \cdot \Delta\sigma \tag{1}$$

$$C = \frac{n^2}{2E}(1+\upsilon)(p_{12} - p_{11}) \cdot \Delta\sigma \tag{2}$$

Where B is response birefringence, c is the piezo-photonic constant, Δσ is the stress difference between x and y direction(as figure1 showed) at fiber core, n is the effective refractive index of fiber, E is Young's modulus of the fiber, υ is Poisson ratio, p11 , p12 are components of strain-optic tensor. For typical side-hole fiber, the Young's modulus E is 7.83×10^{10}, Poisson ratio ν is 0.186, the components of strain-optic tensor P_{11}=0.121,P_{12}=0.27,the effective refractive index of fiber n=1.468. The intrinsic fiber birefringence caused by core ellipticity and fiber residual stresses is ignored in calculation. At the center of optic fiber core, the stress components are approximatively as average stress values in x and y direction of fiber core [14]. By ANSYS software, the finite element method is used in simulation, for calculating the stress distribution at different temperature. Then the birefringence value is got by stress calculation data.

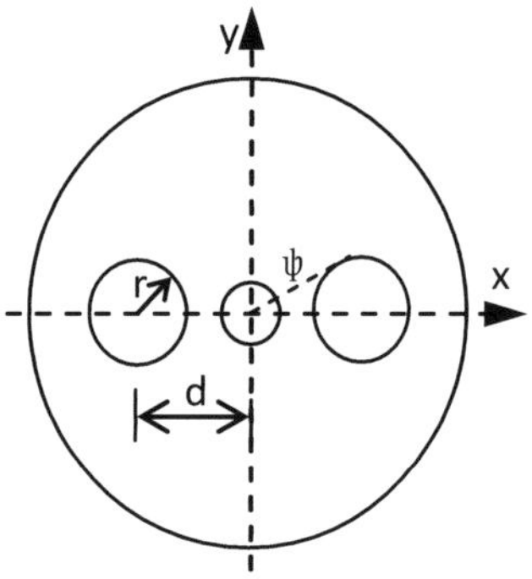

Figure 1. The diagram of side-hole fiber

2.1 Linear Response Curves

The simulations focused on the influence to response birefringence with different geometry parameters. For simulation, the typical hole-core distance and hole radius value are set as 26.6μm, 15.6μm, respectively. In ANSYS software calculations, free meshing method is used to get FEM grid. And the unit size is set as 2μm to get FEM meshing. Then the stress distribution is calculated by ANSYS software after setting the loaded temperature. Fig.2 shows the relationship between birefringence and temperature with different hole radius, when hole-core distance value is 26.6μm. The hole radius varied from 11.9μm to 19μm. Fig.3 is the curves of response birefringence versus temperature with different hole-core distance, when the hole radius value is 15.6μm. The hole-core distance varied from 20 μ m to 40 μ m. For two figures, the temperature varied from 20℃ to 200℃.

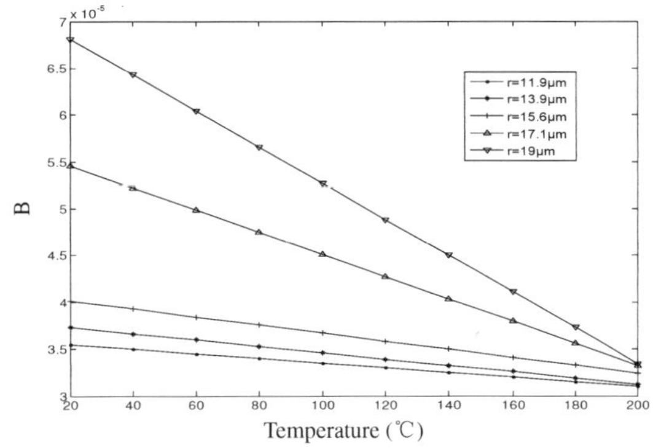

Figure 2. The temperature response curves with different hole-core distance

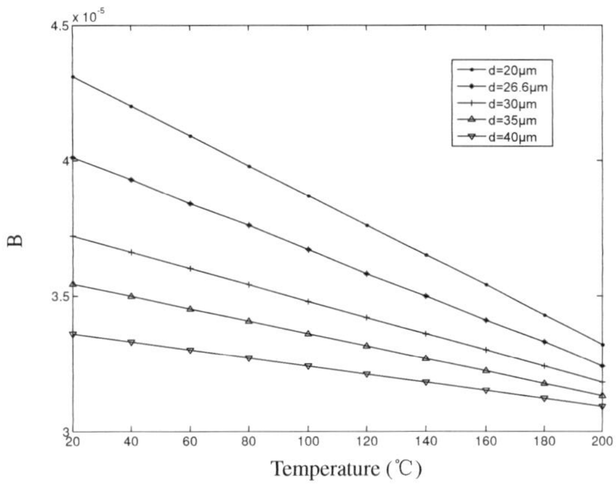

Figure 3. The temperature response curves with different hole radius

As the above figures showed, the response birefringence is always linear with temperature no matter how the geometry changed. But the interesting thing is that it looks that the slope variation of curves is nonlinear with the changing of hole radius and hole-core distance. That means the changing of response sensitivity value is not linear with the fiber geometry parameters.

2.2 Nonlinear Performance Curves

For deeper investigating the response performance, the response sensitivity is also calculated. The temperature response sensitivity is defined as:

$$K_T = \frac{\Delta B}{\Delta T} \tag{3}$$

In calculation, the chosen hole-core distance values are 20μm, 26.6μm, 30μm, 35μm, 40μm, respectively. The hole radius value is set as 3μm, 7μm, 11.9μm, 15.6μm, 19μm, respectively. More than ten values are given to simulate response sensitivity curve to each geometry parameters. The temperature response sensitivity curves are showed in Fig.4 and Fig.5. The geometries choosing must be accordant with the actual structure. That means the hole-core distance value must be more than hole radius value. And the radius value cannot be too small to let the software run well.

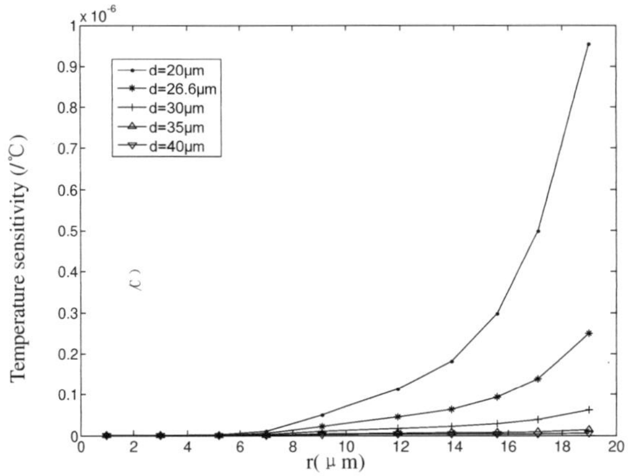

Figure 4. The temperature response sensitivity versus hole radius with different hole-core distance

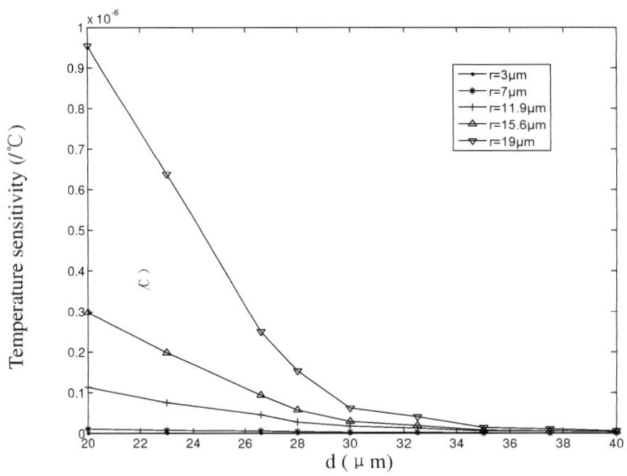

Figure 5. The temperature response sensitivity versus hole-core distance with different hole radius

Obviously the temperature response sensitivity curves are all nonlinear as two figures showed, which is similar to pressure response sensitivity curves at reference [15]. The response sensitivity increased with bigger hole radius or smaller hole-core distance. So large hole radius and small hole-core distance are good to temperature sensing, because of more stronger anisotropy accordingly. Furthermore, the response sensitivity is greatly influenced by hole-core distance when the hole radius value is big. And the sensitivity varied obviously with hole radius value when the hole-core distance value is small. As Fig.4 and Fig.5 showed, the responding is obvious when the hole radius value is more than 7μm, and the hole-core distance value is less than 35μm. That means the hole radius and hole-core distance both have their limit value to get obvious temperature sensitivity.

Some analysis [2, 15] was reported that the pressure response sensitivity is depended lineally on the geometry parameter φ^2, where φ is hole-core separation angle as Fig.1 showed. However, the interesting point is that the temperature response sensitivity changing situation is different by simulations. For deeply investigating this responding performance, the sensitivity curves are simulated with different value of φ^2. For discussing the changing situation, in two simulations, the typical hole-core distance values are uniformly set as 26.6μm, 30μm, 35μm, respectively.

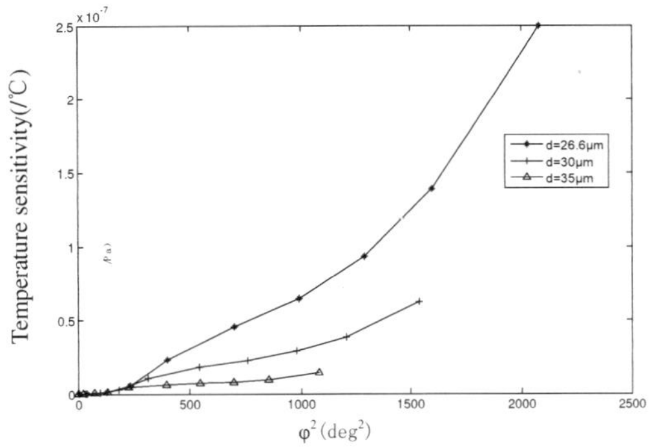

Figure 6. The temperature response sensitivity versus φ^2 with different hole-core distance

As Fig.6 showed, the curves presented the relationship of the temperature sensitivity with φ^2. Obviously the temperature response sensitivity is nonlinear with φ^2. However, according to the results in reference [2], the pressure response sensitivity is linear with φ^2. This interesting difference should be caused by the different influence between the thermal expansion and pressure-induced stress to this special fiber structure.

3. Conclusion

We analyzed the linear and nonlinear temperature response birefringence performance of side-hole fiber in this paper. According to the theoretical analysis, the response birefringence is linearly dependent on external temperature even if with different hole radius and hole-core distance. But the response sensitivity are nonlinear with varied hole radius and hole-core distance. Large hole radius and small hole-core distance are good to improve temperature response sensitivity. And the hole radius and hole-core distance both have their limit value to get obvious temperature sensitivity. Furthermore, different with pressure response sensitivity curves, the temperature response sensitivity is nonlinear with the square of hole-core separation angle. The analysis results are helpful to deeply understand the sensing performance of side-hole fiber, especially the pressure-temperature crossing response performance.

4.Acknowledgements

The authors acknowledge the support of the Fundamental Research Funds for the Central Universities (15CX05055A, 15CX05026A, 15CX05057A), the research project of China Petroleum Science and Technology Innovation Fund (2015D-5006-0604).

5.References

[1] Xie, H. M., Dabkiewicz, P., Ulrich R. and Okamoto K. Side-hole fiber for fiber-optic pressure sensing. Optics Letters, 1986. 11: 333-335.
[2] Clowes, J. R., Syngellakis, S., Zervas, M. N. Pressure sensitivity of side-hole optical fiber sensors. IEEE Photonics Technology Letters, 1998.10(6): 857–859.
[3] Xin, Y., Dong, X. Y., Meng, Q. Q., Qi, F., Zhao, C. L. Alcohol-filled side-hole fiber Sagnac interferometer for temperature measurement. Sensors & Actuators: A. Physical, 2013. 193: 182–185.
[4] Lee, S. H., Kim, B. H., Han, W. T. Effect of filler metals on the temperature sensitivity of side-hole fiber. Optics Express, 2009.17(12): 9712-9717.
[5] Jin, Y. X., Chan, C. C., Zhang, Y. F., Dong, X. Y. Mechanically induced long-period fiber grating in side-hole single-mode fiber for temperature and refractive sensing. Optics Communications, 2010.283(7):1303-1306.
[6] Frazão O, Silva S. O., Baptista J. M., Santos J. L., Statkiewicz-Barabach G., Urbanczyk W., Wojcik J. Simultaneous measurement of multiparameters using a Sagnac interferometer with

polarization maintaining side-hole fiber. Applied Optics, 2008.47(27):4841-4848.

[7] Bock,W. J., Nawrocka, M. S., Urbanczyk, W. Application of side-hole fibers for dynamic pressure measurements. Proceedings of the Society of Photo-Optical Instrumentation Engineers, SPIE, Quebec City, Canada, June, 2000. pp.1175-1179.

[8] Schroeder, R. J.; Yamate, T.; Udd, E. High Pressure and Temperature Sensing for the Oil Industry Using Fiber Bragg Gratings Written onto Side Hole Single Mode Fiber. 13th International Conference on Optical Fiber Sensors and Workshop on Device and System Technology toward Future Optical Fiber Communication and Sensing, Kyongju, Korea, April, 1999.pp.42-45.

[9] Xin,Y., Dong, X. Y., Yuan,J. W., Li,Y., Jin, S. Z., Zhang, S. L. Sensing characteristics of side-hole fiber long-period grating. Advances in Materials Science and Engineering, 2013(11): 2070-2074.

[10] Lee,S. H., Son, D. H., Kim, B. H. and Han, W. T. 2012. Effect of infiltration pressure on the birefringent properties of a side-hole fiber filled with indium. Optics Letters, 2012.37(12): 2322-2324.

[11] Qi F., Xin Y., Dong X. Y. Meng Q., Zhao C.L., Jin S. Z. Sagnac Interferometer Based Temperature Sensor With Alcohol-Filled Side-Hole Fiber. Laser &Optoelectronics Progress, 2013. 50(1): 99-102.

[12] Moon, D.S., Kim, B. H., Lin A., Sun, G., Han, Y. G., Han, W. T. and Chung Y. G. The temperature sensitivity of Sagnac loop interferometer based on polarization maintaining side-hole fiber. Optics Express, 2007. 15(13):7962-7967.

[13] Dong, X. Y., Jin, Y. X., Zhou, W. J., Shum P. Simultaneous Measurement of Strain and Temperature with Side-hole Fiber Based Bragg Grating. Proceedings of the 7th international conference on Information, communications and signal processing, ICICS 2009, Macau, China, December, 2009. pp.611-613.

[14] Yueai, L., Rahman, B. M. A., Grattan, K. T. V. Thermal-stress-induced birefringence in bow-tie optical fibers. Applied. Optics , 1994. 33 (24): 5611-5616.

[15] Wang N., Fu J., Li D., Zhu H., Tian Y., Wang T. Pressure-Induced Birefringence Analysis of Side-Hole Optic Fiber, Journal of Nanoelectronics and Optoelectronics, 2015.10: 795-798.

IOP Conf. Series: Materials Science and Engineering **250** (2017) 012025 doi:10.1088/1757-899X/250/1/012025

Characterization Methods of Subsurface Cracks in Grinding of Optical Elements

Rui Gao, Hongxiang Wang*, Chu Wang, Shunzhi Feng and Benwen Zhu

School of Mechatronics Engineering, Harbin Institute of Technology, Harbin, Heilongjing, 150001, China

*whx@hit.edu.cn

Abstract. In this paper, the subsurface cracks of fused silica introduced in grinding were detected by taper polishing method, and then characterized by the maximum depth, the cluster depth, and the density of cracks. The quantitative relationship between maximum depth and cluster depth of subsurface cracks was determined, and the distribution of subsurface crack density along the depth was studied. The results showed that the abrasive grain size had a great influence on maximum depth and cluster depth of subsurface cracks, while the grinding fluid concentration had little effect on them. The different processing parameters changed maximum depth and cluster depth, but their ratio was almost unchanged. The subsurface crack density decreased exponentially along the depth, and the decrease trend had slowed down significantly at the half of cluster depth.

1. Introduction

The grinding for the optical brittle materials mainly removes the material by brittle fracture, leaving subsurface defects on the final surfaces. The subsurface defects inevitably affect laser induced damage threshold (LIDT) which will shorten the lifetime of optical components [1-2]. In the early subsurface defect detection, due to the lack of advanced detection equipment and reliable detection methods, the subsurface crack depth was mainly estimated by the processing conditions such as the average abrasive grain size. Sabia et al. [3] found the subsurface crack depth was about 5 times to the abrasive size for fixed abrasive grinding, and it was 1~1.8 times for loose abrasive grinding (lapping). Menapace et al. [4] pointed out it was about 3 times in grinding, and Hed and Edwards [5] thought it was 1 time in lapping. Differences in estimated values were caused by differences in processing methods, equipment and parameters. These estimated values of subsurface crack depth were not very accurate, reducing processing efficiency. Tonnellier et al. [6] examined the subsurface cracks during the grinding of fused silica glass, and found that the majority of subsurface cracks clustered together near the machined surface and terminated at cluster depth. With the further increase of depth, there were separate cracks remaining, and the maximum depth of such cracks was much larger than the cluster depth. Researchers of Lawrence Livermore National Laboratory (LLNL) made breakthroughs in the study on characterization methods of subsurface cracks, and broke the limitations of only using maximum depth to characterize. Suratwala et al. [7] utilized magnetorheological finishing (MRF) method to detect subsurface cracks produced by the grinding of fused silica, and obtained the binary images of subsurface crack distribution by image processing technology. The results showed that the subsurface crack density decreased exponentially along the depth. With the continuous progress of science and technology, there have been many advanced detection equipment based on optics, mechanics, acoustics, electromagnetics, quantum mechanics and other fields, promoting the development of subsurface defect detection technology. There are non-destructive detection methods including ultrasonic, eddy current, radiographic, magnetic particle, ultrasonic-infrared phase testing and so on. In addition, there are destructive detection methods including chemical etching method [8-9], taper polishing method, bonded interface technique, and MRF method. Compared with subsurface crack detection technology, the research on characterization of subsurface cracks is still in the initial stage. In order to realize the quantitative characterization of subsurface cracks in optical element, it is necessary to study the density distribution and depth of subsurface cracks. In this paper, the subsurface cracks of fused silica introduced in grinding were characterized by the maximum depth, the cluster depth, and the density of cracks. The distribution of subsurface crack density along the depth was studied.

2. Maximum depth and cluster depth of subsurface cracks

The grinding experiments of fused silica specimens were carried out by grinding and polishing machine (ZYP300), with grinding pressure of 22kPa, grinding time of 30min, spindle speed of 50r/min, and grinding fluid concentration of 8wt.%. The specimen dimension was Φ50mm×6mm and the abrasive grains were W45, W28, and W14, respectively. In order to analyze the relationship between maximum depth and cluster depth, the subsurface cracks were detected by taper polishing method and MRF method, respectively. The detection results are shown in table 1, and the ratio of maximum depth to cluster depth remained between 1.12~1.19.

Table1. Detection results of maximum depth and cluster depth

Abrasive grains	maximum depth (μm)		cluster depth (μm)		ratio	
	taper polishing method	MRF method	taper polishing method	MRF method	taper polishing method	MRF method
W45	26.7	25.2	23.8	21.4	1.12	1.18
W28	15.6	13.7	13.4	12.0	1.16	1.14
W14	11.3	10.9	9.5	9.8	1.19	1.15

The morphologies of subsurface cracks obtained by taper polishing method are shown in figure 1. The subsurface cracks produced by grinding overlapped each other near the machined surface and connected together into dense distribution. The crack density decreased gradually with the increase of depth, and the dense cracks disappeared at a certain depth. This depth was defined as the cluster depth of subsurface cracks, which is the distance between the horizontal line and the machined surface (see figure 1). With the further increase of depth, there were separate cracks remaining under the cluster depth. After reaching a certain depth, there was the non-damaged matrix, and cracks no longer existed. This depth was defined as the maximum depth of subsurface cracks, as shown in the circle (see figure 1). Each stage of grinding for optical elements always introduces a certain depth of subsurface cracks. In the rough grinding stage, the cluster depth is used to characterize the subsurface crack depth. In order to reduce processing time and improve processing efficiency, each operation step should eliminate the dense cracks left from the previous step. In the fine grinding stage, the maximum depth is used to characterize the subsurface crack depth, guaranteeing that the subsurface cracks produced by fine grinding can be completely removed in the subsequent polishing process.

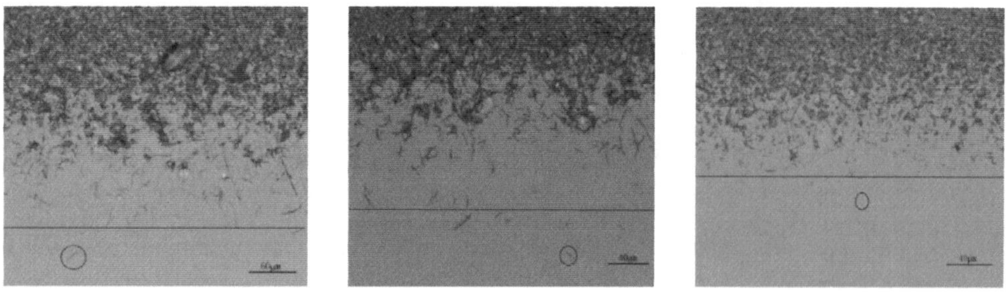

(a) Surface machined by W45 (b) Surface machined by W28 (c) Surface machined by W14

Figure 1. Morphologies of subsurface cracks after grinding with different abrasive grains

3. Subsurface crack density

Grinding is a very complex dynamic random micro-cutting process. With inconsistency of protrusion height, uneven distribution of abrasive grain sizes, and random distribution of subsurface cracks, it is very difficult to calculate the number of cracks at different depths by traditional method. At the near-surface, subsurface cracks overlap each other, so it is more difficult to describe the crack distribution accurately. MATLAB has powerful image processing function, which can realize the evaluation and characterization of subsurface crack distribution at different depths.

Because of the difference of gray scale between cracks and non-damaged matrix in the images, the cracks can be detected from the matrix by setting a suitable threshold *T*. The gray scale of input image is $F(x, y)$ and the gray scale of output image is $g(x, y)$:

$$g(x,y)=\begin{cases}1 & F(x,y)>T\\0 & F(x,y)<T\end{cases} \tag{1}$$

The image processing flow is shown in figure 2. Firstly, the true color image of subsurface cracks acquired by microscope was transformed into a gray scale image. Then, the median filter removed the noise and improved the clarity of crack edge. Subsequently, the appropriate threshold *T* was selected t o turn the gray scale image into a binary image, where cracks were showed black and non-damaged m atrix was showed white. Finally, the pixel number of cracks at a certain depth and all the pixel number of binary image were counted. The ratio of them was defined as the subsurface crack density at this de pth. The original image obtained by optical microscopy, the gray scale image, and the binary image pr ocessed by MATLAB are shown in figure 3.

Figure 2. Image processing flow

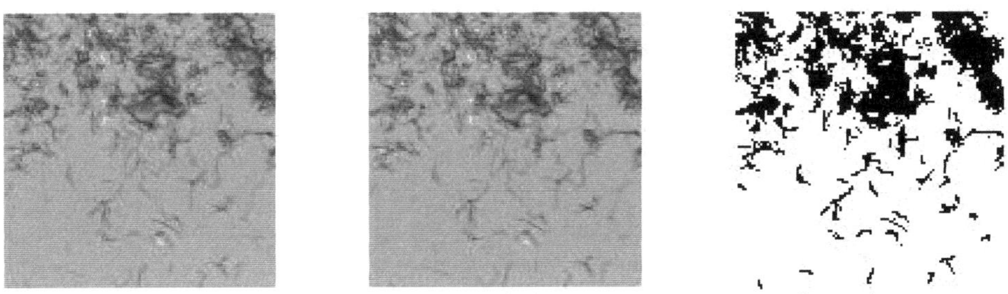

Original image Gray scale image Binary image

Figure 3. Original image, gray scale image, and binary image in image processing

4. Distribution of subsurface crack density along the depth

The distribution of the subsurface crack density along the depth depends on instantaneous distribution of cracks, overlapping of cracks at different grinding moments, and continuous removal of cracks. The subsurface crack distribution at a certain depth introduced at time t is assumed to conform to the normal distribution, and denoted as $f_0(c)$. After time Δt, the material removal increment is Δ, and it is assumed that the same crack distribution as time t is produced on the new surface at this time, which is denoted as $f_0(c+\Delta)$. In the same way, when the material removal increment is $i\Delta$, the distribution is denoted as $f_0(c+i\Delta)$ [7]. The generation and removal of subsurface cracks in grinding process indicate

IOP Conf. Series: Materials Science and Engineering **250** (2017) 012025 doi:10.1088/1757-899X/250/1/012025

that the subsurface crack density decreases gradually with the increase of subsurface depth (see figure 4).

When i is large enough, the generation and removal of subsurface cracks reach a dynamic equilibrium (the specimen surface is adequately grinded), and the function of subsurface crack distribution $f_c(c)$ [7] can be expressed as:

$$f_c(c) = f_0(c) + f_0(c+\Delta) + \ldots + f_0(c+i\Delta) = \sum_i f_0(c+i\Delta) \tag{2}$$

The distribution of subsurface crack density along the depth $F_c(c)$ can be obtained by integrating $f_c(c)$:

$$F_c(c) = \int_c^{\infty} f_c(c) d_c \tag{3}$$

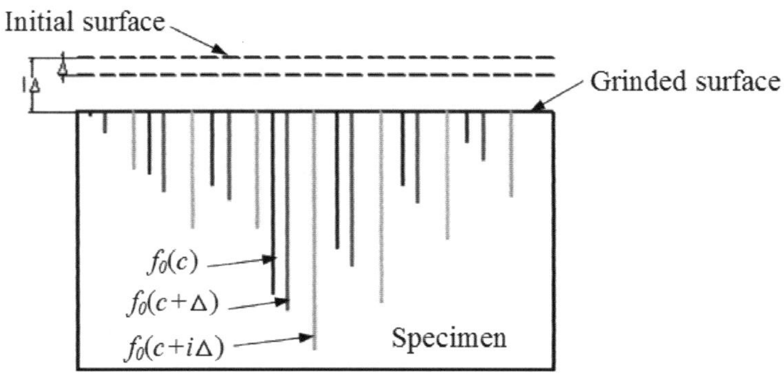

Figure 4. Generation and removal of subsurface cracks in grinding process

In the actual grinding process, abrasive grains are randomly distributed on the grinding plate, and the space between abrasive grains is also random. The interactions between abrasive grains will inevitably affect the subsurface crack depth. In order to determine the relationship between cluster depth and maximum depth, the pre-polished fused silica specimens with dimension of Φ20mm×5mm were used in the experiment. The abrasive grains were 320#, 400#, 600#, and 800#, respectively. The grinding fluid concentrations were 5wt.%, 8wt.%, 12wt.%, and 15wt.%, respectively. Other processing parameters remained constant. Maximum depth and cluster depth of subsurface cracks detected by taper polishing method were calculated by MATLAB, and their relationships with abrasive grains are shown in figure 5.

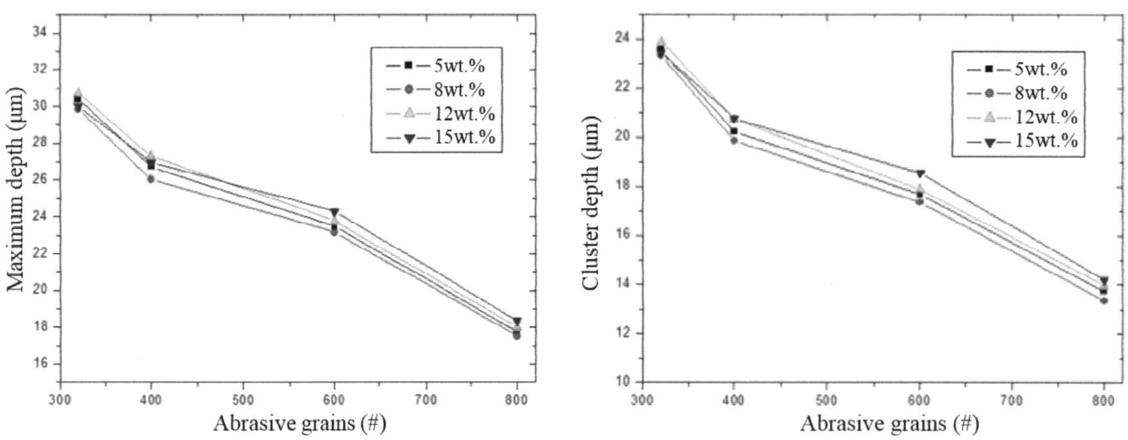

(a) (b)

Figure 5. Relationships between maximum depth and cluster depth with abrasive grains under different grinding fluid concentrations

As shown in figure 5, 1) with the decrease of abrasive grain size, maximum depth and cluster depth of subsurface cracks were reduced. The smaller the abrasive grain size was, the more the effective abrasives involved in the grinding process. In the case, with the same grinding pressure, the effective load of each abrasive grain was smaller, leaving the decrease in maximum depth and cluster depth. 2) When the abrasive grain size was constant, with the increase of grinding fluid concentration, maximum depth and cluster depth changed very little. The reason for this phenomenon was that the effective abrasive grains were alternately updated in the actual grinding process, resulting in the actual grinding fluid concentration tended to be consistent. 3) Although the different processing parameters changed maximum depth and cluster depth, their ratio was almost unchanged. This is because the random distribution of space between abrasive grains existed during the grinding, and the crack propagation depth reached a dynamic equilibrium after a long period of grinding.

According to the experimental data, the variations in subsurface crack density along the depth are shown in figure 6. The subsurface crack density decreased exponentially along the depth, and the decrease trend had slowed down significantly at the half of cluster depth. With the continuous processing of grinding, the irregular pyramid-like abrasive grain tips were gradually passivated or broken. The subsurface crack distribution corresponding to each material removal increment Δ was changed, and inconsistent with the theoretical assumptions, which was the root cause of the differences between experiment and theoretical analysis.

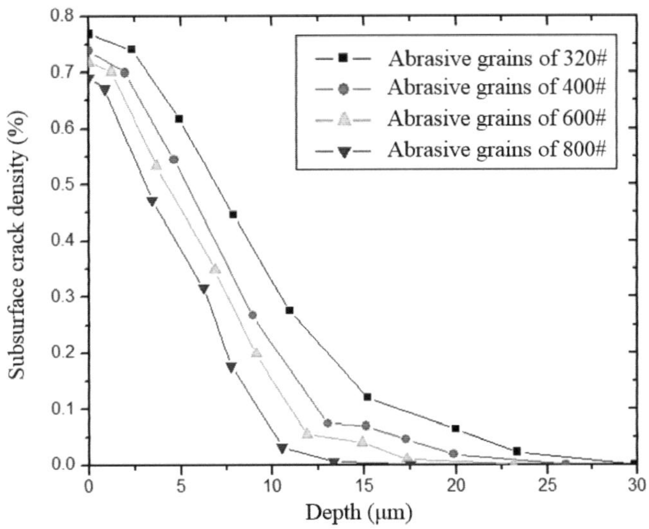

Figure 6. Variations in subsurface crack density along the depth under different abrasive grains

5. Conclusions

The present work studied the characterization methods of subsurface cracks in grinding of optical elements. Following conclusions can be drawn from this investigation:

1) The subsurface crack distribution in grinding of fused silica was detected by taper polishing method. The quantitative relationship between maximum depth and cluster depth was determined, and their ratio remained at 1.12~1.19.

2) The subsurface crack density at a certain depth was calculated through image gray processing, median filter, and image thresholding method. The subsurface cracks were characterized by maximum depth, cluster depth, and crack density.

3) For maximum depth and cluster depth of subsurface cracks, they were reduced with the decrease of abrasive grain size, while the grinding fluid concentration had little effect on them. Although they were changed with the processing parameters, their ratio was almost constant.

4) The subsurface crack density decreased exponentially along the depth, and then the decrease trend had slowed down significantly at the half of cluster depth.

6. Acknowledgments
This research was financially supported by National Natural Science Foundation of China (Grant No.51475106) and Science Challenge Project (Grant No.JCKY2016212A506-0503).

7. References
[1] Jiang Y, et al. Effect of residual stress on laser-induced damage characterization of mitigated damage sites in fused silica 2015 *Journal of Non-Crystalline Solids* **410** 88-95.
[2] Lv D, Wang H, Zhang W and Yin Z Subsurface damage depth and distribution in rotary ultrasonic machining and conventional grinding of glass BK7 2016 *International Journal of Advanced Manufacturing Technology* **86** 2361-2371.
[3] Sabia R, Stevens H J and Varner J R Pitting of a glass-ceramic during polishing with cerium oxide 1999 *Journal of Non-Crystalline Solids* **249** 123-130.
[4] Menapace J A, Suratwala T I and Miller P E MRF applications: measurement of process-dependent subsurface damage in optical materials using the MRF wedge technique 2005 *Proceedings of SPIE - The International Society for Optical Engineering* (2005)599103-599103-11.
[5] Hed P P and Edwards D F Optical glass fabrication technology. 2: Relationship between surface roughness and subsurface damage 1987 *Applied Optics* **26** 4677.
[6] Tonnellier X, Morantz P, Shore P, Baldwin A, Evans R and Walker D D Subsurface damage in precision ground ULE® and Zerodur® surfaces 2007 *Optics Express* **15** 12197-12205.
[7] Suratwala T, Wong L, Miller P, Feit M D, Menapace J, Steele R, Davis P and Walmer D Subsurface mechanical damage distributions during grinding of fused silica 2006 *Journal of Non-Crystalline Solids* **352** 5601-5617.
[8] Neauport J, Ambard C, Cormont P, Darbois N, Destribats J, Luitot C and Rondeau O Subsurface damage measurement of ground fused silica parts by HF etching techniques 2009 *Optics Express* **17** 20448-56.
[9] Zhou Y, Funkenbusch P D, Quesnel D J, Golini D and Lindquist A Effect of Etching and Imaging Mode on the Measurement of Subsurface Damage in Microground Optical Glasses 2010 *Journal of the American Ceramic Society* **77** 3277-3280.

IOP Conf. Series: Materials Science and Engineering **250** (2017) 012026 doi:10.1088/1757-899X/250/1/012026

Human Factor Consideration on Color Consistency in Flight Deck

Ning Yang

Shanghai Aircraft Design & Research Institute, COMAC, Shanghai, 201210, P.R.China

Email:yangning0528@163.com

Abstract: The essential factors for information legibility and the evaluation approach for aviation chromaticity difference and luminance difference measurements are analyzed in this paper. Based on the color difference measurement given by display-related aviation regulation, one algorithm which is applicable for cross-area color consistence is proposed. This algorithm not only provides one evaluation method for different color application area but also effectively reduces human perceptual color difference.

1. Introduction

Color can enhance information understanding, AC 25.1302-1, 5-5.b.(3)(a) states that a common color philosophy across the flight deck is desirable, although deviations may be approved with acceptable justification [1]. It is important that the use of color in all cockpit applications be consistent across all cockpit displays. Color has a high potential for distraction, the overuse of color can increase visual clutter and reduce legibility, the proper uses of color can aid in pilot information recognition and attention-getting and alerting.

Ref.[1] suggests to avoid using many different colors to convey meaning on displays. The large number of colors may actually be counter-productive to organizing information. Color should be applied in a conservative manner, no more than seven colors shall be used for coding applications in small fields, including white, red, green, blue, yellow (amber), cyan and magenta. Considering the color discriminability, AC 25-11A, 31.c.(5)(g) advises 12 color pairs should be avoided [2].

Ref.[2] emphasizes the essentiality of color discriminability which can be affected by a variety of factors, such as the human color vision, lighting conditions, the display performance and viewing angle. Therefore, it appears to be significant for aviation designers to improve the legibility performance and evaluate the perceived color differences with current available measures in flight deck.

In aircraft cockpit, the pilots sense the color from self-luminous display and lighting devices (incandescent or LED). Since these two types color apply different evaluation systems, it is a critical issue how to solve the color consistence among different color standard systems. This paper provides an algorithm to evaluate the cross-area color consistence.

2. Critical Factors for Legibility

2.1 Human Perceptual Ability

Human vision system is the basic limitation to legibility performance. Figure 1 demonstrates the response spectra of three different kinds of cones in the retina of human eyes.

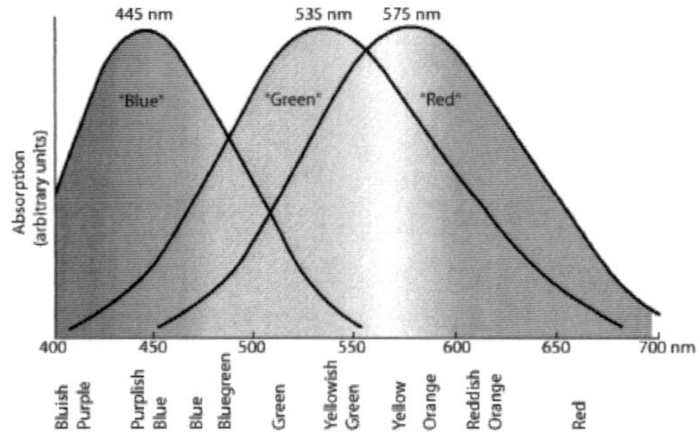

Figure.1 Response spectra of human cones

According to the experimental results, about 64% of the cones are red-sensitive, about 32% green sensitive, and about 2% are blue sensitive. The reason is that blue tends to be focused in front of the retina and has relative scarcity receptors in the fovea compared to red and green.

Another impact of color on legibility is the potential consequences of color illumination. The pilots could perceive the surface colors caused by colored cockpit light or the display self-illuminating color.

2.2 Color Usage Principle

The closer in hue each color will be, the harder will be for pilots to discriminate them. The more colors used the higher errors probability the pilots will make in assigning the intended meaning to them [3]. Table 1 gives an example of absolute discrimination.

Table 1 Absolute discrimination error probability

Number of Colors	Percent Errors
10	2
12	4
15	5
17	28

The poorly organized and crowded display arrangement attributes the clutter which generally increases interpretation processing time, resulting in reducing the legibility. Luminance contrast is also considered one of the most important factors for the legibility. For optimum legibility, character contrast should be between 6:1 and 10:1.

3. Color Difference Measurement

With the certain period usage of display, color would be varied in terms of luminance and chromaticity which are mainly caused by pixel degradation. However, color degradation should be obvious and should not preclude the pilot from interpreting the remaining display information referring to AC 23.1311-1C, 22.6. The means of distinguishing the displayed symbols from each other in all ambient conditions fall into two categories: luminance differences and/or chromaticity differences.

3.1 Chromaticity Difference

Chromaticity is an objective specification of a color regardless of its luminance, that is, as determined by its hue and saturation (or chroma, or intensity, or excitation purity). In CIE system (not take luminance into account, such as 1931 and 1976 CIE systems), the more widely spaced apart two colors are, the more different they appear from one another in terms of hue (wavelength) and saturation (excitation purity).

Currently, no single color difference method is available to give an accurate measure for all tasks involving the use of color. New metrics CIE ΔE*00, as reported in Ref.[4], may be a better predictor of color discrimination performance. Unfortunately, it could not have adequate metrics for symbol legibility and some problems still exist with this method that precludes its use as a completely precise measurement.

Although there is no adequate basis for specifying a minimum acceptable difference at this time, color difference tolerances are still recommended to prevent color confusion when observing several displays on the same flight deck. Ref.[5] believes the most credible color difference system is 1976 CIE ΔE* (CIELUV, considers luminance as well as chromaticity) which perceptually weights both luminance and chromaticity components in determining color difference.

The color difference tolerances are partitioned into the following three separate categories.

Color uniformity gives the color difference between any two points on the screen of the same color when viewed from a fixed point that should not exceed 24.

$$\Delta C^*_{\text{fixed}} = \sqrt{\Delta u^{*2} + \Delta v^{*2}} \tag{1}$$

where, Δu^* (Δv^*) are the difference between measured color CIE 1976 u^* (v^*) values; u^* and v^* are defined as below.

$$\begin{cases} u^* = 13L^*(u' - 0.1978) \\ v^* = 13L^*(v' - 0.4684) \end{cases} \tag{2}$$

where, L^* is defined as below.

$$L^* = \begin{cases} 116\sqrt[3]{\frac{L_{\text{meas}}}{Y_{\text{n}}}} - 16, & L_{\text{meas}} > 1\text{fL} \\ 9.03 \times L_{\text{meas}}, & L_{\text{meas}} < 1\text{fL} \end{cases} \tag{3}$$

where, L_{meas} is measured luminance; Y_{n} is measured luminance of display white reference at screen center taken from DEP. D6500(u′=0.1978,v′=0.4684) is recommended as the reference to insure uniformity from flight deck to flight deck.

DEP color tolerance expresses the color difference between the specified color and measured color that should not exceed 24.

$$\Delta C^*_{\text{dep}} = 13L^*_{\text{meas}}\sqrt{\Delta u'^2 + \Delta v'^2} \tag{4}$$

where, $\Delta u'(\Delta v')$ is difference between specified u' values and u' measured at the display center; L^*_{meas} is defined as below.

$$L^*_{\text{meas}} = \begin{cases} 116\sqrt[3]{\frac{L_{\text{meas}}}{Y_{\text{n}}}} - 16, & L_{\text{meas,DEP}} > 1\text{fL} \\ 9.03 \times L_{\text{meas,DEP}}, & L_{\text{meas,DEP}} < 1\text{fL} \end{cases} \tag{5}$$

where, L_{meas} is measured luminance; $L_{\text{meas,DEP}}$ is measured design eye luminance.

Viewing envelope color tolerance should be no greater than 26.

$$\Delta C^*_{\text{fov}} = 13L^*_{\text{meas}}\sqrt{\Delta u'^2 + \Delta v'^2} \tag{6}$$

where, $\Delta u'$ ($\Delta v'$) is difference between u' (v') measured at any point in each viewing envelope and u' (v') measured at the design eye position; the definition of L^*_{meas} is same as formula (5).

3.2 Luminance Difference

Luminance is considered as the amount of light emitted, transmitted or reflected from the display, lighting source, or other object.

Ref.[3] states that the maximum and minimum luminance of transmissive display (e.g. LCD), based on 0.6 mrad line width, should be at least 257 cd/m2 (from DEP) or 171 cd/m2 (anywhere) and no greater than 0.343 cd/m2, respectively. Meanwhile, it is certainly needed to balance the luminance and line width to achieve the same apparent brightness (attribute of visual sensation). The relationship between luminance and apparent brightness is shown below.

$$Apparent\ Brightness = K_{\mathrm{n}} \times Luminance \tag{7}$$

The relative luminance difference between two areas can be defined in several different ways. Two important formulas, background uniformity and background color, are recommended in Ref.[3].

Background uniformity defines the total color difference at any two locations commanded to their dark state, within the usable display surface. This value should not exceed 12.

$$\Delta E^* = \sqrt{\Delta L^{*2} + \Delta u^{*2} + \Delta v^{*2}} \tag{8}$$

where, ΔL^*, Δu^* and Δv^* are the difference between background color CIE 1976 L^*, u^* and v^* values, respectively.

Background color is the average background CIE 1976 color, measured across the utilized display surface should be less than or equal to 26 anywhere within the display's designed viewing envelope.

$$C^*_{\mathrm{back}} = 13L^*_{\mathrm{back}}\sqrt{\left(u'_{\mathrm{back}} - 0.1978\right)^2 + \left(v'_{\mathrm{back}} - 0.4684\right)^2} \tag{9}$$

where, u'_{back} and v'_{back} are CIE 1976 u' and v' of background, respectively; L^*_{back} is defined as below.

$$L^*_{\mathrm{back}} = \begin{cases} 116\sqrt[3]{\frac{L_{\mathrm{meas}}}{Y_{\mathrm{n}}}} - 16, & L_{\mathrm{back}} > 1\mathrm{fL} \\ 9.03 \times L_{\mathrm{meas}}, & L_{\mathrm{back}} < 1\mathrm{fL} \end{cases} \tag{10}$$

where, L^*_{back} is background luminance.

4. Cross-area Color Consistence Algorithm

Most avionics companies have established internal color standards in order to both reduce errors and to make the product be performed comfortable. Table 2 gives an example of color standards.

Table 2 Color standards with CIELUV coordinates

color	u'	v'	%Lum
black	0.2189	0.4943	0.58
white	0.2238	0.4952	100
red	0.4393	0.5295	30.21
yellow	0.2311	0.5551	88.12
green	0.1407	0.5652	62.43
cyan	0.1386	0.4423	45.89
blue	0.1430	0.2339	8.42
magenta	0.3226	0.4117	37.93

Display panels, even though the same type displays, are still slightly different from each other in terms of chromaticity and luminance because of the manufacture process. Therefore, it is necessary to measure and set up the primary colors table for a certain type display. Primary colors are being used specifically and only for the purposes of verifying compliance with ARP4256 section 4.2.4.1. Table 3 shows one color setting of primary colors.

Table 3 Primary colors table

color	u'	v'	L^*	ΔC^*
black	0.2188	0.4939	5.69	0.030499
white	0.2227	0.4943	100	1.847647
red	0.4390	0.5293	60.95	0.285686
yellow	0.2301	0.5549	94.92	1.258397
green	0.1408	0.5655	82.67	0.339853
cyan	0.1389	0.4404	63.85	1.596633
blue	0.1435	0.2346	36.73	0.410752
magenta	0.3211	0.4133	68.34	1.948457

By using formula (1), all the available display color could be validated base on the primary color as shown in table 3.

Figure 2 plots u′ and v′ coordinates of the full scale targets for red, green and blue on the CIE 1976 which represents the gamut for display, where point a is the white point.

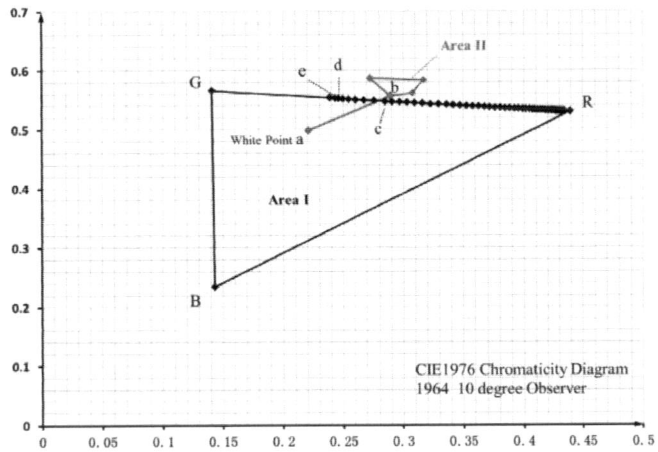

Figure 2 Display gamut in CIE 1976

Assume that the amber area of lighting is defined in Area II, the critical value (point b) obviously falls outside the display gamut. In order to meet the cross-area color consistence in cockpit, it is necessary to determine one closest available approximation of the color that lies inside the display gamut.

By connecting point a and point b, point c is observed be the closest one from red to green along all the available grayscale settings. The CIE LUV coordinate of point c is tested as:

$$(u', v', L^*) = (0.2843, 0.5462, 78.36)$$

Based on formula (4), the color difference between point c and primary yellow color from DEP is:

$$\Delta C^*_{\mathrm{dep(c)}} = 13L^*_{\mathrm{meas}}\sqrt{\Delta u'^2 + \Delta v'^2} = 55.92 \tag{11}$$

Point c is not the desired point as $\Delta C^*_{\mathrm{dep}}$ is greater than 24. The color chosen was the one near the green side of the intersection since the value is the most "yellow".

According to the analysis above, the color difference of point d and point e are:

$$\Delta C^*_{\mathrm{dep(d)}} = 13L^*_{\mathrm{meas}}\sqrt{\Delta u'^2 + \Delta v'^2} = 27.35 \tag{12}$$

$$\Delta C^*_{\mathrm{dep(e)}} = 13L^*_{\mathrm{meas}}\sqrt{\Delta u'^2 + \Delta v'^2} = 22.98 \tag{13}$$

Therefore, point e is the optimum color to guarantee amber consistence for different application area in cockpit.

5. Summary

This paper analyzes the essentiality of color application based on the requirement of related aviation regulation. The critical factors for flight information legibility in cockpit are also summarized. Combined with the color difference measurement method, an effective algorithm is proposed to solve the cross-area color consistence issue in cockpit and reduces human perceptual color difference.

6.Reference

[1] FAA AC 25.1302-1. Installed systems and equipment for use by the flightcrew [S]. 2013.
[2] FAA AC 25-11A. Transport category airplane electronic display systems [S]. 1987.
[3] ARP 4256A. Design objectives for liquid crystal displays for Part 25 (Transport) aircraft [S]. 2001.
[4] Commission Internationale de L'Eclairage (CIE), 2001, Improvement to Industrial Colour-Difference Evaluation [C], Technical Report CIE 142-2001, Vienna, Austria.
[5] Supplement No. 2 to CIE Publication No. 15, Recommendations on uniform color spaces color difference equations - Psychometric Color Terms.

IOP Conf. Series: Materials Science and Engineering **250** (2017) 012027 doi:10.1088/1757-899X/250/1/012027

Structural andOptoelectronicProperties of a-SiO_x:Ag Films Used for Ag/SiO_x/p-SiMemristor

YanhaoGou[1], WeiHou[1],DongyangLi[2], Wenlin Meng[1], Yicheng Chen[1]and Wei Li*[2]

[1]School of Optoelectronic Information, University of Electronic Science and Technology of China, Chengdu, 610054 China

[2]State Key Laboratory of Electronic Thin Films and Integrated Devices, University of Electronic Science and Technology of China, Chengdu, 610054 China

*Email: wli@uestc.edu.cn

Abstract. Silicon basedmemristorhas attracted much attention due to its potential applications not onlyinelectrical storagebut also in electro-optical modulator.No matterwhat applicationsare mentioned, the device performance is closely related to the characters of the materials used.In this paper, a-SiO_x:Agthin films aredepositedon quarzsubtratesby co-sputtering, and theirmicrostructural andoptoelectronic properties are investigated by using XRD, Ramanand ellipsometry, respectively.Finally, amemristor with Ag/a-SiO_x/p-Si structure is fabricated successfully, and the"ON/OFF"performanceofresistivities aremeasuredby voltammetry. The results show that a-SiO_x is a potential alternate of insulator layer in the system of "Ag/insulator or semiconductor/p-Si"memristors which could be usedin optical modulators in the near future.

1.Introduction

Memristor has been paid much attention in microelectronic field as a promising candidate for the next-generation nonvolatile memory with excellent miniaturization potential, high on/off ratio, fast programming speed and high-endurance[1–3]. Generally, a memristorcomprisesmetal/insulator/metalormetal/semiconductor/metal sandwich structure.The operating principle is resistance switching between high and low states under different excitation voltages.The switching of resistanceis realized by metal filament formation (or dissolution) in the insulator (or in the semiconductor)at positive (or negative) applied voltage. During this resistance switching procedure, optical properties of the insulator (or semiconductor) could be adjusted as well,makingthe optical application of memristora potentiallypossible.Given that compatibility of the fabrication with complementary metal oxide-semiconductor (CMOS) technology, Si-base materials are the best choice for the insulator (or semiconductor). Recently, Emboraset al[4-5] demonstrated experimentally that a newoptical switch with atomic sizes and low power consumptioncan be realized by incorporatingmemristor with Ag/a-Si/p-Si (p-typesilicon) structure into an optical waveguide. In the new optical switches,the p-Sisubtrateacts as Si waveguideand the a-Si dielectric layer acts as cladding layerboth inan optical waveguide, respectively. Thus the propagating light in the waveguide would be modulateddue to the optical property changes of the dielectric layerduring switching process.Although the optical switching based on memristorhas already been realized [6], the principle behind the device is still unclear up to now.

Concerned about a new optical switch based on memristor, SiO_x is a better alternate than a-Si.A memristorbased on SiO_xhavesome uniquecharacteristics.More importantly, SiO_x is more suitable than a-Si used as cladding layer of an optical waveguide due to its lower refractive index, and it is well known that silicon oxide (SiO_2) has been widely used asthe cladding layer of optical waveguide.Accordingly, the performance of the optical switchbased on SiO_xmight be expected to have a surprising improvement.

In addition,device performance is closely related tothe characters of the materials used.It is noted that several literatures have investigated the surface plasma resonance and optical absorption of SiO_x:Ag thin films[7-8].So, for Ag/SiO_x/p-Si memristor, the SiO_x layer consistingof silver filament could be seen asa kind ofSiO_x thin film embedded with Ag nanoparticles, and the study of silicon oxide alloyed by silver(SiO_x:Ag) thin film will help us to well understand the optoelectronic properties of SiO_xmemristor.

In this paper, a-SiO_x:Ag thin films are prepared by co-sputtering method, and the structural, optoelectronic properties of the thin films are characterized by using XRD, Raman, voltammetry and ellipsometry methods, respectively. Finally, amemristor with Ag/a-SiO_x/p-Si structure is fabricated and the resistivity ON/OFFbehaviouris studied based on I-V measurement.

2. Experimental Procedure

Thea-SiO_x:Agthin films were deposited on the quartz substrates by RF magnetron co-sputtering. The chamber was pumped to a base pressure of 7×10^{-4}Pa and the substrates were heated to 250℃ before deposition. The RF power wasset to 200W. Argon(Ar)flow served as working gas was kept at 200sccmto keep the chamber pressure in 0.5 Pa. At the same time, oxygen (O_2) flow was kept at 3 sccm as the reaction gasduring sputtering. To introduce Ag atoms, several Agchips(99.99%)were fixed on theSitarget(99.999%) andH_2 flowwasnotintroducedhere.The nominal Ag concentration, varying from 0.8% to 6.4%, was estimated by the target coverage proportion ofAg chips and the sputtering yields.PanalyticalX'pert HighScore XRD instrumentwas applied to measure the X-ray diffraction spectra, in which a conventionalCuKα radiation (λ=1.5418Å) was used, and the angle ofincidence was set at1°. The amorphousnetworkof a-SiO_x:Ag thin filmswere analysedbya RENISHAWinViaRamanMicroscopeatawavelength of 514nm.Thepowerwassetbelow 5mWtoavoidlaser-inducedcrystallization.Optical constantsof the a-SiO_x:Ag thin films were extractedby SENTECH SE850 ellipsometer.The film resistivitieswere measured by a Keithley6517A semiconductor characterization system (SCS). Finally,a memristormade of Ag/SiO_x/p-Si wasfabricated andthe I-V performance was carried out by a Keithley4200ASCS.

3. Results and Discussion

Figure1 shows the XRD spectra of a-SiO_x:Ag thin films with different Ag concentrations. The four peaks belong to Ag nanocrystals including 38.4° assigned to Ag (111), 44° assigned to Ag (200), 65° assigned to Ag (220) and 77° assigned to Ag (311)with Ag concentration at about 6.4%, respectively. For samples with lower Ag concentrations (3.2% and 1.6%), only single peaks of Ag (111) are observed, but when Ag concentration is lower than 1.6%, this typical peak can not be found in XRD spectra. This could be explained by the detection limit of XRD methodowing to the relatively low fraction of Ag nanocrystals. Mean sizes of Ag nanocrystals could be calculated by Scherrer formula and the relationship between mean grain size and Ag concentration is shown in the insert of Figure.1. It can be seen thatwhen the Ag concentration is increased from 1.6 % to 6.4 %, the Ag nanocrystal size goes up from 1.48 nm to 2.43 nm. These sizesareclose to those of Ag nanopartcles formed in the Ag/SiO_2/Ptmemristor[9].

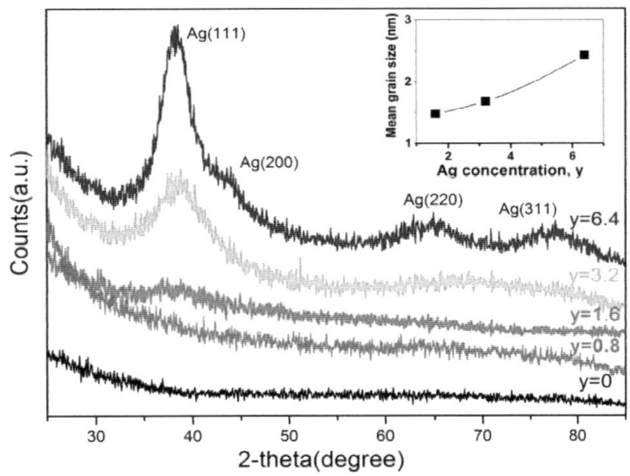

Figure1.XRD spectra of SiO_x:Agfilms and mean size of nanocrystals with Ag concentration.

Figure2(a~d) shows Raman spectra andthe variationsof Raman parameter of the testing thin films with Ag concentration.It can been seen apparently that these Raman spectra aretypical forms of amorphous semiconductor materials.As shown in figure2(b) and (c),when Ag concentration y increases from 0 to 6.4, the transverse optical peak(ω_{TO}) shifts from 492.5 cm^{-1} to 450.7 cm^{-1} and the peak widthΓ_{TO}increases from 52.4 cm^{-1} to 92.6 cm^{-1}, indicatingthat a-SiO_x:Ag thin films present a large short-range disorder with the increase of Ag concentration[10-11].From Figure2(d), one can see thatthe I_{TA}/I_{TO}ratios increase from 0.63 to 1.66with the increase of Ag concentration, implying a more disorder amorphous network in medium-range.

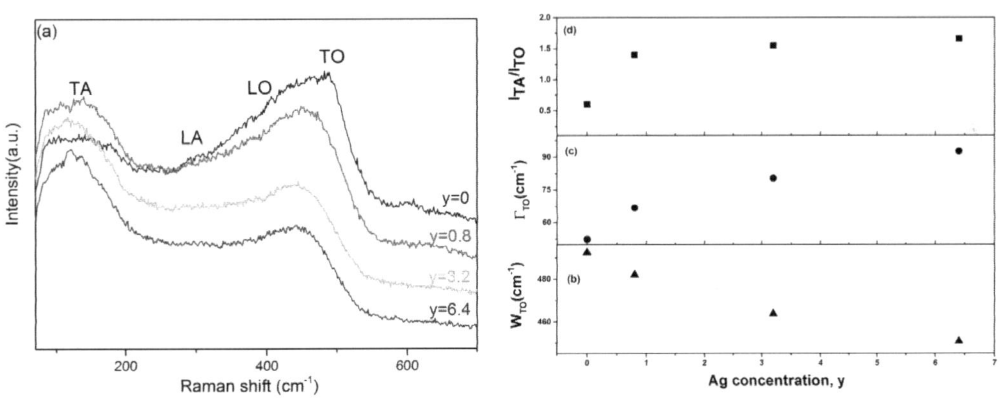

Figure2. Raman spectra (a) and the parameter variation (b), (c) and (d) of a-SiO_x:Ag thin films.

Figure3(a) shows the resistivityvariationofa-SiO_x:Ag thin films withdifferent Ag concentrations.It can be seen that with the increase of Ag from 0 to 6.4%, the resistivity decreases from ~10^6 (Ω.cm)to 10^{-1} (Ω.cm), which spansalmost 7 orders of magnitude.The metallic conducting tunneling formed by Ag nanocrystals could be responsible for this large gap of resistivity. And this gapis large enough to cover therange between On and Off states of an Ag/a-SiO_x/p-Si memristor. In order to gain more insights into the resistance switching behavior of an Ag/a-SiO_x/p-Si memristor, we have measured the I-V characteristicsof this memristoras shown in figure 3(b). The bias voltage is applied to the top metal (Ag), while the back contact (p-Si) is grounded, and the area of Ag electrode is about 1mm^2. A high on/off ratio is observedfrom the I-V curves.Whenthe voltage is increased above a threshold voltage, the memristor isturned on and the conductance is increased very rapidly. It can be seen that the current increases from ~10^{-8} A to 10^{-3} A, which spans almost 5 orders when the bias

IOP Conf. Series: Materials Science and Engineering **250** (2017) 012027 doi:10.1088/1757-899X/250/1/012027

voltageexceeds the threshold voltage. Thisswitching behavioris attributed to the formation of silverfilament as in Ag/a-Si/p-Si memristor[12-13]. It is considered that when the memristoristurnedONata positiveappliedvoltage,Agcationsmigrateintoa-SiO_xlayer, and the layer could be taken as a kind ofa-SiO_x thin film embedded with Ag nanoparticles. Therefore, the large variationin resistivity of a-SiO_x:Ag thin films might be related to the resultthat the experimental Ag/a-SiO_x/p-Si memristorexhibits over 5 orders of On/Off ratio. The switching threshold voltage and the compliance current observed in the present study are3.2V and 1mA, respectively, whicharealmost equal to the values of Ag/a-Si/p-Si memristorpublished elsewhere [14].Another point needs to pay attention is the repeatability. It can be seen in Figure3(b), moreover, that in the first circle, the On/Off ratio reaches 5 orders, while after 10th circle, the On/Off ratio decreases to less than 3 orders. Thisunsatisfactory repeatability can be attributed to the limits ofmanufacturing conditions inthe fabricationof Ag/a-SiO_x/p-Si memristor.

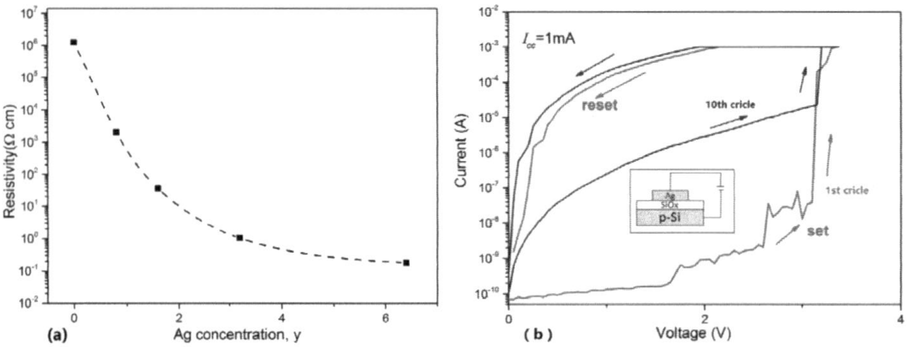

Figure3. Resistivity of SiO_x:Agfilm with Agconcentration(a) and I-V curves of Ag/SiO_x/p-Si memristor(b).

Figure4shows the dispersionrelations about the refractionindex and the extinction coefficientofa-SiO_x:Ag thin films with different Ag concentration. As shown in figure4(a), one can see that changes of sliver concentration have an obviously effect on the variation of refraction index.The refraction index exhibits anomalous dispersion characteristics as the wavelength increases from 300 nm to 580 nm, but exhibitsnormal dispersion characteristics from 580 nm to 800 nm.Thisanomalous dispersioncharacteristic is due to the fact that resonant oscillation of the conductionelectrons would be replaced by the single electron transitions of sliver nanocrystals[15-16]. In the normal dispersion range from 580 nm to 800 nm, the refraction index increases with silver concentration.As shown infigure4(b), theextinction coefficientincreaseswith the increase of Ag concentration.Givena pure a-SiO_xfilm, the refraction index is much lower than that of c-Si,the transmission light can be confined inside the Si waveguide, ifana-SiO_xfilm is used as the claddinglayer of Si waveguide.More importantly, the refraction indexof a-SiO_x:Ag thin film developsas the increaseofAg concentration, which meansthat the optical parametersof a-SiO_x thin film could be heavily adjustedby introducing Agnanoparticles into thea-SiO_xthin films. These interesting results imply that$SiO_x$is a potential alternate of a-Sias in Ag/a-Si/p-Si memristorto realize an applicable optical switch.

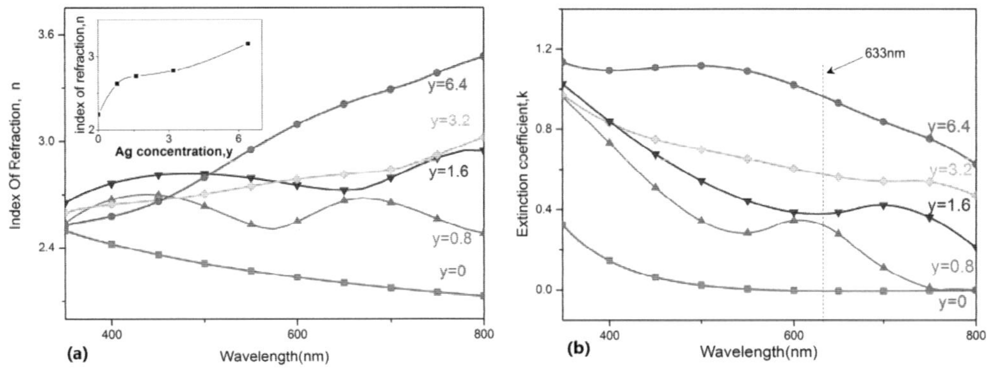

Figure4. Refraction index (a) and extinction coefficient (b) of a-SiO_x:Ag thin films with Ag concentration. The inset of (a) shows the n valuesat λ=633nm.

4. Conclusions

In summary, a-SiO_x:Ag thin films are deposited by co-sputtering method, and the structural, optoelectronic properties of the thin filmsare characterized byusing XRD, Ramanand ellipsometry, respectively.Amemristor with Ag/a-SiO_x/p-Si structure is fabricated and the I-V curves are measured.

Ag nanocrystalsaresuccessfully embedded ina-$SiO_x$and the sizes of them are close to those of nanopartcles formed in the Ag/a-Si/Ptmemristor.With the increaseofAg concentration,the amorphous network of thea-SiO_x:Agfilms exhibitsan increase in short-range disorder and a decrease in medium-range order. The resistivity of thea-SiO_x:Agfilmsdecreases from ~10^6 (Ω.cm) to 10^{-1} (Ω.cm) with the increaseof Ag concentration, which is high enough to cover the gap between ONandOFF statesin the Ag/a-SiO_x/p-Simemristor.Our resent research on dispersion relation of a-SiO_x:Ag thin filmsdemonstrates thata-SiO_xfilm is a potential alternate of insulator layer in the system of "Ag/insulator or semiconductor/p-Si" memristors which could be used in optical modulators in the near future.

5. References

[1] Schindler C, Weides M, Kozicki M N and Waser R 2008*J. Applied Physics Letters***92** 122910.

[2]Huang R, Zhang L, Gao D, Pan Y, Qin S, Tang P, Cai Y and Wang Y2011*J.Applied Physics A* **102**927-931.

[3] Dong Y, Yu G, McAlpine M C,Lu W and LieberC M 2008 *J.Nano Letters***8** 386-391.

[4]Emboras A, Goykhman I, Desiatov B, MazurskiN, Stern L, Shappir, J and LevyU 2013*J.Nano Lett***13** 6151-5.

[5]Emboras A, Niegemann J, Ma P, Haffner C, Pedersen A, Luisier M and Hafner C 2016 *J.Nano Lett***16** 709-14.

[6] Li D, Guo A and Song Q 2016*J.Proceedings of the Spie***19**1001909.

[7] Güner S, Budak S, GibsonB and Ila D 2014*J.Applied Surface Science***310** 180-183.

[8]Ye YH, Jiang YW, Tsai MW, Chang YT, Chen CY, Tzuang DC, WuYT and Lee SC 2008*J.Applied Physics Letters* **93** 033113.

[9] YangY, Gao P, Gaba S, Chang T, Pan X and Lu W 2012 *J.Nat Commun***3** 732.

[10] Wei W, Xu G, Wang J and Wang T 2007 *J.Vacuum***81** 656-662.

[11] Guo A, Li W, Jiang X, Wang C and Jiang Y 2015 *J.Journal of Raman Spectroscopy***46** 619-623.

[12] Jo S H and Lu W 2008*J.Nano letters* **8** 392-397.

[13] JoS H, Kim K H and Lu W 2009*J.Nano Letters* **9** 870-874.

[14] Liu Y, Gao P, Jiang X, Bi K, Xu H and Peng W 2014*J.Applied Physics Letters***104** 043502.

[15]Roy R K, Mandal S K, Bhattacharyya D and Pal A K 2003 *J.The European Physical Journal B - Condensed Matter***34** 25-31.

[16] Guo A, Li D, Li W, Gu D, Jiang X and Jiang Y 2016 *J.Materials Letters***185** 5-8.

IOP Conf. Series: Materials Science and Engineering **250** (2017) 012028 doi:10.1088/1757-899X/250/1/012028

The Device Design of a Rural Domestic Sewage Treatment Performance

Zuhan Liu[1*], Lili Wang[2] and Chengzhi Deng[1]

[1]Jiangxi Province Key Laboratory for Water Information Cooperative Sensing and Intelligent Processing, Nanchang Institute of Technology, Nanchang Jiangxi 330099, China
[2]School of Science, Nanchang Institute of Technology,Nanchang Jiangxi 330099, China
E-mail: lzh512@nit.edu.cn

Abstract.A kind of device for removing pollutant from rural domestic sewage isstudied in this paper, which belongs to the field of sewage treatment technology. The device include anaerobic pool, aerobiotic pool and aerating apparatus. Specifically, the aerobiotic pool is a sealed rectangular body with a rectangular groove on its top cover. The fixed wall is established on one side of the rectangular groove, and the wall is located in the middle of the top cover. The anaerobic pool is opposite to the fixed wall. And there is a aerating apparatusbetween the anaerobic pool and fixed wall, and the apparatus is situated right above the rectangular groove. The design is simple in structure and low in manufacturing cost. The biochemical ratio to the sewage could be improved through the anaerobic pool, and then the sewage is sufficiently aerated by means of natural ventilation and height of water. Theaerated sewage enter into aerobiotic pool, and then the organic matter is degraded into Carbon Dioxide and water under the action of aerobic bacteria, of which function is that the ammonia nitrogen is oxidized into nitrate and nitrite. The water that is far from the aerating apparatus inside of the aerobiotic pool enters into oxygen-deficient environment, and the nitrate is converted into nitrogen. After the effluent is checked qualified,the water is discharged into the drainage ditch or utilized for irrigation.

1. Introduction

The environmental pollution caused by rural domestic sewage is not only the potential safety hazard in rural water source, but itintensifies the crisis of freshwater resource, thus, it severelyimpacts the development of the rural area and the improving of the material living standard [1]. Compared with the urban water consumption,the river water, well water and tap water are generallyintegrated usage in rural water use with characteristic ofa small amount of sewage rate, the inhomogeneity of sewage discharge, the obliviouswater volume change, the relatively hightransient variation and the small daily variation coefficient[2].But beyondthat,there is little difference on the quality of domestic sewage in most rural areas, the favorablebiodegradabilityand containtiny amountsofnitrogen, phosphorus, pathogenic bacteria and so on. With the improvement of the standard of living, heavy metal, toxic and harmful substance may be contained in some domestic sewage, that the consequences arethe markedly increasedoil component,more and more organic pollutants [3-5], so that the risk of organic matter, nitrogen and phosphorus increased as the above pollutants rose.

The biological treatment is employed to most rural domestic sewage in China [6]. These methods could be divided into anaerobic and aerobic biological treatment in terms of the treatment process. Andfromthetreatment mode, they are divided into centralized and decentralized treatment. It's worth mentioning that centralizedtreatmentmostly adopt aerobic biological method, of which its advantage of reliableoperation and good treatment result, but a series of problems still exist, such as high costs, enormous power consumption, high operation expenseanddifficulty in surplus sludge treatment etc. [7]. As for decentralized treatment, anaerobic biological treatment is widely adopted. The technique has the characteristic of small investment and running expense, easy manipulation, however there are many shortagesinitsapplication for a largepartof treatment facilities, such as instability treatment effect and instability for the quality of processed water [8].

From the above analysis we could see, the research and development on better treatment result, lower construction and operating cost, easier management and higher efficiency energy saving technology are the key to solving the problems of rural domestic sewage treatment [9-10]. For this, the paper proposed a new device for removing pollutant from rural domestic sewage and the removal approach. The process was characterized by steady operation, low operating cost, simple structure and low manufacturing cost.

2. The Device Design

There are many domestic sewage sources in our rural regions. Besides human feces and waste water from kitchens, the sewage could also be generated from household cleaning, garbage stacking and percolating.With the development of the economy and the improvement of the peasants'living standard and changes in life style, the domestic sewagealso will increase. Aiming at the rural sewage treatment, the device is generally constructed at low-lying and remote places. The problem presented by this device is lack of power equipment and there is no professional technician to maintain it. Moreover, it also requires that there is the factor of lower construction cost and convenience of installation and service, so it is very hard for common treatment device to meet the requirement.

The device for removing pollutant from rural domestic sewage provided by this design includes anaerobic pool, aerobiotic pool and aerating apparatus. The mentioned aerobiotic pool is a sealed rectangular body, a rectangular groove has been set on its top cover and at the edge of its one side there is fixed wall, which is located in the middle of the top cover. The mentioned anaerobic pool locates on one side of the aerobiotic pool and is opposite to the fixed wall. Between the mentioned anaerobic pool and the fixed wall there is an aerating apparatus, which situates right above the rectangular groove. Several aeration units are contained in the mentioned aerating apparatus, and each of them incorporates crossing rectangular groove, as well as the first wallboard and second wallboard that are parallel to each other. The aeration holes are set on both the first and second wallboards mentioned above. One end of the first wallboard and one end of the second wallboard mentioned above are vertically connected to the fixed wall, and the other ends to both are some distance away from the mentioned anaerobic pool. The connecting plate is equipped on top of the first and second wallboards mentioned above. Because such plate is inferior to the top of the first and second wallboards, the drainage groove has been formed. The connecting water channel is set on one side of the mentioned aeration unit and it is connected through the mentioned drainage groove. Meanwhile, both of the sidewalls of the connecting channel are higher than those to the drainage groove. The bottom of the mentioned anaerobic pool is no lower than the relative height to the top of the mentioned aerobiotic pool. The overflow opening has been established transversely on the mentioned anaerobic pool and one sidewall that is opposite to the fixed wall mentioned above. Several aeration units are connected with the overflow opening via the connecting water channel.

The device design is shown in figure 1-4.It includes anaerobic pool 1, aerobiotic pool 2 and aerating apparatus 3 (figure 1). The aerobiotic pool 2 is a sealed rectangular body with a rectangular groove 201 on its top cover (figure 2). And one side of the rectangular grooveestablishthe fixed wall 5 located in the middle of the top cover of the aerobiotic pool 2. The anaerobic pool 1 islocated the side of the pool 2 and its antithesis is the fixed wall 5. There is an aerating apparatus 3 between the anaerobic pool 1 and the fixed wall 5, the rectangular groove 201 of which is at the top of the aerating apparatus 3 (figure 2 and 3). The aerating apparatus 3 includes several aeration units and eachunit cover the crossing rectangular groove 201, as well as the first wallboard 301 and second wallboard 302are mutual parallel and one side of both wallboard. The aeration holes are set on both the first and second wallboards 301 and 302. One end of the first wallboard are vertically connected to the fixed wall 5, and there is a certain spacing betweenthe other side and anaerobic pool 1. The connecting plate is equipped on top of the first and second wallboards 301 and 302. Such plate is inferior to the top of 301 and 302 toform a drainage groove 303. The connecting water channel 305 is set on one side of the aeration unit. The connecting channel 305 is connected to the drainage groove 303, and both of sidewalls of the connecting channel 305 is higher than those of the drainage groove 303 (figure 4). The bottom of the anaerobic pool 1 is no lower than the relative height to the top of the aerobiotic pool 2. The overflow opening 101 has been established transversely on the anaerobic pool 1 and one sidewall that is opposite to the fixed wall 5. Several aeration units are connected with the overflow opening 101 via the connecting water channel 305.

The above-mentioned device mainly consists of anaerobic pool 1, pool 2 and aerating apparatus 3. With the simple structure made of cement, it is firm and durable. The collected rural domestic sewage passes through the anaerobic pool via the inlet pipeline. The pollutant in the anaerobic pool that is difficult to be degraded could be hydrolyzed and acidified into the substance that is easily degradable under the function of anaerobic bacteria. Thus the biochemical ratio to the sewage could be improved. Then under the circumstance of natural ventilation, the sewage after hydrolytic acidification flows into connecting channel and drainage groove successively via the overflow opening 101. Due to both of sidewalls of the connecting channel are higher than those of the drainage groove, the overflow weir 304 is formed. After the sewage that is flown into the drainage groove reaches to the set height, it spills through both of the sidewalls of the drainage groove via the outer lateral wall of the first and second wallboards, enters into the aerobiotic pool via the rectangular groove and aerates by means of natural ventilation and height of water. The sewage after being sufficiently aerated enters into aerobiotic pool, and then the organic matter is degraded into carbon dioxide and water under the function of aerobic bacteria and the ammonia nitrogen is oxidized into nitrate and nitrite. The water that is far from the mentioned aerating apparatus inside of the aerobiotic pool enters into oxygen-deficient environment, and then the nitrate is converted into nitrogen. After the effluent is checked qualified it is discharged into the drainage ditch or utilized for irrigation. Due to the sewage stays for a long duration inside of the device, the sludge completes self-decomposition and there is hardly any surplus sludge.

It should be explained that there is certain space between the anaerobic pool 1 and aeration device 3, and thus the ventilation gallery has been formed. Meanwhile, there is space between every two aeration units so that the ventilation gallery has also been formed. In addition, the above-mentioned first and second wallboards are made from stainless steel or glass reinforced plastic. The height of the first and second wallboards is 3~5m and the thickness is 2~10mm. The aeration holes with the diameter of 1.5~3cm have been set on the first wallboard 301 and the second wallboard 302. The aperture ratio to the first and second wallboards is 30~42%. Consequently the ventilation condition to the aeration device could be improved greatly, which is beneficial for the sewage after hydrolytic acidification to be sufficiently aerated.

For the sake of enhancing the effect of natural ventilation, as a specific and appropriate way the ventilation gallery with the width of 1.5~3m could be formed between the ends of the first and second sideboards 301 and 302 as well as the anaerobic pool 1. As the optimized plan, in order to further enhance the effect of natural ventilation to the aeration device 3, several aeration holes are set on the fixed wall 5.

It should be further explained that the inlet pipeline is installed on the anaerobic pool 1 and the opposite sidewall of the overflow opening 101. The outlet end of such pipeline is connected to the bottom of the sidewall. The inlet end is higher than the height setting of the lower edge to the overflow opening 101 on the anaerobic pool 1. The incoming water flows into the anaerobic pool 1 via the bottom waiting to be processed by virtue of the height difference.

Furthermore, because some floating impurities will be mingled in the initial sewage, as the optimized approach the filter screen is set on the inlet end of the pipeline for filtering the floating impurities. As the preferable plan, the operation hole 4 has been set on the top cover of the aerobiotic pool 2, and the hole cover has also been set on the operation hole 4 for the convenience of entering for maintenance. After the device is completed construction and put into use, the treatment effect is very

IOP Conf. Series: Materials Science and Engineering **250** (2017) 012028 doi:10.1088/1757-899X/250/1/012028

good. Moreover, unattended operation, non-power consumption, zero-operation expense and micro-common sense could be realized, and it is generalized for using in rural areas.

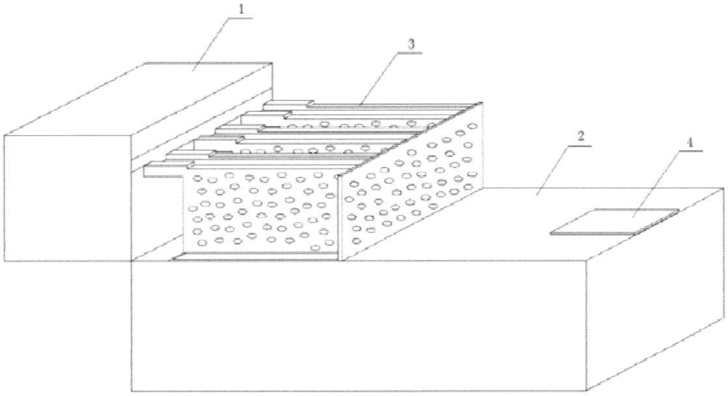

Figure 1. The structural diagram of the removal device

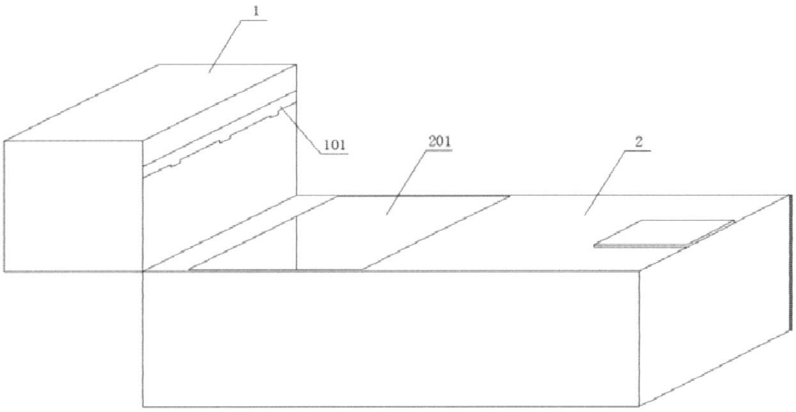

Figure 2. The structural diagram of the removal device's main body

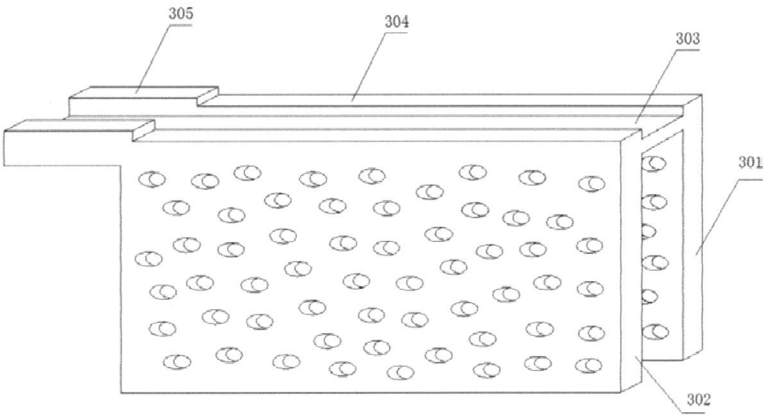

Figure 3. The structural diagram of the aeration apparatus to the removal device

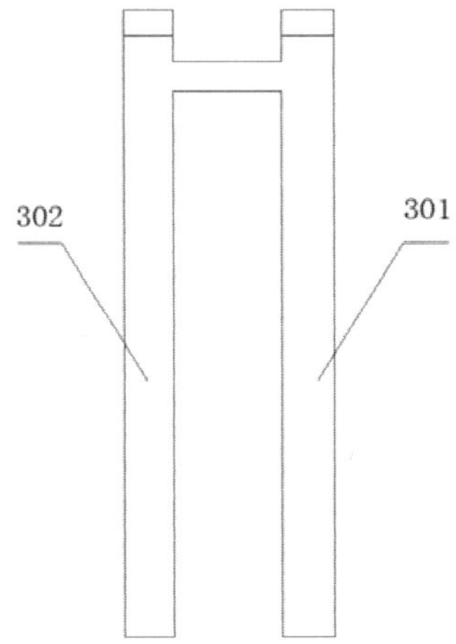

Figure 4. The lateral view of the aeration apparatus to the removal device

3. Conclusion

A kind of device for removing pollutant from rural domestic sewage had been designed in this paper.Itscharacteristic lies in that it included anaerobic pool 1, aerobiotic pool 2and aerating apparatus 3. The mentioned aerobiotic pool 2 was a sealed rectangular body, a rectangular groove 2had been set on its top cover, the fixed wall 5 had been established on one side of the rectangular groove 2, and meanwhile the wall was located in the middle of the top cover. The mentioned anaerobic pool 1 located on one side of the mentioned aerobiotic pool 2, and was opposite to the mentioned fixed wall 5. The aerating apparatus 3 had been equipped between the mentioned anaerobic pool 1 and the fixed wall 5, and meanwhile it situated right above the rectangular groove 201.

The mentioned aerating apparatus 3 included several aeration units. Each aeration unit incorporated crossing rectangular groove 201, as well as the first wallboard 301 and second wallboard 302 that was parallel to each other. The aeration holes were set on both the first and second wallboard 301 and 302 mentioned above. One end of the first wallboard 301 and one end of the second wallboard 302 mentioned above were vertically connected to the fixed wall 5, and their other ends had certain gap to the mentioned anaerobic pool 1. The connecting plate was equipped on top of the first and second wallboard 301 and 302 mentioned above. Such plate was inferior to the top of the wallboard 301 and 302 and drainage groove 303 had been formed. The connecting water channel 305 was set on one side of the mentioned aeration unit. Such channel and the mentioned drainage groove 303 were linked together, and meanwhile both of the sidewalls of the connecting channel 305 were higher than those to the drainage groove 303.

The bottom of the mentioned anaerobic pool 1 was no lower than the relative height to the top of the mentioned aerobiotic pool 2. The overflow opening had been established transversely on the mentioned anaerobic pool 1 and one sidewall that was opposite to the fixed wall 5 mentioned above. Several aeration units were connected with the overflow opening 101 via the connecting water channel 305.

4.Acknowledgments

This work was supported by the Key Project of Jiangxi Provincial Department of Science and Technology (20161BBF60061),Natural Science Foundation of Jiangxi Province(20161BAB202040), the Open Research Fund of Jiangxi Province Key Laboratory of Water Information Cooperative Sensing and Intelligent Processing (2016WICSIP012 and 2016WICSIP014) and the Science and Technology Project of Jiangxi Provincial Department of Education (GJJ151109 and GJJ161097).

IOP Conf. Series: Materials Science and Engineering **250** (2017) 012028 doi:10.1088/1757-899X/250/1/012028

5. References

[1] Shen FJ, Zhang KQ, Li JX, Huang ZP, Zheng XQ and TangJW2014Evaluation method for engineering technology of rural domestic sewage treatment based on fuzzy integral model*Trans. Chinese Soc. Agricultural Eng.***30**272-280

[2] Parnell PE2003The effects of sewage discharge on water quality and phytoplankton of Hawai'ian coastal waters *Mar. Environ. Res.* **55**293

[3] Mounaouer B, Nesrine Aand Abdennaceur H2014Identification and characterization of heavy metal-resistant bacteria selected from different polluted sources*Desalin. Water Treat.***52**7037-7052

[4] Chen Hand Zhang M2013Occurrence and removal of antibiotic resistance genes in municipal wastewater and rural domestic sewage treatment systems in eastern China*Environ. Int.* **55**9-14

[5] Wan YS, Zhang P, Li DL and Ma JF2011Analysis on selection of domestic sewage treatment method in rural area*Agr. Sci.Technol.***32**44-47

[6] Li XN, Li XA, Jiang B and Lv XW2007Research on optimization of splashing oxygenated biological filter for rural domestic sewage treatment*China WaterWastewater***23**15-18

[7] Wang LM, Guo FG, Zheng Z, LuoXZ and ZhangJB2011Enhancement of rural domestic sewage treatment performance, and assessment of microbial community diversity and structure using tower vermifiltration*Bioresource Technol.* **102**9462-9470

[8] Zheng W, Yang YM, Li YL and Zheng JQ2014Treatment of rural domestic sewage by solar ozonic ecological sewage treatment plant*Adv. Mater.Res.* **955-959**3393-3399

[9] Sun P, Li Y, Kong FL, Qu LL2013Comment on the rural domestic sewage treatment technology *Jour. Qingdao Technol. Univ.***34**71-75

[10] Tonetti AL, Coraucci Filho B, Bertoncini EI, Oliveira RA and StefanuttiR2010Evaluation of a simple system of wastewater treatment aiming at use in rural areas*Rev. Bras. Eng. Agr. Amb.* **14**227-234

IOP Conf. Series: Materials Science and Engineering **250** (2017) 012029 doi:10.1088/1757-899X/250/1/012029

Application Research of Quality Control Technology of Asphalt Pavement based on GPS Intelligent

Min Wang[1,2], Bo Gao[1], Fei Shang[1] and Tao Wang[1]

[1] Chongqing Zhixiang Paving Technology Engineering Co.,Ltd., No.33, Xuefu Avenue, Nan'An District, Chongqing, 400067, China
[2] China Merchants Chongqing Communications Research & Design Institute, No.33, Xuefu Avenue, Nan'An District, Chongqing, 400067, China
Email: panda9496@163.com

Abstract. Due to the difficulty of steel deck pavement asphalt layer compaction caused by the effect of the flexible supporting system (orthotropic steel deck plate), it is usually hard and difficult to control for the site compactness to reach the design goal. The intelligent compaction technology is based on GPS control technology and real-time acquisition of actual compaction tracks, and then forms a cloud maps of compaction times, which guide the roller operator to do the compaction in accordance with the design requirement to ensure the deck compaction technology and compaction quality. From the actual construction situation of actual bridge and checked data, the intelligent compaction technology is significant in guaranteeing the steel deck asphalt pavement compactness and quality stability.

1.Introduction

At present, gussasphalt is usually adopted on steel deck pavement with thickness of 6cm~8cm, which is constructed in two layers. Due to thin thickness of pavement and high construction requirement, it belongs to very delicate project, in which every application detail is very sensitive, and minor construction defects can bring serious subsequent damage.

Compaction is one of key procedures of steel deck pavement construction, and the under-compaction and uneven compaction caused by traditional roller and control procedure can be one of the main factors triggering early damage of steel deck pavement. The under-compaction and poor stability of steel deck pavement can cause cracks or potholes etc., which generates additional repairing and maintenance charges. Therefore, not only compactness but also the evenness of compactness must be guaranteed. Considering that the complicated compaction procedure of steel deck pavement requires more flexible and economic application technique, the intelligent compaction technology has better feasibility and prospect, and the real-time control of construction procedures such as compaction times can effectively improve the steel deck pavement quality and reliability. This paper has done the analysis and research on the application of intelligent compaction technology in the steel deck pavement.

2.The Intelligent Compaction Technology Principles and Composition

2.1 Working Principle

The intelligent compaction technology is a kind of new compaction quality control technology which integrates the sensor technology, embedded system and computer technology together. It could guide the roller to automatically judge and adjust the compaction performance parameters in accordance with the rolled material and mechanical operation status in the compaction process, so as to achieve the best compaction effect.

The intelligent compaction equipment can measure and record the compaction quality and automatically judge and adjust the compaction performance parameters in accordance with the rolled material and mechanical operation status with the going of compaction process, so as to achieve the best compaction effect. This system can automatically check the stiffness and surface temperature of compacted asphalt surface layer and integrate the checked road compaction quality with automatic vibration adjustment. Besides, it uses the map based on the GPS to indicate the parameter situation and compaction quality in the compaction process and realizes the automatic storage of documents with specialized software. This technique solves the problems such as miss-compaction, under-compaction and over-compaction fundamentally so as to ensure the evenness of compactness. The working principle of controlling rolling times, automatic frequency conversion and amplitude variation control is showed in Figure 1.

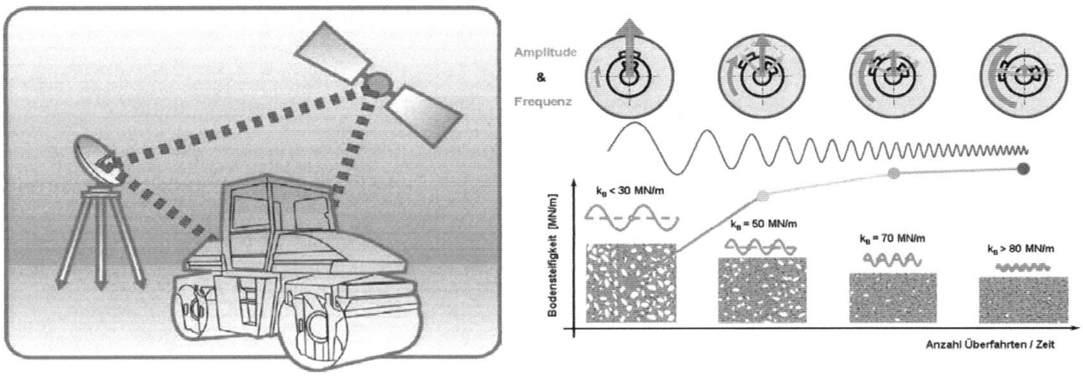

Figure 1. The Working Principle of Controlling Rolling Times, Automatic Frequency Conversion and Amplitude Variation Control

The intelligent compaction technology was used in subgrade construction. Because the compaction effect of subgrade is obvious and identification of the compaction degree is relatively easy. However, when it comes to asphalt pavement, the application becomes more difficult. The application of intelligent compaction technology in this research for the asphalt pavement is a technological breakthrough.

2.2 System Composition

CCSFlex asphalt compaction control system with GPS real-time dynamic location control technology and compaction sensor monitoring technology will be adopted in the test, and CCS900 application software is used to real-time deal with acquired correct location and posture information of roller steel wheel, and rolling times, road surface temperature, heading direction and speed are shown in forms of figures, values and sound signal etc. to lead the operator to do work.

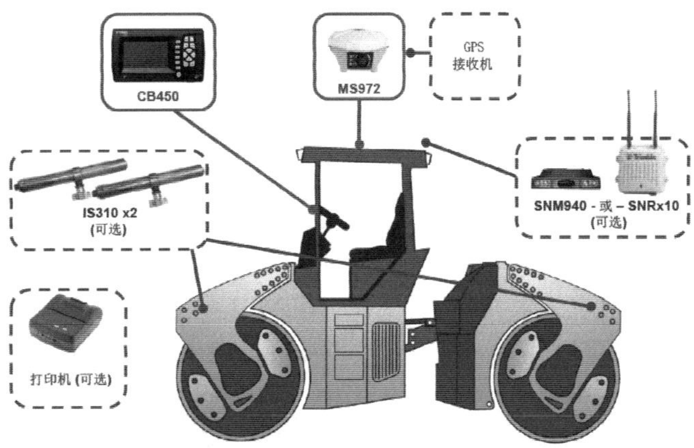

Figure 2. CCSFlex Asphalt Compaction Control System Composition

From Figure 2, it is shown that CCSFlex asphalt compaction control system is composed of the following parts: MS990 receiver, CB450 control box, SNR410 data transfer radio, IS310 temperature sensor (optional), printer (optional), GNSS base station.

3.Site Implementation and Data Analysis

3.1 Implementation Scheme

It is shown in Figure 3 and Figure 4 that, in September 2013, intelligent compaction system was tested in Guangxi Liuzhou Guangya Bridge construction, and three sets of intelligent compaction systems (one for the first rolling and two for the second rolling) were installed on three rollers to check rolling times of each roller and finally, the data of three rollers was combined to confirm the accumulated rolling times.

Guangxi Liuzhou Guangya Bridge steel deck pavement project adopts the pavement scheme: TOPEVER® waterproof binding layer + gussasphalt + high-elastic modified SMA, as shown in Figure 3. Gussasphalt is formed by fluidity and doesn't need rolling and the test focuses on the high-elastic modified SMA mixture (thickness 35mm) compaction process. The pavement structure is shown in Table 1.

Table 1. Guangxi Liuzhou Guangya Bridge Steel Deck Pavement Structure

Layer	Detailed parameters
Pavement upper layer	High-elastic asphalt SMA, thickness: 35mm Scatter modified emulsified asphalt, dosage: 300~500g/m^2
Pavement lower layer	Gussasphalt concrete GA10, thickness: 35mm: scatter 5~10mm pre-mixed debris
Waterproof binding layer	Responsive binder, dosage: 100~200 g/m^2 TOPEVER (two layers) Polymer waterproof coating, total dosage: 2500~3500g/m^2 Anti-corrosive primer, dosage: 100~200 g/m^2
Steel plate	Sand blasting and de-rusting, cleanness: Sa 2.5, roughness: 50~100μm

At the same time, two comparison tests are done to verify the control effect of intelligent compaction system. In two tests, intelligent compaction systems are installed to check the rolling times and temperature. For the first construction (left deck), the operator does rolling according to the previous experiences and just records rolling tracks and times; for the second construction (right deck), the operator is guided by real-time shown tracks and times to set up rolling times control goal to do rolling.

Figure 3. Install the Intelligent Compaction System

IOP Conf. Series: Materials Science and Engineering **250** (2017) 012029 doi:10.1088/1757-899X/250/1/012029

Figure 4. Collect Data in Construction Process

After the recorded data was input into the computer, rolling result was showed in Figure 5 and Figure 6. By putting the rolling times of three rollers together we get the accumulated rolling times of each position. (From red to green represents the rolling times from more to less, and the green stands for once, and the red stands for over 20 times).

Figure 5. Rolling Result of Using the Traditional Method (Left Deck)

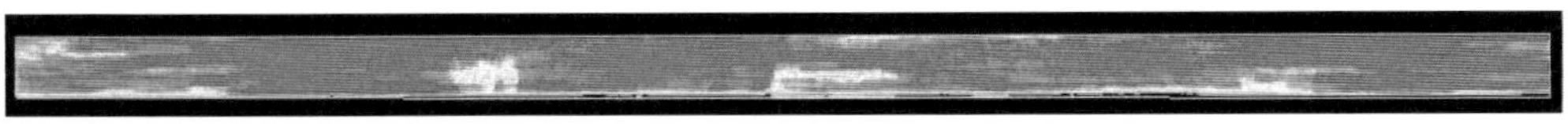

Figure 6. Rolling Result of Using the Intelligent Compaction Equipment (Right Deck)

Figure 5 and Figure 6 show that, after applying intelligent compaction system to control rolling times, the rolling evenness is better.

Through the site compaction engineering we found that the intelligent compaction technology had obvious advantages. Compared with traditional rolling procedure, the intelligent compaction is more convenient in field management and more efficient. Because in the traditional compaction, roller operator is relatively blind. More controllers are needed on site and over-compaction often happened.

3.2 Data Analysis

After the completion of SMA on the whole bridge, pave track TM Plus is used to inspect the compactness of SMA which is shown in Figure 7. Test data is obtained from every 3m on both left and right width of the deck.

Figure 7. Using Pave Track TM Plus to Collect the Data of Compactness

Table 2. The Test Results of Guangya Bridge Compactness

Chainage No.	Left Deck Compactness/% (Traditional Rolling Procedure)			Right Deck Compactness/% (Controlling Rolling Times)		
k0+20	98.9	100.0	100.6	99.2	98.7	97.9
k0+40	99.4	96.9	101.0	99.3	97.8	99.1
k0+60	99.8	97.1	103.0	99.4	99.0	98.2
k0+80	99.4	99.2	99.6	100.2	99.1	99.0
k0+100	99.7	99.9	97.0	97.8	97.5	99.7
k0+120	95.9	98.3	99.6	100.3	99.5	98.4
k0+140	97.8	96.1	97.8	96.2	98.6	97.8
k0+160	97.9	97.3	96.1	100.0	99.1	98.4
k0+180	97.3	97.8	98.3	98.1	97.6	99.3
k0+200	97.9	99.6	98.1	98.6	98.6	98.0
k0+220	98.9	101.8	97.4	97.7	98.6	98.1
k0+240	98.4	98.6	98.5	99.2	97.7	97.5
k0+260	99.4	99.3	99.7	98.6	98.4	99.9
k0+280	96.8	98.5	99.5	99.6	98.9	96.7
k0+300	99.4	96.8	97.1	98.8	98.1	97.9
k0+320	96.8	97.5	98.4	99.5	99.5	98.2
k0+340	97.2	99.0	96.5	98.3	96.5	99.5
k0+360	98.3	97.8	97.5	99.3	100.9	98.7
k0+380	97.6	99.1	98.0	98.8	98.4	98.6
k0+400	99.0	96.1	98.0	98.2	101.2	98.5
k0+420	95.9	100.8	97.8	97.5	97.8	99.5
k0+440	98.2	92.7	98.0	97.7	98.4	99.0
k0+460	95.6	98.1	97.0	97.9	97.8	99.6
k0+480	97.0	97.6	97.3	99.5	101.1	98.3
k0+500	97.6	99.4	97.9	98.9	98.8	99.4

Test results of Guangya Bridge compactness is shown in Table 2. In asphalt pavement project, the compactness variation coefficient and standard discrepancy are usually used to reflect the discretion degree of gussasphalt compactness and larger variation coefficient and standard discrepancy indicates higher data discretion degree. In accordance with the compactness analysis method stipulated in *Technical Specifications for Construction of Highway Asphalt Pavement*, the variation coefficient and standard discrepancy of compactness of two decks are analyzed.

The average compactness, standard discrepancy and variation coefficient of tested road section can be calculated by equation (1), equation (2) and equation (3).

$$K_0 = \frac{K_1 + K_2 + \ldots + K_n}{N} \tag{1}$$

$$S = \sqrt{\frac{(K_1 - K_0)^2 + (K_2 - K_0)^2 + \ldots + (K_n - K_0)^2}{N-1}} \tag{2}$$

$$C_V = \frac{S}{K_0} \tag{3}$$

Where K_0 is the average compactness of that tested road section, %; S is the standard discrepancy of that tested road section, %; C_V is the variation coefficient of that tested road section, %; K_1, K_2, …, K_N is the compactness of every tested point in that tested road section, %; N is the total number of tested points in that tested road section, and DOF is N-1.

The analysis results are shown in Table 3.

Table 3. The Calculation Results of Standard Discrepancy and Variation Coefficient

Parameters	Average Compactness/%	Standard Discrepancy/%	Variation Coefficient/%
Left Deck (Traditional Rolling Procedure)	98.2	1.53	1.56
Right Deck (Control Rolling Times)	98.7	0.96	0.96

The data in Table 3 shows that the compactness standard discrepancy and variation coefficient after applying the intelligent compaction system to control rolling times are obviously lower than that of the layer which doesn't control rolling times and the compactness after applying the intelligent compaction system to control rolling times indicates that the intelligent compaction system can better control rolling times and achieve better construction quality control result.

4.Conclusions

By the application of intelligent compaction equipment in the steel deck pavement, the following conclusions can be obtained:

(1) The intelligent compaction system can real-time record the walking tracks of pavement roller and feedback the rolling times and temperature to the operator so as to effectively control the steel deck pavement layer compaction quality.

(2) For steel deck pavement with the intelligent compaction system, the rolling times of asphalt mixture can be better controlled than that without the intelligent compaction system.

(3) The compactness standard discrepancy and variation coefficient used the intelligent compaction system to control rolling times were obviously lower than those didn't use intelligent compaction system to control. Moreover, the compactness using the intelligent compaction system to control rolling times is higher. The intelligent compaction system plays an important role in the quality control.

5.References

[1] Yang Xueru and Bi Ruqiang 2011 *Pavement Machinery and Construction Mechanization* **06** 43-45
[2] Xie Weiping and Zhang Guoqian 2011 *Shanxi Architecture* **09** 154-156
[3] Li Jianjun 2011 *Construction Mechanization* **05** 60-63
[4] Liu Wei 2011 *The Railroad Architecture Technology* **03** 65-68
[5] Shen Peihui and Yang Ping 2011 *Highways & Automotive Applications* **03** 118-120
[6] Zhang Zhiming 2005 *Construction Machinery and Equipment* **03** 5-8
[7] Yang Lu and Feng Zhanqiang 2011 *The Engineering Machinery Digest* **01** 50-53
[8] Ma Xueliang, Sun Zuwang and Long Shuigen 2008 *Pavement Machinery and Construction Mechanization* **11** 81-84

IOP Conf. Series: Materials Science and Engineering **250** (2017) 012030 doi:10.1088/1757-899X/250/1/012030

Feasibility of Using Dredged Mud for Prepared the Permeable Brick

Chaoqun Zhou, Xiaosu Cheng, Lingke Zeng*, Hui Wang and Jing Chen

School of Materials Science and Engineering, South China University of Technology,Guangzhou 510641, China

Abstract. Through experimental analysis found that the chemical composition of the dredged mud is similar to clay and the dredged mud does not leach heavy metals. Using the dredged mud in the preparation of permeable bricks reduced the quantity of incineration dredged mud buried in landfills, and the exploitation and consumption of natural sandstone. The dredged mud needs to be checked by the validation criteria when the second use, so we used the TCLP test to identify hazardous materials. Its leaching of heavy metals was in line with industry standard. And the basic formula of permeable brick were prepared, its performance was in line with national standards. The use of dredged mud preparing eco-friendly permeable bricks, not only solves the problem of environmental pollution, but also gets some economic and social profit.

1. Introduction

In recent years, the harbor waterway expansion and the coastal dredging engineering maintenance result in dredged mud increased gradually. Proper handling of dredged mud has become a top priority. At present, the treatment of dredged mud is mainly in landfill, marine dumping, filling and so on.

Over the past decades, extensive studies dealing with characterization and stabilization of dredged marine soils have been reported, and dredged material has been recognized as a promising marine resource for the back-fills used for land reclamation [1]. In China, dredged materials have always been treated as geo-waste and directly dumped into the ocean, resulting in unforeseen large volume of dredged materials accumulated in the ocean[2]. It is statistically reported that the annual production of dredged materials in Shanghai region alone exceeds 70 million m^3, and so far more than 2.5 billion m^3 of dredged materials had been already dumped into the surrounding ocean [3, 4]; In South Korea, about 729 million cubic meters of coastal dredging has occurred between 2001 and 2008 with some 31 million cubic meters of dredged sediment dumped in the open sea [5]; Annually, 5.9 million cubic meters of dredged sediments is excavated and placed in several disposal areas in Jasper County, South Carolina [6]. In addition to marine landfill, the dredged mud also can use in the other way. Over the past decade, most people are vigorously promoting green and environmentally friendly energy materials. The dredged mud has a huge output annually, and the second use of materials bear the brunt of a large number of researchers attracted attention. For distance, the preparation of light weight aggregates used the dredged mud as aggregates [7]; the preparation of fired brick sediment used the harbor river as the primary raw material [8]. In China, many cities are calling the concept of sponge city, permeable brick meets this concept. For ordinary bricks, it has good strength, but poor water permeability leads to a series of problems, such as urban waterlogging, heat island effect, ground subsidence and etc. The permeable brick with light and porous features is different from other ordinary brick. The porosity of permeable brick is between 10% and 20%, so that the permeable brick have good water permeability. The permeable brick has many advantages: alleviate the city waterlogging and heat island effect, replenishment of groundwater timely and absorb the noise. Based on the rough surface, it can increase the wear resistance to increase the safety of the road. In recent years, many researchers used various materials to prepared permeable bricks and achieved good results. Such as: preparation of water permeable brick using fly ash as main raw material. The compressive strength and coefficient of water permeability of the fly ash water permeable brick was 30.6 MPa and 1.08×10^{-2} cms^{-1} [9]. The prepared water permeable brick by discarded ceramics has a compression strength of 48.9 MPa, a bending strength of 7.8 MPa and a permeation coefficient of 0.0312 cms^{-1} [10]. And the water permeable brick from red mud has the compressive strength of 35.32 MPa with permeability coefficient of 0.028 cms^{-1} and wear length of 27.35 mm [11].

2. The Characteristics of Dredged Mud

The dredged mud was obtained from the river near Cencun, Guangzhou. And the dredged mud was sampled in three different places, to avoid the difference of the dredged mud composition in the same river. The dredged mud was pretreated firstly. The dredged mud was placed on the tray, then was oven-dried at 105℃ for 2 hours. Then the dry dredged mud was grinded for 2h in the ball mill machine, the obtained powder sieved to a granular size <0.546 mm to removed other impurities.

2.1 XRD and XRF Analysis of Dredged Mud Composition

Table 1 shows the X-ray fluorescence (XRF) analysis of the dredged mud. The main ingredients of the dredged mud are SiO_2, Al_2O_3 and Fe_2O_3 similar to natural clay. All materials present a significant loss of ignition which is likely associated to carbonate and sulphate decomposition and also to burn-off of organic matter that is usually adsorbed in three kinds of materials [12]. Therefore, dredged mud was used instead of natural clay as a low temperature binder to prepare the permeable brick. Figure 1 shows the X-ray diffraction (XRD) patterns of the dredged mud, it demonstrated that the dredged mud mainly contains quartz, kaolin and silica phase.

Table 1 The chemical composition of dredged mud (%)

Sample	SiO_2	Al_2O_3	Fe_2O_3	K_2O	TiO_2	CaO	P_2O_5	Na_2O	Total	LOI
P1	61.12	21.19	4.76	2.63	0.84	0.58	0.27	0.24	99.72	8.09
P2	54.54	20.34	5.58	3.01	1.07	1.05	0.21	0.22	99.75	13.73
P3	59.34	22.02	4.34	2.52	1.02	0.87	0.19	0.27	99.6	9.03

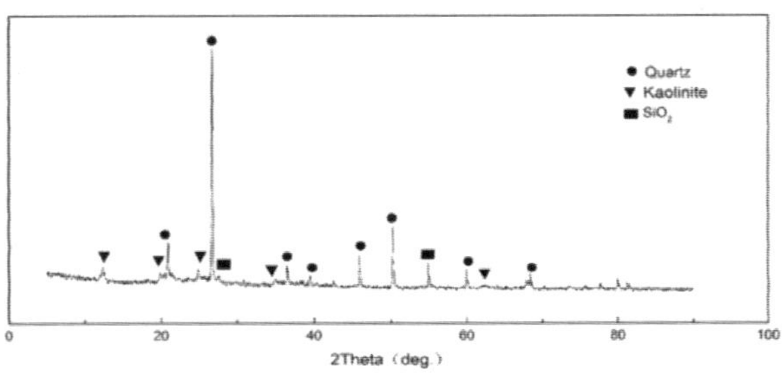

Figure 1. The XRD pattern of the dredged mud.

2.2. The Analysis of the Ash of Dredged Mud

The ash of dredged mud refers to the percentage of inorganic solid oxide in the dredged mud. The organic matter residue completely was burned, and water was completely discharged (including structured water).

A certain amount of dry dredged mud(D_1= the weight of dry dredged mud and the crucible) was weighed into the crucible, then put into the muffle furnace for 30 min from room temperature slowly heated to 500 °C, and then 30 min to 815 °C and 815 °C for 1 h. Then the crucible removed to the asbestos plate and cooled to room temperature. And the above procedure was repeated until the difference in mass between the two measurements was less than 0.002 g, meanwhile the mass (D_3) was recorded. D_2 is the weight of the crucible. The content of ash in the dredged mud was calculated according to equation (1).

$$A_{sh} = \frac{D_3 - D_2}{D_1 - D_2} * 100\% \quad (1)$$

Where A_{sh} is the content of ash in the dredged mud (%), D_3 is the weight of ash and crucible (g), D_2 is the weight of crucible (g), D_1 is the weight of dry dredged mud and crucible (g).

The test results are shown in **Table 2**. It can be seen from the Table 2 that dredged mud was mainly ash. Three points of dredged mud's A_{sh} were more than 94%. The dredged mud has high content of organic matter.

Table 2 The results of ash content of dredged mud

Sample	D_1(g)	D_2(g)	D_3(g)	A_{sh}(%)
N_1	135.19	84.77	132.72	95.10
N_2	141.43	88.15	138.55	94.59
N_3	163.30	109.28	160.63	95.06

2.3. TCLP Leaching

Many countries use TCLP leaching test to identify hazardous materials. If the concentration of leached heavy metals exceeds the regulatory standard, the dredged mud is considered hazardous waste, and must be further treated to reduce the risk of leaching of heavy metals into the environment. The dredged mud leaching test was carried out according to the Taiwan EPA standard testing method for solid waste. The dredged mud and a leaching solution (CH_3COOH) at a weight/volume ratio of 1/20 were mixed and agitated at 30rpm for 18h. The leachate was then filtered. The heavy metal concentration was analyzed by atomic absorption [13, 14]. The experiment of this study was conducted dredged mud whose heavy metal dissolution met the regulatory standard.

Table 3 Dredged mud TCLP test results (mgL^{-1})

Sample	P_1	P_2	P_3	Taiwan limits
Cu	0.031~7.86	0.012~0.21	0.017~3.9	15
Cr	0.8~1.83	0.19~0.24	0.18~1.03	5
Pb	0.01~0.047	0.005~0.05	0.032~0.14	5
Cd	0.0005~0.0023	<0.0005	<0.0005	1

Table 3 lists the result of TCLP analysis for the dredged mud from three sampling points and all samples met the QA/QC requirements that generated the data ranges. So the dredged mud in place of clay to prepare the permeable brick met the required regulations and standards, and it can not to leave a negative heritage to future generations [15].

2.4 Preparation of Permeable Bricks with Dredged Mud

The dredged mud powder sieved to a granular size <0.546 mm, the particle size of glass powder was 0.154 ~ 0.197 mm, and the particle size of waste porcelain was 2.675 ~ 3.53 mm. The permeable brick was prepared according to the proportions of the samples in **Table 4**. In the laboratory, the permeable brick was prepared according the flowed chart (**figure 2.a**) and the samples were fired according the firing systems (**figure 2.b**).

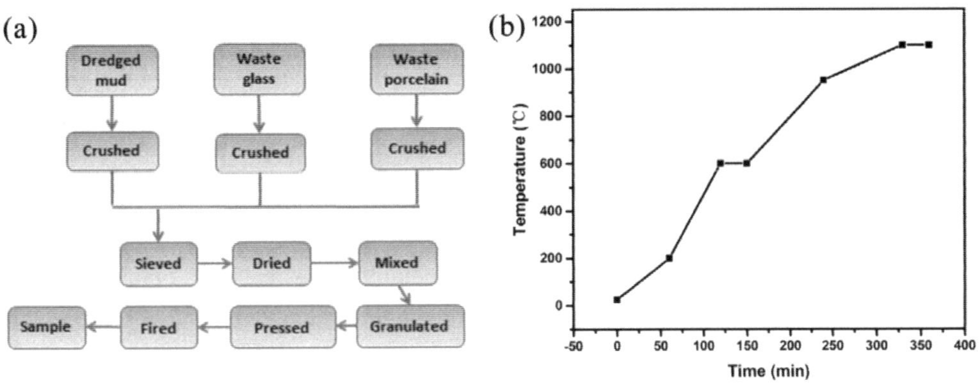

Figure 2.a The preparation of process flow chart. **2.b** The firing systems.

Table 4 The composition of the permeable brick

Sample	MG(mud to glass)	Waste ceramics (%)	Density (gcm^{-3})	Porosity (%)	Strength (MPa)	Permeability (cms^{-1})
1#	1:3		1.866	5.92	159.37	0.00205
2#	2:2	60	1.885	7.25	94.46	0.0294
3#	3:1		1.859	23.48	75.84	0.0405
4#	4:0		1.857	24.41	64.44	0.0532

Figure 3. The permeable bricks under different MG ratios.

As can be seen from figure 3, the surface color of the permeable bricks is very different. With the content of the dredged mud increases, the sample gradually deepened the color. Because the dredged mud contains Fe_2O_3, the sample's color is brighter with the increasing of the dredged mud. In life, the color of the permeable bricks can be changed by controlling the content of dredged mud. The four samples of the body density and porosity can be seen from the Table 4. Increasing the content of the dredged mud causes increases in the porosity, due to the combustion of organic matter in the dredged mud. And the density of the permeable brick is greater than 1.8 ~ 1.9 gcm^{-3}. The water permeability of re-prepared permeable brick are greater than the national standard 1×10^{-2} cms^{-1}.

3. Conclusion

The possibility of using dredged mud as the primary material making permeable bricks was investigated in this study. Through the composition analysis, the ash analysis and heavy metal dissolution analysis of dredged mud, it determined that the dredged mud could to prepare the permeable brick and the performance of the permeable bricks was better. The TCLP testing shown that the heavy metal concentration of the dredged mud were fully in line with industry standards. Based on these preliminary analyses, it can be concluded that studied the dredged mud can be used as secondary raw materials for the production of permeable bricks. In the preparation of permeable bricks, the content of heavy metal is in line with industry standards and can also contribute to the construction of sponge city.The experiment using the dredged mud to prepare permeable bricks will be carried out in the follow-up study.

4. Acknowledgment

This work is funded by a grant from the Science and Technology Planning Project of Guangzhou (Grant no. 201604020116).

5. References

[1] W. Shi, Q. Chen, S. Nimbalkar, W. Liu, A new mixing technique for solidifier and dredged fill in coastal area, Marine Georesources & Geotechnology, **35** (2017) 52-61.

[2] W. Liu, Q. Chen, G. Chiaro, H. Jiang, Effect of a cement-lignin agent on the shear behavior of Shanghai dredged marine soils, Marine Georesources & Geotechnology, **35** (2017) 17-25.

[3] H.M. Jiang, W.B. Liu, Experimental study on deformability and shear strength of solidified dredged soil, Applied Mechanics and Materials, Trans Tech Publ, 2013, pp. 1755-1762.

[4] X. Jiang, J. Cheng, Q. Tang, H. Gao, Z. Jiang, X. Zhao, Analysis of resource utilization of dredged mud in Shanghai, Environmental Science & Technology, **36** (2013) 86-89.

[5] K.-H. Eom, D.-I. Lee, D.-S. Park, G.-Y. Kim, Diagnosis for status of dredging and ocean disposal of coastal sediment in Korea, Journal of Environmental Impact Assessment, **18** (2009) 185-193.

[6] A. Mezencevova, N.N. Yeboah, S.E. Burns, L.F. Kahn, K.E. Kurtis, Utilization of Savannah Harbor river sediment as the primary raw material in production of fired brick, Journal of environmental management, **113** (2012) 128-136.

[7] K. Hamer, A. Hadeler, T. Muschalla, J. Schröter, G. Timmer, Light weight aggregates made from dredged harbour sediments, Journal of Soils and Sediments, **3** (2003) 284.

[8] K. Hamer, V. Karius, Brick production with dredged harbour sediments. An industrial-scale experiment, Waste management, **22** (2002) 521-530.

[9] L. RAO, J. CAO, H. ZHANG, L. YANG, Y. PANG, Y. LIU, J. WANG, Study on preparation of fly ash water permeable brick [J], New Building Materials, **7** (2006) 017.

[10] J.-f. WU, J.-g. CHEN, X.-h. XU, M. LIU, B.-z. FANG, H.-l. LI, Research on Preparation of Water Permeable Brick from Discarded Ceramics [J], Journal of Wuhan University of Technology, **19** (2009) 007.

[11] L.G.W. Ping, Study on Preparation and Performances of the Water Permeable Brick from Red Mud [J], Metal Mine, **12** (2009) 040.

[12] M. Romero, A. Andrés, R. Alonso, J. Viguri, J.M. Rincón, Sintering behaviour of ceramic bodies from contaminated marine sediments, Ceramics International, **34** (2008) 1917-1924.

[13] M.-H. Wu, C.-L. Lin, W.-C. Huang, J.-W. Chen, Characteristics of pervious concrete using incineration bottom ash in place of sandstone graded material, Construction and Building Materials, **111** (2016) 618-624.

[14] F.-Y. Chang, M.-Y. Wey, Comparison of the characteristics of bottom and fly ashes generated from various incineration processes, Journal of hazardous materials, **138** (2006) 594-603.

[15] S.-i. Sakai, M. Hiraoka, Municipal solid waste incinerator residue recycling by thermal processes, Waste Management, **20** (2000) 249-258.

Environment Impact Analysis of Shield Passing Alongside Bridge Pile Platform Using Three Dimensional Numerical Simulation

Yanliang Shang [1,2], Wenjun Shi [2], Tongyin Han [3] , Zhichao Qin [2]and Shouji Du[1]

[1] Department of civil engineering, Shanghai Jiao Tong University, Shanghai 200240, China; 2. School of civil engineering Shijiazhuang TieDao University, Shijiazhuang 050043, China ;3.Langfang Teachers University, Langfang ,065000 , China

E-mail: sty1980926@163.com

Abstract: The shield method has many advantages in the construction of urban subway, and has become the preferred method for the construction of urban subway tunnel. Taking Shijiazhuang metro line 3 (administrative center station - garden park station interval) Passing alongside bridge as the engineering background, double shield crossing the bridge pile foundation model was set up. The deformation and internal force of the pile foundation during the construction of the shield were analyzed. Pile stress caused by shield construction increases, but the maximum stress is less than the design strength; the maximum surface settlement caused by the construction of 10.2 mm, the results meet the requirements of construction.

Keywords: shield; under wear; bridge pile; deformation; internal force

1. Introduction

Shield is one important method of tunnel construction. It has many merits such as little impact on the surroundings, less ground operation, high automation degree, fast construction speed and low labor intensity. Due to the obvious advantages, shield is widely used in city subway tunnel construction. In the process of shield construction, it is a problem worth paying close attention to when it meets the surrounding structures which are sensitive to deformation.This paper takes Shijiazhuang metro line 3 (administrative center station - garden park station interval) Passing alongside bridge as the engineering background to study the surface deformation law of shield pile passing through bridge pile platform.

Administrative Center Station ~ Garden Park Station is located in Zhengding New District, starting from the administrative center station, north along the planning of Beijing South Street (formerly known as Metro Avenue) below the laying, side wear and under the comprehensive pipe gallery to park garden station. The administrative center station to park station interval by shield construction ,starting mileage of the right line of the shield is K31+593.450, terminate the mileage for the K32+682.090, the total line length is 1088.640 m; the left line of the shield chainage K31+593.450, terminate the mileage for the K32+682.090, the total line length of 1088.640 M . The soils within the exploration range of the project are divided into artificial accumulation layer (Qml), newly deposited layer (Q4al), Quaternary alluvial (Q4al + pl), Quaternary system, Q3al + pl) 4 layers. Shield close to the existing Zhou Hanhe bridge bridge pile, the impact is relatively large. First of all, the outer surface

IOP Conf. Series: Materials Science and Engineering **250** (2017) 012031 doi:10.1088/1757-899X/250/1/012031

of the bridge piles and the distance between the subway tunnels is relatively close, so the control of the shield tunneling axis coordinate requirements are higher.[1~2] Second, the shield excavation squeeze the soil, and will produce friction, the two together on the pile will make the internal force increases, so the bridge under the pile when the more dangerous. Many scholars have done a lot of research [3~9]

2. Establishment of Three - Dimensional Numerical Model of Shield Crossing Construction

In this chapter, we use the method of numerical simulation to establish the model of shield tunneling through Zhouhanhe bridge The following first describes the calculation of the model assumed, shield tunneling process simulation:

(1) calculation assumption

Pile foundation, pipe material, grouting layer as a linear elastic material.

(2) calculation parameter selection

The parameters of segment and pile are shown in table 1.

Table 1. Parameters of segment and pile

structure	Material name	Severeγ/ (kN·m-3)	Elastic ModulusE /MPa	Cohesion /kPa	Internal friction angle/°	Poisson's ratio
Segments with piles	C50 Concrete	2 400	3 450	—	—	0.20

In the K32+100~K32+150 section under wear shield piles, pile depth of about 20 m, pile bottom tunnel vertical minimum distance is 5.4 m, the minimum horizontal distance is 1.5 m, the tunnel center average depth is 28 m, thickness 0.3 m, every excavation 1.2 m, the 3D model along the tunnel axis is 42 m long, 70 m wide and 50 m high, the model shown in figure 1. Figure 2 shows the relative position of the tunnel and the bridge pile. The model consists of 54844 units, 56878 nodes and 82 excavation steps. Due to the pile on the basis of the load to withstand the bridge, so set the surface of its 700 kN concentrated load.

Shield tunneling process: first simulation of the left line of shield tunneling, and other left after the end of the tunnel, the right line of shield began to dig, to the end of the excavation. Excavation first full-section excavation, and then joined the shield, simulated shield support soil, and then add the grouting layer and lining.

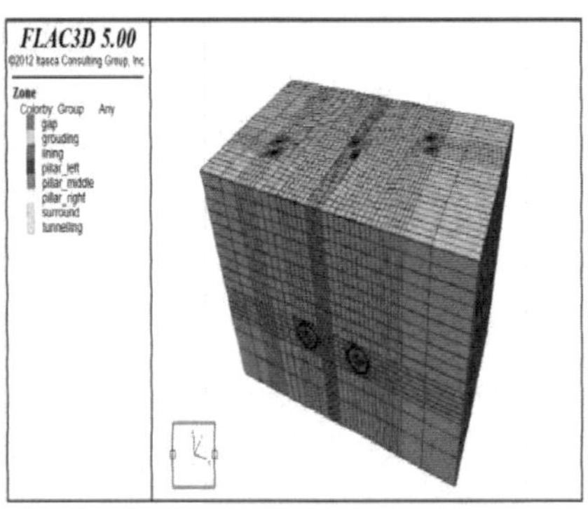

Figure1. Model of tunnel crossing pier

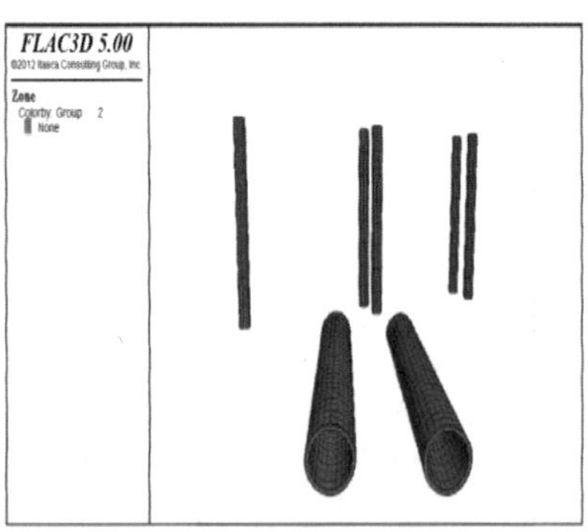

Figure2. Shield tunneling structure and bridge pile foundation structure

3. Analysis on the Influence of Shield Tunnel's Influence on Construction of Zhouhanhe Bridge

Based on the model and calculation results of the shield tunnel and bridge pile, analysis of shield tunnel Undercrossing weeks pile surface deformation and bridge deformation and internal force characteristics of the Han River Bridge Pile foundation.

3.1 Surface Deformation

Figure3 is the vertical displacement nephogram after shield excavation. Figure4 horizontal displacement nephogram after excavation. As can be seen from the diagram, the maximum settlement of the ground surface is 10.2 mm, and the maximum horizontal displacement of the ground surface is 3.7 mm when the surface subsidence is formed by shield tunnelling.

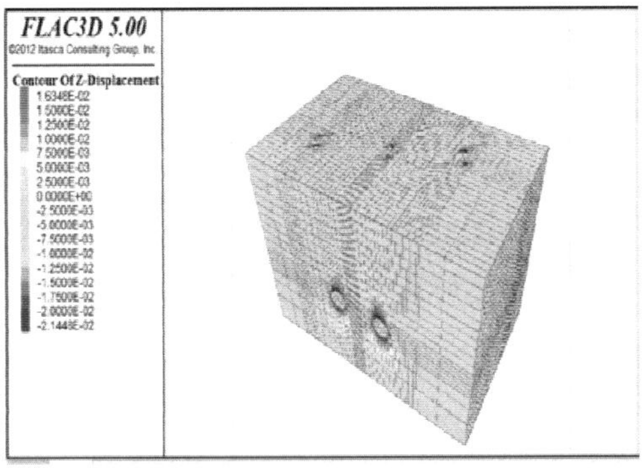

Figure 3. Vertical displacement diagram

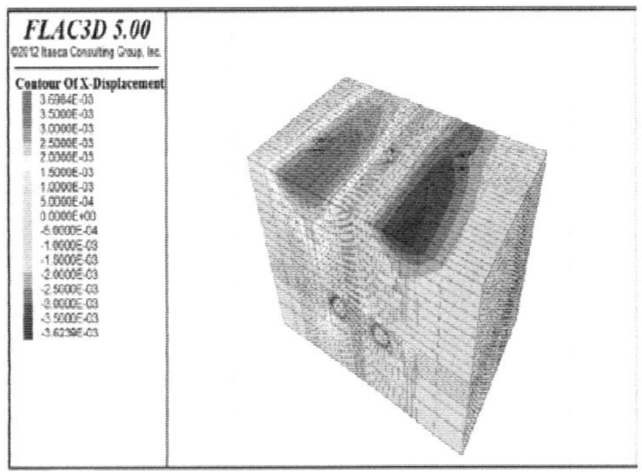

Figure 4. Horizontal displacement diagram

Figure 5 is a pile of vertical displacement, it can be seen from the chart No. 2 pile settlement is 9.35 mm, the biggest, the other two vertical displacement of pile is small, 3 mm. Figure 6 shows the pile horizontal displacement nephogram, it can be seen from the chart the horizontal displacement of No. 2 bridge pier is less than 1 mm, the horizontal displacement of No. 1 bridge is 2.95 mm, the horizontal displacement of No. 3 pier is 2.6 mm.

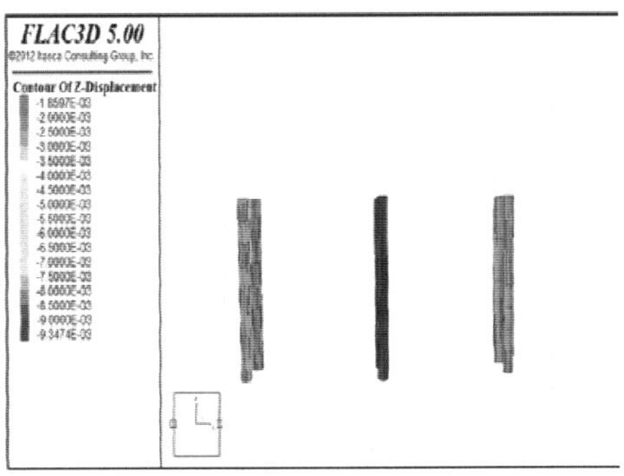

Figure 5. Vertical displacement diagram of bridge piles

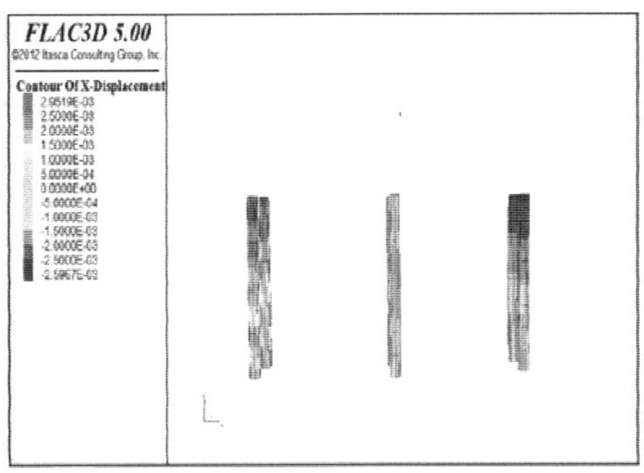

Figure 6. Horizontal displacement diagram of bridge pile

According to the set of monitoring points, you can get the end of the construction and the end of the two lines are the end of the settlement. It can be seen from the figure that the subsidence is smaller and the maximum settlement is 4.52 mm after the excavation of the left line. After the excavation of the right line, the surface settlement is obviously increased and the maximum settlement is 8.74 mm.

3.2 Internal Force of Bridge Piles

In order to study the effect of shield tunneling on piles, so analysis of the internal force and displacement of pier No. 1 and No. 2 bridge pier. In the five stages of the excavation process, the internal force and displacement are monitored respectively. The stage 1 is the initial excavation of the left shield, the second stage is the left shield excavation, the stage 3 is the left line shield tunneling completion, the stage 4 is the right line Shield tunneling half, stage 5 for the right line shield tunneling end.

The internal force and displacement of five stages of No. 1 bridge piles results as shown in figure 7～8. It can be seen from Figure7 and Figure8 that the axial force of the bridge pile is proportional to the depth of the pile at the same stage. With the construction of the shield tunneling, the axial force of each part of the pile tends to increase, the value is relatively small, and so in the shield tunneling construction process, the pile shaft force has a small impact. The bending moment of the bridge pile is different. With the construction of the shield tunneling, the pile top and the pile bottom tend to decrease gradually, and the middle of the bridge pile is gradually increased with the excavation construction. 12.5M is the largest bending moment, 20.2 kN • m.

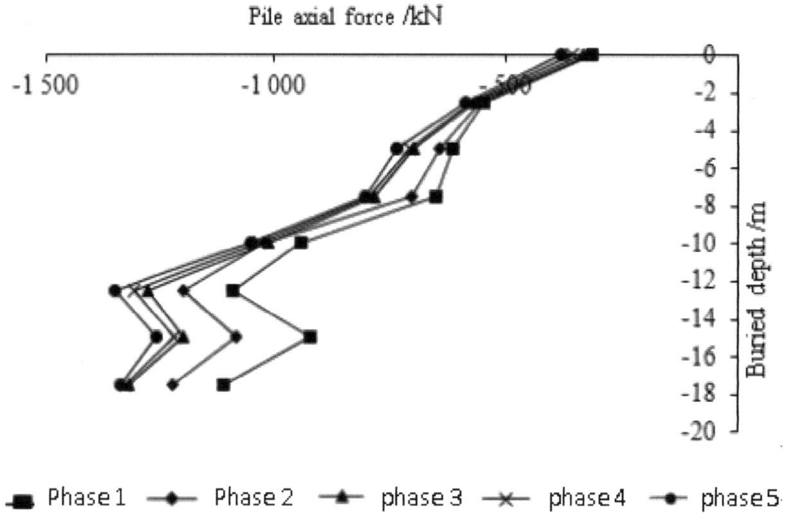

Figure 7. Axial force diagram

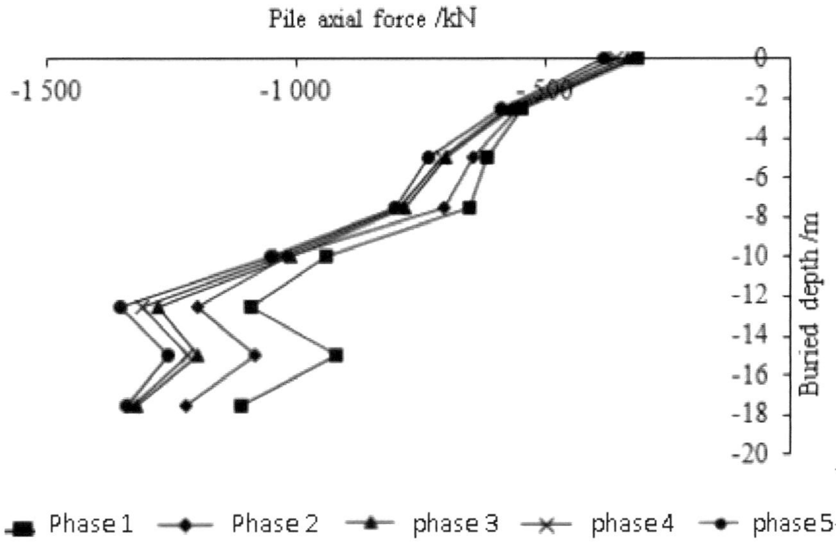

Figure 8. Bending moment diagram

According to the relevant specifications, C50 reinforced concrete axial compressive strength design value of 22.4 MPa, calculated from the results can be seen, the bridge pile stress is much smaller than its compressive strength

4. Conclusion

This paper mainly through the establishment of three-dimensional numerical simulation (FLAC3D) method to simulate and analyze the impact of the construction process on the bridge piles in the administrative center station and the garden road station, and evaluate the influence degree of the shield tunneling, Numerical calculation, the impact of shield excavation on the bridge pile, the main conclusions are as follows:

(1) Shield on Han River pier week case, the maximum pressure of pile shield construction caused by stress is 1.96 MPa, the stress is less than the pile of concrete strength.

(2) After the completion of the left line shield tunneling, the maximum subsidence of the surface is 5 mm. After the right line shield tunneling is completed, the maximum subsidence of the surface is 10.2 mm.

5. Acknowledgments

This study was supported by the Science and Technology Project of Hebei Province (Grant No. 16215408D),and also by Hebei Provincial Disaster Prevention and Mitigation Collaborative Innovation Center of Large Infrastructure Projects.

6.References

[1] Li Hongan. Beijing Metro Line 10 shield under the bridge pile pile settlement alculation formula application [A]. China Civil Engineering Society. Underground construction and harmonious development of the environment - the fourth China International Tunnel Engineering Symposium [C]. *China Civil Engineering Society*: 2009: 5.

[2] Li Yigang. Finite element simulation analysis of shield tunnel under the tunnel [J] .*Guangdong Building Materials, 2011*, (**03**): 24-26.

[3] Li Song, Yang Xiaoping, Liu Tingjin. Effect of Shield Tunneling on Pile Foundation of Viaduct Bridge in Guangzhou Metro [J]. *Railway Construction*, 2012, (**07**): 74-78.

[4] Wu Yanlian. Bridge pile foundation load test [J]. *Small and medium-sized enterprise management and technology (later)*, 2009, (**10**): 147.

[5] Deng Tuo. Shield tunnel under the bridge pile foundation deformation characteristics of the study [D]. *Hubei University of Technology*, 2015.

[6] Wang Bao, YI Dai-bing, YU Hai-ying, GAI Wen.Numerical analysis of influence of subway shield construction on Zhengxi high-speed rail multi-span high-speed bridge and its influence on surface [J]. *Construction Technology*, 2016,45 (**S1**): 443-446. [2017-08-18].

[7] Wang Meng-lin.Effects of Interference Shield Construction on Zhengzhou Special Bridge in Zhengkai Intercity Railway [J]. *Construction Technology*, 2016,45 (**S1**): 435-438. [2017-08-18].

[8] Xiao Li, Zhang Qinghe. Numerical simulation of ground subsidence caused by track shield construction [J]. *Journal of Tongji University (Natural Science Edition)*, 2011,39 (**09**): 1286-1291. [2017-08-18]

[9] Liang Rongzhu, Xia Tang Dynasty, Lin Cungang, Yu Feng. Shaped Propulsion Causes Surface Deformation and Horizontal Displacement Analysis of Deep Soil [J / OL]. *Chinese Journal of Rock Mechanics and Engineering,* 2015,34 (**03**): 583-593. (January 9, 2015) [August18, 2017].

Http://kns.cnki.net/kcms/detail/42.1397.o3.20150115.1810.027.htmlDOI:10.13722/j.cnki.jrme.2015.03.016

"Wave - Particle Duality" and Soil Liquefaction in Geotechnical Engineering

Demin Wang

Liaoning Provincial Building Design & Researching Institute geotechnical engineering company , Liaoning Shenyang ,110005.

E-mail: wangdemin24741925@163.com

Abstract : In the disaster situation of multi-earthquake, with the phenomenon of vibrating phenomenon and the occurrence of cracks in the surface soil , the collapse of the buildings on the ground are caused. The author tries to explain the phenomenon of earthquake disaster in this geotechnical engineering by using the wave-particle duality theory of sunlight. And proposed the sun in the physics of the already high frequency of the weak light superimposed into the low frequency of the low light wave volatility, once again superimposed, superimposed as a lower frequency of linear light, the energy from low to high. Sunlight from weak light into a strong sunlight, that is, the sun near the observation may be weak light or black sunspots is composed of black holes. By long distance, the convergence of light becomes into a dazzling luminous body. Light from the numerous light quantum and an energy line form a half-space infinite volatility curve , and the role of light plays under the linear form of particles. When the night is manifested of l black approaching unconnected light quantum. The author plays the earth as the sun, compared to the deep pressure of low-viscosity clay soil pore ,water performance is complex. Similar to the surface of the sun's spectrum, saturated silty sand is showed volatility, Ground surface high-energy clay showed particle properties. Particle performance is shear strength.

1.Introduction

In recent years of our country ,more frequent earthquakes lead to the earthquake disaster with the phenomenon of high-rise buildings collapse. Roads and other hard soil staggered cracks. The hard mesons transfer most of the energy in the process. However, the soft daughter meson consumes energy and delivers less energy.. The hard delevering meson is like deleverimg the energy of the quantum collision, but the soft sub-meson shows the volatility of light consuming of a lot of energy. The less energy is more prominent for the volatility to consume energy. However, The stronger the energy is more prominent to the quantum of light, that is, the more powerful the more difficult to dissipate the energy. And the lower the energy the more easily dissipated for the performance of high-energy material particles. And low power shows object volatility. As the time as earthquake liquefaction,, the water transfers energy best and the energy dissappers best. So it shows the performance of volatility like water ripples. The surface of area soil is small, and the collision surface is just a local range. The transmission of energy is not easy, which shows the performance of the particle.. The energy line of lighet is like the performance of capillary water in the soil, which can lead to high frequency fluctuations. Soil particles are the most important factor for making the curve waveform. Capillary water and soil particles are made into a half-space infinite body, with the

IOP Conf. Series: Materials Science and Engineering **250** (2017) 012032 doi:10.1088/1757-899X/250/1/012032

performance of the volatility of movement. Just like buddha beads. Saturated soil as the buddha beads material in medium frequency fluctuations in the energy line, and the capillary water is made into a low energy concentration of the moderate energy line. Sand is a common material on the surface of the land with mostly non-capillary water and usually more than the performance to the energy of particles,. When an earthquake occurs, the phenomenon of flow sand leads to the surface of the water outflow. The lower part of the saturated sandy sandy soil occurred shaking, there is a water outflow that the surface of the convergence of energy is very strong. If the upper part are the high-rise buildings, the soil generally does not have water outflow, and more energy gathered to high-rise buildings. The higher the convergence of the greater more energy, so in the earthquake-prone areas, the underlying buildings are bullded.

2.The Principle of Energy Convergence In Physics

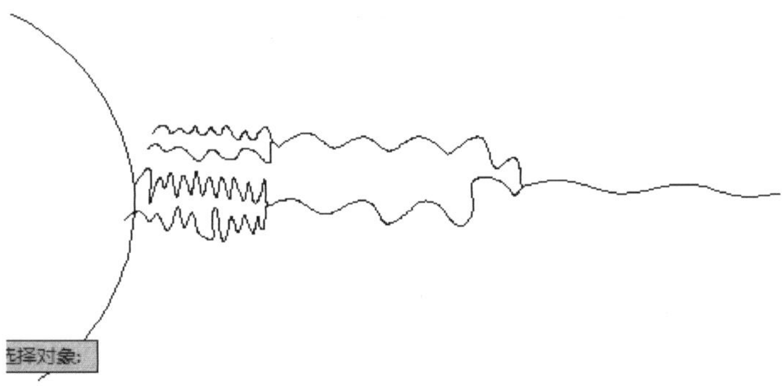

Figure 1. Principle of spectral convergence of sunlight

The energy of the sun in the process of dissipation shows the total convergence, near away from the sun the light energy is weak, with rapid dissipation as the performance for high frequency fluctuations. In the distance of the middle of the sun when the energy occurred convergence. Long distant from the sun it occurrs convergence. The emergence of low-frequency pop is similar to the linear state. Its energy is quite large in a dazzling state. To the distance near the sun it is for the volatility, the long distance from the sun, it is the main performance of the particle.

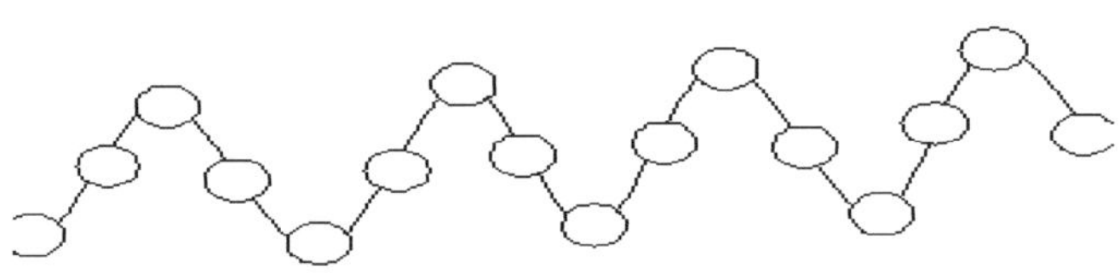

Figure 2 The curve of energy line and light quantum

Light quantum is the existence of Ethernet tiny transparent particles, and the movement is connected by the energy line with the composition of the law of volatility. When the energy is very difficult to dissipate when the performance of a linear collision process appears.

Similar to the flashlight poke, strong external force is put on the wave of light, strong energy first performs the particle, after the dissipation and then performs the volatility. This is the experimental nature of Wave - particle duality.

3. Wave-Particle Duality in Geotechnical Engineering

In geotechnical engineering, the center of the earth is equivalent to the body of the sun, and the soil particles are equivalent to the light quantum of the space, and the void water is equivalent to the energy line in the sun light.

In the event of earthquakes, that is, when the crust is moving, the distribution of more frequent and weak energy, or else the volcano will not erupt to the surface and turn to the crustal surface accumulation of volcanic lava. In the role of groundwater, the energy continues to converge to the surface.. Sand liquefaction occurs under the action of shaking vibrating sand. The author publishs papers on the foundation soil liquefaction damping research. If the liquefaction does not occur, the destructive power between the soil transfers to the surface of the building, and further passes to the high-level part so that the violent shaking. occurs.

This study presents several assumptions:

(1) the soil in the crustal energy dissipation process changes from the weak capillary water into sufficient groundwater, and then it expresses as a strong particle collision mechanism, that is, the shear strength of soil . That is, the convergence principle of energy , it convergences from low energy to high energy quantum..

(2) in the process of energy convergence , the soil particles may be broken.

(3)The thickness of the capillary water channel determines the level of energy convergence, like the pool coming together into the same river.

(4) in the surface of the soil, mostly capillary water is existed just like no volatility, fully understood as the particle. that is the linear collision state.

(5) light is formed of the light quantum and energy lines, the particle collision is changing from volatility into a straight line.

(6) the soil is the composition of soil particles and capillary water. There is no capillary water when the energy is completely transferred to the soil particles, that is, the structural principle of the soil..

In the ether space, the energy of each light quantum is E = MC2, and the kinetic energy is low so the vibration frequency is high . At last, the performance is fluctuating; when the kinetic energy is high, the vibration frequency is low, and the particle is linear.

In the soil, the distance of the soil particles from the center of gravity is larger, the the potential energy G is greater. so G is positive. E = 2 \ 1 Mv2 + G , soil particles moves to the center of the earth, which are consolidated. The closer to the center of the earth, the soil reflecs the complete consolidation of the state. Farther away from the ground and more close to the surface, the soil showed more loose state. Under the groundwater level, the water is now fluctuating, hanging up in the capillary water, volatility weakens, to no capillary water, which performs he linear relationship between particle collision. In other words, the energy in the transmission process is always volatile, but in longer distance, the volatilefrom performs liner state. The weaker the amplitude of the beat, the the frequency is lower.

Particles are only relatively speaking, volatility is permanent. .Einstein's black box theory is realized in the local space, if the light in a broader space will bend, there is a slight trend of bending, in a limited local range, the light is always along the straight line. Soil particles are consolidated.

In the soil analogy analysis, more close to the ground, the capillary water performs more complex chaotic. Under the groundwater level, the water body in the sand or silty sand shows regularity. By the Darcy law V = K i, the hydraulic gradient performs laminar flow state, that is, the law state. The greater the hydraulic gradient, the more turbulent state, that is, irregular. Of course, there are two reasons that the flow of water is not easy. One for the confining pressure of the water channel compacted closes water channel, two is that channels for the water is too thick, there is no enough water to fill the channel. When the confining pressure is too large, that is, in the deeper soil, the soil crowding degree, the soil energy iseasy into water kinetic energy, which is easy to laminar flow trend. In thc surface, the soil pore channel is large, the kinetic energy of the soil is completely transformed into the shear strength of the soil particles. At this time, the water in the soil can not fill the pore channel, so the kinetic energy of the water in the pore is not big. From the effective stress principle, p = u + a, p is the total energy, u is the excess pore water pressure, and a is the effective stress.. At the surface, the water body u = 0. a =p. The performance of the collision of soil particles is friction, that is, the shear strength of soil. At this time the performance of soil is particles.

In the volatility curve, the waterline only provides the energy transferration, the degree of damage depends on the size of the natural energy of the soil particles. If the surface the energy of the earth ,maybe it is huge enough to shake a building.. In the process of energy transmission, the energy of the soil particles is getting bigger and bigger, and finally the fragmentation of the soil particles and the collapse of the building are caused. In the light I-line or water flow connection, the energy from one end of the half-space infinite body to the other end of the infinite distance, the performance of energy more and more concentrates, and the performance of light is linear propagation,far away. That is, the collision process of light quantum.

4.The Relationship Between the Liquefied Liquefaction and the Two Particles

In the sand, when the buoyancy generated by the void water is equal to the gravity of the soil particles, the soil particles are in the suspended state. There is no friction among the soil particles,and it can be seen as a liquid fluid state, but it is just a lot bigger than water molecules. It can flow, which is sand liquefaction. When the sand has liquefaction, the water body shows volatility,and the energy transfers out, the occurrence of liquefaction is entirely due to the volatility of water molecules . The reason for the occurrence of vibration liquefaction is the vibration of the earthquake. Water molecules are full of fluctuations, and no cohesion is among the soil particles. Soil particles among the existence of water is reflected the volatility. No pore water performs the particles.

Particle directly lead to earthquake damage, volatility is mainly used to dissipate energy, so in the application, the use of particle volatility is to reduce the earthquake damage. The Particle particles are as weak as possible. .In the earthquake-prone areas, in the treatment of the foundation, to the groundwater level, the clay will be filled into sand, saturated sand in the vibration state is prone to sand liquefaction, In the middle of the ground we build a layer of saturated sand, we take the use of vibration liquefaction to reduce damage. At the surface, for effective rolling, the soil compaction becomes smaller. The use of rolling soil particles make the pore become smaller, the soil particles dense, compaction. Increase the effective stress,and increase the friction, control the movement of soil particles. In this case, the liquefaction will dissipate the main energy, the energy delivered to the surface is minimal, the surface seismic capacity is strong, because the surface is hard soil. The surface soil particles are compact, reducing the collision motion of the particles, so it is not exercise, which is the equivalent of no energy transmission.. The energy is enclosed in the sand liquefaction zone. For example, as if in the iron drum in the agitation of water, water constantly moves, which has the constant consumption of energy, but the iron has always no movement. This is like a liquefied soil layer is set to hard soil layer of protective layer role. Liquefied office access to the fine sand for the useless soil, the cost is very low. we can use waste and greatly improve the use of fine sand, turning waste into treasure.

5.Iron Bucket Effect

Put an iron bucket at the ground, the bucket fills with water, with the utensils or hand kept stirring the water inside the bucket, but the barrel does not move, everything outside the barrel seems to have nothing happened. Water is the damper of the earthquake, and now the water for the dampers is taken place of saturated fine sand, which can occur liquefaction.. Iron bucket is buckled under the surface, the barrel is filled with liquefied saturated fine sand.. We can study how much the effect of iron drum on the shock absorption, How does the effect of saturated fine sand liquefaction on the effect of reducing vibration. Analysis of shock absorption from two perspectives: (1) dissipation of vibration energy, take the use of volatility to reduce the seismic energy, that is, the method of dredging; (2) with the resistance method of shock absorption, the iron is very good line of very dense soil, there is a certain degree of stiffness, with less prone to deformation. The volatility of the earthquake and the particle are completely separated from the outside world. Use the volatility and then hinder the particle transferring energy. The dissipated energy is separated by an isolator. That is, the energy is completely consumed by volatility, and the energy is not allowed to pass through the iron drum, and the seismic energy is shielded with a hard and dense barrier. The inverted buckets can be analogous to the support piles of the foundation pit, and the support pile behind the backfill is similar to the outer wall of the drum.

As Figure below:

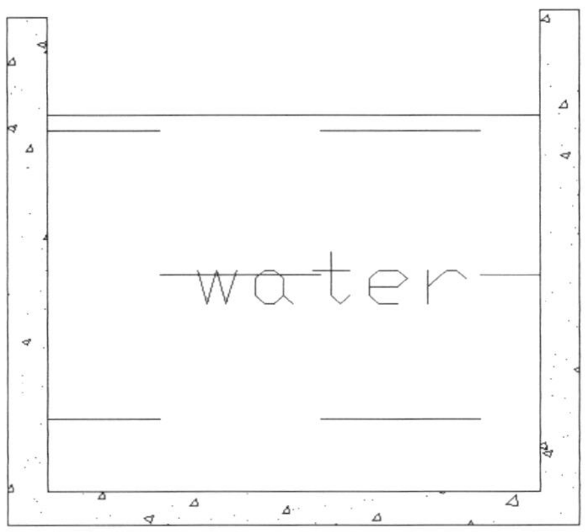

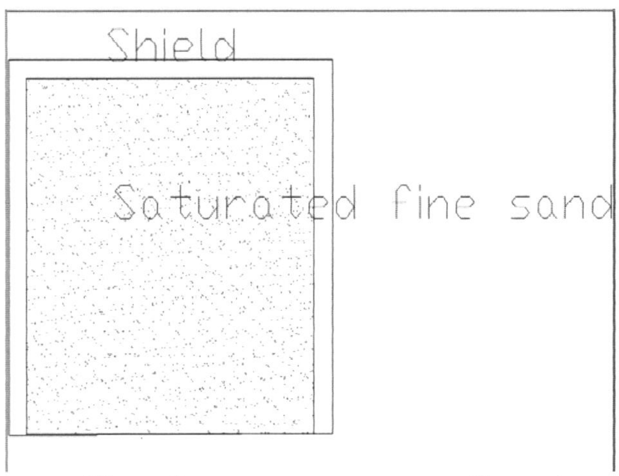

Figure 3 **Figure 4**

6. Conclusions

In the role of seismic load, the soil may occur liquefaction damping effect , and it may also not meet the conditions of soil liquefaction, and the energy transfers further to the upper high-level structure. In the process of energy transferration, seven conclusions are summarized as follows:

1) In the process of soil energy transferration, the soil particles are equivalent to the light quantum, the water is equivalent to the light line, and the light quantum wears into a curve which can be volatile material;

2)Volatility is eternal, only the strength of the fluctuations is high or low, the particle is weak volatility, the particle is relatively speaking. That is, the linear propagation of light is relatively speaking;

3)In the light transmission process, the energy from the low energy level to high energy level, from weak to high energy, it is mainly because of the superposition of energy concentration;

4)Water is a volatility, the water is the earthquake dampers, with cheap saturated powder sand as a damper, so that fine sand liquefaction is used to consume energy to reduce shock;

5)Iron bucket effect reflects the fine liquefied sand can be used to reduce shock, shield can be effectively isolated earthquake from our buildings;

6) The inverted buckets can be analogous to the support piles of the foundation pit, and the support pile behind the backfill is similar to the outer wall of the drum.

IOP Conf. Series: Materials Science and Engineering **250** (2017) 012032 doi:10.1088/1757-899X/250/1/012032

7.References

[1] Wei Li, Demin Wang, Chao Li, Qinghui Zeng.Exlperimental Research on Settlement Distribution of Composite Foundation Configured With Walled Soil Improvement. ISSN:1660-9336.

[2] Feng Yong-gen, Wang Bin, Wei Dong.Discussion on the Mechanism of Reinforcing Foundation by Dynamic Consolidation Method. Exploration Science and Technology. 1988 (2): 27-30

[3] Qian Jiahuan, Qian Xuede. Theory and Practice of Dynamic Consolidation. Journal of Geotechnical Engineering. 1986 (6): 25-28

[4] Ye Weimin, Tang Yiqun, Yang Lin.Study on the Effect of Dynamic Consolidation on Reinforcement of Saturated Soft Clay Foundation. Rock Mechanics. 1998 (3): 72-76

[5] Wang Demin. Foundation soil liquefaction damping research. Urban and rural construction economy yearbook.

IOP Conf. Series: Materials Science and Engineering **250** (2017) 012033 doi:10.1088/1757-899X/250/1/012033

Numerical Simulation on Cuttings Transport with Drillpipe Rotation in Extended Reach Well

Xiaole Guo and Wenyin Li

School of oil and gas engineering, Chongqing University of Science and Technology, 20th East Road Street, University Town, Chongqing, P.R. China.
Email: alpha2003 @126.com

Abstract Computational Fluid Dynamics (CFD) is used to simulate the solid-liquid two-phase flow in the annulus. The results show that the mixture of liquid and solid moves in spiral in horizontal annulus, and the solid phase distributes asymmetrically in the annulus. The distribution of solid volume fraction and the velocity of liquid phase and solid phase in annulus are given. The thickness of cuttings bed of rotating drill pipe is 1/3 of that of non-rotating drill pipe. Main factors which influence the cuttings transport in the annulus are analyzed, and the following results are got: the thickness of cuttings bed changes sensitively when the rotation speed of drill pipe ranges from 80rpm to 120rpm, the rotation of drill pipe disturbs the flow of liquid and solid in annulus, improving the cleaning of annulus.

1. Introduction

The biggest difference between Extended Reach Wells (ERWs) and conventional wells is that it has a long horizontal displacement. Cuttings take a long time to reach the wellhead from the bottom, which makes it very difficult to remain hole cleaning because of the formation of cuttings bed.

Many scholars have carried out a lot of research on cuttings transport in ERWs [1], and established a number of models [2-3]. These models can be broadly divided into two categories, one is theoretical model including steady models like it in [4-8] and second is the experimental regression model like it in [9]. The regression model is generally based on the experimental data regression, so it can be used to calculate the influence of drill pipe rotation on the cuttings bed height, but it can not explain the mechanism of the impact of drill pipe rotation on the cuttings transport [10]. The theoretical models can be subdivided into two layers steady model like it in [4], three layers steady model like it in [5], two layers dynamic model like it in [6] and three layers dynamic model like it in [7]. Most of these theoretical models ignored or simplified the effect of the drill pipe rotation on cuttings transport. The reason is that the rotation of the drill pipe can make the two-phase flow in annular become more complicated and difficult to solve. As a result, these theoretical models can not be used in the field design because of the big error and cannot be used to analyze the influence of drill pipe rotation on the cutting transport either.

Thus, a numerical model of cuttings transport solved by Fluent software is presented to simulate the cuttings transport in fluid in this paper. The change laws of concentration, velocity and pressure of liquid or cuttings with or without pipe rotation are analyzed. The simulation result has a very great theoretical and engineering value to hole cleaning technology of ERWs.

2. Numerical Model

Euler multiphase flow model is adopted to analyze the problem. The mathematical model is shown as follows.

（1）Continuity equation

$$\frac{\partial}{\partial t}(\alpha_q \rho_q) + \nabla \cdot (\alpha_q \rho_q \bar{v}_q) = \sum_{p=1}^{n} \dot{m}_{pq} \tag{1}$$

（2）Momentum equation

$$\frac{\partial}{\partial t}(\alpha_q \rho_q \vec{v}_q) + \nabla \cdot (\alpha_q \rho_q \vec{v}_q \vec{v}_q) = -\alpha_q \nabla p + \nabla \cdot \overline{\overline{\tau}}_q + \sum_{p=1}^{n} (\vec{R}_{pq} + \dot{m}_{pq} \vec{v}_{pq}) + \alpha_q \rho_q (\vec{F}_q + \vec{F}_{lift,q} + \vec{F}_{Vm,q}) \tag{2}$$

In the above formulas, ρ is density, v is velocity, α is the volume fraction, $\vec{F}$ is the external volume force, $\vec{F}_{lift}$ is the lift force, $\vec{F}_{Vm}$ is the virtual mass force, $\vec{R}$ is the interphase force, p is the pressure, $\dot{m}$ represents mass transfer between two phases, $\overline{\overline{\tau}}$ is the pressure strain tensor, subscript p and q represents a certain phase respectively.

3. Model Verification

According to the actual situation of extended reach drilling, this paper chooses the 215.9mm diameter of hole and the 127.0mm diameter of drill pipe with an eccentricity of 0.5 as the annular border. Considering the capability of the computer CPU and memory, the axial length is set to 20m. Through a lot of trial, the final grid scheme is determined as 56x8x360=161280. The grid division diagram of annular section is shown in Figure 1 below. After the meshing work done, the software of Fluent was used to complete the rest simulation works.

The comparing result between the numerical model in this paper and Wang's model [9], which is regressed from experimental data, is shown in Figure 2. The Wang's experimental parameters are used in the simulation and comparison. We can see that the two calculation results are close. But there are also some differences. The result of Wang's model is approximate to a straight line, while our numerical model is more like a parabolic curve. The result of cuttings bed height calculated by our model is higher when the flow rate is small and lower when the flow rate is high than it using Wang's model.

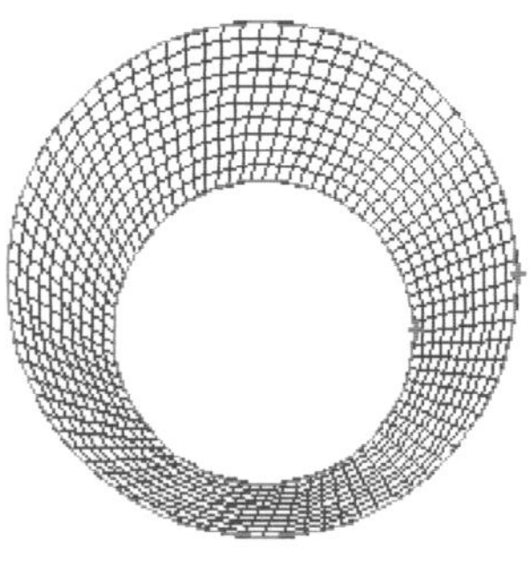

Figure 1. Grid division diagram

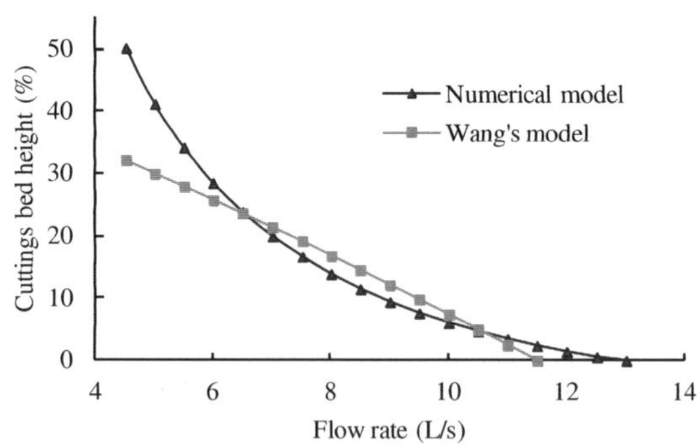

Figure 2. Comparison with Wang's model

IOP Conf. Series: Materials Science and Engineering **250** (2017) 012033 doi:10.1088/1757-899X/250/1/012033

4. Simulation Result

The default values used by simulation are as follows. The pump rate is 30 L/s, the drillpipe rotation speed is 100 rpm the eccentricity is 0.5, the viscosity of fluid is 25cP the inlet volume fraction of cuttings is 5%, and the cuttings diameter is 1mm. The section where the cuttings volume fraction is bigger than 70% is defined as fixed bed area and the **section** where the cuttings volume fraction is between 10% and 70% is defined as mobile bed area.

4.1. Cuttings Distribution in Annular

Cuttings distribution with drillpipe rotation and without rotation are shown in Figure 3. It can be seen that the distribution of cuttings is not symmetrical because of the drillpipe rotation. In the right side of annular, the fluid flows upward drived by rotating drillpipe and carries more cuttings into the upper flow area, which leads to a loose cuttings bed in large range. But the left side is the opposite. In comparison with the integration of cuttings volume fraction without rotation, the cuttings amount is reduced by nearly 1/3 with drillpipe rotation, which shows the important effect of drillpipe rotation on cuttings transport.

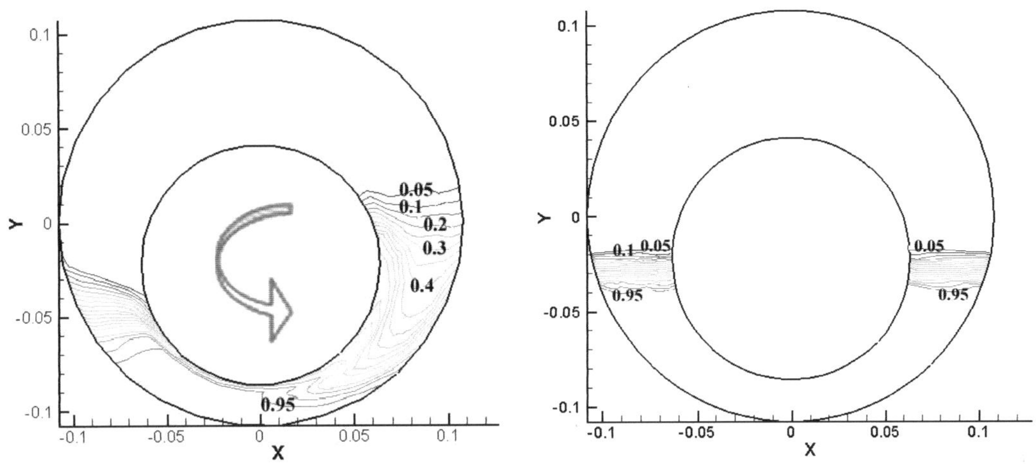

Figure 3. Cuttings volume fraction distribution with (left) and without (right) pipe rotation

The solid velocity distribution in the annular space with and without pipe rotation are shown in Figure 4. It can be seen from the figures, due to the rotation the cuttings bed in lower part of annular becomes loose and part of the cuttings jump into the upper liquid flow area, which makes the cuttings concentration of upper liquid flow region increases and the fluid flow energy decreases. As a result, the velocity of solid phase in upper layer decreases, but the velocity of solids in cuttings bed of lower region increases significantly, which makes the average velocity of solids in annular increases.

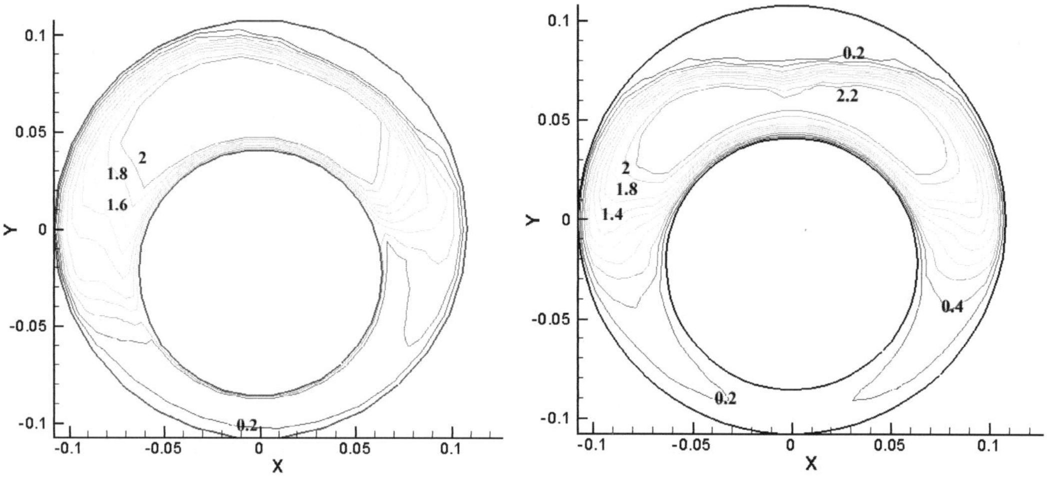

Figure 4. Cuttings velocity distribution with (left) and without (right) pipe rotation

4.2. Effect of Pipe Rotation and Eccentricity on Cuttings Bed Height and Velocity
Curves of the height of cuttings bed and the average velocities of each area of the annular at different rotation speeds are shown in Figure 5. As can be seen, with the increase of rotation speed, more cuttings are carried into the upper part of the liquid flow area and taken away. As a result, the cuttings bed height is declined, the velocity of fixed cuttings bed and mobile cuttings bed is increased, while the rate of upper area of liquid flow is decreased. It can also be seen that the decreasing speed becomes small when the rotation speed is more than 100 rpm. From Figure 6 it can be seen that, with the rising of eccentricity, the height of the cutting bed increases. But the influence of eccentricity is more complex, which is related to the height of the bed height and the geometrical size of the annular.

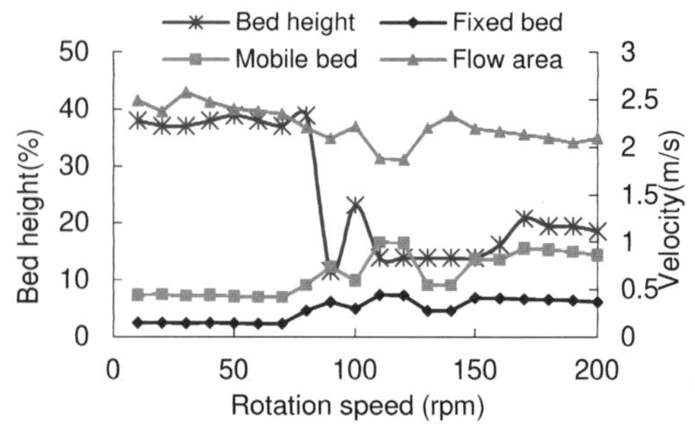

Figure 5. Cuttings bed height and solid-phase velocity vs. pipe rotation speed

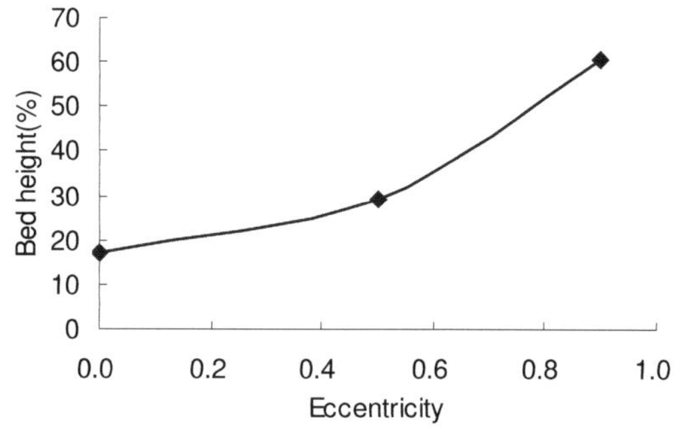

Figure 6. Cuttings bed height vs. pipe eccentricity

4.3. Effect of Flow Rate and Fluid Properties on Cuttings Bed Height and Velocity
Curves of the height of cuttings bed height and the average velocities of each area of the annular at different flow rates are shown in Figure 7. As can be seen, the bed height declines and the velocities increases with the increase of flow rate. The changing speeds reach maximum while the flow rate is between 30～40L/s. When the flow rate reach 50 L/s, almost all the cuttings are moved up and there is no cuttings deposited. The influence of fluid viscosity and density on the height of cuttings bed is shown in Figure 8. It can be seen that, with the increase of the viscosity and density of the fluid, the height of the cutting bed is reduced.

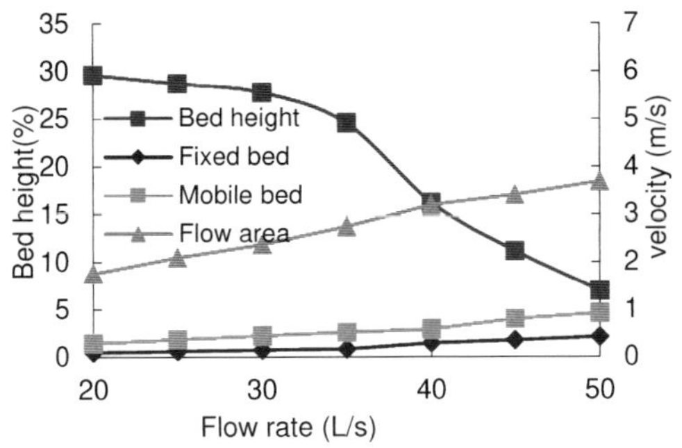

Figure 7. Deposition bed height and solid-phase velocity vs. flow rate

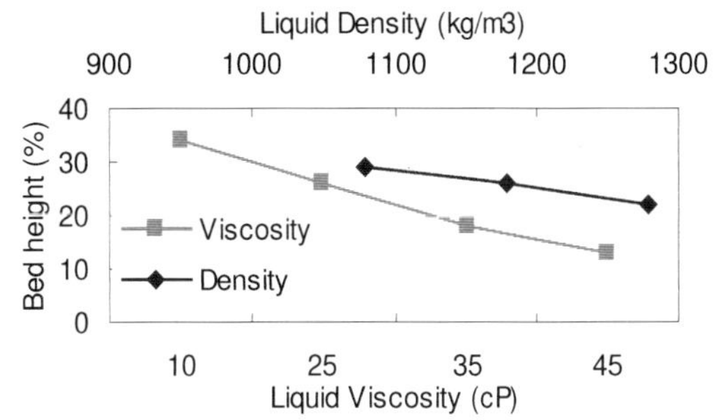

Figure 8. Deposition bed height vs. liquid viscosity and density

4.4. Effect of Cuttings Diameter and Inlet Concentration on Cuttings Bed Height

The influence of cuttings diameter on the height of cuttings bed is shown in Figure 9. It can be seen that with the increase of cuttings diameter the height of the cutting bed is increased. The ROP increases with the rising of inlet concentration of cuttings, and thus the inlet concentration of cuttings is used instead of ROP in this paper. From Figure 10, it can be seen that the height of the cuttings bed increases with the rising of the inlet concentration of cuttings.

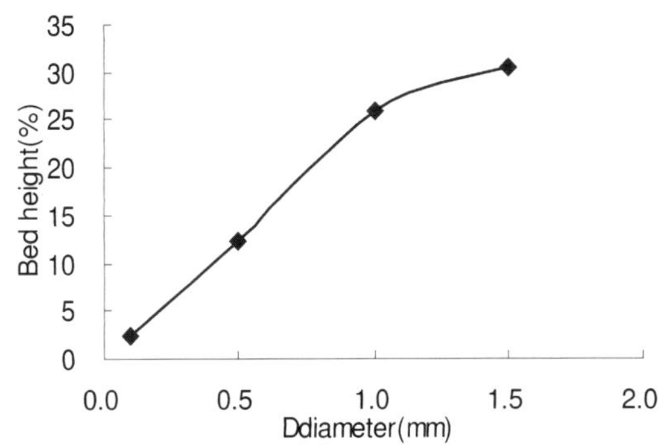

Figure 9. Deposition bed height vs. cuttings diameter

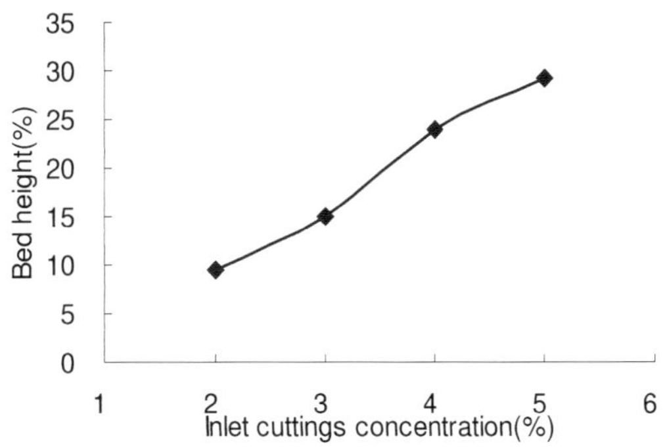

Figure 10. Deposition bed height vs. inlet concentration of cuttings

5. Conclusion

(1) Drillpipe rotation turns the axial motion of cuttings in annular into the helical motion and the distribution of cuttings concentration and velocity become symmetrical in annular.

(2) Drillpipe rotation can not only reduce the cuttings concentration in annular, but also increase the cuttings velocity, and thus promotes the hole cleaning condition. But it also increases the annular pressure drop.

(3) The cuttings bed height decreases effectively with the rising of flow rate. There is a sharply declination of cuttings bed at the rotation speed of 80 to 120 rpm. When the rotation speed is more than 120 rpm, the cuttings bed height does not change obviously. Thereby, the speed of more than 80 rpm is recommended in fields to enhance the important role of drillpipe rotation on hole cleaning.

(4) With the increase of the viscosity and density of the fluid, the cuttings bed height decreases. With the rising of cutting diameter, ROP and eccentricity, the cuttings bed height increases.

6. Acknowledgment

This research is funded by project supported by Scientific and Technological Research Program of Chongqing Municipal Education Commission (Grant No. KJ1401318) and project 51404050 supported by National Natural Science Foundation of China.

7. References

[1]. Li, J., Luft, B., 2014, Overview of Solids Transport Studies and Applications in Oil and Gas Industry - Experimental Work. SPE Russian Oil and Gas Exploration & Production Technical Conference and Exhibition, 14-16 October, Moscow, Russia, SPE-171285-RU.

[2]. Li, J., Luft, B., 2014, Overview Solids Transport Study and Application in Oil-Gas Industry-Theoretical Work. Int. Petroleum Technology Conference, 10-12 December, Kuala Lumpur, Malaysia, IPTC-17832-MS.

[3]. Nazari, T., Hareland, G., & Azar, J. J., 2010, Review of Cuttings Transport in Directional Well Drilling: Systematic Approach. SPE Western Regional Meeting, 27-29 May, Anaheim, California, USA, SPE-132372-MS.

[4]. Wang ZhiMing, Zhang Zheng, 2003, Study on Cuttings Transportation Tendency of the Dual Layer stabilizing model in highly inclined well, Oil Drilling & Production Technology, vol. 25, no.4, pp8-10.

[5]. Guo Xiaole, Wang Zhiming, 2007, Study on Cuttings Transport in ERW, Proc. 20th National Conference on Hydrodynamics, 23-25 August, Tai Yuan, Shan Xi, China.

[6]. Wang Zhiming, Zhang Zheng, 2004, A Two-Layer Time-dependent Model for Cuttings Transport in Extended-reach Horizontal Wells, Journal of Hydrodynamics, vol. 19, no. 5, pp676-681.

[7]. Guo Xiaole, Wang Zhiming, Long Zhihui, 2010, Study on Three-Layer Unsteady Model of Cuttings Transport for Extended-Reach Well, Journal of Petroleum Science and Engineering, vol. 73, pp171-180.

[8]. Guo Xiaole, Wang Zhiming, Long Zhihui, 2011, Transient Cuttings Transport Laws through All Sections of Extended Reach Well, Journal of China University of Petroleum, vol. 35, no. 1, pp72-76.

[9]. Wang Haige, Liu Xisheng, Ding Gang, 1993, The Model of Cuttings Bed Thickness in Horizontal Well Section, Journal of the University of Petroleum, China, vol. 17, no. 3, pp25-32.

[10]. LOCKETT, T.J., Richardson, S.M., Worraker, W. J., 1993, The Importance of Rotation Effects for Efficient Cuttings Removal during Drilling", SPE/IADC Drilling Conference and Exhibition, 22-25 February, Amsterdam, Netherlands, USA, SPE-25768-MS.

IOP Conf. Series: Materials Science and Engineering **250** (2017) 012034 doi:10.1088/1757-899X/250/1/012034

Study on Frost Heaving Characteristics of Canal Foundation Soil in Hadashan, China Irrigation Area

Zuozhou Li[1], Donghe Ma[2, 3], Chengcheng Zheng[1]and Rui Wang[2, 3]

[1]ShanXi ZhenAn Pumped Storage Co.,Ltd,State Grid Xinyuan Company, Fengning, China

[2] Chin Water Northeastern Investigation, Design & Research Co., Ltd., Changchun, China

[3] Cold area engineering technology research center of Ministry of water resources , Changchun, China

Abstract: Frost heaving characteristics of canal foundation soil in HaDashan area are analyzed and studied by indoor frost heave test and field test. The research results show that, under the experimental conditions, the frost heave of the canal soil increases with time and temperature, but after a certain period of time, the amount of frost heave tends to be stable; the frost heaving process is divided into 4 stages: unstable freezing, continuous freezing, stable freezing and thawing. The research results have certain guiding significance for the construction of large channel in HaDashan area.

1. Introduction

In cold regions such as northeast, northwest and high altitude areas in China, cold weather in winter caused freezing of different depths of crustal surface soils, so that many kinds of engineering and structures which are located on the soil base and water are destroyed by ice pushing, frost heaving and freezing thawing, resulting in the destruction of different parts, forms and degrees. The water conveyance channel has the characteristics of long lines, complicated geological conditions, diverse soil quality of filling materials and light quality of channel lining, as well as different winter water conditions and freezing effects. The freezing damage is more likely to occur than other hydraulic structures.

The main water conveyance canal of Hadashan water control project is located in the cold area. and the channel soil is single, basically clay and loess loam, and the preliminary analysis is frost heaving soil.Therefore, it is necessary to carry out indoor and field frost heave observation test to define the frost heave type of foundation soil and provide scientific basis for design and construction.

2. Methodology

2.1 Indoor Frost Heave Test

The indoor frost heaving test was carried out, taking into account the high groundwater level and supplying the frost heaving front, the open system test was used to test the frost heave under the condition of external water supply [1-3]. The frost heave test of the open system is designed with 4 groups, and the test results are shown in Figure 1 and table 1:

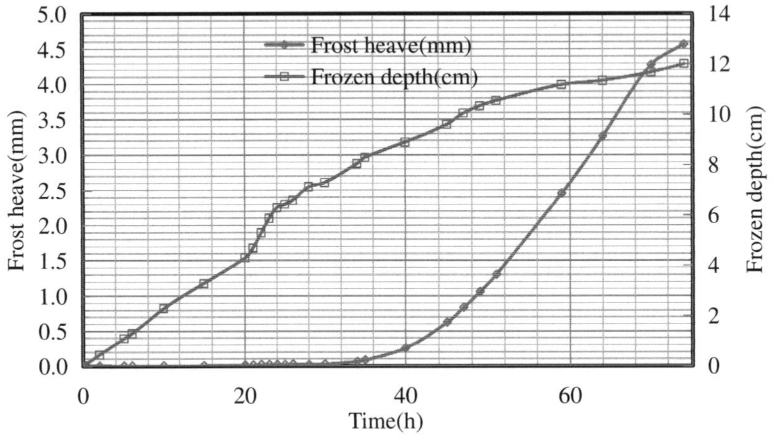

Figure 1. frost heave volume and frozen depth curve

Table 1 HaDaShan drainage base soil frost heave test results table

Number	Water content (%)	Frost heave(mm)	Frost heaving rate(%)
1-1	17.4	4.57	3.95
1-2	18.0	5.23	4.37
1-3	17.2	5.50	4.56
1-4	16.5	4.80	4.03
average value	17.3	5.03	4.23

Test results show that the frost heave of canal foundation soil increases with the freezing depth; under the condition of test, the frost heave of canal foundation soil increases with time and temperature decrease, but reach a certain time after the frost heave stabilized; frost heave of canal foundation soil increases with the increase of water quantity.

2.2 Freezing Process and Freezing Depth of Soil in Site

In the seasonally frozen soil region, the temperature is the upper boundary condition of soil freezing. With the seasonal variation of temperature, the frozen soil melts alternately[4-5]. According to the weather data for many years in the region, the temperature began to be negative in late October to early November, When the soil temperature is lower than the freezing temperature of soil water, soil water particle surface freezes and the soil to form one form of frozen soil, as the temperature decreases, the freezing front also gradually to the lower layer forward until it reaches the maximum frozen depth. After that, the temperature began to thaw, and the frozen layer began to melt in two directions. On the one hand, the temperature rose and the soil on the upper crust melted. On the other hand, under the influence of heat flow in the ground, the upper layer is ablated from the lower layer, and the frozen soil layer is connected between the middle of May and the end of May.

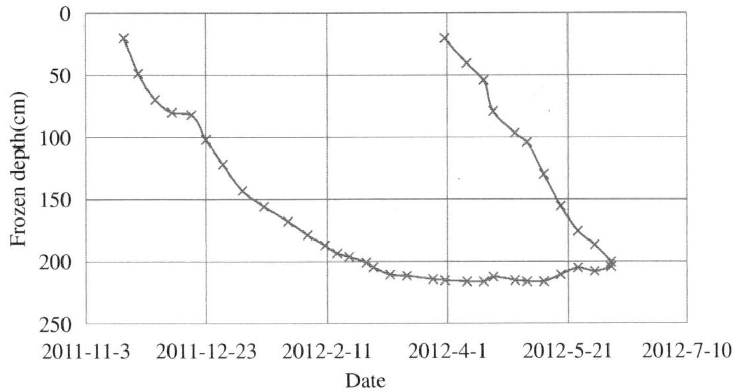

Figure 2. development process of freezing depth

It can be seen from Fig. 2 that the freezing-melting process of winter soil in the test site can go through the following stages:

The first stage is not stable freezing stage: generally occurs in late October to early November, the temperature fluctuate around 0 ° C, the surface soil freezing and thawing day and night, freezing thawing frequently, can not form stable frozen layer.The second stage is the continuous freeze stage: generally occurred in late November, and when the average temperature dropped to a steady negative temperature, the freezing depth began to increase, and reached the maximum freezing depth in late February.In this stage, although the freezing depth fluctuates briefly with the fluctuation of the temperature, but the overall trend is gradually increasing.. The third stage is the stable freezing stage: generally occurs in late February to late March, when the frost depth reach or close to the maximum, there are more than a month of time, due to the freezing interface in the heat balance, the freezing front is relatively stable, that is basically no development or reduced freezing depth.The fourth stage is the thawing stage: when the average temperature in late March reached the positive temperature, the surface began to melt from top to bottom. Thereafter, the bottom of the frozen layer began to melt from bottom to top until the whole layer of the melting layer disappeared. It can be seen that the melting of the frozen layer is carried out in both directions.

3. Results and discussion

The maximum frost heave and frost heave rate of the test ground in each year are shown in table 2. Seen from the table, in the test field, whether it is cold or warm year, surface amount of frost heaving are strong frost, according to the 《freezing resistance design specification》 (GB/T50662-2011) in division level are frost heave grade IV~Ⅴ.

Table2 The maximum frost heave of the surface in each year

Year	2008-2009	2011-2012
Maximum frost heave（cm）	16.0	19.3
Frost heaving rate （%）	12.7	10.7

4. Conclusion

(1) Channel soil frost heave increases with the development of the deep freeze; under the condition of test, the frost heave of canal foundation soil increases with time and temperature increased, but reach a certain time after the frost heave stabilized; frost heave channel soil increased with the increase of water supplement.

(2) In the test field, whether it is cold or warm year, surface amount of frost heaving are strong frost, according to the 《freezing resistance design specification》 (GB/T50662-2011) in division lcvcl arc frost heave grade IV~Ⅴ.

(3) The frost heave test results show that the freezing-thawing process in winter can be divided into four stages: unstable freezing, continuous freezing, stable freezing and thawing.

IOP Conf. Series: Materials Science and Engineering **250** (2017) 012034 doi:10.1088/1757-899X/250/1/012034

5.References

[1] Luo Xianhu, Xu Xing, Shi Xiaobing, Cheng Zongheng. Principle and Method of Submarine Sediment Thermal Conductivity Measurement in Laboratory- with TK04 Thermal Conductivity Measurement System as Examle[J].OCEAN TECHNOLOGY, 2008, 27(6):89-91.

[2] Ji Lan, Xiong Zhiwei, Cheng Jia, Cai Hancheng. Experimental study on frost heave performance improvement of coarse grain filling in high speed railway subgrade [J] .Railway Engineering,2016,(2):102-105.

[3] Han Chunpeng, He Dongpo, Jia Yanmin, Cheng Peifeng.Frost heave properties of lime modified subgrade soil[J].Journal of Chang'an University(Natural Science Edition), 2013, 33(3):27-29.

[4] Jiang Haibo, Tian Yan. Test of frost heaving damage mechanism of rigid-soften composite trapezoidal canal in seasonally frozen ground region [J]. Transactions of the Chinese Society of Agricultural Engineering, 2015, 31(16):145-148.

[5] Wang Jiangwei. Experiment research of frost heave on U-shaped canal with concrete linging and discussion of frost heaving mechanics model[D]. Ningxia University,YinChuan,2016.

IOP Conf. Series: Materials Science and Engineering **250** (2017) 012035 doi:10.1088/1757-899X/250/1/012035

Kinetic Analysis of the Main Temperature Stage of Fast Pyrolysis

Xiaoxiao Yang, Yuying Zhao, Lanshu Xu and Rui Li*

College of materials science and technology, Beijing Forestry University, 100083, Beijing, China

Email: 13020021531@163.com

Abstract. Kinetics of the thermal decomposition of eucalyptus chips was evaluated using a high-rate thermogravimetric analyzer (BL-TGA) designed by our research group. The experiments were carried out under non-isothermal condition in order to determine the fast pyrolysis behavior of the main temperature stage (350-540°C) at heating rates of 60, 120, 180, and 360°C min^{-1}. The Coats-Redfern integral method and four different reaction mechanism models were adopted to calculate the kinetic parameters including apparent activation energy and pre-exponential factor, and the Flynn-Wall-Ozawa method was employed to testify apparent activation energy. The results showed that estimation value was consistent with the values obtained by linear fitting equations, and the best-fit model for fast pyrolysis was found.

1. Introduction

In view of the important role that fast pyrolysis plays in thermal conversion process of biomass, a great need exists for robust comprehensive biomass pyrolysis mechanism that could predict product speciation and yields as well as optimize process conditions. Non-isothermal thermogravimetric analysis (TGA) is the most common method to investigate the kinetics , but heating rates often confined to lower than 120 °C min^{-1} due to commercial instrument limitation, which is quite different from fast biomass pyrolyisis case. In the present work, we investigated fast pyrolysis behavior of eucalyptus chips tested by a high-rate thermogravimetric analyzer (BL-TGA) designed by our group at different heating rates (60, 120, 180, and 360 °C min^{-1}) under nitrogen atmosphere, and determined the kinetic parameters of the main temperature stage through the Coats–Redfern integral method and four different reaction mechanism models (RO1, RO2, R2, D1). The Flynn-Wall-Ozawa method was used to testify apparent activation energy and determine the best-fit model.

2. Experimental

2.1. Materials

Eucalyptus obtained from Guangxi Province, China were air-dried and sliced into sheets with a area of 50 mm×10 mm and thickness of 2 mm. which has the chemical composition as Table 1.

Table 1. Chemical composition of eucalyptus chips.

Sample	Moisture (%)	Ash (%)	Cellulose (%)	Hemicellulose (%)	Lignin (%)
Eucalyptus chips	5.8	0.3	41.02	20.21	30.43

2.2. *Fast Pyrolysis Kinetic Analysis*

Under non-isothermal conditions, the actual temperature can beexpressedin terms of the initial temperature T and the heating rateβas follows:

$$\frac{d\alpha}{dT} = \frac{A}{\beta} f(\alpha) \exp\left(-\frac{E}{RT}\right) \quad (1)$$

WhereT is the temperature, A is the pre-exponential factor,E is the apparent activation energy, and R is the universal gas constant(8.3145Jmol^{-1} K^{-1}), α is the mass-loss rate, expressed by $\alpha = M_0 - M_i / M_i - M_\infty$. $M_0$is the initial mass of the sample, M_i is the mass at a given time t, andM_∞ is the final mass of the sample and f(α)is the reaction model.

The integral function was defined:

$$G(\alpha) = \int_0^\alpha \frac{d(\alpha)}{f(\alpha)} \quad (2)$$

Combination of (1) and (2) gives,

$$G(\alpha) = \frac{A}{\beta} \int_{T_0}^{T} \exp\left(-\frac{E}{RT}\right) dT \quad (3)$$

Through the approximate induction of integration with respectto temperature, Coats-Redfern equation is derived asfollows and E and A can be calculated through fitting the linercurve:

$$\ln\left[\frac{G(\alpha)}{T^2}\right] = \ln\left[\frac{AR}{\beta E}\left(1 - \frac{2RT}{E}\right)\right] - \frac{E}{RT} \quad (4)$$

Flynn-Wall-Ozawa formula was obtained by the transformation of equation(3):

$$\log\beta = \log\frac{AE}{RG(\alpha)} - 2.315 - 0.4567\frac{E}{RT} \quad (5)$$

Flynn-Wall-Ozawa method can directly determine valueE, which avoidedcalculation errorcaused by different assumptions of reaction mechanism models select in Coats-Redfern equation.

Different reaction mechanism models G(α) elucidating fast pyrolysis process of eucalyptus chips werepresented in Table 2,which will be put into equation (4). The best-fitmodel was selected by usingthe Flynn-Wall-Ozawa method through the comparison of apparent activation energies obtained by two methods

Table 2. Different reaction mechanism models of pyrolysis with various functions of G(α)andf(α).

Reaction model		G(α)	f(α)	Reaction mechanism
Chemical reaction	RO1	$-\ln(1-\alpha)$	$1-\alpha$	First-order reaction
	RO2	$(1-\alpha)^{-1}-1$	$(1-\alpha)^2$	Second-order reaction
Phase boundary reaction	R2	$1-(1-\alpha)^{1/2}$	$2(1-\alpha)^{1/2}$	Two-dimensional, shrinking cylinder
Diffusion-controlled reaction	D1	α^2	$1/2\alpha$	One-dimensional diffusion, parabolic law

3. Results and Discussion

3.1. Fast Pyrolysis Behaviour of the Main Temperature Stage

Figure1 shows TG mass-loss curves and corresponding DTG analysis curves atdifferent heating rates: 60, 120, 180 and 360 °Cmin^{-1}. As shown in Figure 1,many peaks could be distinguished from the DTG curves at around 350-540 °C, which indicated that main mass loss stages during fast pyrolysis. From DTG curve, we can conclude that the process could be mainly divided into four stages: moisture loss, the early stage of mass loss, the main stage of fast pyrolysisand charring. Table 3 showed the main temperature points of decomposition observed in DTG curves. Both curves and data could obviously illustrated that the most rapid mass loss stage shifted to higher temperature corresponding with increasing heating rates,which is the joint outcome of mainly different reaction conversion and slightly physically uniform heat transfer.

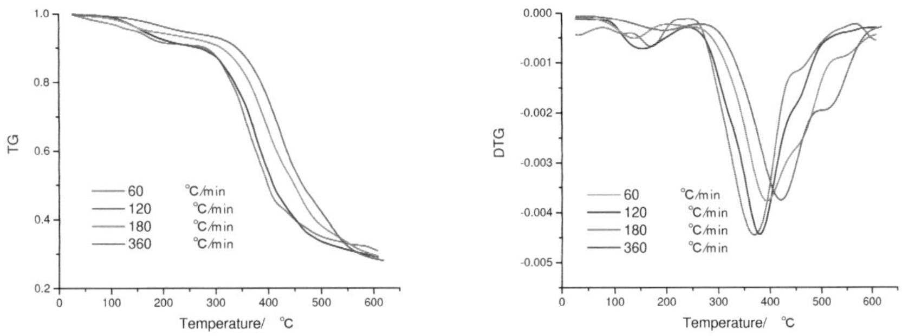

Figure1.TG (left) and DTG (right) curves of eucalyptus chips at different heating rates.

Table 3. The main temperature points of decomposition.

Heating rates β(°C min-1)	60	120	180	360
First temperature point T (°C)	348.2	350.4	366.3	414.7
Second temperature point T (°C)	382.4	401.7	434.1	468.9
Third temperature point T (°C)	409.8	431.0	458.1	494.6
Fourth temperature point T (°C)	479.8	493.7	509.6	533.0

Moreover, as shown in Table 4, when the temperature reached the main stage of fast pyrolysis (350 °C), the weight proportion at the heating rate of 60 °Cmin^{-1} was the lowest observed, about 0.8174 and the 360 °Cmin^{-1} was the highest, about 0.9284. It indicated that samples had been consumedmuch more in low-temperature stage of lower heating rates, and consequently produce higherresidue yield ,about 0.3074 in case of 60 °Cmin^{-1}. This is attributed to that slow pyrolysis of low heating ratescan lead to the formation of a large amount of carbon. From Figure1.we can see the higher heating rates could heat the samples up to required temperature in a short time and reveal fast pyrolysis mechanism more perfectly.

Table 4. Mass loss process of fast pyrolysis.

Final weight / Temperature T (°C) / β(°C min⁻¹)	200	350	540	650
60	0.9240	0.8174	0.3475	0.3074
120	0.9343	0.8360	0.3342	0.2905
180	0.9507	0.9038	0.3910	0.2987
360	0.9723	0.9284	0.4622	0.2945

3.2. Kinetic Analysis of the Maximum Mass-loss Points

Coats-Redfern integral method and four kinds of reaction mechanism models (RO1, RO2, R2, D1) described in section 2.2 were applied to estimate the kinetic parameters of the maximum mass-loss points. The fitting straight lines of various reaction mechanism models at different heating rates were showed in Figure 2. and the corresponding kinetic parameters were summarized in Table 5.

As shown in Figure 2, the fitting straight lines had significant difference and their intercepts were utterly inconsistent at different heating rates. However, the lines simulated by the four different reaction mechanism models (RO1, RO2, R2, D1) had similardistribution order and theircorrelation coefficients were very high. The finding from A. V.Bridgwater also confirmed that the integral methods gave better fitting results. The specific parameters were shown in Table 5, the highest apparent activation energy values were obtained by the diffusion-controlled reaction model (D1), for which pre-exponential factor Aalso had the highest value. On the contrary, the corresponding values obtained by the phase boundary reaction model (R2)were thelowest, and the chemical reaction mechanism models (RO1, RO2) hadsimilar apparent activation energy values. The changes of the apparent activation energy values calculated by the four different reaction mechanism models (RO1, RO2, R2, D1) at the maximum mass-loss points of different heating rates can be obviously observed in Figure 3. The trend of the change could be vividly presented and the apparent activation energies obtained by the chemical reaction mechanism models (RO1, RO2) had a slight change. The variationcould be attributed to the mass and heat transfer limitations at different temperature points and heating rates.

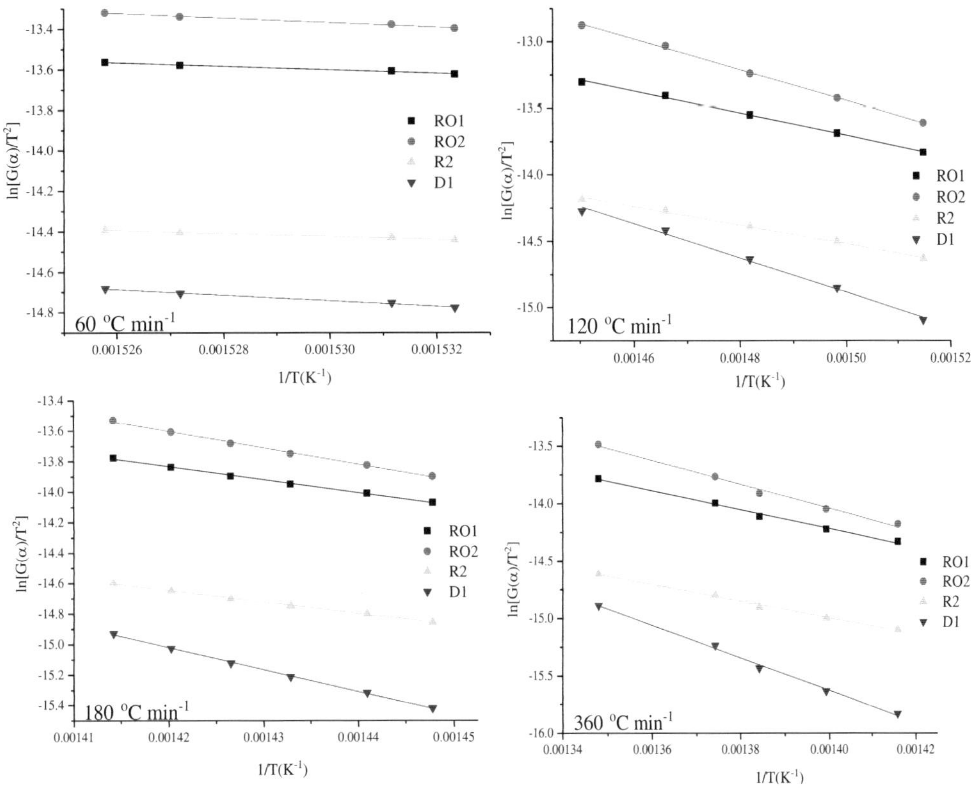

Figure 2. Fitting straight lines ofvarious reaction mechanism models at different heating rates

Table 5. Kinetic parameters of various reaction mechanism models under different heating rates.

Heating rates β (°C min^{-1})		60	120	180	360
Maximum mass-loss points T (°C)		382.4	401.7	434.1	468.9
Fitting straight line equations	RO1	y = -8631.9x - 0.3934	y = -8379.8x - 1.1345	y=-8547.2x - 1.6956	y = -8213.2x - 2.7187
	RO2	y = -11294x + 3.9122	y = -11553x + 3.8918	y=-10802.0x + 1.7373	y = -10380x + 0.4922
	R2	y = -7362.4x - 3.1584	y = -7005.9x - 4.0085	y=-7494.0x - 4.0039	y = -7238.6x - 4.8611
	D1	y = -13729x + 6.2628	y = -12917x + 4.4907	y=-14440.0x + 5.4852	y = -14120x + 4.1427
Correlation coefficient R2	RO1	0.9903	0.9976	0.9975	0.9935
	RO2	0.9909	0.9988	0.9966	0.9914
	R2	0.9898	0.9961	0.9980	0.9943
	D1	0.9911	0.9950	0.9987	0.9958
Apparent activation energy E (kJ mol^{-1})	RO1	71.77	69.67	71.06	68.28
	RO2	93.90	96.05	89.81	86.30
	R2	61.21	58.25	62.31	60.18
	D1	114.14	107.39	120.05	117.39
Pre-exponential factor A	RO1	6.8673E+03	6.4241E+03	5.6378E+03	3.9672E+03
	RO2	6.3896E+05	1.2819E+06	2.1187E+05	1.1888E+05
	R2	3.8062E+02	3.1517E+02	5.0558E+02	4.2294E+02
	D1	7.9638E+06	2.5727E+06	1.1578E+07	5.9615E+06

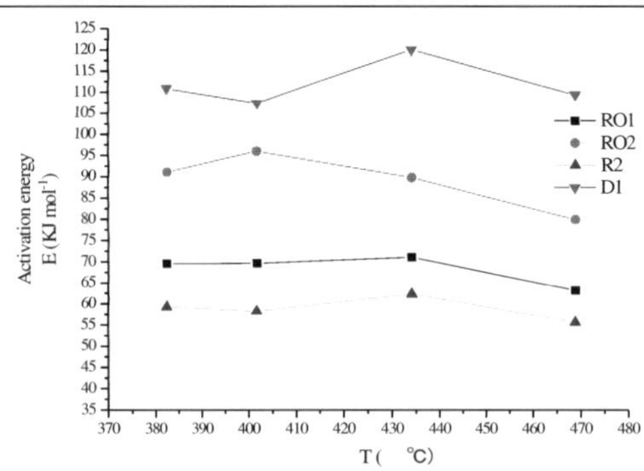

Figure 3. Variation of the apparent activation energy for fast pyrolysisusing four kinds of reaction models

In order to find the best-fit models ofbiomass fast pyrolysis, the Flynn-Wall-Ozawa method was employed to verify the apparent activation energy. The specific parameters of the Flynn-Wall-Ozawa methodwereshown in Table 6 and the apparent activation energy E was 74.97 $kJmol^{-1}$, which was similar to the apparent activation energiesobtained by the first-order reaction model(RO1) of the Coats-Redfern method. This result accorded with other works presented in the literature comparing kinetic parameters determined by TGA and calculatedby different methods. Moreover, as shown in Table 5, with the increase of the heating rates, the pre-exponential factor of RO1 slightly decreased, which could be caused by the heat transfer error between samples at different heating rates. The further study of heat transfer isin progress.

Table 6. Kinetic parameters of the Flynn-Wall-Ozawa method

method	Fitting straight line equation	Apparent activation energyE (kJ mol^{-1})	Correlation coefficient R2
Flynn-Wall-Ozawa	y = -4118.4x + 8.108	74.97	0.9730

4. Conclusions

In this study, the fast pyrolysis of eucalyptus chips had been investigated under nitrogen atmosphere by means of BL-TGA at different heating rates (60, 120, 180 and 360℃min^{-1}). The fast pyrolysis behavior and reaction mechanism of the main temperature stage (350-540 ℃) were investigated, which indicated that kinetic analysis of fast pyrolysis under the condition of higher heating rates would be more comprehensive.

The maximum mass-loss points occurred at 382.4, 393.4, 434.1 and 468.9℃ respectively. Kinetic parameters were calculated by the Coats-Redfern integral methods using four reaction mechanism models and verified by the Flynn-Wall-Ozawa method.

5. Acknowledgements

The authors gratefully thank for special fund for Beijing common construction project.

6. References

[1] M. A.Lopez-Velazquez, V.Santes, J.Balmaseda, E.Torres-Garcia, *Pyrolysisof orange waste: a thermo-kinetic study,* Journal of Analytical and Applied Pyrolysis, 2013

[2] Y.F.Huang,W.H.Kuan, P. T.Chiueh, S.L. Lo, *A sequential method to analyze the kinetics of biomass pyrolysis,* Bioresource Technology, 2011

[3] A. M. Cortés, A.V. Bridgwater,*Kinetic study of the pyrolysis of miscanthus and its acid hydrolysis residue by thermogravimetric analysis,*Fuel Processing Technology,2015

[4] S. S.Kim, H. V.Ly, J.Kim, J. H.Choi, H. C.Woo,*Thermogravimetric characteristics and pyrolysis kinetics of Alga Sagarssum sp. biomass,* Bioresource Technology, 2013

[5] X. Huang, J. P. Cao, X. Y. Zhao, J. X. Wang, X. Fan, Y. P. Zhao, X. Y. Wei, *Pyrolysis kinetics of soybean straw using thermogravimetric analysis,* Anesthesiology, 2016

[6] R. Zhou, H. W. Lei, J.Julson,*The effects of pyrolytic conditions on microwave pyrolysis of prairie cordgrass and kinetics,*Journal of Analytical and Applied Pyrolysis, 2013

[7] S. Kim, H. V.Ly, J.Kim, J. H.Choi, H. C.Woo,*Thermogravimetric characteristics and pyrolysis kinetics of Alga Sagarssum sp. biomass,*Bioresource Technology, 2013

[8]K. Wu, J. Liu, Y. L. Wu, Y. Chen, Q. H. Li, X. Xiao, M. D. Yang,*Pyrolysis characteristics and kinetics of aquatic biomass using thermogravimetric analyzer,*Bioresource Technology, 2014

[9] C. D.Blasi, *Modeling chemical and physical processes of wood and biomass pyrolysis,* Progress in Energy & Combustion Science,2008

[10] J. Q. Zhou, P. Gao, C. Q. Dong, Y. P. Yang,*TG-FTIR analysis of nitrogen conversion during straw pyrolysis: A model compound study,*Journal of Fuel Chemistry and Technology, 2015

[11] A. G. Barneto, J. A. Carmona, J. E. M. Alfonso,*Simulation of the thermogravimetry analysis of three non-wood pulps,*Bioresource Technology, 2010

[12] J.A.Caballero, J.A.Conesa, R.Font, A.Marcilla, *Pyrolysis kinetics of almond shells and olive stones considering their organic fractions,*Journal of Analyticaland Applied Pyrolysis, 1997

[13] S. M. Liu, M. Q. Chen, Q. S. Hu, J. Wang, L. H. Kong, *The kinetics model and pyrolysis behavior of the aqueous fraction of bio-oil,*Bioresource Technology, 2013

[14] S. Wang, X. Guo, L. Tao, Z. Yan, Z. Luo,*Mechanism research on cellulose pyrolysis by Py-GC/MS and subsequent density functional theory studies,*Bioresource Technology, 2012

[15] P. R.Patwardhan, D. L.Dalluge, B. H.Shanks, R. C.Brown,*Distinguishing primary and secondary reactions of cellulose pyrolysis,*Bioresource Technology, 2011

[16]D. Liu, Y. Yu, H. Wu,*Differences in Water-Soluble Intermediates from Slow Pyrolysis of Amorphous and Crystalline Cellulose,*Energy Fuels, 2013

[17] F. X. Collard, J.Blin,*A review on pyrolysis of biomass constituents: Mechanisms and composition of the products obtained from the conversion of cellulose, hemicelluloses and lignin,*Renewable and Sustainable Energy Reviews, 2014

[18] G. N. Richards,*Glycoaldehyde from pyrolysis of cellulose,* Journal of Analytical and Applied Pyrolysis, 1987

IOP Conf. Series: Materials Science and Engineering **250** (2017) 012036 doi:10.1088/1757-899X/250/1/012036

Research on Performance and Microstructure of Sewage Pipe Mortar Strengthened with Different Anti-Corrosion Technologies

Song Mu [1,2], Huaxin Zhou [1,2], Liang Shi [1,2], Jianzhong Liu [1,2], Jingshun Cai [1,2] and Feng Wang[1]
[1] Jiangsu SOBUTE New Materials Co, Ltd., Nanjing 211103;
[2] State key laboratory of high performance civil engineering materials, Nanjing 211103.
Email: musong@cnjsjk.cn

Abstract. Mostly urban underground sewage is the acidic corrosion environment with a high concentration of aggressive ions and microbe, which resulted in performance deterioration and service-life decrease of sewage concrete pipe. In order to effectively protect durability of the concrete pipe, the present paper briefly analysed the main degradation mechanism of concrete pipe attacked by urban underground sewage, and proposed that using penetrating and strengthening surface sealer based on inorganic chemistry. In addition, using index of compressive strength, weight loss and appearance level to investigate the influence of the sealer on corrosion resistance of mortar samples after different dry-wet cycles. Besides, comparative research on effect of the sealer, aluminate cement and admixture of corrosion resistance was also addressed. At last, the SEM technology was used to reveal the improvement mechanism of different technologies of corrosion resistance. The results indicated that the sealer and aluminate cement can significantly improve corrosion resistance of mortar. Besides, the improvement effect can be described as the descending order: the penetrating and strengthening surface sealer > aluminate cement > admixture of corrosion resistance. The mortar sample treated with the sealer displayed the condensed and sound microstructure which proved that the sealer can improve the corrosion resistance to urban underground sewage.

1. Introduction

Sewage pipe, as an important component of urban sewerage system, effectively assures the green and sustainability of urban development. At present, the sewage pipes are mainly made of concrete materials in China. Mostly organic and inorganic substances from municipal sewage are corrosive for concrete. In addition, the corrosion phenomenon developed slowly in a long term, therefore the damage resulted from the corrosion is slight in a short duration. However, it is worthy of notice that the difficulty of maintenance and rehabilitation for concrete pipe under the sewage [1,2]. Based on deterioration mechanism of concrete pipe under sewage, the corrosion could be categorized into sulfate and acid attack.

As for the deterioration mechanism of concrete pipe, sulfate attack is a complicate corrosion which is composed of physical salt crystallization destruction under the role of drying and wetting cycles, expansion destruction from formation of ettringite and gypsum crystal, and decomposition destruction of C-S-H gel transformed into thaumasite at the condition of cold and moisture (the environmental temperature is lower than 15 ^{o}C) [3,4].

Acid corrosion is also a key factor for deterioration of concrete pipe under the sewage. Particularly, acid corrosion will accelerate precipitation of calcium ions and contribute to destructive reaction from sulfate attack on concrete. Based on the resource of acid corrosion, acid substance originated from polluted water in the sewage or from microbiological corrosion. The former acidity, originated from acetic acid, fatty acid, citric acid, lactic acid, etc. from industrial waste and domestic sewage, could result into concrete destruction at the bottom and lateral side of the sewage pipe. Although the organic acid mentioned above is weak, the corrosion mechanism from the weak acid is similar to the strong acid which reacted with the cement hydration products into dissolved salt. Finally, the calcium ion flowed away from the hydrated products and resulted in concrete destruction. The later acidity referred in particular to that hydrogen sulfide corrosion on concrete pipe resulted from microbiological role [5]. The microbiological corrosion can be categorized into as follows: firstly, sulfide-rich waste water are released into sewage pipe and resulted in microbiological corrosion; secondly, sludge precipitated on pipe bottom resulted in the corrosion. When the organic and inorganic substances precipitated on the bottom of concrete pipe and transformed into sludge, sulfate ion was reduced by sulfur reducing bacteria to produce hydrogen sulfide gas. Under the role of biochemistry, the gas transports into upper space of concrete pipe unsaturated with water, and the sulfuric acid by the gas oxidation resulted in corrosion destruction of upper part of concrete pipe [6].

Based on corrosion mechanism and deteriorated characteristic of concrete exposed to sewage, it is suggested that increasing densification and reducing porosity of concrete and decreasing transport rate of aggressive ions in concrete can increase resistance of concrete to sewage corrosion. At present, the main technical solution of improving concrete resistance to acid and sulfate attack is corrosion resistance concrete and surface protection materials. The corrosion resistance concrete is mainly made by corrosion resistance admixture or aluminate cement with an excellent ability to resist sulfate attack. Surface protection materials are composed of film-forming coating, hydrophobic liner and pore-blocker [7-9]. In addition, the pore-blocker is a liquid solution consisting of highly reactive components which can effectively penetrate into concrete surface and react with hydrated products and anhydrate cement particles so as to strengthen the concrete surface and protect from the corrosion. Based on the reaction mechanism, pore-blocker can effectively improve the corrosion resistance of existing concrete, such as concrete pipe. Besides, the pore-blocker mainly made by inorganic substances can form a firm barrier to aggrieve substance in concrete surface without ageing and adhesive problem of organic coating.

Mortar is the main material at the internal surface of concrete pipe prepared by centrifugal compacting process. Considering corrosion characteristic of concrete pipe under sewage, the mortar is easily corroded by the aggressive substance from sewage water. Therefore, the present research mainly focused on effect of corrosion resistance on mortar sealed with the pore-blocker based on modified silicate technology, and the mechanism is also be discussed..

2. Raw Materials and Experimental Method

Considering the acid corrosion characteristic of concrete pipe under sewage, the present research conducted comparative research on effect of corrosion resistance admixture, aluminate cement for sewage corrosion, and modified-silicate based pore-blocker on corrosion resistance of mortar exposed to acid environment.

2.1. Raw Materials

Cement: Standard silicate cement conformed to the China National Standard GB 8076-2008 "Concrete admixture".

Sand: Standard sand conformed to the China National Standard GB/T 17671-1999 "Method of testing cements-Determination of strength".

Water: Tap water.

Corrosion resistance admixture (CY referred in the present research): CY admixture, yellow appearance and powder, conforms to the Chinese Industrial Standard JC/T1011-2006 "Sulfate corrosion resistance admixtures for concrete", provided by an anti-corrosion company from Tian Jing city. Based on the product specification, the CY admixture is composed of surfactant, inorganic salt, organic promoter and carrier. The specification also claimed that the CY admixture, replaced cement content by 8 to 12 %, significantly improved the corrosion resistance of cementitious materials exposed to sulfate environment with content of 10000～15000 mg/l. In addition, CY admixture is recommended to use in some severe environment, for instance, salty soil, sewage.

Aluminate cement (KS referred in the present research): KS cement, grey appearance and conforms to the Chinese National Standard GB/T 201-2015 "Calcium Aluminate Cement", provided by an anti-corrosion company from Tian Jing city. The specification of the cement indicated that KS cement is a hydraulic binder specially developed for concrete pipe, which contains approximately 50% alumina.

Modified-silicate based pore-blocker sealer (NF referred in the present research): NF sealer, as a translucent liquid solution with a pH value of 12, produced by Jiangsu SOBUTE New Materials Co., Ltd. In addition, density of the sealer is 1.15g/cm3, and measured solid content is about 26.1%. The sealer conforms to Chinese Industrial Standard JC/T1018-2006 "Water based capillary inorganic waterproofer".

2.2. Mortar Preparation

Standard mortar sample: Mixture of standard mortar is described as follows: standard cement: standard sand: water = 1:3:0.5 (by weight). According to the Chinese National Standard GB/T 17671-1999 "Method of testing cements-Determination of strength", the freshly mixed mortar was casted into steel mold with a size of 40×40×160 mm and vibrated until compacting. After 24 hours curing, the demolded mortar samples were put into standard curing room for 28 days.

Anti-corrosion mortar sample prepared by CY admixture (CY mortar mentioned in the present research): using the same mixture as the standard mortar, the standard cement replaced with CY admixture in the weight of 8 %. Besides, the CY mortar also adopted the same curing condition.

Anti-corrosion mortar sample prepared by KS aluminate cement (KS mortar mentioned in the present research): using the same mixture as the standard mortar, the standard cement replaced with KS aluminate cement in the weight of 100 %. Besides, the KS mortar also adopted the same curing condition.

Anti-corrosion mortar sample prepared by NF sealer (NF mortar mentioned in the present research): using the same mixture as the standard mortar, the NF sealer was firstly brushed the surface of standard mortar with usage of 0.5 kg/m2 and then cured at a room temperature condition for 8 days. Before the sealer brush, the standard mortar should be put into standard curing room for 20 days, and then wiped off the water on the surface of mortar samples until reaching the surface-dry condition.

2.3. Corrosion Test

Considering corrosion characteristic of concrete pipe under sewage, acid solution of sulfuric acid and acetic acid with pH value of 2 was prepared separately. When reached surface-dry condition after 28 days curing, the testing samples were kept into a room with the condition of 20±2℃and RH=65% for drying and wetting cycles of acid corrosion. The corrosion cycle persisted 24 hours as described as follows: drying for 16 hours at 50 °C, cooling down for 2 hours at 20 °C, and then immersing into acid solution for 6 hours. After reaching different cycles, the mortar samples were taken out and dried at room temperature, and then performance was evaluate with index of compressive strength, weight loss and appearance damage. Index of compressive strength was defined that the ratio of compressive strength of corroded mortar exposed to acid solution to that of the standard mortar cured at standard room at the same duration.

3. Results

3.1. Index for Compressive Strength

The results from Figure 1 indicated that the sealer NF and aluminate cement KS significantly improved corrosion resistance of mortar to sulfur acid with the pH value of 12. After 90 days of corrosion cycles, NF and KS mortar exhibited a good result of index of compressive strength higher than 0.75 (required value for sulfate attack from the China National Standard GB/T 50082 "Standard for test methods of long-term performance and durability of ordinary concrete") and almost reached at 0.9, which meant high resistance to attack of sulfur acid on mortar. When mortar brushed with NF sealer, the reactive component reacted with hydration products and anhydrate cement particles to block the transport channel of aggressive ions and protect from acid attack on mortar. With increasing number of corrosion cycles, the continuous hydration of mortar sample will also improve its ability to resist corrosion by means of microstructure optimization. Therefore, the improved role of pore-blocker sealer NF and aluminate cement KS became more significant when increasing the exposure duration. By contrast, The CY admixture firstly increased the index for compressive strength of mortar but then decreased. The reason could be in that incorporation of CY admixture had a poor resistance to inhibit transportation of sulfate ions into the mortar samples, which could contributed to formation of ettringite resulted from reaction between sulfate ions and hydrated products. The ettringite improved compressive strength of the mortar at the early age of corrosion cycles, which appeared as increased index of compressive strength. However, hydration products could be decomposed in that increasing acidity with increased corrosion cycles deteriorated alkality environment which is important for stability of cement hydration products. Finally the decomposition of hydration products decreased the index of compressive strength.

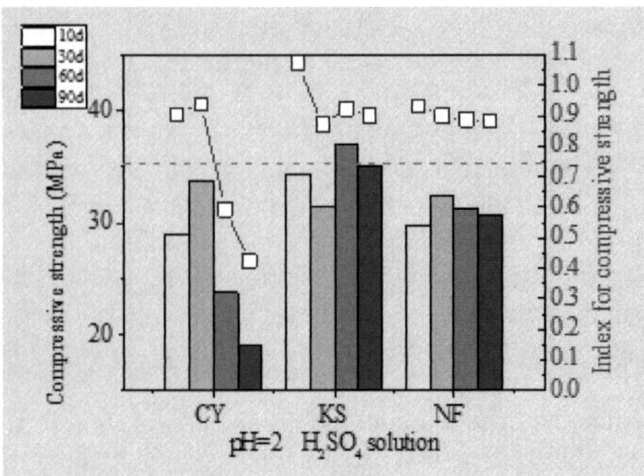

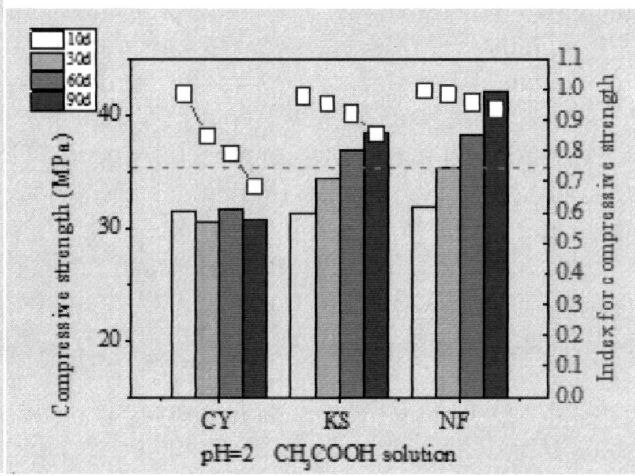

Figure 1. Index of compressive strength for mortar exposed to sulfur acid condition at pH=2.

Figure 2. Index of compressive strength for mortar exposed to acetic acid condition at pH=2.

3.2. Weight Loss

Figure 3 shows that NF and KS mortar also effectively decreased weight loss of the mortar samples exposed to sulfur acid solution, but the CY mortar had a significant weight loss. The weight loss of aluminate cement KS could ascribe to crystallization transfer of hydration products which also resulted in decrease of strength of aluminate cement.

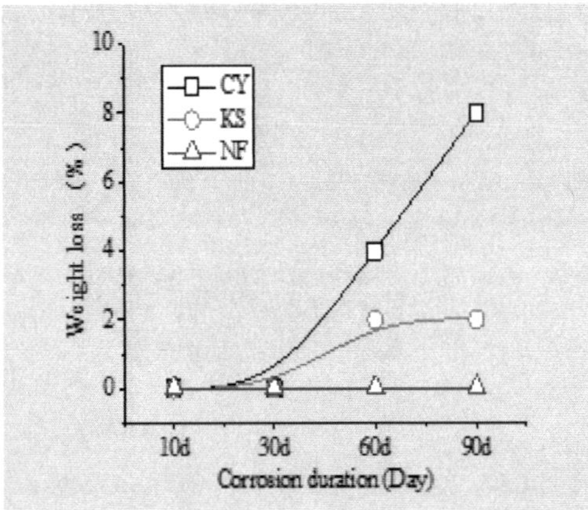

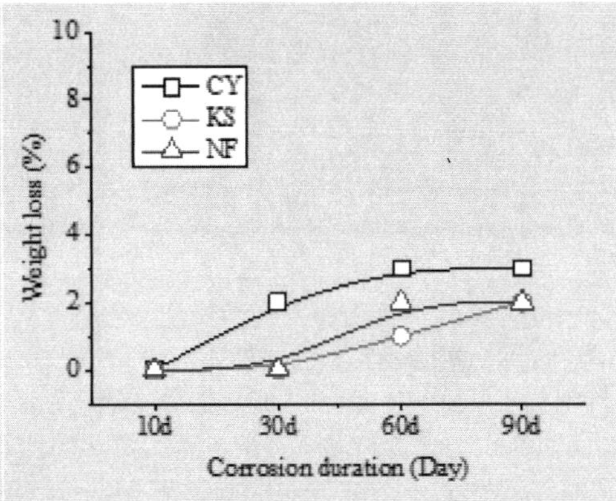

Figure. 3 Weight loss for mortar exposed to sulfur acid condition at pH=2.

Figure. 4 Weight loss for mortar exposed to acetic acid condition at pH=2.

The results from Figure 4 indicated that CY admixture, KS aluminate cement and NF sealer had a good performance to resist corrosion from acetic acid with the pH value of 2. When the corrosion cycles reached at 90 days, weight loss of mortar sample was lower than 5 % which is regarded as the critical value for weight loss of concrete exposed to frost condition according to the China National Standard GB/T 50082. Therefore, the acetic acid with a pH value of 2 had a slight influence on mortar performance.

Appearance of mortar samples exposed to sulfur acid solution was showed in Figure 5. By contrast, mark of NF mortar can be observed from the appearance, but the marks of KS and CY mortar were disappeared. In addition, it has been observed that peeling and sand leakage. Therefore, the results demonstrated that NF sealer significantly improved corrosion resistance of mortar samples than KS aluminate cement and CY admixture.

(A) Mortar appearance

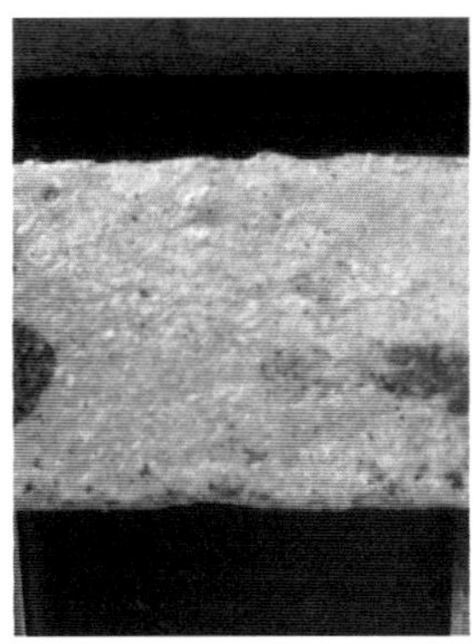

(B) Mortar appearance protected by NF sealer

(C) Mortar appearance protected by aluminate cement KS

(D) Mortar appearance protected by corrosion resistance admixture CY

Figure. 5 Corrosion appearance for mortar exposed to sulfur acid condition at pH=2.

3.3. SEM

FEI Quanta 250 Scanning Electron Microscope was used to observe morphology of mortar surface after 90 days of corrosion in sulfur acid solution. The SEM images of CY, KS and NF mortar were showed in Figure 6 to 8 separately. The SEM results indicated that hydration product of CY mortar was loosed and porous, and CPS value of calcium ion was lower than that of KS and NF mortar with a magnification of 1000 times. Therefore, acid solution with the pH value of 2 resulted in significant decomposition and destruction. By contrast, KS mortar displayed a denser microstructure than CY mortar. However, some micro-cracks were observed on KS mortar surface, which indicated that sulfur acid or crystallization transformation of hydration products resulted in destruction of KS mortar. NF mortar has a denser microstructure than other mortar, no cracks and large pores were observed on the mortar surface. In addition, high value of CPS proved that NF sealer effectively improved microstructure of mortar and inhibited calcium leaching from hydration products.

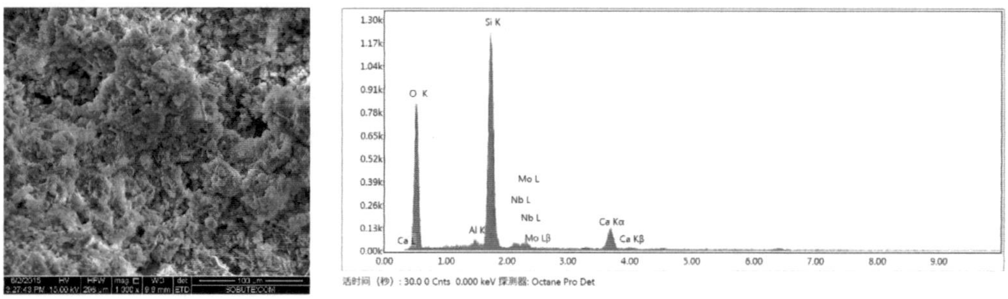

Figure. 6 Microstructure morphology and chemical composition of mortar appearance protected by corrosion resistance admixture CY (×1000).

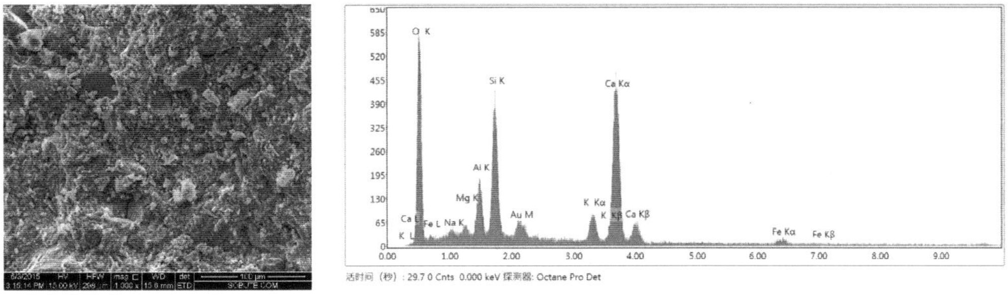

Figure. 7 Microstructure morphology and chemical composition of mortar appearance protected by mortar appearance protected by aluminate cement KS (×1000).

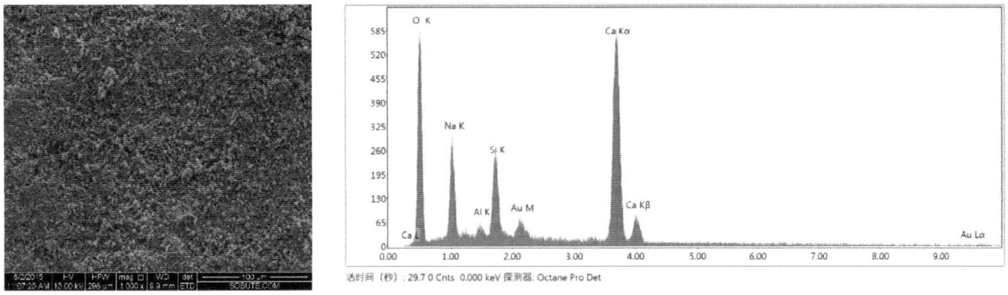

Figure. 8 Microstructure morphology and chemical composition of mortar appearance protected by Mortar appearance protected by NF sealer (×1000).

4. Conclusions

(1) In the present research, the mortar sample sealed with modified silicate-based pore-blocker NF or the mortar sample made by aluminate cement KS displayed an excellent resistance to acid solution with a pH value of 2 respectively. However, corrosion resistance admixture CY, conforms to the Chinese Industrial Standard JC/T1011-2006 "Sulfate corrosion resistance admixtures for concrete", had a poor performance to improve corrosion resistance of mortar exposed to acid solution. In addition, the corrosion resistance could be described in the following descending order: NF>KS>CY

(2) Results from SEM images also indicated that the mortar coated with NF sealer displayed denser microstructure than CY and KS mortar. In addition, no cracks and large pores were observed on NF mortar, and EDAX results demonstrated that calcium leaching is slight when compared with other mortars mentioned above.

5.Acknowledgement

Authors appreciate the financial supports from the "National Key R & D Program of China" under the contract No. 2017YFB0309904, the Natural Science Foundation of Jiangsu Province under the contract No. BK20141012 and BK20162011.

6.Reference

[1] Zhou F C 2002 *J. Chongqing Jiaotong Institute* **21** 125-8
[2] Zhou X Z, Song P D, Ni G H 2008 *Concrete* **4** 100-1
[3] Han Y D, Zhang J and Gao Y 2011 *Concrete* **1** 52-6
[4] Ma B G, Gao X J, He Z M and Dong R Z 2004 *J. Chin. Ceram. Soc.* **32** 1219-24
[5] Tang X, Jia X, Feng C, Huang H and Chen L 2007 *J. CHIN. SOC. CORROS. RROT.* **27** 373-8
[6] Jiang G, Keller J and Bond P L 2014 *Water Research* **65** 157-169
[7] Bentur A, Diamond S and Berke N S 2005 *Steel corrosion in concrete: fundamentals and civil engineering practice* (E & FN Spon)

IOP Conf. Series: Materials Science and Engineering **250** (2017) 012036 doi:10.1088/1757-899X/250/1/012036

[8] Yang Y and Li M 2014 *Shanghai Coatings* **52** 40-2
[9] Bin L I, Chang L I and Wang X 2001 *Materials Protection* **34** 33-4

IOP Conf. Series: Materials Science and Engineering **250** (2017) 012037 doi:10.1088/1757-899X/250/1/012037

Preparation of Shrinkage Compensating Concrete with HCSA Expansive Agent

Changcheng Li and Fujia Jia

China Building Materials Academy, Guanzhuang Dongli 1#, Chaoyang District, Beijing, China.
Email: cbmali@126.com

Abstract. Shrinkage compensating concrete (SCC) has become one of the best effective methods of preventing and reducing concrete cracking. SCC is prepared by HCSA high performance expansive agent for concrete which restrained expansion rate is optimized by 0.057%. Slump, compressive strength, restrained expansion rate and cracking resistance test were carried out on SCC. The results show that the initial slump of fresh SCC was about 220mm-230mm, while slump after 2 hours was 180mm-200mm.The restrained expansion rate of SCC increased with the mixing amount of expansive agent. After cured in water for 14 days, the restrained expansion rate of C35 and C40 SCC were 0.020%-0.032%. With the dosage of expansive agent increasing, restrained expansion rate of SCC increased, maximum compressive stress and cracking stress improved, cracking temperature fell, thus cracking resistance got effectively improvement.

1. Introduction

Concrete will crack when internal tensile stress exceeds the tensile strength of concrete under restrained condition. Due to the emphasis on early strength, increase of concrete strength grade and cement fineness, the cracking caused by thermal shrinkage and drying shrinkage are the most common in actual projects. And these will inevitably result in a series of performance degradation and hence influence the durability and service life of concrete structure. The expansion generated by hydration of expansive agent establishes 0.2MPa-0.7MPa pre-stress in shrinkage compensating concrete (SCC). Thus, thermal shrinkage and drying shrinkage can be effectively compensated by the pre-stress. SCC has been widely used to prevent and reduce cracking of concrete in engineering.

HCSA is a double expansive sources high performance expansive agent for concrete. Calcium oxide is the major expansive sources, and gypsum and calcium sulphoaluminate are the assistant expansive sources. HCSA has the advantage of high expansion (restrained expansion rate is 0.17% in water), fast expansion speed, anhydrous expansion, short expansion period, small drying shrinkage, and can activate the activity of mineral admixture [2]. HCSA can effectively compensate drying shrinkage and thermal shrinkage, thus improve crack resistance and anti-seepage of concrete. HCSA has been successfully used in concrete structural self-waterproofing, jointless constructing of the excessive-length construction, and cracks prevention of mass concrete [3].

It is well known that waterproofing is the premises to ensure the normal use of underground engineering. Once the concrete structure cracks, the leakage is unavoidable. So crack resistance is more important than anti-permeability of concrete in underground engineering. The object of this study is preparing SCC with 0.020%-0.032% restrained expansion rate by HCSA high performance expansive agent.

IOP Conf. Series: Materials Science and Engineering **250** (2017) 012037 doi:10.1088/1757-899X/250/1/012037

2. Experimental

2.1 Materials

Ordinary Portland cement (P.O 42.5), S95 ground granulated blast furnace slag (GGBS), Class F II grade flay ash (FA) and calcium sulphoaluminate-calcium oxide expansive agent (HCSA) were used in this study. The physical properties of cement and performance index of HCSA expansive agent are listed in Table 1 and Table 2, respectively. The restrained expansion rate of HCSA is shown in Figure 1. River sand with a fineness modulus of 2.60, crushed stone with a continuous gradation of 5mm-25mm were used as fine and coarse aggregate. Polycarboxylate superplasticizer with a water-reducing rate of 28% was also used. Tap water was used as mixing water.

Table 1. Physical properties of cement.

Density (g/cm^3)	Specific surface area (m^2/kg)	Setting time (min)		Flexural strength(MPa)		Compressive strength(MPa)	
		Initial setting	Final setting	3d	28d	3d	28d
3.15	345	151	216	6.0	8.5	29.3	49.6

Table 2. Performance index of HCSA expansive agent.

Items			Index	Measured value
Fineness	Specific surface area (m^2/kg)	≥	200	325
	Residue on sieve (1.18 mm mess,%)	≤	0.5	0
Setting time (min)	Initial setting	≥	45	195
	Final setting	≤	600	238
Restrained expansion rate (%)	Water curing 7days	≥	0.06	0.056
	Air curing 21days	≥	−0.010	0.011
Compressive strength (MPa)	7days	≥	20.0	26.7
	28days	≥	40.0	48.2

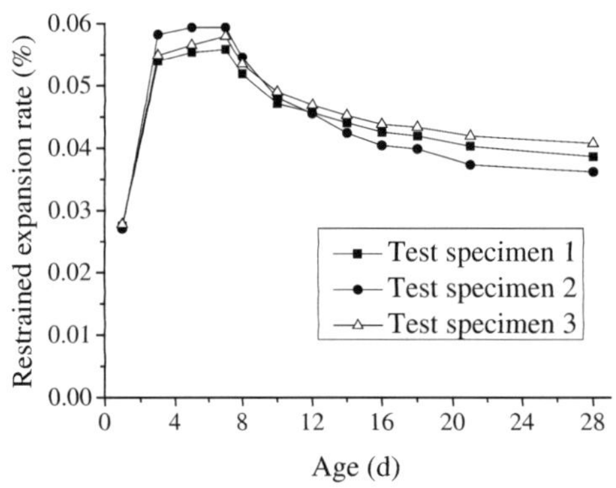

Figure 1. Restrained expansion rate of HCSA expansive agent

Table 3. Concrete mixture proportions (kg/m^3).

Sample	Cement	FA	GGBS	HCSA	Fine aggregate	Coarse aggregate	Water	Superplasticizer
C35B	230	117	40	35	633	1175	175	8.34
C35W	230	112	40	40	633	1175	175	8.34
C40E	255	91	50	45	595	1209	170	9.53

2.2 Mix Proportions
Based on the comprehensive consideration of strength, restrained expansion rate, workability and durability of concrete, the optimized mixture proportions of SCC are shown in Table 3. Samples were designated as C35B (SCC for bottom plate), C35W (SCC for external wall and top plate) and C40 (SCC for expansion reinforcing band), respectively.

Slump, slump loss and density of fresh concrete were tested conforming to the Chinese national standard GB 50080-2002. Compressive strength of 7days and 28days was tested with cube specimens of 100mm×100mm×100 mm conforming to the Chinese national standard GB/T 50081-2002.

Restrained expansion rate of SCC were tested by the field measuring device for restrained expansion rate of concrete which invented by China Building Materials Academy. This device can be used to detect the restrained expansion rate of concrete specimen, which was placed in the prism mould measuring 100mm×100mm×400 mm with vertical limiter, both in water and in air. The restrained expansion rate of SCC was calculated by from equation (1).

$$\varepsilon = \frac{L_1 - L}{2L_0} \times 100 \tag{1}$$

Where ε is the restrained expansion rate of SCC, %; L_1 is the specimen length measuring results at the measured age, mm; L is the initial specimen length measuring results, mm; L_0 is the reference length, 300 mm.

The schematic diagram of restrained expansion rate measuring device for SCC is shown in Figure 2.

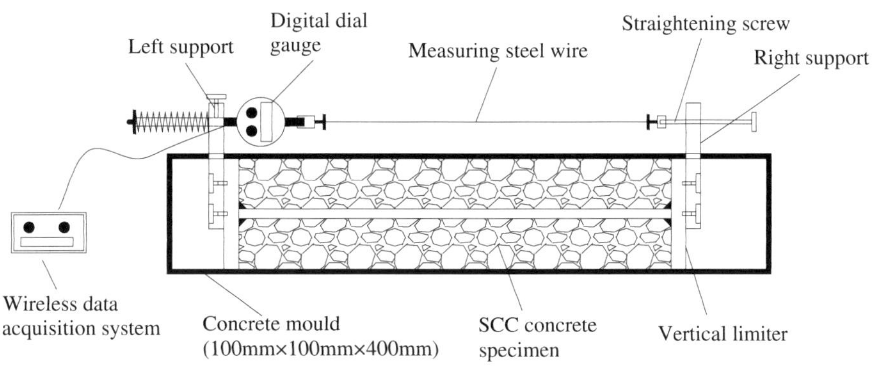

Figure 2. Restrained expansion rate measuring device for SCC

Cracking resistance of SCC was tested by a temperature stress testing machine (TSTM) under fully restrained condition. The TSTM-specimen was shaped like a dog-bone with the dimension of 1000mm×150mm×150 mm in the center part and 150mm×280mm×150 mm at its ends. The imported temperature rise-cooling curve was obtained from the engineering field measurement. The maximum compressive stress, crack stress, and crack temperature of SCC were measured to evaluate the cracking resistance.

3. Results and Discussion

3.1 Property of Fresh Concrete
Slump and slump loss directly decide the both constructability and operability of fresh concrete. To ensure the strength, restrained expansion rate and durability of concrete, the slump of concrete should be as low as possible when the concrete can be vibrated well. Considering execution conditions and the quality of Chinese construction workers, (180±30) mm is appropriate for the control value of concrete slump. The performance test results of fresh concrete are listed in Table 4.

IOP Conf. Series: Materials Science and Engineering **250** (2017) 012037 doi:10.1088/1757-899X/250/1/012037

As shown in Table4, the initial slump of fresh SCC was about 220mm - 230mm, while slump after 2 hours was 180mm to 200mm. The concrete construction from mixing, transportation to placing can be completed in 2 hours. Thus the slump of fresh SCC simultaneously satisfied the demands of construction, durability and cracking resistance of concrete. Besides, 2400-2410kg/m^3 density were also in accordance with the specification requirements.

Table 4. Properties of fresh concrete.

Sample	Slump (mm)		Density (kg/m^3)
	initial	2h	
C35B	225	190	2410
C35W	230	200	2400
C40E	220	180	2405

3.2 Compressive Strength

Table 5 shows the compressive strength of SCC. According to the Chinese national standard JGJ 55-2011, the mixing strength of C35 and C40 strength grade SCC should be not lower than 41.6MPa and 46.6MPa, respectively.

Table 5. Compressive strength of SCC.

Sample	Compressive strength (MPa)			Compressive strength growth rate (%)		
	3d	7d	28 d	3d	7d	28 d
C35B	23.8	32.6	45.5	52.3	71.6	100
C35W	23.6	35.1	47.1	50.1	74.5	100
C40E	28.9	41.8	54.5	53.2	76.8	100

It could be seen that the strength of C35B and C35W SCC at 28days were 45.5 MPa and 47.1MPa, while the strength of C40E at 28 days was 54.5MPa. Thus all the SCC had reached the design mixing strength requirements. Compressive strength growth ratio at 3 days was about 52%, and the ratio reached 75% at 7 days. The strength development was reasonable, long-term strength continued to increase. It is the best developing period for restrained expansion rate of SCC when the compressive strength increased from 5 MPa to 20MPa [1]. So the early strength of SCC not only met construction schedule, but also matched the expansion process. And this was beneficial to improve the cracking resistance.

3.3 Restrained Expansion Rate

Restrained expansion rate is an important design parameter for SCC, because it directly decides the compensating thermal shrinkage and drying shrinkage of SCC. Due to external wall and roof suffered more restraint degree, these parts are easier to crack. So, the restrained expansion rate of external wall and roof is higher than that of bottom plate [4]. Post-pouring joint can be replaced by expansion reinforcing band to release deformation and stress. The theory is that the higher pre-stress established in expansion reinforcing band can effectively strengthen the compensating ability of SCC in both sides. And therefore, both restrained expansion rate and strength of expansion reinforcing band are enhanced a rank. The restrained expansion rate of SCC was shown in Figure 3.

The restrained expansion rate of SCC increased with the mixing amount of expansive agent. After cured in water for 14 days, the restrained expansion rate of C35B, C35W and C40E were 0.021%, 0.027% and 0.032%, respectively. Meanwhile, the restrained expansion rate of SCC still remained at 0.006%-0.008% after 28 days' air cured at 60±5% of RH. This illustrated that the SCC had been kept in press condition at all times. And there had established enough pre-stress for compensating early thermal shrinkage and later drying shrinkage. So the SCC had excellent crack resistance.

3.4 Cracking Resistance

TSTM can simulate different temperature gradient of concrete under fully-constrained condition. The test procedures cover actual temperature development of early concrete, the development process of mechanical property, thermal property and deformation performance of concrete as well. Therefore, the evaluation results of TSTM are more comprehensive and objective.

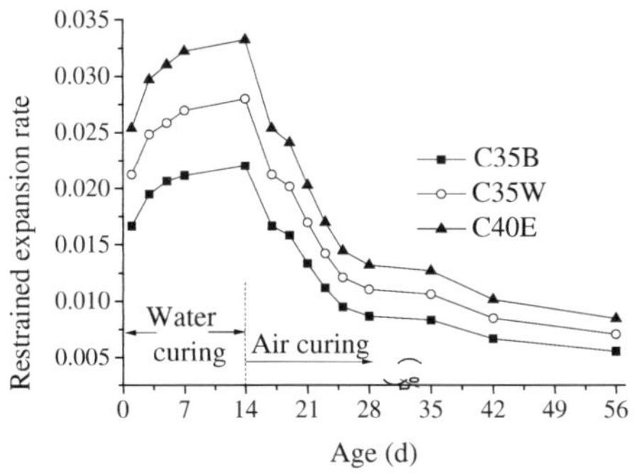

Figure 3. Restrained expansion rate of SCC

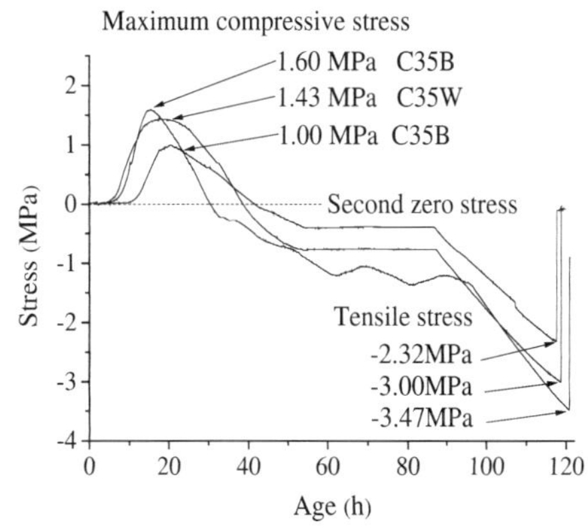

Figure 4. Stress curve of SCC

Based on the pre-stress generated by the hydrated expansive products of expansive agent, SCC can offset the tensile stress caused by shrinkage. And this is advantageous for improving the cracking resistance of concrete. The stress developing curve of SCC was shown in Figure 4, and the characteristic parameter of temperature-stress test were listed in Table 6.

Table 6. The characteristic parameter of concrete temperature-stress test.

Items	C35B	C35W	C40E
Maximum compressive stress (MPa)	1.00	1.43	1.60
Maximum compressive stress time (h)	15.42	17.41	20.38
Second zero stress temperature (°C)	41.88	44.00	50.25
Second zero stress time (h)	41.50	38.55	30.38
Cracking stress (MPa)	-2.32	-3.00	-3.47
Cracking temperature (°C)	-18.5	-13.4	-7.6

Maximum compressive stress, cracking stress and cracking temperature can comprehensive reveal the cracking resistance of SCC. The greater value of maximum compressive stress and cracking stress indicates a better cracking resistance of concrete. The lower value of cracking temperature, the better supporting temperature shock of concrete gets. With the dosage of HCSA expansive agent increasing from 35kg/m^3 to 40 kg/m^3 and 45kg/m^3, the maximum compressive stress of SCC at temperature rise period increased from 1.00MPa to 1.43MPa and 1.60MPa. Meanwhile, the cracking stress at temperature-fall period also increased from 2.32MPa to 3.00MPa and 3.47MPa. These results further indicated that there was enough pre-stress to compensate shrinkage. In addition, cracking temperature decreased from -7.6℃ to -13.4℃ and -18.5℃. As a result, the thermal cracking tendency of concrete at early stage descended. The cracking resistance of SCC enhanced with the dosage increasing of HCSA expansive agent. The results accorded with restrained expansion rate growth law in chapter 3.3.

4. Conclusion

SCC prepared in this study had the characteristic of good workability, compressive strength increase steadily, coordinated growth of restrained expansion rate and compressive strength.

With the dosage of expansive agent increasing, restrained expansion rate of SCC increased, maximum compressive stress and cracking stress improved, cracking temperature fell, thus cracking resistance got effectively improvement.

IOP Conf. Series: Materials Science and Engineering **250** (2017) 012037 doi:10.1088/1757-899X/250/1/012037

5. Reference

[1] Zhao S Z and You B K 2010 *Shrinkage compensating concrete crack control and its application* (Beijing: China Building Industry Press) p65

[2] Zhao S Z, Liu L, Zheng W L, Jia F J and Li CC 2016 *Expansive Agents & Expansive Concrete* 1-5

[3] Zhao S Z, Liu L, Zheng W L and You B K 2009 *Expansive Agents & Expansive Concrete* 2-5

[4] Feng N Q, Gu Q X and Hao T Y 2008 *Cracking in Concrete Structure and its Countermeasures* (Beijing: China Machine Press) p202

IOP Conf. Series: Materials Science and Engineering **250** (2017) 012038 doi:10.1088/1757-899X/250/1/012038

The Effect of Freeze - thaw Cycles on Slope Anchorage and Preventive Measures of Research

Quanchao Bi[a] , Xuelei Wang, Mingxiu Zhang, Qiang Rao, Zhixin Fu and Jianyao Su

Hebei Institute of Architectural Engineering, Zhangjiakou City, Hebei Province, China, 075000

E-mail: [a] hbjzdxbqc@sina.com

Abstract. Freeze-thaw cycles have great harm to slope anchorage structure, and the theoretical system has not been formed because there are too many constraints and interrelations. This paper analyzes the failure principle and appearance of the slope under the freeze-thaw cycles and puts forward the prevention and control measures. On the whole, at this stage there is no way to raise the anchoring technology as a system to do the whole study. It is urgent to establish a complete concept and overall evaluation mechanism of rock and soil anchoring system, because it can give full play to the function of the favorable elements in the anchoring system and weaken the harm of the unfavorable elements so that the system can achieve the best working condition.

1. Introduction

Slope is a natural or artificial formation of the slope and it is one of the most basic geological environment of human engineering activities as well as the construction of the most common form of engineering [1]. The anchoring technology is a reinforcement technology to reinforce the rock and soil by using the anchor (cable). It is widely used in rock and soil reinforcement, which has the advantages of improving the stability of the rock and soil itself, faster, economical and reliable and so on. No matter what kind of anchoring system, theoretically the way of transmission of force is from the anchor to the grouting body, and then by the slurry into the surrounding rock and soil.

There are many factors that affect the durability of slope engineering; a factor that can not be ignored is the freeze-thaw cycles. In the area where the temperature is affected by the freeze-thaw cycles, the slope engineering is prone to freeze-thaw damage, and the stability of the slope is greatly affected, even the instability of the slope is caused.

2. Analysis of Axial Force of Bolt

Yang Shuangkuo etal.[2] proposed the concept of the first and second critical deformation of the anchor and revealed the three-stage characteristics of the anchoring force of the anchor with the deformation of the anchor - the anchoring force strengthened and deformed, and the anchoring force remained constant stage and anchoring force weakening deformation stage. And the characteristics of three kinds of anchoring forces with the variation of deformation are introduced in detail. The numerical simulation of the whole of the anchor is carried out by Luo Yiping[3] through using FLAC3D. It is confirmed that the above conclusions are different, but their essence is the same.

IOP Conf. Series: Materials Science and Engineering 250 (2017) 012038 doi:10.1088/1757-899X/250/1/012038

3. Analysis of Shear Stress at the Interface between Bolt and Mortar

Micro-wrinkle in the anchor body surface plays an important role in the anchoring force. The cohesive force between the anchor body and the grouting material is the main way of transmitting force, but if the drawing force is continued to be increased, the displacement between the anchor body and the grouting material will destroy the cohesive force. The main form of force transmission is the friction [4-5].

Results of research at home and abroad [6-9] show that the shear stress distribution is not immutable, and it gradually changes with the increase of drawing force. When the drawing force is small and the anchoring interface is in the elastic bond state, the shear stress at the interface of the bolt-grouting material is the largest at the end and decreases gradually in the shape, and its shape is roughly distributed according to the hyperbolic function or exponential function. As the drawing force continued to increase, bolt-grouting material interface shear stress in the end reduced and its maximum value to the lower part of mobile, big in the middle and small at both ends of distribution; When drawing force continues to increase, the end of the anchored body interface damage occurs, if the design on the interface end of bolt and grouting material, the friction resistance between the interfaces of the depths of the maximum shear stress to pass to anchor solid stability to provide the required pulling force; As the drawing force increases again, bolt-grouting material interface maximum shear stress to the anchor still deep solid, until the bolt-grouting material interface is broken.

4. Analysis of Interface Shear Stress of Mortar-rock Mass

It's generally believed that the friction between the anchor and the hole wall determines the effective anchorage length [10], the average surface friction decreases with the increase of bolt length. When vertical drawing force effect on anchoring body, anchoring body and rock mass around the friction on the contact interface, due to the friction resistance is from top to the bottom of the anchored body development gradually, the friction resistance at different anchorage depth is different. The upper end of the anchor first reaches the limit value. When the drawing force is greater than the extreme value of the friction resistance, the interface between the anchor and the surrounding rock mass softens into the partial slip state. With the increase of drawing force, the softening of the interface between the anchor and the surrounding rock mass will develop toward the bottom of the anchor until the interface between the anchor and the surrounding rock mass is all in the slip state and the anchoring structure is destroyed[11,12]. The destruction process is shown in Figure 1.

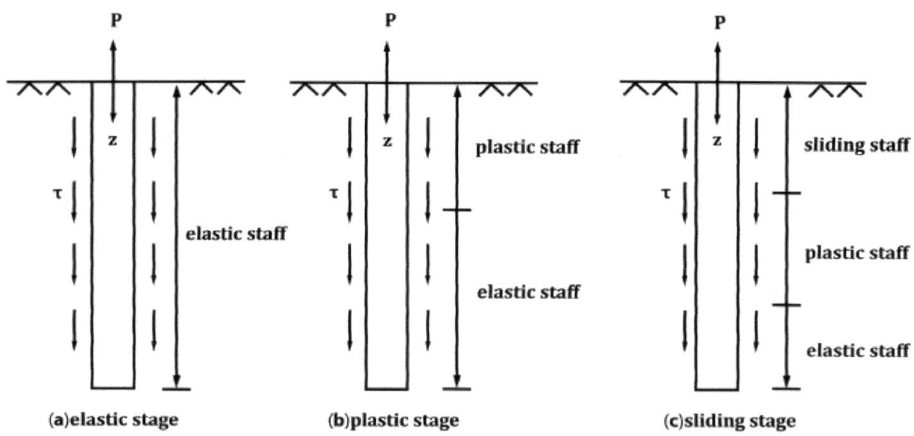

Figure.1 Interaction of mortar and rock mass

5. Influence of Freeze-thaw Cycling on Slope Anchorage Structure

Affected by ambient temperature change, water saturated rock mass and supporting structure in the transition between solid and liquid. The volume of water caused by changes of internal crack and the change of pore size, and the anchoring structure stress state changes so that produce frost heaving and thawing settlement. When the rock mass structure is subjected to frost heave, the structure will have frost heave deformation, so that the internal cracks and pores of the structure to further develop. And the deformation caused by frost heave is not fully recovered during the melting process of the rock mass. The rock mass structure is subjected to repeated frost heaving or melting by the external temperature. Because the influence mechanism of the rock mass and the supporting structure is different, the influence degree is different, so it will be in the rock mass and support. The interaction of the structure will be the surface of the stress distribution changes, which induced the interaction surface cracking, dislocation or even off the phenomenon seriously affected the slope protection engineering safety [13, 14].

6. Protective Measures for Freezing and Thawing of Slope Supported by Anchorage

For non-support of the natural slope, in order to prevent the occurrence of landslides and other disasters under the action of freeze-thaw cycles, it must take reinforcement support measures.
For example: ①retaining wall; ②anti-slide pile; ③soil nails;
④prestressed anchor (cable); ⑤anti-slip plug; ⑥slope protection;
⑦shot anchor support; ⑧cutting slope lightening; ⑨drainage and seepage and so on.

The anchoring section can increase the bond strength of the interface at the interface of the anchor and increase the anchoring length of the bolt. The increase of the anchorage length can also improve its ultimate pull force. However, beyond the limit anchorage length, the tensile capacity of the bolt can not be improved. In addition to increasing the length of anchoring, the following methods can be used in the project, such as single-hole composite anchoring (a)and post-high pressure grouting anchoring(b). The expansion of the anchorage(c) can also improve the pull-out force of the anchor in the weak or complex rock and soil layer, but also increase the shear strength of the rock mass around the anchor and improve the overall stability of the slope.

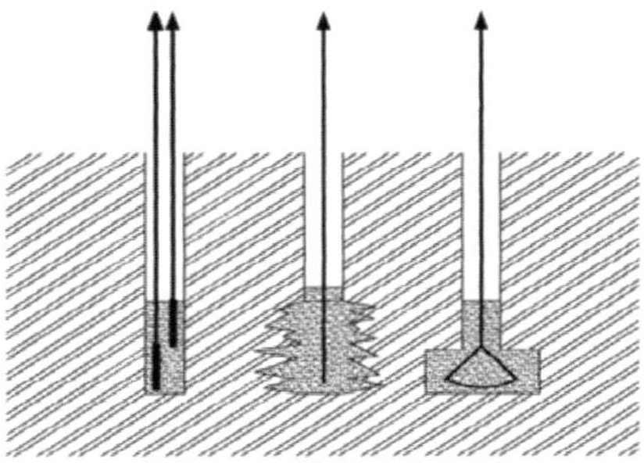

Figure.2 Method to improve the capacity of anchor rods

7. Conclusion

Freeze-thaw cycles have an adverse effect on the slope anchoring structure, and this effect can not be avoided. This effect can only take favorable measures to minimize its harm. The anchorage works can obviously improve the rock and soil properties, and the lattice structure can enhance the slope protection. The plant slope has many advantages and the effect of the above method can not be neglected. Slope protection measures can take the combination of prestressed anchored lattice beam and vegetation slope protection, which can reinforce the slope and green the environment. But it still needs a complete theoretical system. It is urgent to establish a complete concept and overall evaluation mechanism of rock mass anchoring system so that the function of each favorable element in the anchoring system can be fully exploited, thus weakening the harm of the unfavorable elements so that the system can achieve the best working condition.

IOP Conf. Series: Materials Science and Engineering **250** (2017) 012038 doi:10.1088/1757-899X/250/1/012038

8. References

[1] Jian-lin LI. Slope Engineering [M].Chongqing: Chongqing University Press.
[2] Shuang-suo YANG,Jian-ping CAO. Evolution Mechanism of Anchoring Stress and Its Correlation with Anchoring Length [J]. Journal of Mining and Safety Engineering, 2010, 027 (001):1-7.
[3] Yi-ping LUO,Sheng SHI, Zhi-xin YAN. Numerical Simulation of Vertical Uplift of Bolt [J]. Journal of China & Foreign Highway, 2016, 036(002):6-10.
[4] Hansonn W. Influence of Surface Roughness of Prestressing Strand on Band Performance [J]. Journal of Prestressed Concrete Institute, 1969, 14 (1):32–45.
[5] GotoY. Cracks Formed in Concrete around Deformed Tension Bars [J].Journal of American Concrete Institute, 1971, 68(4):244–251.
[6] Phillips S.H.E. Factors Affecting the Design of Anchorages in Rock[R]. London:Cementation Research Ltd.1970.
[7] Ji-ru ZHANG,Bao-fu TANG. Hyperbolic Function Model to Analyze Load Transfer Mechanism on Bolts [J]. Chinese Journal of Geotechnical Engineering, 2002, 24(2):188–192.
[8] Guo-jin CAO,Hong-dao JIANG,Hong-mei XIONG. Calculation Method of Support Length for Stretched Bolts [J]. Chinese Journal of Rock Mechanics and Engineering, 2003, 22(7):1141–1145.
[9] Jian-chao ZHANG,Jian-qing HE,Xin JIANG. Mechanical Analysis for Anchorage Segment of Tension-type Anchor Based on Kelvin Solution [J]. Mining and Metallurgical Engineering, 2012, 32(4):16–19.
[10] Wen-xiang PENG,Ming-hua ZHAO,Hai-ping YUAN,etal. Parameters Analysis of Grouted Bolts by Lagrangian Difference Method [J]. Journal of Central South University (Science and Technology), 2006, 37(5):1002−1007.
[11] Yi-ping LUO,Sheng SHI, Zhi-xin YAN,etal. Shear Interaction of Anchorage Body and Rock and Soil Interface Under the Action of Uplift Load[J]. Journal of China Coal Society,2015,40(1):58-64.
[12] Ya-lin LIU,Shu-tong YANG,Liang LIU,etal. Mortar Anchor Adhesive Performance Test Research [J]. China Concrete and Cement Products, 2014, (2):61-63.
[13] HORI M. Micromechanical Analysis of Deterioration Due to Freezing and Thawing in Porous Brittle Materials [J]. International Journal of Engineering Science, 1998, 36(4):511–522.
[14] Guang-miao XU,Quan-sheng LIU. Analysis of Mechanism of Rock Failure Due to Freeze-thaw Cycling and Mechanical Testing Study on Frozen-thawed Rocks [J]. Chinese Journal of Rock Mechanics and Engineering, 2005, 24(17):3076-3082.

IOP Conf. Series: Materials Science and Engineering **250** (2017) 012039 doi:10.1088/1757-899X/250/1/012039

Experimental Research on Seismic Performance of Four-Element Variable Cross-Sectional Concrete Filled Steel Tubular Laced Columns

Zhijing Ou [a], Jianmao Lin [b], Shengfu Chen [c] and Wen Lin [d]

College of Civil Engineering, Fujian University Of Technology, Fuzhou 350118, China

Email: [a]sina99@163.com, [b]
924697140@qq.com, [c]280460976@qq.com,[d]linw617@qq.com

Abstract. A total of 7 experimental tests were conducted to investigate seismic performance of four element variable cross-sectional Concrete Filled Steel Tubular (CFST) laced columns. The experimental parameters are longitudinal slope and arrangement type of lacing tubes. The rules on hysteresis loop, ductility, energy expenditure, and stiffness degradation of specimens are researched. Test results indicate that all specimens have good seismic performance; their hysteresis loops are full without obvious shrinkage. With the increase of longitudinal slope, the horizontal carrying capacity increases, energy dissipation capacity improve, and there is slightly increase in stiffness degradation. The influence of arrangement type of lacing tubes on displacement ductility of specimens is big.

1. Introduction

Variable cross-sectional Concrete Filled Steel Tubular (CFST) laced column consist of CFST longitudinal elements tied together using steel lacing tubes, the section size of column is gradually increasing from top to bottom. Compare with equal sectional laced column, variable cross-sectional laced column is advantageous in relatively high compressive strength capacity, relatively high horizontal bearing capacity, good overall stability, excellent deformation capacity and attractive appearance. Furthermore, this type of hybrid column has more excellent earthquake resistance for its center of gravity is lower and the reaction of the column under earthquake function is decreased [1]. So it is believed that there is great potential to apply variable cross-sectional CFST laced columns in high-pier bridges, equipment frame and large-span CFST truss arch bridges located in strong earthquake region.

Examples of such structures include Ganhaizi super large bridge in Sichuan province in China [2]. It crosses seismic belt (Magnitude=9), the overall length is 1811m, and it is the longest total CFST truss girder bridge in the world. In the substructure of the bridge, CFST laced column and CFST hybrid columns are used according to different pier height. In which No. 26 pier is 67.29 meters high, the longitudinal slope is 1:50, and it is the tallest variable cross-sectional CFST laced column pier in the world.

In recent years, several investigations have been performed to quantify the dynamic behavior of CFST columns. Some researchers have focused on the seismic performance of CFST single longitudinal column and same cross-sectional CFST laced columns [3]-[11]. However, seldom research has been conducted on aseismic behavior of variable cross-sectional CFST laced columns, and additional experimental results are needed to quantify the hysteretic properties, ductility, failure mode and horizontal load-carrying capacity of these CFST laced columns.

IOP Conf. Series: Materials Science and Engineering **250** (2017) 012039 doi:10.1088/1757-899X/250/1/012039

In this paper, the pseudo-static test of 7 specimens was conducted to investigate seismic performance of four-element variable cross-sectional CFST laced columns, with the experimental parameters of longitudinal slope and arrangement type of lacing tubes. The rules on hysteresis loop, skeleton curve, ductility, energy expenditure, and stiffness degradation of specimens are researched, then failure mode failure mechanism are discussed. These experimental results can provide research basis for dynamic characteristics analysis and seismic design theory of variable cross-sectional CFST laced columns.

2. Specimens

Seven 1/8-scale specimens were designed and fabricated taking No.9 pier of Ganhaizi super large bridge as the prototype, include six laced column with inclined lacing tube and one laced column with flat lacing tube, as shown in Table.1. They had same section dimension and same length of 2.5 meter, outer center distance of longitudinal element is 700mm, inner center distance of longitudinal element on the top section of column is 0.3, inner center distance of longitudinal element on the bottom section of column is decided according to longitudinal slope. The axial compression ratio is 0.15.

Each specimen was composed of four longitudinal CFST members (φ114×2mm steel tubes filled with concrete C50) and several lacing steel tubes ofφ48×2mm. Seven specimens designated as BP-1 to BW-2, corresponding to experimental parameter of longitudinal slope and arrangement type of lacing tubes, which were listed in Tab.1. Longitudinal slope is set at 1:40 and 1:20, the layout of the tubes is respectively made up of flat, K, N and W, as shown in Tab.2.

Table.1 Characteristics of Test Specimens

Specimen	arrangement type of lacing tube	longitudinal slope	Peak load (kN)
BP-1	Flat	1:40	80.11
BK-1	K-type	1:40	201.09
BN-1	N-type	1:40	179.23
BW-1	W-type	1:40	213.39
BK-2	K-type	1:20	209.02
BN-2	N-type	1:20	208.21
BW-2	W-type	1:20	236.39

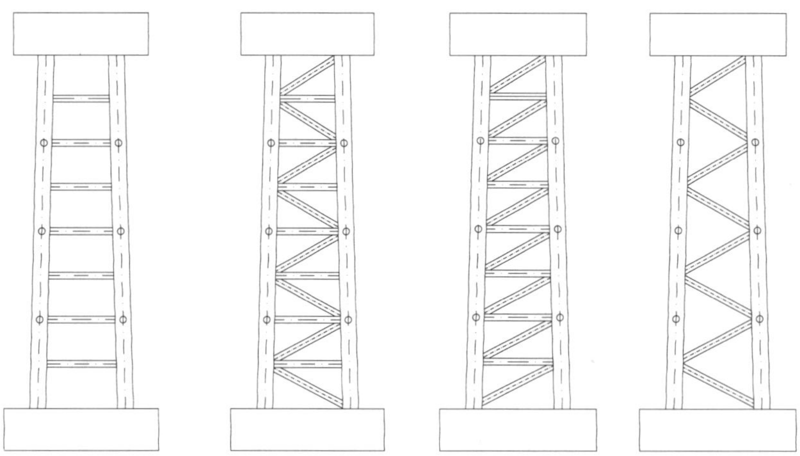

(a)Flat (b) K type (c) N type (d) W type

Figure.1 arrangement type of lacing tube

From the tests of the materials used in the specimens, the yielding strength f_s of the longitudinal steel tube and lacing tube are 365MPa and 384MPa, respectively, and elastic modulus E_s is

2.06×105MPa, the standard value of curb strength of the concrete fck is 59.4MPa, its elastic modulus Ec is 3.45×104MPa.

3. Experimental method

All specimens are tested by MTS servo loading system. Pseudo static loading scheme is adopted, the constant vertical load is applied by hydraulic jack on top of pier to control axial load ratio of the specimen. Then horizontal load (low reversed cyclic displacement) is acted by horizontal jacks fixed on the counterforce wall. The testing set-up is shown in Fig.2. Displacement control is adopted in load mode. After the specimen yielded, load step was changed to multiples of yield displacement. When cyclic loading dropped to 85% of peak load, test was stopped. The specimens were instrumented to measure (a) strain of longitudinal member at the bottom and mid-height section in both longitudinal and transverse direction; (b) strain at lacing tubes' middle section;(c) deformation in longitudinal and lateral axial.

Figure.2 Test device

4. Experimental results

4.1 Load-displacement hysteretic loop

The load-displacement hysteretic loops of typical specimens are shown in figure 3. The hysteretic curves of each specimen are relatively plump fusiform and there is no obvious pinch, it shows that variable cross-sectional CFST laced columns have good seismic performance, they can absorb and dissipate seismic energy well. Comparing these curves, it can be found that fullness degree of CFST laced column with flat lacing tube's hysteresis loop is evidently bigger than that of CFST laced column with inclined lacing tube.

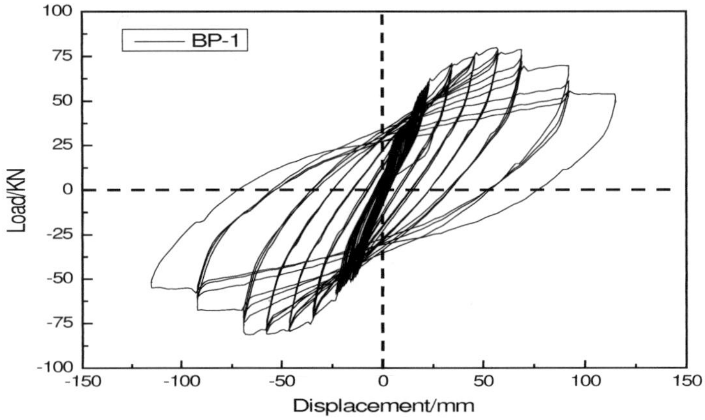

(a)BP-1 specimen

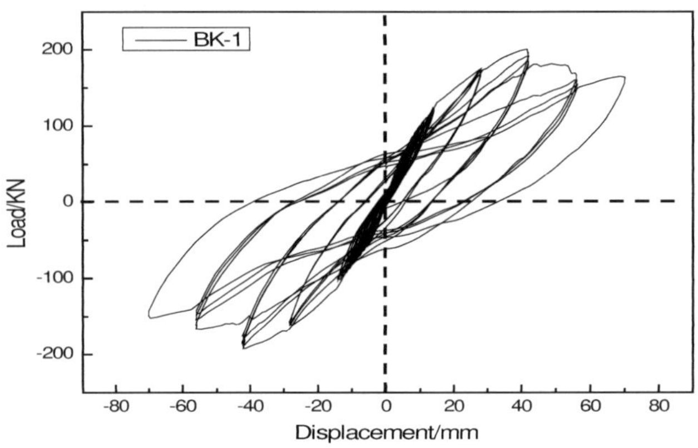

(b) BK-1 specimen

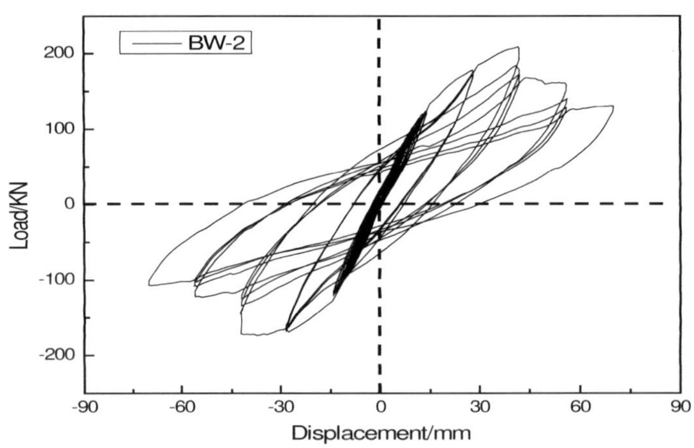

(c) BN-2 specimen

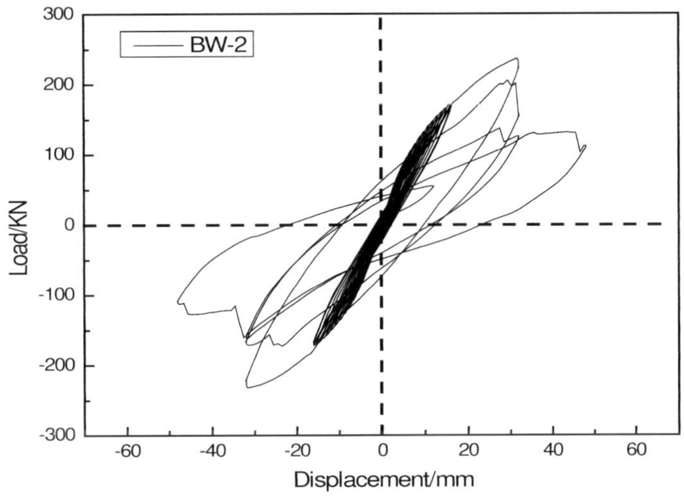

(d) BW-2 specimen
Figure.3 Comparison of load-displacement hysteresis loops

4.2 Load-displacement skeleton curve

The load-displacement skeleton curves of typical specimens are shown in figure 4, these laced columns have same longitudinal slope (1:40) and different arrangement type of lacing tubes. From Fig.3, it can be found that elastic stiffness and horizontal peak load of CFST laced column with inclined lacing tube (specimen BK-1, BN-1, BW-1) are obviously larger than that of CFST laced column with flat lacing tube (specimen BP-1). Because the lacing tubes and longitudinal elements form a stable triangle relationship in variable cross-sectional CFST laced column, thus the overall performance of the specimen evidently improved, and the stiffness and bearing capacity increase accordingly.

The skeleton curves of specimens with N-type lacing tubes are shown in figure 5. It can be found that with the increase of longitudinal slope from 1:40 (specimen BK-1) to 1:20 (specimen BK-2), the elastic stiffness and horizontal peak load increase 16.7% and 16.2%, fall-period stiffness and ultimate displacement are respectively nearly.

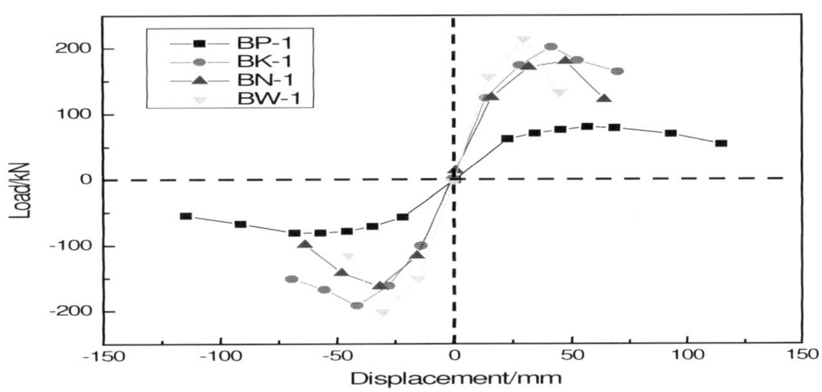

Figure.4 Skeleton curve with different lacing tube type

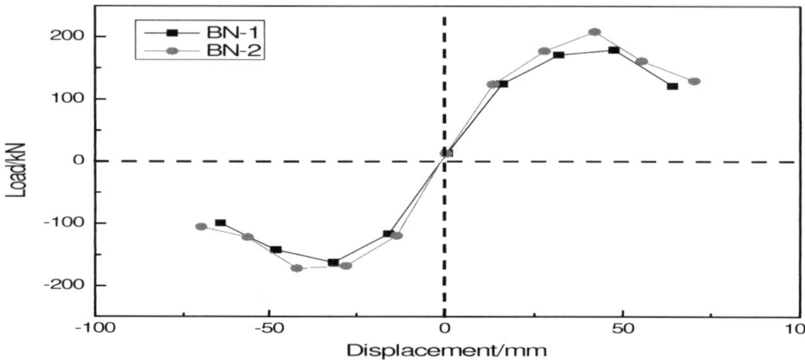

Figure.5 Skeleton curve with different longitudinal slope

4.3 Ductility

Ductility is one of the important indexes to evaluate seismic performance, it can be expressed by displacement ductility coefficient μu, the value of μu is one of the important indexes to evaluate seismic performance, it can be expressed by displacement ductility coefficient μu. The value of μu is larger, the ductility is better. Displacement ductility coefficient (μu) of specimens is shown in Table. 2. From Tab.2, it shows that arrangement type of lacing tubes has a great influence on displacement ductility of variable cross-sectional CFST laced column, the value order of μu as follow, K-type>flat>N-type >W-type. Meanwhile, the influence of longitudinal slope on variable cross-sectional CFST laced column is respectively small.

Table.2 Displacement ductility coefficient (μu) of Specimens

Speci-ments	BP-1	BK-1	BN-1	BW-1	BK-2	BN-2	BW-2
μ_u	4.12	4.39	3.47	2.38	4.06	3.78	2.24

5. Conclusions

Test results indicate that the hysteretic curves of specimen are relatively plump fusiform, it shows that variable cross-sectional CFST laced columns can absorb and dissipate seismic energy well and they have good seismic performance. Arrangement type of lacing tube and longitudinal slope has large influence to seismic behaviors of variable cross-sectional CFST laced columns. Compare with CFST laced column with flat lacing tube, elastic stiffness and horizontal peak load of CFST laced column with inclined lacing tube enlarge obviously, the value order of ductility as follow, K-type>flat>N-type >W-type. With the increase of longitudinal slope, the elastic stiffness and load carrying capacity increase gradually, and energy consumption improved consequently.

6. Acknowlegement

This research work has been conducted under the grant number 51408128 of the national natural science foundation of China ,the grent number 2017J01471 of the natural science foundation of Fujian province and the grant number GY-Z15108 of the program for new century excellent talents in Fujian Province university. This support is sincerely appreciated.

7. References

[1] CHEN Shengfu, OU Zhijing. The Research Summary of Seismic Performance on Concrete Filled Steel Tubular Laced Columns.Fujian Architecture & Construction, ,(2):48-52(2016)

[2] Zhijing Ou .The Practice of Concrete Filled Steel Tube Piers to Bridges: A review, Applied Mechanics and Materials, Vols.405-408,pp1602-1604, (2013)

[3] LV Xilin, Lu Weidong. Seismic Behavior of Concrete-filled Rectangular Steel Tubular Columns Under Cyclic Loading.Journal of Building Structures, 21(2):2-11(2000)
[4] HAN Linhai. The concrete filled steel tubular structures from theory to practice [M],Beijing: Science Press, (2004). (in Chinese)
[5] CHEN Bowang, Zou Yanhua , Tang chu,etc. Contrast research on square and circular CFST laced columns pseudo-static test. China civil engineering journal, 47(S2):108-112(2014)
[6] CAO Yan,Chen bowing,He ran,etc.Self-compacting concrete filled square steel tubes limb lattice column quasi static test research. Journal of Hunan institute of engineering,24(1):77-81.(2014)
[7] DENG Xuanyi. Study on the seismic performance of concrete filled steel tubular lattice column. Central South University, (2012)(in Chinese)
[8] LUO Yao. Studies on the seismic performance of four-tube concrete filled steel tubular laced columns. Central South University,(2013)(in Chinese)
[9] LV Yinhua.Studies on hysteretic behavior of uniform section concrete filled steel tubular laced columns with flat lacing tube.Fuzhou University, (2015)(in Chinese)
[10] Abdelgadir Elzien, Boha Ji; Zhongqiu Fu; Zhengqing Hu .The behavior of lightweight aggregate concrete filled steel tube columns under eccentric loading . Steel and Composite Structures, 11(6): 469-488(2011)
[11] Chung, Kyung-Soo, Kim Jin-Ho; Yoo Jung-Han. Experimental Study on the Structural Behavior of High Strength Concrete-filled Square Steel Tube Subjected to Flexural Loading. Steel and Composite Structures, 14(2): 133-153(2013)

Numerical Flexibility Determination Method of Stress Intensity Factor for Concrete

Yongjin Wu[1,a], Hu Chen[2] and Xiangdong Wang[3]

[1] Key Laboratory of Geotechnical Mechanics and Engineering of Ministry of Water Resources, Yangtze River Scientific Research Institute, Wuhan 430010, China

[2] College of Civil and Transportation Engineering, Hohai University, Nanjing 210098, China

[3] College of Mechanics and Materials, Hohai University, Nanjing 210098, China

Email: [a] wuyj@hhu.edu.cn

Abstract. Flexibility method is a commonly used method to determine fracture toughness. In the experiment, it is necessary to prepare specimens with different crack lengths and other exactly same conditions, and carry out a large number of repeated experimental works. Given the above problems, this paper develops a flexibility determination method based on numerical simulation method to calculate the stress intensity factor and fracture toughness of concrete specimens. The results of the test example show that the use of the numerical simulation experiment method is practically feasible and effective to obtain the relation curve between the flexibility C and the crack length a of the concrete specimens and further get the stress strength factor and the fracture toughness under the ultimate load.

1. Introduction

Concrete is a kind of quasi-brittle material with low tensile properties. Under the external factors such as load, temperature change and chemical action, the occurrence of cracks in a concrete structure is unavoidable. Under the action of certain external factors (load, temperature change and so on), the internal defects of the concrete structure will produce a high degree of stress concentration, and gradually expand the development to form micro-cracks in the concrete body. Besides, the joint surfaces between phases will be separated under the influence of external factors to form interface cracks and develop into micro-cracks. Micro-cracks continue to expand, converge and pass through to for macro-cracks. The macro-cracks continue to expand, and ultimately may lead to complete failure of concrete structure [1-2]. Crack is the main reaction of aging and lesion of the concrete structure, which is very harmful to its safety. Both the failure modes of complex structures such as high arch dam and engineering fracture problems of the conventional buildings such as piers, foundations, and chimneys are urgent and challenging problems to be solved in design. In the concrete structures there are many examples of cracks that affect the operation and influence the benefits.

At present, the stability analysis of cracks is mainly from two aspects. Firstly, according to the influencing factors, establish a mathematical monitoring model using the monitoring data and analyze the development trend of cracks. Secondly, calculate the stress strength factor at the crack tip or other fracture parameters using the fracture mechanics principles and methods, and determine whether the cracks will expand according to the corresponding fracture criteria [3]. In determining whether the cracks will expand, the fracture criteria of crack-tip in the structure criteria is essential. The stress intensity factor K of the fracture parameters is a useful parameter to characterize the strength of the stress field near the crack tip. The methods of determining the stress intensity factor K include the analytical method, numerical method, and experimental method. Flexibility method is commonly used in the experiment to determine the fracture toughness. The specimen with the different crack lengths and other exactly same conditions should be prepared in the experimental process, and a large number of repeated experimental works should be carried out. Given the above problems, this paper develops a flexibility determination method based on numerical simulation method to calculate the stress intensity factor and fracture toughness of concrete specimens.

2. Flexibility Expression of Stress Intensity Factor

Flexibility is the amount of displacement caused by the unit force, only related to the structural geometry, E and ν. The expression of flexibility is:

$$C = \Delta / P. \tag{1}$$

The energy release rate G is an important index of fracture mechanics related to the external load and the structure type (including the length, shape, position and structural geometries of the crack). The dimension is [force] × [length] -1, which can be regarded as the necessary force of unit length for crack propagation. There is a particular relationship between the stress intensity factor K and the energy release rate G, assuming that the crack expands a certain length Δa along its propagation line, calculate the stress relaxation work of the crack tip, we can get the relationship between the energy release rate and stress intensity factor [4-5]:

$$G = \frac{1}{E'}\left(K_{\mathrm{I}}^2 + K_{\mathrm{II}}^2\right) + \frac{1}{2\mu}K_{\mathrm{III}}^2 \cdot \tag{2}$$

Where E is the elastic modulus, E'=E during the plane stress state, $E' = E/(1-\nu^2)$ during the plane strain state; μ is the shear modulus.

The energy release rate is independent of the nature of the loading system. Whether it is constant displacement or constant load, G is equal to $dU_a/(B \cdot da)$. Then the relationship between the energy release rate G and the flexibility C is deduced as follows:

$$G = \frac{P^2}{2B}\frac{dC}{da}. \tag{3}$$

Formula (3) is often used for the energy release rate of experimental determination. dc/da can be determined by experimental determination of flexibility of multiple specimens with different crack lengths and other exactly same conditions.

The stress intensity factor K of the specimen with certain geometric type and crack shapes can be obtained by combining formula (2) and formula (3), and furthermore, the ultimate load Pmax obtained by the experiment can determine the fracture toughness Kic of the specimen. Fracture toughness is a mechanical property parameter of the material and a toughness index in which the material resists the crack [2].

3. Implementation Process of Numerical Flexibility Method

It is necessary to carry out a large number of repeated experimental works to obtain dc/da through the experimental method, including the preparation of specimens with different crack lengths and other exactly same conditions. Each specimen should be loaded to obtain P-Δ relation curve, consuming a lot of time and materials. With the advancement of computer technology and numerical method, it is possible to carry out the numerical simulation of the fracture process of concrete materials. Under the approximate physical conditions of numerical simulation structure and actual structure, the numerical simulation method saves time, effort, without instrument limit, and may repeat and done a lot of experiments. It not only can simulate complex loading paths that are difficult to implement by many loading devices but also is immune to the effects of various factors in the experiment. Therefore, it is of importance to study the mechanical properties of concrete by numerical simulation method [6-7].

The load - displacement curve of the concrete specimen under different crack lengths during fracture failure process is obtained by numerical simulation experiment method, and the relation curve between the flexibility C and the crack length a is established to obtain dC/da. The numerical simulation of this paper is carried out based on ANSYS. The solid65 element in ANSYS is a three-dimensional 8-node solid isoparametric element which is usually used to simulate concrete structure. It is characterized by plastic deformation, which can be cracked in three directions and crushed. The detailed process of the numerical implementation is as follows:

(1) Establish the geometric model of the specimen in the ANSYS, the crack length of the model is a.

(2) Define the element, material properties and divide the grid. The solid65 element is used as the element, and the failure criterion of William-Warnke five parameters is adopted. The followings should be defined: uniaxial tensile strength ft, uniaxial compressive strength fc, shear strength reduction coefficient of open and closed slip surface, two-way compression strength fcb, and uniaxial compressive strength f1 and biaxial compressive strength f2 under a certain confining pressure.

(3) Add the boundary condition and the external load to solve and extract the displacement Δ of the load point.

(4) Change the crack length a, repeat steps (1) ~ (3).

(5) Establish the relation curve between the flexibility C and the crack length a, and get dC/da through derivation.

(6) Calculate the stress intensity factor K, for example complete type I cracks:

$$K_{\mathrm{I}}^{2}=E^{'}G=\frac{E^{'}P^{2}}{2B}\frac{dC}{da}\,. \tag{4}$$

4. Applications

4.1 Complete Type I Crack Problem

The geometry of the three-point bending beam specimen and loading method is shown in Fig.1. As shown in Fig.1, there will be only normal stress on both sides of the crack including the crack surface, so it's a complete type I crack problem. The specimen size is L × h × b=2100mm×500mm×200mm, S=2000mm, span ratio s/h=4. Elastic modulus E=30GPa, Poisson's ratio v=0.194, compressive strength fc=25.2MPa, tensile strength ft=2.37MPa.

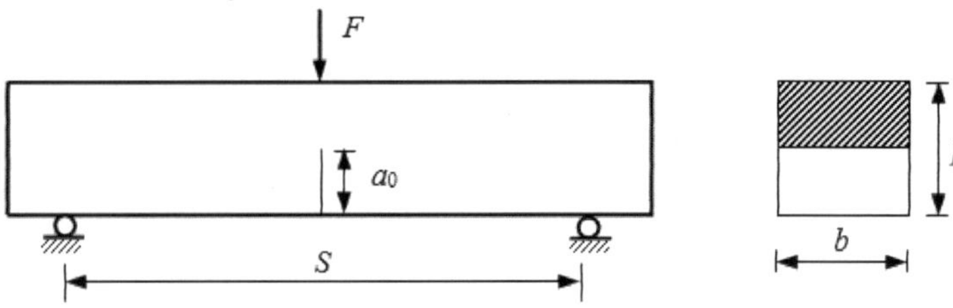

Figure.1. Three-point bend beam specimen.

The results obtained according to the method in Section 3.2 are shown in Fig.2. It is observed that the C-a curve of the complete type I crack is gentle while the seam height ratio is less than 0.4, but the amplitude of change is significantly increased when the ratio is greater than 0.4. At this point, the specimen has been approaching the destruction state. This is consistent with the change of the formula f(a) proposed by Chen Chi[8s]. The relationship between C and a of this three-point bending beam is obtained by quadratic polynomial fitting:

$$C=2\times10^{-6}a^{3}-8\times10^{-7}a^{2}+2\times10^{-7}a-6\times10^{-9}\,. \tag{5}$$

By deriving the formula (5), we can get dC/da. Substituting into formula (4), the value of the stress intensity factor KI can be obtained. The comparison results between numerical flexibility method and Chenchi's formula are shown in Fig.3. It can be seen that the error is basically less than 15%.

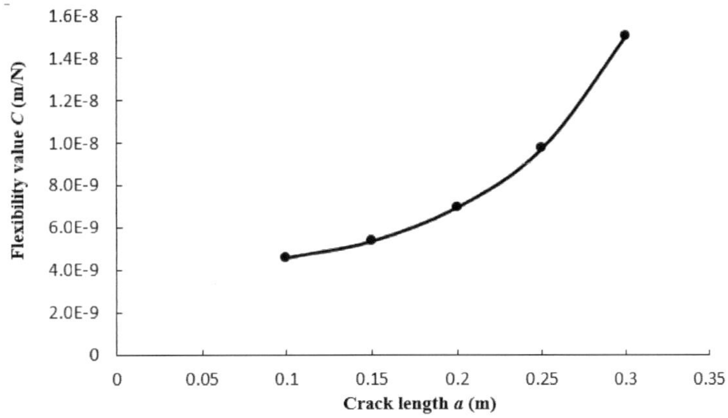

Figure.2. Curve of the relationship between C and a of the complete type I crack obtained by the numerical flexibility method.

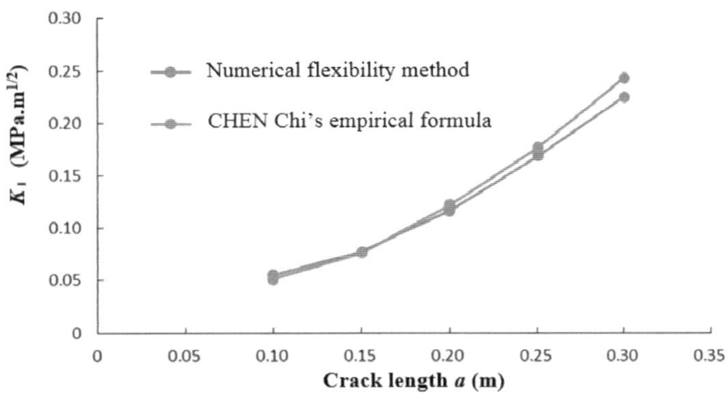

Figure.3. Comparison of calculation results of the complete type I crack by different methods.

4.2 Complete Type II Crack Problem

The geometry of the four-point shear specimen and loading method is shown in Fig.4. According to the illustrated loading method, there will be only shear stress on both sides of the crack including the crack surface, and the bending moment on the crack plane is zero. The specimen is a complete type II crack under this loading method. The specimen size is L×h×b=500mm×100mm×100mm, S=200mm, S'=20mm. Elastic modulus E=30GPa, Poisson's ratio ν=0.17, compressive strength fc=25.2MPa, tensile strength ft=2.37MPa.

The flexibility C corresponding to the crack length a obtained by numerical simulation is shown in Fig.5. Compared with complete type I crack, the C-a curve of the complete type II crack is quite gentle. This also proves that type I cracks are the most dangerous. The relationship between C and a of this four-point shear specimen is obtained by quadratic polynomial fitting:

$$C = 0.0719a^6 - 0.0156a^5 + 0.0014a^4 - 6\times10^{-5}a^3 + 2\times10^{-6}a^2 - 2\times10^{-8}a + 2\times10^{-9}. \qquad (6)$$

By deriving the formula (6), we can get dC/da. Then the value of the stress intensity factor KII can be obtained. The comparison results between numerical flexibility method and the finite element displacement method are shown in Fig.6. It can be seen that under the same computing conditions the error is basically less than 10%.

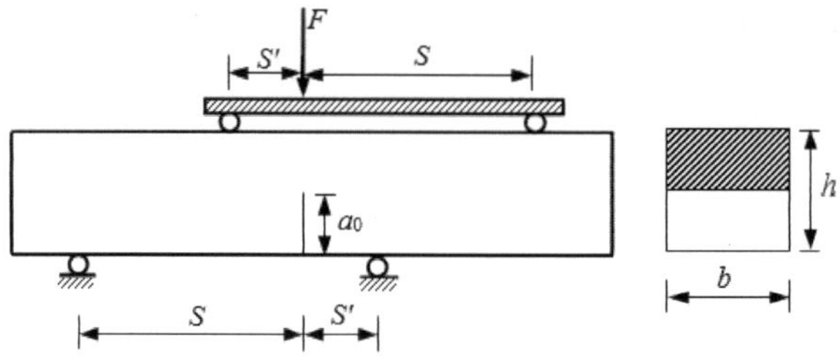

Figure.4. Four-point shear beam specimen

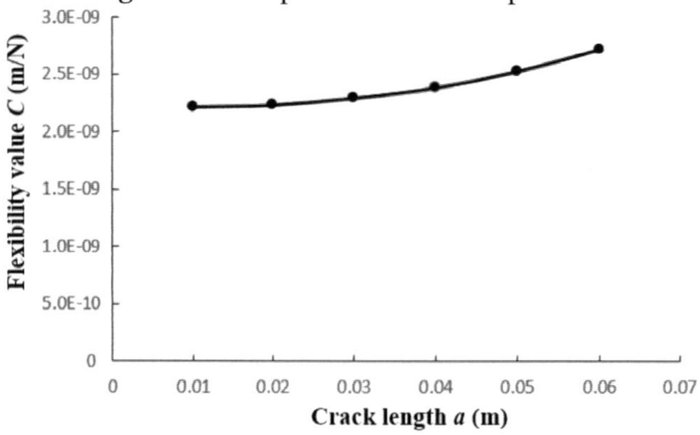

Figure.5. Curve of the relationship between C and a of the complete type II crack obtained by the numerical flexibility method.

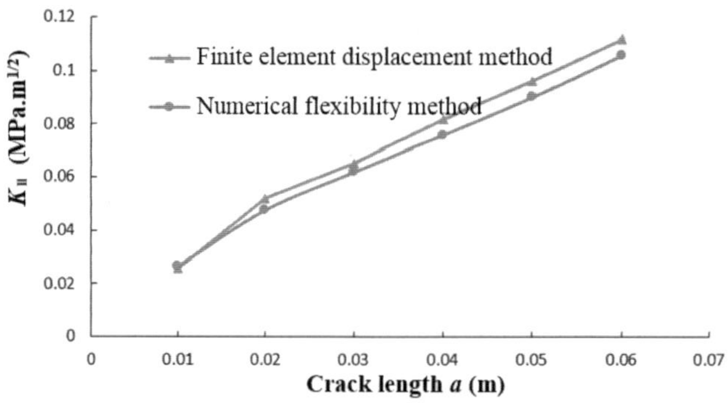

Figure.6. Comparison of calculation results of the complete type II crack by different methods.

5. Conclusions

The results of the test example show that the use of numerical simulation experiment method is a practically feasible and effective method to obtain the relation curve between the flexibility C and the crack length a of the concrete specimens, and further get the stress strength factor and the fracture toughness of the specimen under the ultimate load. Compared with the experimental determination method, it is not necessary to carry out a large number of repetitive experimental works, and compared with the traditional finite element method; it also can meet the precision needed for the engineering without dividing the grid of the crack tip region into small scale. But the concrete is a heterogeneous composite material, including coarse aggregate, fine aggregate, cement hydrate, pores, and cracks, etc. The numerical simulation of the loading failure process has certain limitations, and the accurate numerical simulation has room for further improvement.

6. Acknowledgements
Financial support from the National Natural Science Foundation of China (Nos. 51539002), the National Key Research and Development Program of China (Nos. 2016YFC0401802, 2016YFC0401803, 2016YFC0401804, and 2016YFC0402008), and the Basic Research Fund for the Central Research Institutes of Public Causes (No. CKSF2015041/YT) are greatly acknowledged.

7. References
[1] Wang, D.: Fracture Mechanics (Volume 1). Harbin: Harbin Institute of Technology Press. (1989)
[2] Yu, X.Z., Qiao, C.X., Zhou, Q.L.: Rock and Concrete Fracture Mechanics. Changsha: Central South University of Technology Press (1988)
[3] Xu, S.L., Reinhardt H.W.: Crack extension resistance and fracture properties of quasi-brittle softening materials like concrete based on the complete process of fracture. Int. J. Fracture. 92, 71--99 (1998)
[4] Irwin, G.R.: Analysis of stresses and strains near the end of a crack traversing a plate. J. App. Mech. 24, 361--365 (1957)
[5] Orown E.: Fracture and strength of solids. Reports on Progression Physical, the Physical Society.12, 185--232 (1949)
[6] Deng, A.M., Cao L., Xu D.Y.: Studies on Size Effect of Concrete Fracture Toughness. J. Yangzhou University. 10, 63--66 (2007)
[7] Deng A.M.: Study on Fracture Criterion Series of Concrete Considering the Influence of Fracture Process Zone. Nanjing: Hohai University. (2007)
[8] Chen C.: Research on Metal Fracture. Beijing: Metallurgical Industry Press. (1978)

Sustainability-based Study on the Development of Human Settlement in Traditional Villages

Yali Chen[1], Liping Huang[2] and Qi Lu[3]

[1]School of Architecture, South China University of Technology, PH.D School of Design, South China University of Technology, Lecture
[2]shanghai normal university, PH.D and Lecture
[3]School of Architecture, South China University of Technology, Professor
Email: 340075865@qq.com

Abstract: The purpose of this study is to investigate and analyze the status quo of the living environment in traditional villages, and to identify current issues and developing strategies based on data analysis. It is proposed that comprehensive sustainable strategies for land use should be designed, defined and used as guidelines for the constructions of traditional villages in the process of rapid urbanization. Such sustainable strategies should be applicable for remediation and development of traditional villages. This will promote the coordinated development in society, economy and environment in traditional villages.

1. Introduction

1.1 Research Background

The protection of living environment in traditional villages is a long-term dynamic process, which requires dynamic monitoring and adjustment throughout the entire process. The effectiveness of village protection measures, which is manifested in the quality of achievements, depends on the research methods of management relating to sustainable development. Therefore, sustainable development strategies for living environment must be established first, including living environment design with local villagers' involvement, hierarchical and phase-by-phase continuous management system, and a instructive and achievable sustainable development model.

Langtou Village, locating in Huadu District, Guangzhou, a traditional village with characteristic traditional Academy of classical learning and ancestral hall, was chosen as a representative example in this study.

2. Literature Review

Currently, domestic research on living environment in traditional villages mainly focuses on improvement and optimization, through adjustments of land and business structures, from a business planning, ecological planning and tourism planning point of view. Back in 2001, Wu Liangyong [1] made scientific research on human settlement, and ten years later, in 2011, he published a book titled Scientific Research Progress on Human Settlement (2002-2010), which showed that human settlement should employ different adaptive criteria and methods to keep up with the times. Shan Jixiang [2] (2010), the Director of State Administration of Cultural Heritage, emphasized the importance of village human settlement protection and expounded on its classification and renovation targets. He proposed that village human settlement protection should stick to the principles of strengthening community functions, hold on to traditional culture and boost regional economy. In House Form and Culture, Amos Rapopport [3] probed into the design process of vernacular architecture and maintained that the process should be the one during which architects and users perfectly understand each other, because local architecture is closely related to the real life and geography of villages, which also embodies the relation between space form and lifestyle. The basic requirements and function of traditional rural settlements include producing and servicing function, Eco-environmental function, and social function. [4] Sustainable human settlements are organic, dynamic and growing. Thus, it should be considered carefully in the design of land use and eco-environment. [5]

3. Research Method

With Langtou Village in Guangzhou as a sample, the researcher undertook a questionnaire survey that covered the planning and design of new villages, with a view to obtaining data concerning basic information and current housing condition of individual villagers and their families, their satisfaction with life, their suggestions for future village construction, etc. All respondents are native villagers. Altogether 50 questionnaires were handed out, and 43 were retrieved. Data analysis based on public participation is adopted in this research. Then, statistical analysis was performed to obtain the result.

4. Data collection and analysis

4.1 Basic personal and family situations

The main sources of income for respondents' families and the types of work in which respondents are currently engaged.

Figure 1 shows that income source is mainly divided into three parts including agricultural income (36%), wage income (34%) and collective dividend (20%).

Figure 2 shows that 40% of the villagers mainly work in agriculture, and 38% of the villagers work in factories. The data analysis was proceeded from the main income source of villagers and types of work they are currently engaged in.

Respondents’ age is between 14 to 61 years old, a relatively balanced residential structure.

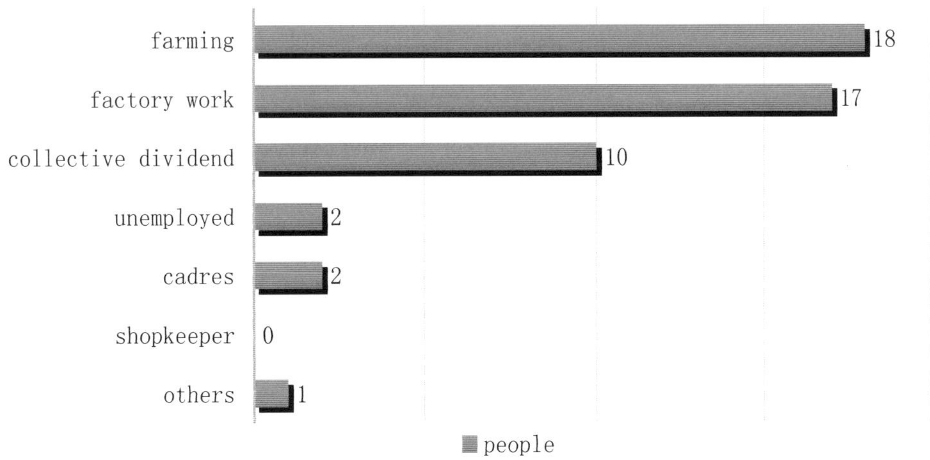

Figure.1 Respondents’ major income source

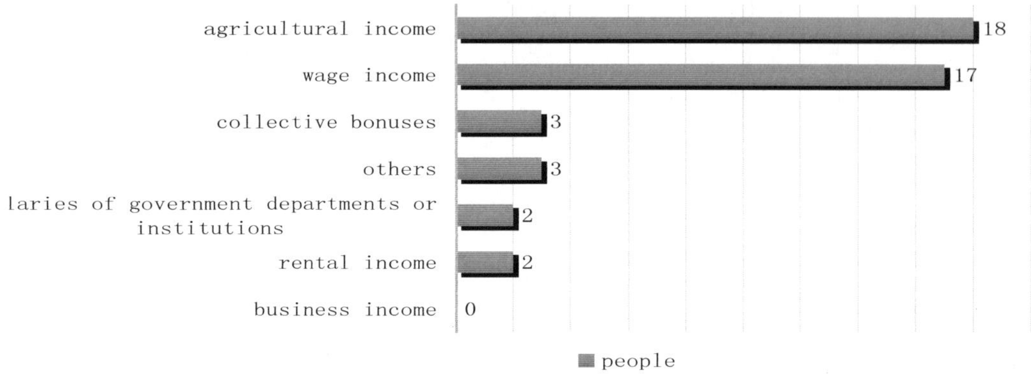

Figure.2 Respondents' current work category

4.2 The housing situation

It includes the building time and size of the house.

As shown in Figure 3, 100% of villagers own a house. Most of the houses were built between 1991 and 1995 with one floor.

As shown in Figure 4, 55% of villagers own a house measuring 80-90 square meters.

The desire of the villagers for the residential housing for the future planning years is related to the housing improvement type.

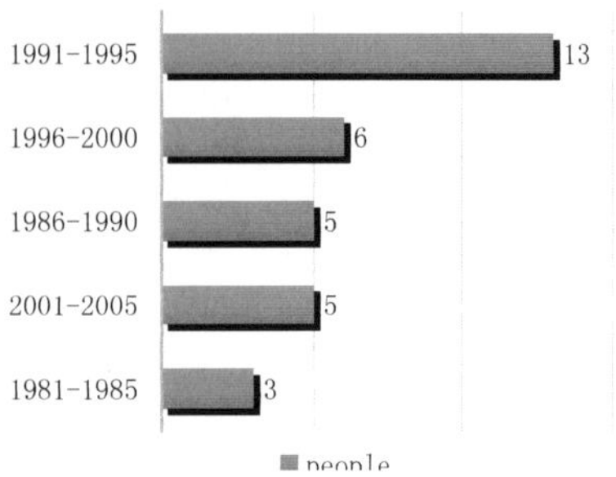

Figure.3 The Building time of the respondents

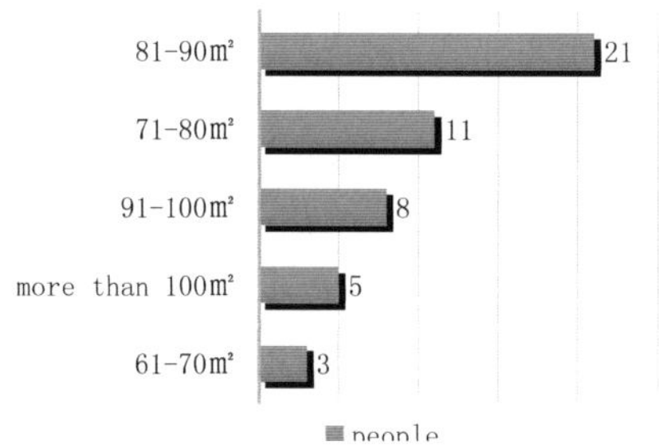

Figure.4 Housing area of the respondents

4.3 Satisfaction on rural infrastructure

It includes respondents' satisfaction on rural infrastructure and environmental remediation.

Fig.5 data states that 60% of villagers are satisfied with the greening, just at an average leve.

Figure 6 expounds that 60% of the villagers are dissatisfied with the construction leisure sites and facilities in the village, and look forward to the improvement of the living environment of the public facilities in the village. Among the problems for environmental renovation listed in the recycling questionnaire, 20% of the villagers think that it is most urgent to dismantle the dilapidated houses, and 15% of the villagers believe that the most urgent problem is centralized sewage disposal.

Based on the proportion of villagers' demands for public service facilities, the call for improvement on existing environment is intensively reflected in sustainable strategies.

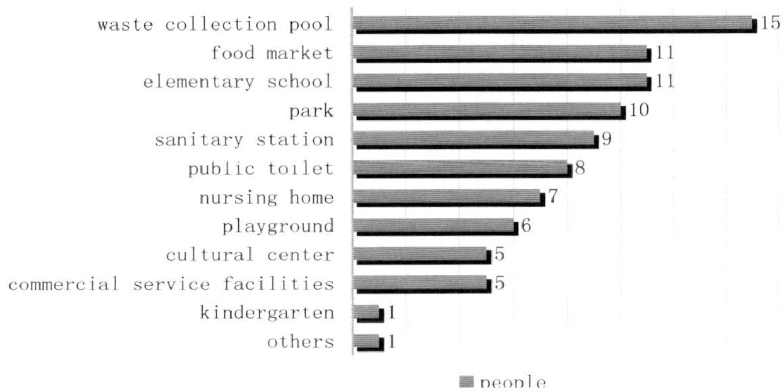

Figure. 5 Respondents' satisfaction on rural infrastructure

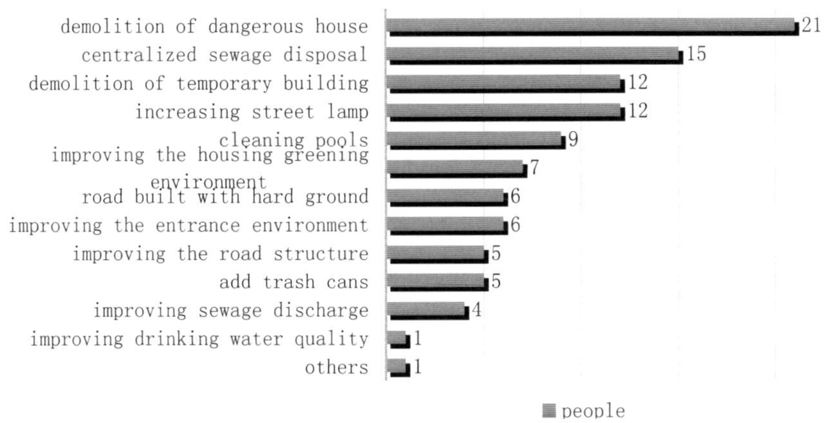

Figure.6 Respondents' satisfaction on the current environmental remediation

4.4 Proposals for new rural development

Concerns by respondents include several aspects, (I) the primary problem that needs to be addressed in constructing a new countryside, (II) the landmarks to represent village, (III) the respondents' willingness to devote a certain amount of funds for reinvention living environment, (IV) the respondents' willingness to accept the demolition of their own houses.

As shown in Figure 7, 32% of the villagers think that the improvement of the living environment is the primary issue needed to be solved in the process of new rural construction. 30% of the villagers consider construction capital security as the most primary thing, 20% of the villagers believe that making scientific planning is most important.

Figure 8 shows the local identity data of villagers. 38% of the villagers think the landmarks most likely to represent their village are the ancestral temple, memorial arch. 30% of the villagers give priority to the hometown of historic celebrities, and 18% of the villagers prefer to local-style dwellings.

Figure 9 shows that 44% of percent of villagers are not very motivated to invest some money for the improvement of their living environment. 34% of the villagers feel hard to accept the burden of self-financing in the new countryside, and 33% think they could not accept it.

Figure 10 shows that 28% of the villagers refuse to accept the demolition of their houses, and 35% of the villagers are not sure. 38% of the villagers are under the obligation to build the new countryside, and 27% of the villagers are not sure.

It can be learned that the sustainable strategies for the improvement of the living environment in traditional villages lies in the scientific measure of multi-cooperation and multi-party construction.

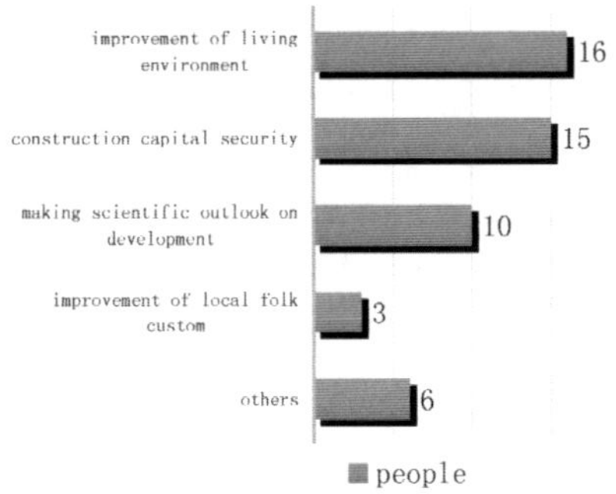

Figure.7 respondents' suggestions for the primary issue that needs to be addressed in new rural construction

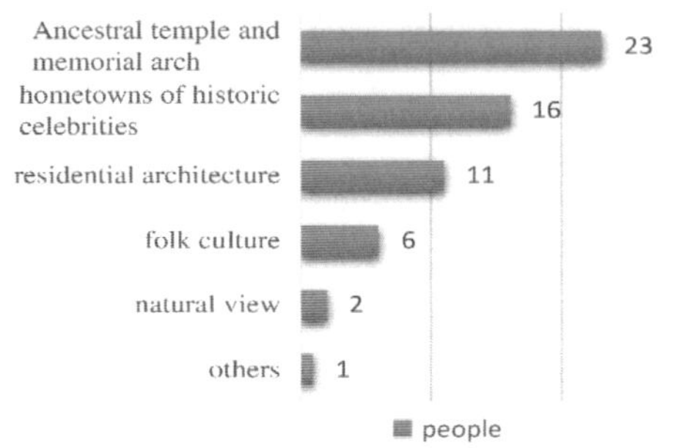

Figure.8 respondents' suggestion for the symbol most likely to represent village in new rural construction

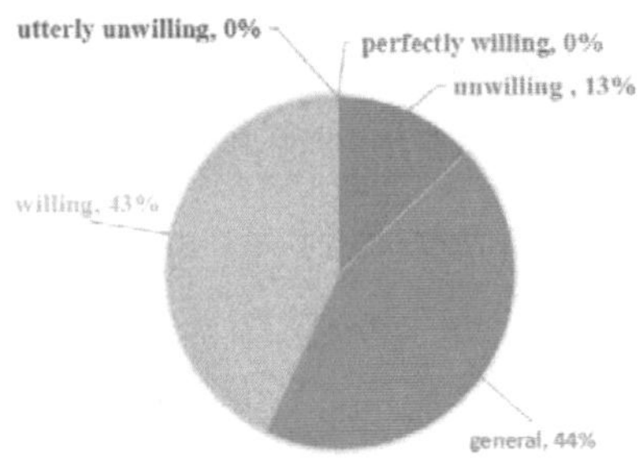

Figure.9 The willingness of respondents to invest a certain amount of money in improving the living environment

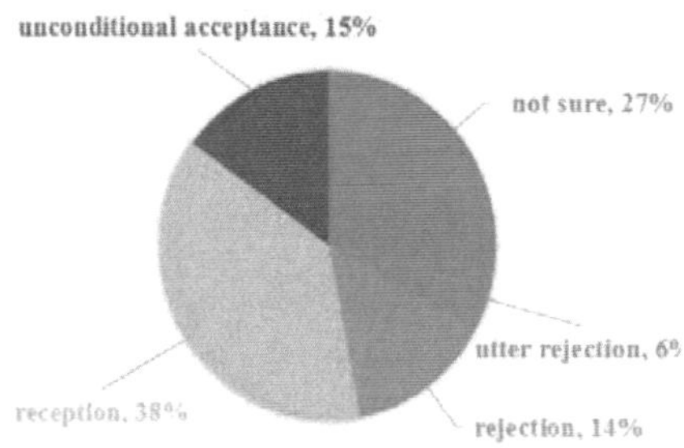

Figure.10 interviewees' reception of duty in the new rural construction

5. Discussion

5.1 The sustainability of traditional village human settlements

Local economic development should be coordinated with fund raising capacity according to the principle of land economy, energy conservation, water and material saving. Importance should be attached to regional characteristics, protection of the ecological environment, natural landscape, traditional architecture, historical and cultural relics, with respect for local folk customs and the protection of the natural texture of the village. Improvement of the production and living conditions of the villagers should be conducted by intensive construction of agricultural housing renovation, village infrastructure and public service facilities. The core is to increase the villagers' income and to enhance the life quality of the villagers.

5.2 Control of the population size of the planned years

The population predication is carried out through time series to obtain scientific statistics for future village capacity. Prediction of the future population growth is based on the historical and present-day change. The purpose of population capacity prediction is to achieve an appropriate balance between the native population quantity and alien population quantity, and to control the capacity of the village scientifically.

Formula: $Q=Q1(1+K)^{a}+P$

In formula: Q - population prediction (human);

Q1 -- the current number of registered population (people);

K - natural growth rate of agricultural population during the planning period (‰);

a- planning term (year);
P- mechanical growth of agricultural population during the planning period

5.3 Improvement of living and landscape habitat

Planning villagers' residential land. It mainly relies on the adjustment and renovation of the existing village residential land: focusing on the remediation of qualified residential land; adding new village land at reasonable location; arranging and utilizing public facilities and infrastructure in a centralized manner.

Improving the quality of land used for public facilities. Land should be reserved for village-level facilities for the needs of the villagers to make a reasonable complement for rural life. Such facilities include administrative facilities, commercial and financial facilities, cultural and recreational facilities, sports facilities, health care facilities and educational facilities.

Multi-cooperation and guidelines for sustainable development of traditional villages. The government guidance and coordination should be strengthened to provide organizational guarantee for planning execution, including the system protection of the public; the democratic management to provide human resources support for planning implement; the fund raising through various channels to provide financial security.

6. Conclusion

A standard of sustainability in this study is the residents' satisfaction to living conditions. Langtou Village was chosen as a representative example in this study. The data and results obtained from Langtou Village can be used as important standards in sustainable development of traditional villages.

The goal of rectification should be according to local conditions, with targeted and optional construction standard for each phase. Also, it is necessary to give an overall consideration of housing by villagers and the construction of public facilities and infrastructure. The smooth implementation of democracy between the government and the village can contribute to the equal cooperation for villagers, grassroots government and diversified cooperative groups. On the whole, a sustainable development of human settlement in traditional villages can make a reality through multi-cooperation and coordination. Consequently, the balance between nature and social ecological system in rural areas can be maintained through taking into consideration the balance of multiple parties' interests and finding the resonance point in the law of market economy.

Base on the data analysis, the research in this study confirmed again that the sustainable development of traditional villages should be efficiently design according to organic, dynamic and growing model.

7. Acknowledgement

Acknowledgement: Project supported by the National Nature Science Foundation of China, Project No. 51278194

8. References

[1] Wu Liangyong, Introduction to Human Settlements, China Architecture & Building Press, 2001
[2] Shan Qixiang, Workshop Report for the State Administration of Cultural Heritage, 2010
[3] Amos Rapoppor, House Form and Culture, 2012
[4] Liming Liu, History of rural landscape planning and its prospects in China. , 2001, 17(1):53.
[5] Yihui Sun, Tian Chen, Yuncai Wang, The progress of regional cultural landscape in traditional rural regions, 2008, 27(6):1.

The Quasi-static Properties of Natural Marine Clay under Tidal Low Frequency Cyclic Loading in Yangtze Estuary, China

Dan Li[1] , Nianqing Zhou[1, a] , Xiaonan Wu[2] and Jiachun Yin[2]

[1]College of Civil Engineering, Tongji University, Shanghai, 200000, China
[2]Communications Construction Contracting Co. Ltd., Shanghai, 200000, China
Email: [a]nq.zhou@tongji.edu.cn

Abstract. The enclosure engineering of tidal flats behind Changxing Island is taken as the research background. In order to explore the undrained cyclic response of natural marine clay on tidal level periodic changing, the stress-controlled low frequency cyclic tri-axial tests and static creep tests are conducted on natural K0-consolidated Shanghai clay. In the series I tests, samples are cyclically sheared by low-frequency cyclic loading until stability or failure, the accumulative behavior and reversible strain are studied. The effects of confining pressure and stress ratio are also investigated. Significantly, loading frequency is a key factor in controlling the cyclic behavior. In the series II tests, special attention is given to the low frequency cyclic vibration effect on the undrained cyclic tests by comparison with the static creep tests. The critical stress ratio of low frequency cyclic behavior is less than static creep strength.

1. Introduction

In the marine environment, soil foundations are subject to cyclic loads generally induced by waves, tides or vibrations of super structures over a long period of time (Ling et al., 2002). Accordingly, seabed soils are expected to experience cyclic stresses of different amplitudes and frequencies, which will lead the soils to residual settlements and result in a decrease in strength or even damage (Fan et al., 2017).

As for the response of soft clay under cyclic loading, investigations on clay behavior under cyclic loading have been done by many researchers (Matsui et al.,1992; Yasuhara et al.,1994; Soroush et al.,2009). These excellent works provide fundamental tools for further understanding of the cyclic behavior of clays, which is of significance in practical design methods for the stability of nearshore and off-shore structures. In addition to experimental studies, some researchers have evaluated the undrained cyclic behavior of clays using elastic-plastic models (Karim et al.,2013; Graham et al.,2015).

At present, contraposing the mechanics properties of soft clay under the wave loading, which is a kind of alternating load of low amplitude and long period(Zhu et al.,2006), many researchers have constantly deepened the study, and also have applied the theory research to practice. However, the above researches of the soft clay are almost aimed at the high frequency cyclic loading experiment, and few involved the dynamic cyclic creep characteristics of low frequency cyclic loading. There are many common types of low frequency cyclic loading in practical engineering, such as underground water level cyclic change, the load change caused by the amount of storage in tanks or granary, the periodic ballast of tidal level fluctuation on marine soil and so on (Boulanger et al.,2007).The role of low frequency cyclic loading on the foundation soil can be regarded as a long period process of compression, resilience and re-compression(Chu et al.,2008).As everybody known, the tidal period is so long (normally for 12 hours) ,as well as the tidal cyclic loading so weak, that the tidal cyclic loading on the soil can be treated as a kind of quasi-static loading(Graham et al.,2015).

In order to study the dynamic cyclic effect of the rise and fall of tidal level on natural marine soil of Yangtze Estuary, applying the Britain's GDS static tri-axial apparatus, two series of undrained triaxial tests are carried out to study the low frequency cyclic behavior and the effect of the confining pressure, cyclic stress ratio and loading frequency on K0-consolidated natural marine clay. Firstly, based on the natural clay cyclic test results, the constitution of typical low frequency cyclic curve is discussed, and the accumulative deformation and reversible strain are also analyzed in detail. Secondly, the low frequency cyclic behavior is investigated deeply by altering variables, and it is observed that the loading frequency is a key factor in controlling the undrained low frequency cyclic behavior. Thirdly, the special attention is given to the low frequency cyclic vibration effect on the basis of the undrained static creep tests. Compared with static loading tests, the quasi-static properties under low frequency cyclic loading are studied.

2. Experiment Program

2.1 Test Instrument and Soil Sample

This research adopts the high precision static tri-axial apparatus produced by GDS company for creep tests. The system is mainly composed of triaxis pressure chamber, the axial deformation measurement system, the axial force loading system, data acquisition unit and computer, which can monitor the whole process of tests, conduct the high-speed data acquisition and data storage. All the pressures are exerted by stress controllers, and axial displacements are measured by a GDS axial actuator. Excess pore pressure is measured at the center of the bottom of the specimens.

The soil samples used for tests are soft clay of marine origin, taken from Changxing Island shallow waters in China, which is typical saturated soft soil in Shanghai area with high water content, high sensitivity, large void ratio, high compressibility, low strength and relatively uniform, including visible number of detrital mica and shells, as well as the thixotropic and creep characteristics. The in-situ soft clays are normally consolidated in general, and exhibit obvious viscosity, initial stress-induced anisotropy and some degree of soil structure. All of the samples are quite homogeneous and the index properties are shown in table 1.

Table.1. Basic physical and mechanical parameters of Shanghai marine soft clay

W_0/ %	ρ_0/ (g/cm^3)	G	S_r/%	e_0	W_L/ %	W_P/ %	$a_{0.1\text{-}0.2}$/ Mpa^{-1}	$E_{s0.1\text{-}0.2}$/ Mpa^{-1}
32.0~ 52.1	1.69~1. 88	2.71~ 2.75	94.2~98 .8	0.917~ 1.475	34.5~ 46.7	20.2~ 26.0	0.41~ 1.25	1.92~ 4.67

2.2 Test Scheme and Procedure

To simulate the impact of the tidal level lifting on the creep characteristics of soft soil foundation, long period of low-frequency cyclic creep tests were conducted, started with the soil sample under K0-consolidated state more than 24 hours so as to restore the initial stress state of the soil.

Block samples were taken with thin-walled cubic stainless steel boxes from about 9.0m below the original ground level and were then trimmed for the tri-axial tests. Detailed test steps are described as follows. The specimens are nominally 3.91cm in diameter by 8cm in height and trimmed with their axes of symmetry coincident with the in-situ vertical direction. All specimens were firstly vacuum saturated, and B-values of more than 0.95 were observed. Then, each specimen was consolidated for 24 h under the K0-consolidated condition. The final effective axial stress of consolidation was set to 89kPa according to the in-situ depth, and the radial stress during K0-consolidation was automatically controlled by GDS system. After consolidation, the average value of K0 was 0.55, and each specimen was in normally consolidated state in which the initial mean effective stress and the initial deviator stress.

After K0-consolidation, two kinds of tri-axial tests with load control were performed under undrained conditions. According to the characteristics of the common low frequency cyclic loading, a sinusoidal wave cyclic load was applied in the vertical direction while the cell pressure was in constant. Then the series I low frequency cyclic deformation tests and series II static creep tri-axial tests were both conducted.

In the Series I tests, a cyclic deviator stress σ_d was obtained by calculating the weight of water which is caused by the tide range, shown as Figure.1. Afterwards, the cyclic deviator stress was applied under stress-controlled conditions until the normally K0-consolidated sample failed, ruled as to strain change less than 0.05%, or be loaded into the strain damage, with the standard of destroy 15%. Then the accumulative shear strain and excess pore pressure behavior are investigated. To study the effect of loading frequency on cyclic behavior, the loading period chosen in the tests were 0.5 h, 1 h and 2 h for each given cyclic deviator stress. The loading history curve is shown in Figure.2. The series II undrained static creep tests were conducted on normally K0-consolidated specimens with corresponding strain rate until a shear strain of 15% or more was reached.

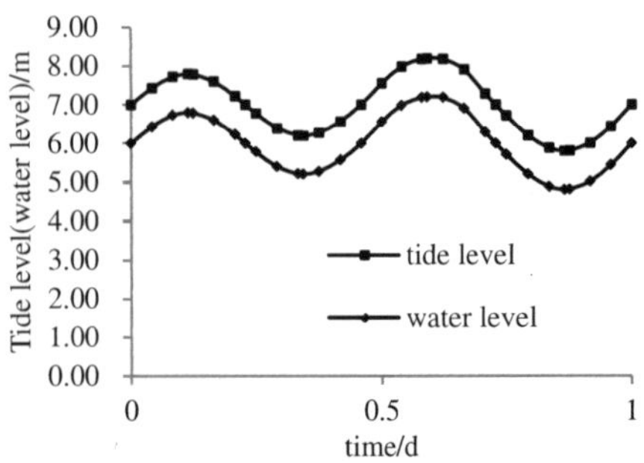

Figure 1 Tide level and water level monitoring curves

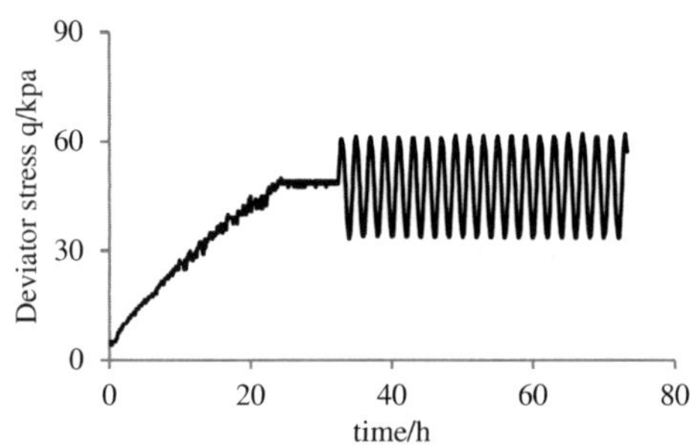

Figure 2 Loading history curve

3. Analysis of Series I and II Test Results

3.1 The Constitution of Typical Low Frequency Cyclic Curve

Due to the specimens with K0-conslidated, special attention was paid to the accumulative behavior of Shanghai clay during undrained low-frequency cyclic shearing. It can be noted that, for a given loading frequency, the development of single amplitude accumulative shear strain is highly dependent on the magnitude of the applied cyclic stress. When the cyclic stress ratio is small, the development of shear strain has the similar creep growing trend, extremely slowly, in an extreme small amount of accumulative strain, which leads the specimen to get into the creep decay stage in a certain cycles, and stabilizes at a small value. A trend of cyclic failure is not observed. However, when the cyclic deviator stress is larger, the creep deformation curve has a relatively large value in a few cycles, and gets into the collapse stage absent of the creep decay stage. There is an inflexion on each accumulative shear strain amplitude curve if the cyclic stress ratio is large enough to induce cyclic failure. If the stress ratio is large to a certain degree, the strain will develop in a sharp trend, which leads the sample to collapse abruptly. The undisturbed clay samples always collapse abruptly in a relatively short period once the shear strain amplitude is larger than that corresponding to the inflexion of the strain curve.

For example, the parameters for the experiment specimen were set as $T=1h$, $\sigma_c=50kpa$, $\sigma_d=15kpa$. According to the measured results, the strain curve can be geared to the decay creep type. As shown in Figure.3, which are creep strain curves of the initial three cycle curve and the last three cycle curve, it can be seen that the increasing rate of strain slows down with loading time, the initial three cycle made the 0.54% of strain amount. While in the last loading period, creep curve keeps stable, with a little strain increasing, and the strain amplitude kept on about 0.3%.

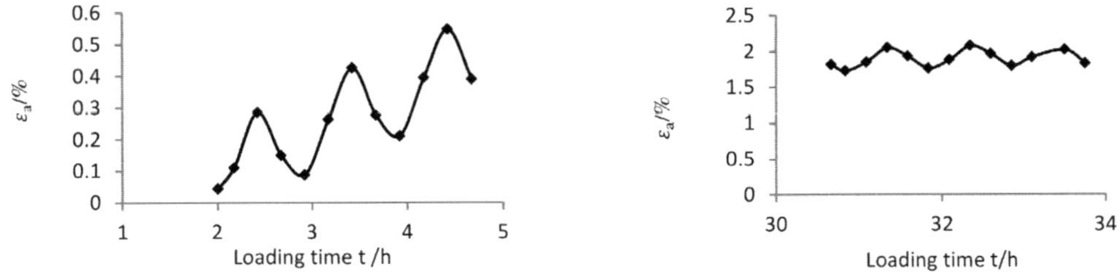

(a) Initial stage curve (b) Final stage curve

Figure 3 Initial stage and final stage of the decay creep cyclic curves

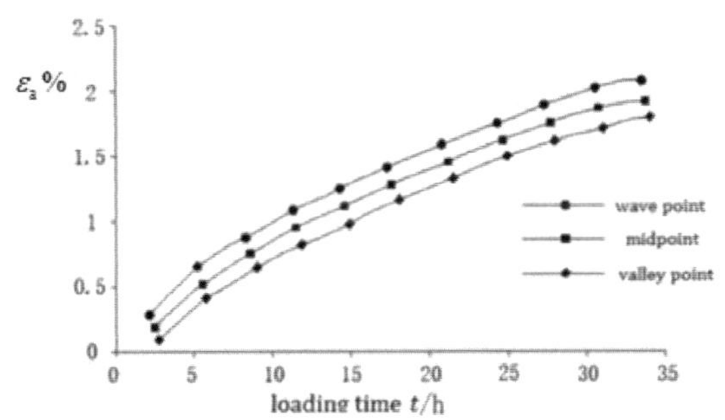

Figure 4 Wave crest curve, wave valley curve and wave midpoint curve (T=1h, σ_c=50*kpa*, σ_d=15*kpa*)

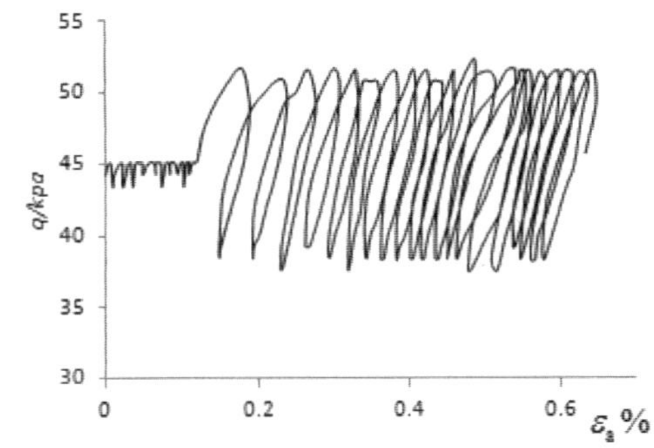

Figure 5 Deviator stress-strain cyclic curve (T=1h, σ_c=50*kpa*, σ_d=15*kpa*)

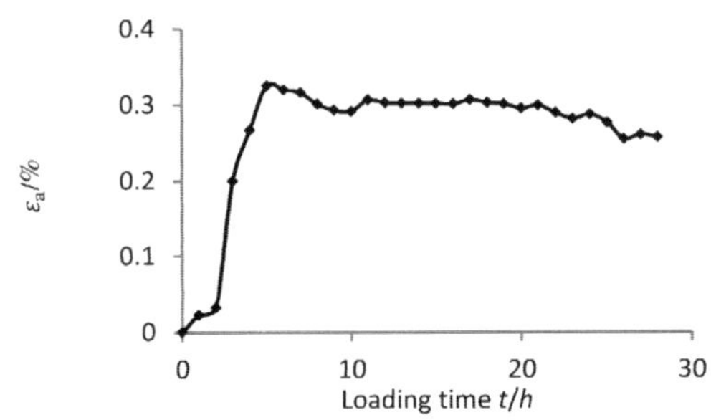

Figure 6 Reversible strain history curve of decay type
(T=1h, σ_c=50*kpa*, σ_d=15*kpa*)

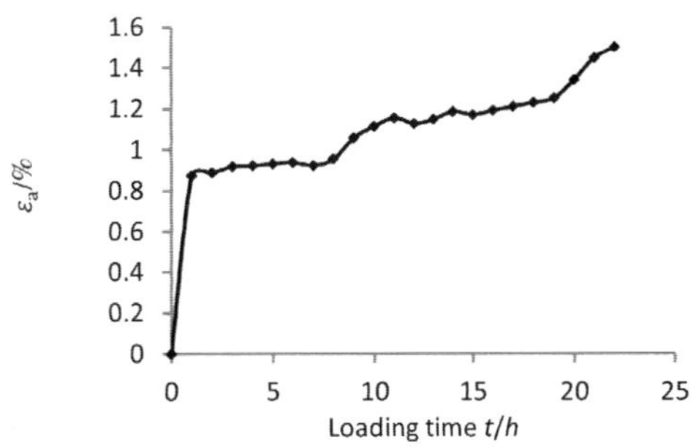

Figure 7 Reversible strain history curve of damage type (*T*=1*h*, σ_c=50*kpa*, σ_d=20*kpa*)
(T=1h, σ_c=50*kpa*, σ_d=15*kpa*)

If the wave crest points, wave valley points and the intermediate points of valueε_0, corresponding to the definition $\varepsilon_0 = 1/2(\varepsilon_a + \varepsilon_b)$ are joined in smooth curves respectively, these three curves are shown in Figure.4. They keep in similar trend of decreasing slope with loading cycles increasing, demonstrating the soil samples keep in stable creep development.

Figure.5 shows the deviator stress-strain variation curve, displayed as nonlinear hysteresis loops, which elapse towards the right more intensely. This phenomenon which is similar to the high frequency cyclic curve (Guo et al., 2013) demonstrates that plastic strain happened and accumulated in the cyclic process with the decrease of the accumulative rate.

In addition, the reversible strain (denoted as $\varepsilon_e = \varepsilon_{max} - \varepsilon_{min}$) increases rapidly during the initial of loading, then declines slowly to stability (illustrated in Figure.6) which is identical with the typical cyclic loading results (Guo et al., 2013).

With regard to the destructive type curve, the reversible strain of experiment specimen set as σ_d=20kpa (other parameters same as above) can be taken as the example, shown in Figure.7. On the contrary, reversible strain increases with the number of cycles till the specimen destroyed.

3.2 Influence of Confining Pressure, Cyclic Stress Ratio

Taking the structural yielding stress range value as the reference basis, for a given cyclic period of one hour per cycle, different cyclic creep tests under different cyclic stress ratios were conducted under two different confining pressure, to explore the soil structural effects on the cyclic creep characteristics. Partial experimental results are demonstrated in the following analysis.

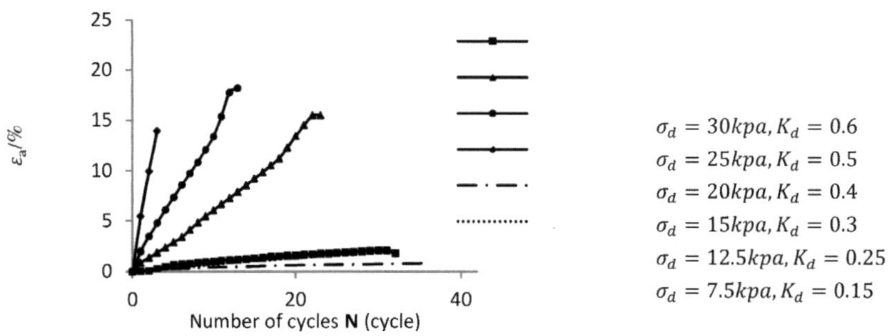

(a) σ_c=50kpa

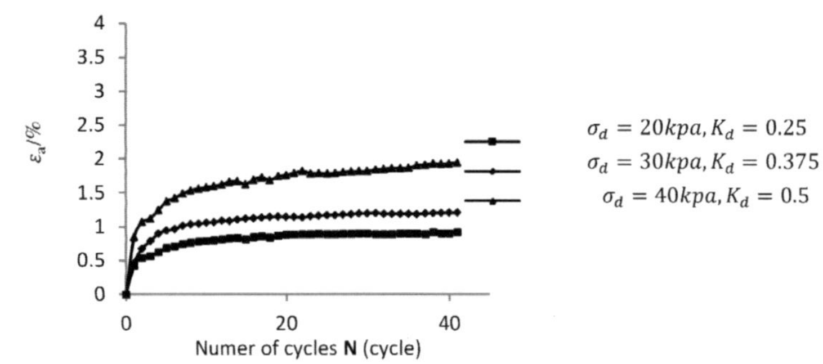

(b) σ_c=80kpa

Figure 8 Low frequency cyclic creep curves under different confining pressure

Figure.8 (a) illustrates cyclic creep curves of different cyclic stress amplitude under condition of 50*kpa* K0-consolidated, denoted as the variable named cyclic stress ratio with the definition of

$K_d = \sigma_d/\sigma_3$. Due to the different effects caused by different cyclic stress ratios on soft clay, the soil samples present different creep characteristics. It can be seen that at the outset of the cyclic loading, axial strain of soil rapidly grows with the increase of cyclic stress amplitude in a very short time. When the stress level is low, with the increase of loading cycles, creep curve is gradually leveling off after reaching a certain number of cycles. The instantaneous deformation is small, although creep deformation increases with loading time increasing, in which only the attenuation creep stage is presented. For the confining pressure is 50*kpa*, the multiple sets of soil samples (the stress amplitude 7.5*kpa*, 12.5*kpa*, 15*kpa*, 20*kpa*, 25*kpa*, 30*kpa*) appear attenuation creep characteristic. As the stress level increases, the deformation of soft clay increases sharply. When the stress level has reached a certain degree, damage occurs in a very short time, such as the value of cyclic stress amplitude set as 20*kpa*, 25*kpa*, 30*kpa*, the loading curve quickly entered the stage of accelerated creep and damage stage. By comparison of the curves in Figure.8 (a), it can be found that the greater cyclic stress ratio is, the greater the total creep deformation is, and the more obvious creep deformation behavior is.

As shown in the Figure.8 (b), the specimens with confining pressure 80*kpa* are used to conduct the tests in the same conditions as above. Obviously, the specimens only go through the creep stage to reach stability. With the confining pressure increasing, the structure of the soil is more powerful, and there will be less possibility to get the specimen to be disruptive.

The results of excess pore pressure of specimens with period 1h are selected to analyze, as shown in Figure.9. It could be seen that the pore pressure rises to a certain value after reaching a stable equilibrium no matter how much the cyclic loading. In the early loading cycles, we find that if the cyclic stress amplitude is greater, the pore pressure rise faster and the peak of the pore pressure is higher. Simultaneously, the curve with the greater stress level achieves stabilization in less cycles, and quickly enters the decay stage. In conclusion, the pore pressure law is consistent with the high frequency dynamic experiment results (Wu et al., 2015).

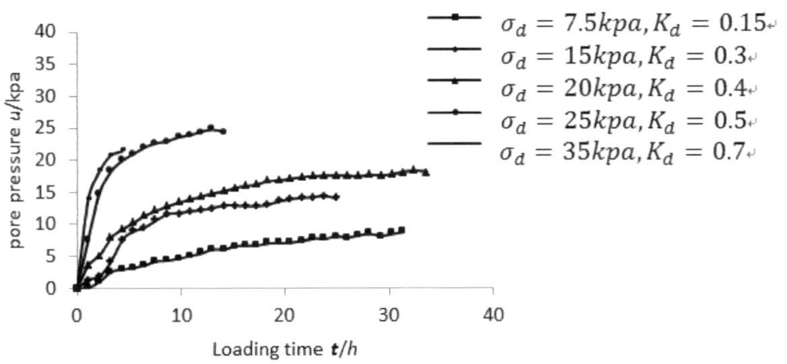

Figure 9 Excess pore pressure time history curves (T=1h,σ_c = 50kpa)

3.3 Influence of Loading Frequency

Keeping the confining pressure at a fixed value, by altering different cyclic frequencies of cyclic tests, the effect of low frequency cyclic frequency on cyclic creep properties of soft clay can be explored. In this test, the confining pressure is set as 50*kpa*, and cyclic stress amplitude alters from 15kpa to 32kpa, with the period set as 0.5*h*, 1*h* and 2*h*, the test results are shown in Figure.9. It discovers that the frequency is a significant role on the deformation of low frequency cyclic loading. With regard to the same confining pressure and stress amplitude, the lower the frequency is, the greater the creep deformation is produced.

When σ_d equals to 15*kpa* and the number of cycles reaches twenty, the creep deformation of period 1*h* is 2.69 times of that with period 0.5*h*, which manifests the strong rheological properties of soft soil. In addition, the critical dynamic load range of different period cyclic loading is roughly same, that is 15*kpa* ~ 20*kpa*.When the cyclic stress ratio exceeds the critical one, the creep curve changes from decay type into damage type. The accumulative effect of plastic deformation with period 0.5*h* appears more apparent than that with period 1 *h* and 2 *h*, of which the cumulative effect of plastic strain is very weak, only after a few number of cycles the curve reaches damage. For the typical dynamic load (such as earthquake load, traffic load), it's noted that usually the higher the load frequency is, the greater the deformation of soft soil is (Guo et al., 2013; Hosseini et al., 2015). On the contrary, the deformation law under low frequency cyclic load appears significantly different from the law of typical dynamic load type. In addition, under low frequency cyclic load, the experimental results show that the lower the frequency is, the more cumulative effect of plastic deformation is more obvious, and more slowly the soil sample get into the attenuation creep stage and reach stable stage of deformation.

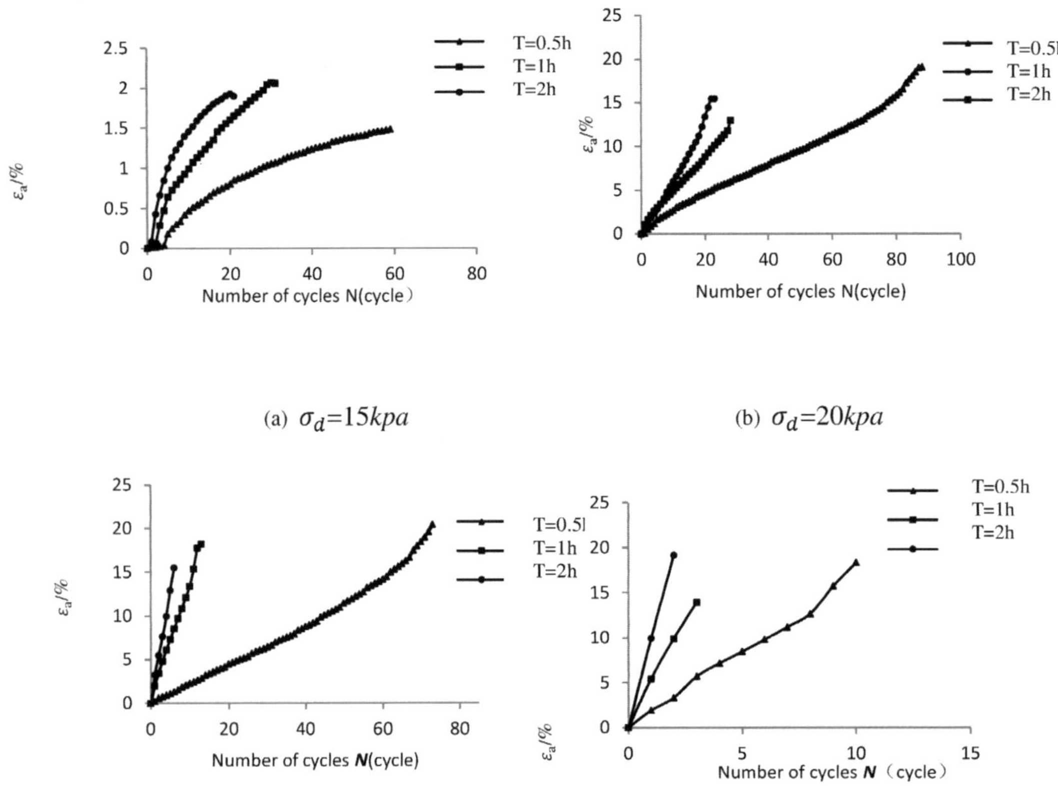

(a) σ_d=15*kpa* (b) σ_d=20*kpa*

(c) σ_d=25*kpa* (d) σ_d=32.5*kpa*

Figure 10 Comparison chart of creep test curves under different frequencies

3.4 Comparison Between Low Frequency Cyclic Properties and Static Creep Properties

By means of setting the parallel tests, that is the static creep tests under the same test conditions, which can be the effective comparison with low frequency cyclic creep tests, the soft clay creep properties under low frequency cyclic loading can be explored further. Static tri-axial fracture strength 90*kpa* under confining pressure 50*kpa* was obtained by K0-consolidated test. Figure.11 illustrates the creep curves of static loading tests in K0-consolidated conditions. Via stepping loading, the static loading creep curves will be obtained by date handling with the "coordinate translation method". For low level of cyclic stress, such as cyclic stress amplitude of 15*kpa*, static load creep curves keep almost the same trend as low frequency cyclic curves, and appear only the attenuation creep stage, in which the accumulated static creep deformation of soil is 57% of the cyclic deformation with period 0.5 h. Static creep deformation of soil samples can reach stability faster. For example, under static loading condition, the stress level is 15*kpa*, the deformation can achieve stability after about 2*h*. Nevertheless, it gradually achieves deformation stability under low frequency cyclic loading after the 60th cycle, with the time lag of up to 30 times to that of static loading. This phenomenon shows that the impact of loading-unloading reciprocating under low frequency cyclic loading plays a significant role on deformation characteristics of soft clay.

With the increase of stress level, static creep deformation gradually keeps rising. For instance, if the cyclic stress amplitude equals 32.5*kpa*, creep curve turns into damage type, and failure strain reaches 15% in 3.6 hour, which demonstrates that the static creep damage stress level keeps between 25*kpa* and 30*kpa*. However, the critical cyclic stress of low frequency cyclic creep curves is in the range of 15*kpa* to 20*kpa*. In conclusion, low frequency cycle vibration intensity is less than the static creep strength of soft soil. The preliminary analysis to this phenomenon can be ascribed to the discharge of pore water in low frequency vibration, and the structure destroying of the soil sample as well (Wu et al., 2015).

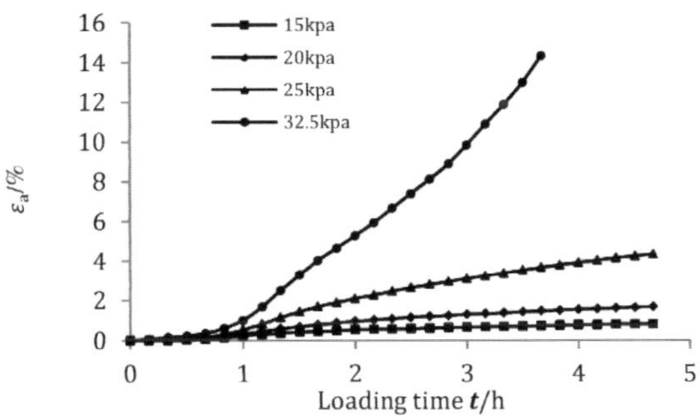

Figure 11 Static load creep curves (T=1h,$\sigma_c = 50kpa$)

4. Conclusion

From the study presented above, the following conclusions can be drawn.

(1) Low frequency cyclic creep curve can be divided into two types: decay type and damage type. The accumulative behavior and reversible strain were studied. The reversible strain of the attenuation curve drops to a stable value after certain cycles. On the contrary, the reversible strain of damage type curve shows the increasing tendency with the increase of cyclic loading time.

(2) The confining pressure and the cyclic stress ratio of low frequency cyclic loading have a significant influence on creep characteristics of soft clay, which reflects the structural influence on its low frequency cyclic creep properties. Tests reveal that the greater the cyclic stress ratio is, the greater the creep deformation of soil is, and the more obvious the creep behavior is;

(3) The cyclic loading frequency is a key factor in controlling the undrained low frequency cyclic behavior .When the number of loading cycles keeps the same, the lower the frequency is, the greater the cumulative plastic deformation is, and the more obvious the cumulative effect is;

(4) The development of the excess pore pressure in low frequency cyclic behavior is similar to the pore pressure properties of typical vibration loading results. The accumulative excess pore pressure can achieve stability after a certain number of cycles, and the cyclic stress ratio makes an effect on stabilizing rate and pore pressure increment;

(5) Comparing the low frequency cyclic loading tests with static creep tests, special attention is given to the low frequency cyclic vibration effect. It has been found that the critical stress ratio of low frequency cyclic behavior is less than static creep strength, the soil samples in low frequency cyclic loading achieve deformation stability more slowly.

IOP Conf. Series: Materials Science and Engineering **250** (2017) 012042 doi:10.1088/1757-899X/250/1/012042

5. References

[1] Smith, T. F., Waterman, M. S., 1981. Identification of Common Molecular Subsequences. J. Mol. Biol. 147, 195-197.

[2] Boulanger, R. W., Idriss, I. M., 2007. Evaluation of cyclic softening in silst and clays.Journal of Geotechnical and Geo-environmental Engineering. 133 (6), 641-652.

[3] Chen, W., Randolph, M. F., 2007. Uplift Capacity of Suction Caissons under Sustained and Cyclic Loading in Soft Clay. J. Journal of Geotechnical and Geo-environmental Engineering. 133(11), 1352 -1363.

[4] Chu, D. B., Stewart, J. P., Boulanger, R. W., 2008. Cyclic Softening of Low-Plasticity Clay and Its Effect on Seismic Foundation Performance. Journal of Geotechnical and Geo-environmental Engineering. 134(11), 1595-1608.

[5] Dinesh, R. K., Tang, J.P., Yazdani, I., 2003. Undrained response of clays to varying strain rate. Journal of Geotechnical and Geoenvironmental Engineering. 129(3), 278-282.

[6] Fan, J., Chen, J., Jiang, D., 2017. Discontinuous cyclic loading tests of salt with acoustic emission monitoring. International Journal of Fatigue. 94, 140-144.

[7] Graham, J., Crooks J. H. A., Bell, A.L., 2015.Time effects on the stress-strain behaviour of natural soft clays.Geotechnique.33 (3), 327-340.

[8] Guo, L., Wang J., Cai Y., 2013.Undrained deformation behavior of saturated soft clay under long-term cyclic loading.Soil Dynamics & Earthquake Engineering.50 (7), 28-37.

[9] Hosseini, S. M. R., Naeini, S. A., Hassanlourad, M., 2016. Monotonic, cyclic and post-cyclic behaviour of an unsaturated clayey soil[J]. International Journal of Geotechnical Engineering.32,1-11.

[10] Karim, M. R., Gnanendran,C.T., 2013. Review of constitutive models for describing the time dependent behaviour of soft clays. Geomechanics and Geoengineering . 9(1), 36-51.

[11] Ling,J.M., Wang,W.,Wu,H.B., 2002. On redisual deformation of saturated clay subgrade under Vehicle load .Journal of Tongji University (in Chinese). 30(11), 1315-1320.

[12] Matsui, T., Bahr, M. A., Ahe, N., 1992.Estimation of shear characteristics degradation and stress-strain relationship of satu-rated clays after cyclic loading. Soils and Foundations. 32(1),161-172.

[13] Soroush, A., Soltanijigheh, H., 2009. Pre- and post-cyclic behavior of mixed clayey soils[J]. Canadian Geotechnical Journal. 46(2), 115-128.

[14] Wu, H. N., Shen, S. L.,Ma, L., 2015. Evaluation of the Strength Increase of Marine Clay under Staged Embankment Loading: A Case Study. Marine Georesources and Geotechnology. 33 (6), 532-541.

[15] Yasuhara, K., 1994. Post cyclic undrained strength for cohesive soils. Journal of Geotechnical Engineering, ASCE. 120(11), 1961-1979.

[16] Zhu, H. H., Chen, X. P., Cheng, X. J., 2006. Study on creep characteristic and model of soft soil considering drainage condition.. Rock and Soil Mechanics (in Chinese). 27(5),693-698.

The Contrast analysis of BRT road and platform section

Shaofei Yin [1], Hui Peng [1], Xuehao Li [2], Xiangzhong Yu [2], Wei Zhang [2] and Jingyan Wang [2]

[1]College of Hydraulic & Environmental Engineering, China Three Gorges University, Yichang, 443002, China
[2]Gezhouba Dam Group First Engineering Co., Ltd. Yichang, 443002, China
Email: 814201831@qq.com

Abstract. Asphalt concrete has been widely used in various fields, and now most of the urban roads have taken asphalt concrete pavement.Because of the way of channelization and the braking force of the vehicle, the road surface of the special road needs to bear greater shear stress, so the road section and the section of the platform usually adopt different pavement structures.In this paper, two kinds of pavement structures are numerically simulated, and the deformation and stress of two kinds of pavement structures under vehicle loading are compared and analyzed.

1. Introduction

As a result of economic development, the number of vehicles in urban areas has increased dramatically. It is easy to cause urban traffic congestion, and BRT has also been used in more and more cities.BRT channelized traffic is more likely to damage Asphalt Concrete Pavements, this paper mainly analyzes the state of the vehicle when the brake is stopped, and compares the stress and deformation of the road structure and the pavement structure. Most of the existing road design methods in China still simplify the vehicle load to static load and assumed to be constant[1]. In the specification[2], the single axial load of 100KN is adopted as the standard axle load, and the ground pressure of the tire is 0.7mpa, The equivalent circle diameter D of single wheel transmission surface is 21.3cm, and the center distance between two wheels is 1.5D.

2. Material and Methods

2.1 Model size and material parameters

Analysis of the data used from the Hubei Province Yichang City Gezhou dam first company, the structure of the road and the structure of the station are shown in table 1 and table 2.

Table.1 Traffic section road structure and parameters.

	Material	Thickness (cm)	Elastic Modulus (MPa)	Poisson's ratio	Density (g/cm3)
Upper layer	SMA-13 type SBS modified asphalt concrete	4	1400	0.35	2.46

IOP Conf. Series: Materials Science and Engineering **250** (2017) 012043 doi:10.1088/1757-899X/250/1/012043

Table.1 Traffic section road structure and parameters.(Continued)

Middle layer	AC-16C medium grained asphalt concrete	6	1200	0.35	2.469
Lower layer	ATB-25 type asphalt stabilized macadam base	8	1000	0.3	2.493
Upper base	Cement stabilized grading gravel base (The representative value of unconfined compressive strength for 7 days is 3 ~ 4.0MPa)	18	700	0.25	2.419
Middle base	Cement stabilized grading gravel base (The representative value of unconfined compressive strength for 7 days is 2.5 ~ 3.0MPa)	18	700	0.25	2.414
Lower base	Cement stabilized grading gravel base (The representative value of unconfined compressive strength for 7 days is2.0 ~ 2.5MPa)	18	700	0.25	2.409
Subbase	Grading gravel base	15	300	0.35	2.24
Subgrade		600	40	0.4	1.8

Table.2 Platform section structure and parameters.

Material		Thickness(cm)	Elastic Modulus(MPa)	Poisson's ratio	Density(g/cm3)
Layer	Black C35 reinforced concrete faceplate	28	31500	0.2	2.43
Middle base	Cement stabilized grading gravel base (The representative value of unconfined compressive strength for 7 days is 3 ~ 4.0MPa)	26	700	0.25	2.419
Lower base	Cement stabilized grading gravel base (The representative value of unconfined compressive strength for 7 days is2.0 ~ 2.5MPa)	18	700	0.25	2.409
Subbase	Grading gravel base	15	300	0.35	2.24
Subgrade		600	40	0.4	1.8

The road surface can be considered as an infinite plane, the length and width of the model are 600cm.

2.2 Load

In the design of the road, double circle vertical load is usually adopted. In fact, the contact surface between the tire and the ground is approximately elliptical. Because of the inconvenience of calculation, the contact area of the ellipse is rarely used [3]. The actual contact area between the tire and the road surface is roughly composed of a rectangle and two semicircles [4], it is approximated as a rectangle in calculation[5], as shown in the figure 1.

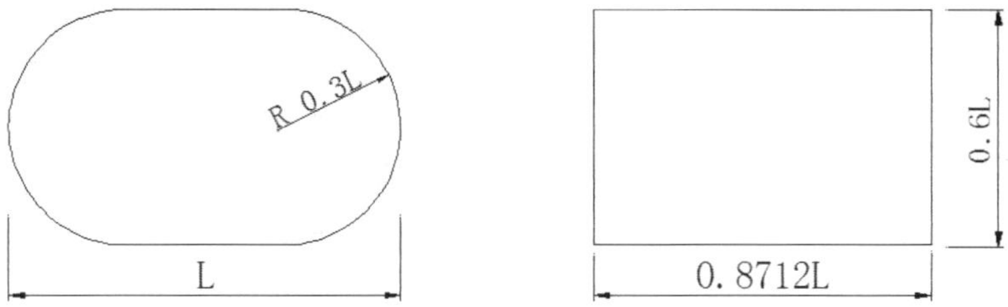

Figure.1 Shape of contact surface.

$$L = \sqrt{\frac{F}{0.5227p}}. \tag{1}$$

In the formula above: F is single wheel load, 25KN; p is tire pressure, 0.7MPa.

The horizontal load acting on the surface of a road; multiplied by the vertical load on the wheel and the coefficient of friction between the wheel and the road.

$$q = f \times p. \tag{2}$$

In the formula above: q is the horizontal load; f is the coefficient of friction.

Refer to "urban road design code" and measured data, When the driver in the case of mental preparation shift and braking, the horizontal force coefficient f is 0.2; when the emergency braking is taken, the f takes 0.5[6]. Consider the most unfavorable conditions, f take 0.5.

2.3 Boundary conditions

The pavement structure model adopts linear elastic solid185 unit, The contact state between the asphalt layer and the semi rigid layer is completely continuous and completely smooth, and all the layers are completely continuous. The boundary conditions are as follows: the bottom surface is completely fixed, there is no X direction displacement between the left and the right, and there is no displacement in the direction of Y. The mesh is shown in Figure 2 and Figure 3.

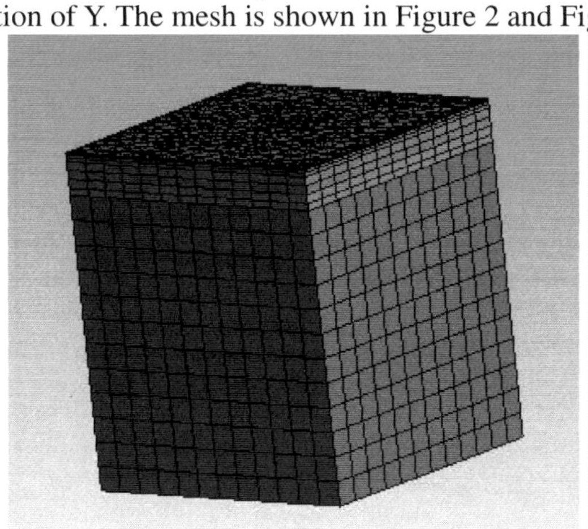

Figure.2 Traffic road structure model.

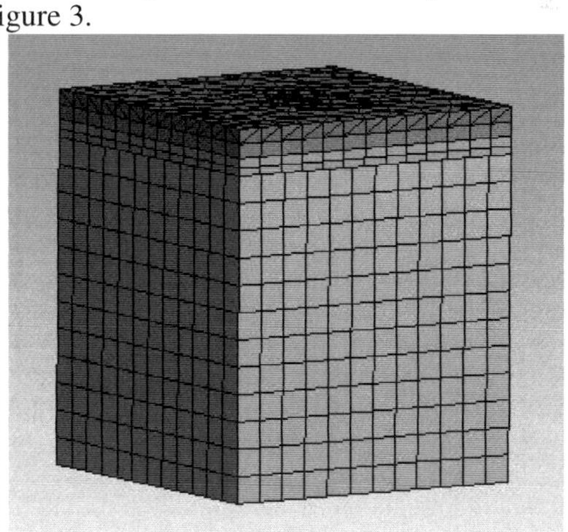

Figure.3 Platform section structure model.

IOP Conf. Series: Materials Science and Engineering **250** (2017) 012043 doi:10.1088/1757-899X/250/1/012043

3. Finite element simulation results

3.1 deformations

Since the load is applied once, the resulting deformation is smaller, the results are shown in Figure 4 and Figure 5.

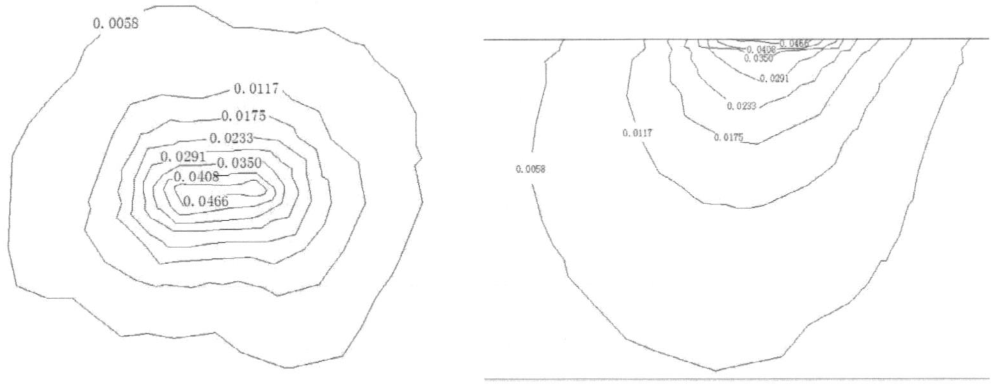

(a) Horizontal section (b) Vertical section

Figure.4 Traffic road deformation contour.

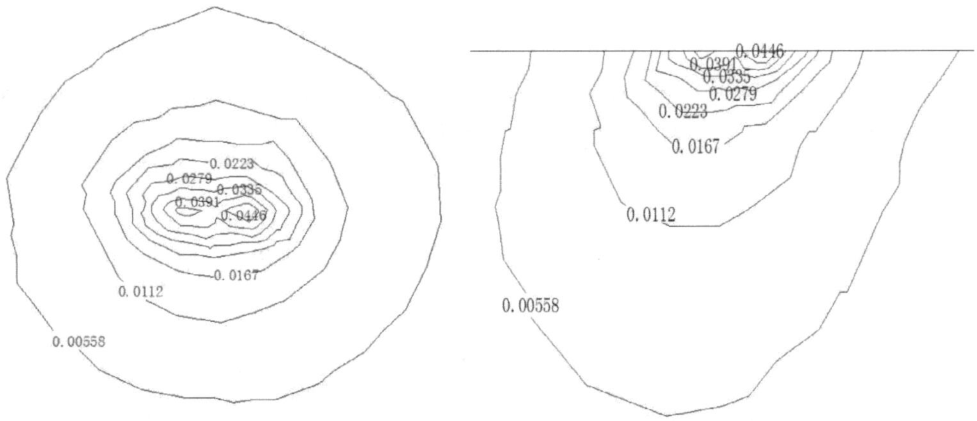

(a) Horizontal section (b) Vertical section

Figure.5 Platform section deformation contour.

3.2 Stress

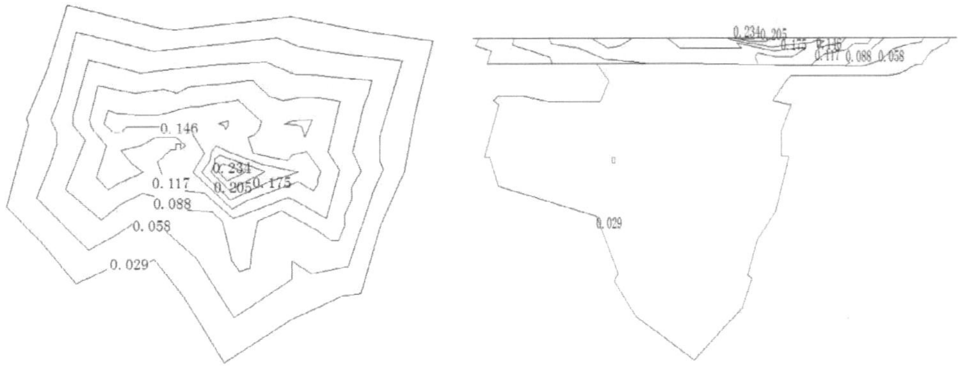

(a) Horizontal section (b) Vertical section

Figure.6 Traffic road stress contour.

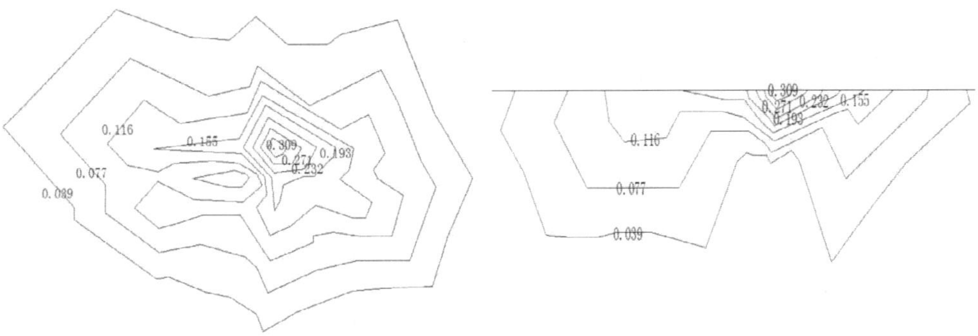

(a) Horizontal section (b) Vertical section

Figure.7 Platform section stress contour.

Table.3 Stress and deformation calculation results.

Structure	Maximum deformation(mm)	Maximum stress (MPa)
Road	0.052487	0.26322
Platform	0.050213	0.3481

4. Conclusions

In this paper, the finite element model of pavement structure is established, and the stress and deformation results are compared. The conclusions are as follows:

(1) From Fig. 4 and Fig. 5, it can be seen that the deformation of the surface layer to the base layer and the bottom layer basically shows a linear change, and the maximum deformation occurs at the center of the contact surface between the wheel and the ground.

(2) By comparing the stress and strain contour maps, it is found that the deformation and stress are not the regular relationship, and stress mutation occurs due to stratification of pavement structure.

(3) In the opposite direction of the road, the below road deformation is larger. Because of the different materials used in the surface layer, the deformation of the platform section is less than that of the traffic section, and the influence of the deformation depth is roughly the same.

IOP Conf. Series: Materials Science and Engineering **250** (2017) 012043 doi:10.1088/1757-899X/250/1/012043

5. References

[1] Tian Fei. Study on the deformation of soft ground under traffic loads.Yangzhou University (2014)
[2] JTG D50-2006, Code for design of highway asphalt pavement(2006)
[3] He Jinlong. Mechanical analysis of rutting causes considering temperature field for asphalt pavement in urban road. Central South University(2014)
[4] Zhang Jiupeng, Huang Xiaoming, Wang Xiaolei. Analysis of Asphalt Pavement Rut Based on Elastic-viscoplastic Theory. Journal of Highway and Transportation Research and Development (2007)Vol.24 No.10
[5] Huang Yangxian. Pavement Analysis and Design. Beijin:China Communications Press(1998)
[6] Chen Bao, Wu Dejun, Liu Zhan. Maximum shearing stress of asphalt pavement structure. Journal of Chang'an University(Natural Science Edition)(2010) Vol.30 No.6

IOP Conf. Series: Materials Science and Engineering **250** (2017) 012044 doi:10.1088/1757-899X/250/1/012044

Technology for High Pure Aluminum Oxide Production from Aluminum Scrap

G N Ambaryan*, M S Vlaskin, E I Shkolnikov and A Z Zhuk

Joint Institute for High Temperatures of the Russian Academy of Sciences, Izhorskaya 13 Bldg 2, Moscow 125412, Russia

*Contact by email: ambaryan1991@gmail.com

Abstract. In this study a simple ecologically benign technology of high purity alumina production is presented. The synthesis process consists of three steps) oxidation of aluminum in water at temperature of 90 °C) calcinations of Al hydroxide in atmosphere at 1100 °C) high temperature vacuum processing of aluminum alpha oxide at 1750 °C. Oxidation of aluminum scrap was carried out under intensive mixing in water with small addition of KOH as a catalyst. It was shown that under implemented experimental conditions alkali was continuously regenerated during oxidation reaction and synergistic effect of low content alkali aqueous solution and intensive mixing worked. The product of oxidation of aluminum scrap is the powder of $Al(OH)_3$. Then it can be preliminary granulated or directly subjected to thermal treatment deleting the impurities from the product (aluminum oxide). It was shown the possibility to produce the high-purity aluminum oxide of 5N grade (99.999 %). Aluminum oxide, synthesized by means of the proposed method, meets the requirements of industrial manufacturers of synthetic sapphire (aluminum oxide monocrystals). Obtained high pure aluminum oxide can be also used for the manufacture of implants, artificial joints, microscalpels, high-purity ceramics and other refractory shapes for manufacture of ultra-pure products.

1. Introduction

High-purity aluminum oxide has a variety of applications and the main are the manufacture of high-quality optics, monocrystal yttrium-aluminium garnets and synthetic sapphires. Synthetic sapphires are widely used for a number of applications such as production of LEDs, substrates for integral microchips, laser diodes, implants, artificial joints, microscalpels, etc. High-purity aluminum oxide is also used for the production of refractory shapes for manufacture of ultra-pure products.

Hydrolysis of aluminum isopropylate [1-2], thermal decomposition of aluminum isopropoxide [3], and thermal decomposition of aluminum nitrate [4] may be used for high-purity alumina manufacturing. Among them, the alcoholate method of aluminum oxide production is the main method which supplies raw material for leucosapphire industry [5]. In this process on the first stage aluminum isopropoxide is synthesized by dissolution of Al in isopropanol. Then aluminum isopropoxide undergoes several purification steps in order to decrease impurity level and then it is hydrolysed by liquid water or water steam. During this process aluminum hydroxide is formed. Further it is calcined to produce alumina. Alcoholate method utilizes aggressive chemicals which cause dangerous pollution.

In all of above mentioned processes, aluminum is used as starting chemical for the production of high-purity aluminum oxide. This is due to the fact that aluminum is a reactive metal, whose industrial production is estimated by tens of millions of tons per year. Aluminum, as a starting material for the production of high-purity aluminum oxide, is more preferable than aluminum oxide, because its purity can be increased from 99.8 % to 99.99 % during the industrial process of three-layer electrolysis (the Hoops process). The energy cost of electrolytic refining does not exceed the energy cost of the primary aluminum production process (the Hall-Heroult process) [6]. When adding only oxygen to aluminum with a purity of 99.99 % (total content of impurities of 100 ppm), it is theoretically possible to obtain oxide with a purity of about 99.995 % (total content of impurities of 50 ppm), since about 1 kilogram of aluminum produces about 2 kilograms of oxide. However, in practice, the direct oxidation reaction in pure oxygen or water without any additives or catalysts is not so simple [7].

Recently, a method has been proposed for the oxidation of granular aluminum in low-percentage alkali solutions with vigorous stirring, in which the alkali acts as a catalyst for the oxidation process [8]. The producing aluminate under the specific experimental conditions quickly decomposed into aluminum hydroxide and alkali, which then participated in the oxidation of the next portion of aluminum. On the example of potassium hydroxide, the optimal mixing speed, the initial concentration of alkali in the solution (0.1 M), and the weight ratio of the aqueous solution to aluminum ($\alpha = 4$) were determined, which ensure a high conversion of aluminum (more than 80%). Recently, it has also been shown the possibility of purifying aluminum oxide during high-temperature vacuum calcination [9-11]. This process is based on the diffusion of impurities to the surface of the crystal (surface segregation) and their subsequent evaporation [12]. The raw material used was a micron powder of aluminum, the oxidation of which was carried out under hydrothermal conditions at a temperature of about 300 °C and a pressure of 10 MPa [13-15].

In this study a simple and environmentally benign method for conversion of aluminum scrap into high-purity alumina is presented. The synthesis process consists of three steps: i) oxidation of high pure aluminum in water at temperature of 90 °C; ii) calcinations of Al hydroxide in atmosphere at 1100 °C) high temperature vacuum processing of aluminum oxide at 1750 °C.

2. Materials and Methods

In oxidation experiments aluminium in the form of granules with aluminium content of not less than 99.99 % has been used (aluminium grade "A99") [8]. Experiments on the oxidation of aluminum were carried out in a pilot plant, those description is shown in [8].

The oxidation reaction of aluminum in aqueous solution of potassium hydroxide proceeds according to the following equation:

$$2Al + 2KOH + 6H_2O = 2K[Al(OH)_4] + 3H_2$$

Produced aluminate is not stable and it is decomposed into a more stable aluminum hydroxide and potassium hydroxide in accordance with the following equation:

$$K[Al(OH)_4] = Al(OH)_3 + KOH$$

Aluminum hydroxide is precipitated at relatively low temperatures. Such precipitation process is used for example in the Bayer process (industrial process producing alumina) when aluminum hydroxide at temperatures about 45-70 °C is precipitated from sodium aluminate aqueous solution. In case of aluminum oxidation in alkaline aqueous solution this reaction leads to the alkali regeneration.

The scheme of method of high-purity α-Al_2O_3 production is shown in Fig.1. After aluminum oxidation the aluminum hydroxide is decanted and dried. Then it is placed into the muffle furnace and heated in order to remove the crystallized water. During the heating the aluminum hydroxide transforms into α-Al_2O_3. At the final step, the aluminum oxide obtained by calcination in muffle furnace is placed into the vacuum furnace to provide high-temperature purification. During this process the end product (α-Al_2O_3) is formed.

A vacuum drying oven VD-53 (Binder GmbH) was used to dry aluminum hydroxide slurry and remove non-bounded water. To remove water from the aluminum hydroxide crystal lattice and to transform it into aluminium oxide a muffle furnace LHT 08/16 (Nabertherm) with molybdenum disilicide heaters was used. To transform the aluminum hydroxide into alumina the 100-ml alumina crucible was used. Crucible with the sample was installed in a central zone of the muffle furnace.

High temperature vacuum purification of aluminium oxide was carried out in a vacuum high-temperature furnace. The heaters of the furnance were made of tungsten, the heat protection shields were made of molybdenum. The crucible with aluminium oxide was placed into the isothermal central part of the chamber. Crucibles were made of tungsten or molybdenum. The useful volume of crucible was about 100 ml (80 mm diameter, 40 mm height, 5 mm wall thickness).

To study the chemical composition of aluminium oxide sample the inductively coupled plasma mass spectrometer ICP-MS, X-2 (Thermo Scientific, USA) was used.

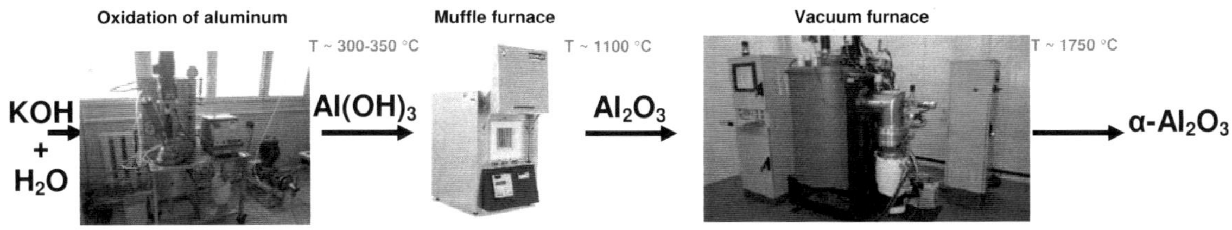

Figure 1. The method of high-purity aluminium oxide production.

3. Results and Discussion

Granulated aluminum was oxidized at 90 °C in an aqueous solution of potassium hydroxide with a concentration of 0.1 M. The degree of conversion of aluminum was about 80 %. After separation of the resulting aluminum hydroxide from unoxidized aluminum, filtration and drying, and high-temperature vacuum treatment were performed.

The results of mass spectrometric analysis of aluminum oxide sample obtained from aluminum are shown in Table 1. It can be seen from the results that when using high purity aluminum, it is relatively easy to obtain alumina with chemical purity of 99.999 %. Total content of detected impurities is no more than 10 ppm. It should be noted that in this case it is possible to significantly reduce the iron content in the final product. Due to the low content of impurities such as Ti, V and Cr in the original aluminum, the content of these impurities in the final product is minimal and meets the modern requirements of consumers of high-purity alumina. Impurities Ti, V, Cr are the most problematic from the point of view of purification of aluminum oxide by the method of high-temperature treatment [10].

The concentration of alkali metals (K, Na, Li), as can be seen from the results of the study, are less than error value of mass spectrometric analysis. This is due to the diffusion of these impurities to the surface of the crystal, which is also shown in [12]. With increasing temperature, the diffusion rate increases. Therefore, a large temperature of heat treatment leads to a decrease in the content of impurities K, Na and Li in the volume of the crystal and an increase in their concentration at the surface and their evaporation.

4. Conclusion

1. Simple and environmentally benign method for conversion of Al scrap of high purity (99.99 wt. %) into high-purity aluminum oxide with the purity of almost 99.999 wt. % (total impurity content of about 10 ppm) is developed and tested. The approach is based on oxidation of aluminum and subsequent high-temperature purification of aluminum oxide.

2. It was shown that aluminum oxide, synthesized by means of the proposed method, in general, meets requirements of industrial manufacturers of high purity alumina single crystals.

3. Described method of high-purity aluminum oxide production was proposed at first. But, in presented work it was tested on the samples in 100 ml crucibles. When the amount of treated sample is increased the evaporated impurities can cause a local extra partial pressure over the sample and so impede its futher purification. Therefore, investigation should be continued in order to scale the proposed method of high-purity aluminum oxide production.

Table 1. The results of mass spectrometric analysis of aluminum oxide sample obtained from high-purity granular aluminum.

Element	Error, ppm	Content, ppm	Element	Error, ppm	Content, ppm
Li	0.1	< error	Sb	0.4	< error
Be	0.03	< error	Te	0.03	< error
B	0.3	< error	Cs	0.004	< error
Na	6	< error	Ba	0.08	< error
Mg	2	< error	La	0.02	0.74
Al	Base	Base	Ce	0.04	0.31
K	1	3.2	Pr	0.004	0.040
Ca	6	< error	Nd	0.04	0.15
Sc	1	< error	Sm	0.01	0.019
Ti	1	< error	Eu	0.002	< error
V	0.6	2.4	Gd	0.02	< error
Cr	1	< error	Tb	0.004	< error
Mn	0.2	< error	Dy	0.01	< error
Fe	9	< error	Ho	0.004	< error
Co	0.1	< error	Er	0.01	< error
Ni	1	< error	Tm	0.004	< error
Cu	1	< error	Yb	0.004	< error
Zn	1	< error	Lu	0.004	< error
Ga	0.3	< error	Hf	0.2	< error
As	0.1	< error	Ta	0.01	< error
Se	2	< error	W	0.03	0.76
Rb	0.04	< error	Re	0.004	< error
Sr	0.09	< error	Ir	0.004	< error
Y	0.02	< error	Pt	0.004	< error
Zr	0.6	< error	Au	0.004	< error
Nb	0.04	< error	Hg	0.06	< error
Mo	0.07	1.2	Tl	0.004	< error
Rh	0.2	< error	Pb	0.06	< error
Pd	0.1	< error	Bi	0.01	< error
Ag	0.04	< error	Th	0.01	0.025
Cd	0.03	< error	U	0.004	< error
Sn	0.3	< error	The sum	36.04	8.86

5. Acknowledgements

The study was supported through a grant of the Russian Science Foundation (project No. 14-50-00124).

6. References

[1] Grinberg E E, Saradzhev V V, Levin Y I and Ryabenko E A 2002 Preparation of Fine Alumina Powders by Hydrolysis of Aluminum Isopropylate *Russian Journal of Applied Chemistry* **75** 245-7

[2] Khishigbayar K E, Moon Y G, Bae E J, Shim K B and Kim C J 2013 Impurity control with the precise measurement of alumina powders synthesized by hydrolysis method *Journal of Ceramic Processing Research* **14** 168-71

[3] Mekasuwandumrong O, Silveston P L, Praserthdam P, Inoue M, Pavarajarn V and Tanakulrungsank W 2003 Synthesis of thermally stable micro spherical χ-alumina by thermal

decomposition of aluminum isopropoxide in mineral oil *Inorganic Chemistry Communications* **6** 930-4
[4] Pacewska B and Keshr M 2002 Thermal transformations of aluminium nitrate hydrate *Thermochimica Acta* **385** 73-80
[5] Fujiwara S, Tamura Y, Maki H, Azuma N and Takeuchi Y 2007 Development of New High-Purity Alumina *SUMITOMO KAGAKU* **2007-I** 1-10
[6] Totten G E and MacKenzie D S 2003 *Handbook of Aluminum* (New York: Marcel Dekker, Inc.)
[7] Shkolnikov E I, Zhuk A Z and Vlaskin M S 2011 Aluminum as energy carrier: Feasibility analysis and current technologies overview *Renewable and Sustainable Energy Reviews* **15** 4611-23
[8] Ambaryan G N, Vlaskin M S, Dudoladov A O, Meshkov E A, Zhuk A Z and Shkolnikov E I 2016 Hydrogen generation by oxidation of coarse aluminum in low content alkali aqueous solution under intensive mixing *International Journal of Hydrogen Energy* **41** 17216-24
[9] Zhuk A Z, Vlaskin M S, Grigorenko A V, Kislenko S A and Shkolnikov E I 2016 Synthesis of high-purity α-Al2O3 from boehmite by high temperature vacuum treatment *Journal of Ceramic Processing Research* **17** 910-8
[10] Vlaskin M S, Grigorenko A V, Zhuk A Z, Lisitsyn A V, Sheindlin A E and Shkol'nikov E I 2016 Synthesis of high-purity α-Al2O3 from boehmite obtained by hydrothermal oxidation of aluminum *High Temperature* **54** 322-9
[11] Lisitsyn A V, Dombrovsky L A, Mendeleyev V Y, Grigorenko A V, Vlaskin M S and Zhuk A Z 2016 Near-infrared optical properties of a porous alumina ceramics produced by hydrothermal oxidation of aluminum *Infrared Physics & Technology* **77** 162-70
[12] Kislenko S A, Vlaskin M S and Zhuk A Z 2016 Diffusion of cation impurities by vacancy mechanism in α-Al2O3: Effect of cation size and valence *Solid State Ionics* **293** 1-6
[13] Bersh A V, Lisitsyn A V, Sorokovikov A I, Vlaskin M S, Mazalov Y A and Shkol'nikov E I 2010 Study of the processes of steam-hydrogen mixture generation in a reactor for hydrothermal aluminum oxidation for power units *High Temperature* **48** 866-73
[14] Vlaskin M S, Shkolnikov E I, Lisicyn A V, Bersh A V and Zhuk A Z 2010 Computational and experimental investigation on thermodynamics of the reactor of aluminum oxidation in saturated wet steam *International Journal of Hydrogen Energy* **35** 1888-94
[15] Shkolnikov E I, Shaitura N S and Vlaskin M S 2013 Structural properties of boehmite produced by hydrothermal oxidation of aluminum *The Journal of Supercritical Fluids* **73** 10-7

IOP Conf. Series: Materials Science and Engineering **250** (2017) 012045 doi:10.1088/1757-899X/250/1/012045

A Simple and Rapid Data Extraction Method for the Precision Aspheric Optical Surface Height

Guohua Xing[1], Yunfeng Peng[2, 3*] and Xing Su[2, 3]

[1] Laboratory on Ship Vibration & Noise, China Ship Development and Design Center, Wuhan, 443003, China

[2]Department of Mechanical and Electrical Engineering, Xiamen University, Xiamen, 361005, China

[3] Shenzhen Research Institute of Xiamen University, Shenzhen 518057, China

Email: pengyf@xmu.edu.cn

Abstract. Nowadays, the application of aspheric optics is becoming more and more popular in the precision optical engineering field. Therefore, it urges the rapid development of the precision machining and measuring technology. Generally, the aspheric optical component is measured by the interferometer. The underlying question is that the figure output by interferometer can't be always recognized by other analysis software or program though the interferometer has its own unique data processing system. In this paper, a robust, rapid and simple method is presented to interpret the surface height data of the precision machined aspheric optical surface. The optical surface is measured by interferometer. The result figure is split into two parts, one of which is the interferogram picture of the whole aspheric optical surface and the other is the colour reference column indicating the height value. The ratios of the red (R), green (G) and blue (B) are analysed based on the middle of the colour reference column, and the corresponding relationship between the colours and surface height is established and looked as a reference data base. Then the interferogram picture of the whole aspheric optical surface is also analysed and divided according to the red (R), green (G) and blue (B) colours. By comparing the ratios and values of RGB colour, the aspheric optical surface height can be extracted approximately. The feasibility of this method was approved by the extraction processing experiment of a polished aspheric optical surface.

1. Introduction

Because of its excellent optical properties, large aspheric surfaces are used in many fields, especially in the astronomical telescope, satellite optical system and laser fusion. The high precision aspheric optics urges the rapid development both of the precision manufacturing and measuring technology. Till now, it is hard to achieve the high precision because of its complex shape and the large size [1-4]. One of the reasons is that it lacks the effective method to evaluate the property of machined optical surface. The machining process of aspheric optics ranges from rough machining, semi-finish machining and finish machining. Each machining process needs their own different measuring method. Generally, polishing is the final procedure to achieve the ultra- precision requirement of the aspheric optics. The polished optical surface has the highest precision degree, and is much close to the theory surface form [5]. Moreover, the aspheric optics is basically transparent, which results the ultra-precision of the polished optical surface needs a specific measuring technique. The optimum equipment to measure the polished aspheric surface is the laser interferometer. The laser interferometer is a type of noncontact measuring method and has its own unique data processing system. But the output result of the laser interferometer is often a colour figure expressed by red, green and blue, which indicates its surface height. In general, researchers can analyse the surface condition by the output colour figure. But unfortunately, the figure output by interferometer can't be always recognized by other analysis software or program, which brings inconvenience to the researcher who doesn't have the interferometer equipment. On the other hand, the analysis with the interferometer is also time consuming and is expensive. Then it is apparently that a data extraction method is necessary to the analysis of polished aspheric optics, by which the surface height can be extracted and dealt as a data base. In this paper, a robust, simple and rapid method is presented to interpret the surface height data of the precision machined aspheric optical surface.

2. Theory of the Surface Height Data Extraction Method

The data processing procedure always runs within the *MATLB*, which was developed by Math Works Corporation in 1984. *MATLB* has many merits in data processing, such as the powerful capability in the matrix calculation, friendly viewing interface, high programming efficiency, opening property and fruitful figure processing function. Now, *MATLB* has become the mainstream in the data and figure analysis field. *MATLB* is suitable for the figure processing operation. The basic research content of *MATLB* includes the figure transformation, figure strengthening, figure analysis and figure condensing. In this paper, the figure processing research is performed based on the *MATLB* software. A figure in *MATLB* may be expressed as a data matrix or a mapping matrix with colour. Based on the matching relationship between the data matrix and the index of figure pixel value, a figure in *MATLB* can be sorted as four types, which are indexing figure, gray figure, binary figure and RGB figure.

The indexing figure deals with the pixel value as the bottom index of RGB colour palette. In *MATLB*, indexing figure includes a data matrix and a colour mapping matrix named as colour palette. The data matrix may be unit 8 and unit 16 or double precision. The colour mapping table is a $m \times 3$m $\times$3 data matrix, and the elements values are double precision floating point data type data and ranges from 0 to 1. Each row of the mapping matrix denotes the colour fraction values of red, green and blue. The indexing figure can map the pixel value to the value of colour palette. The value of each pixel can be achieved by using the value of data matrix's bottom index. For example, value of 1 can be set as the indicator pointing to the first row of the mapping matrix and value of 2 to the second row, and so on. The grey figure only contains the intensity information without the colour fraction information. The grey figure in *MATLB* is also a data matrix. Each element represents a pixel point. The data matrix of grey figure can be double precision and its elements vary from 0 to1. The data matrix of grey figure can also be the unit 8 type and its value ranges from 0 to 255. The value of 0 of both cases denotes the colour of black, and the 1 of double precision type and 255 of in unit 8 which 0~~in unit 8which 0~~ means the colour of white. The binary figure can be looked as either a special grey figure that only contains black and white colours or an indexing figure including two different colours. The matrix of binary figure contains only 0 and 1. The matrix can be saved as data arrays of double precision type or unit 8 type. *RGB* figure is also called as the true colour figure. Liking the indexing figure, it is expressed with three different colours (red, green and blue), which are red (*R*), green (G) and blue (*B*) and is called as the *RGB* figure. The data of *RGB* figure matrix can be either double precision floating type, unit 8 or unit 16. The combination of three colours can get any forms of colours. Different ratio of three colours denotes different surface height.

By comparing the characteristics of figure types mentioned above, it can be found the *RGB* figure is more appropriate to the expression of surface height. If a colour figure can be identified in terms of different colours' element fraction, the mapping relationship between the surface height and colour element fraction can be achieved. Therefore, the *RGB* figure was chosen as the researching object to extract the surface height of polished aspheric optical component. The proposed extraction method of surface height can be described as follows,

2.1 To Read the RGB Interferogram Figure of Polished Aspheric Optics into MATLB

The format of function calling of *RGB* figure in *MATLB* is
x=imread ('filename').

2.2 To Split the Figure into Two Parts

One is the interferogram picture of the whole aspheric optical surface, and the other is the colour reference column indicating the height value. The format of figure trimming is

'imcrop ('filename')'.

Then the split figure is read separately into the *MATLB* with the format of 'imread (*)'. Generally, the size of the read whole aspheric optical surface figure is $426 \times 437 \times 3$, which means that the figure contains 426×437 pixels. Each pixel's color is expressed with the color of red, green and red stored in the third column. The value of RGB ranges from 0 to 255. While the size of color reference column is $365 \times 27 \times 3$, which means that it contains 365×27 pixels. The colour of each pixel is expressed with the color of red, green and red stored in the third column.

2.3 To Build the Reference Data Base

The RGB values of the middle line of the reference column is extracted and stored in a matrix, whose size is 365×1 and its elements ranges from 0 to 255. The reference data is extracted from the middle line of the colour reference column. Through the analysis of the middle line, the fraction of different colours mixed in each pixel can be calculated. The weight component of different colour in each pixel is fitted with polynomials in *MATLB*. In this paper, the red component is fitted with a quadratic polynomial. The green one is fitted with 12 times polynomial and the blue one is fitted with fourteen times polynomial.

In this paper, an assumption is made that the surface height of middle line of the reference column is proportional with the numbers of the combination of different colours. Then the reference data base that maps the surface height of middle line and the numbers of pixel of different colours can be built up.

2.4. To Extract the Surface Height of Aspheric Optical Component

In order to get the surface height of the aspheric optical component, a comparison method is adopted as depicted in Figure. 1. The fraction of different colours in each pixel is firstly calculated, which denote as R_i, G_i, B_i. Then R_i, G_i, B_i are compared with R, G, B stored in the reference data base, by which the deviation errors between, R_i, G_i, B_i and R, G, B are evaluated. The R, G, B with the minimal errors will be thought as the value of R_i, G_i, B_i, and so does the height corresponding to the value of R, G, B. With this method, the surface height of the whole aspheric surface can be achieved.

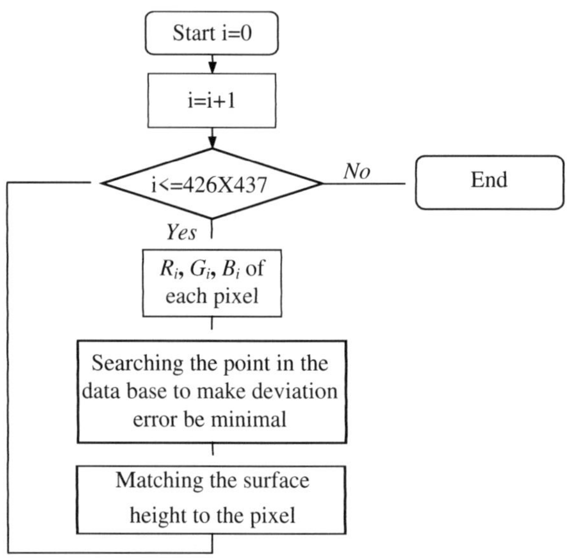

Figure. 1 Flowchart to extract the surface height by comparison

IOP Conf. Series: Materials Science and Engineering **250** (2017) 012045 doi:10.1088/1757-899X/250/1/012045

3. Verification of the Surface Height Data Extraction Method

In order to verify the feasibility of the proposed method, the extraction procedure was performed on one polished aspheric optical component. A polished aspheric optical component, of which the diameter is 100mm, was chosen as the rescarch object. Figure. 2 is the interferogram picture of the polished surface output by the interferometer. The interfergram picture was read into *MATLB* and shown in Figure. 3.

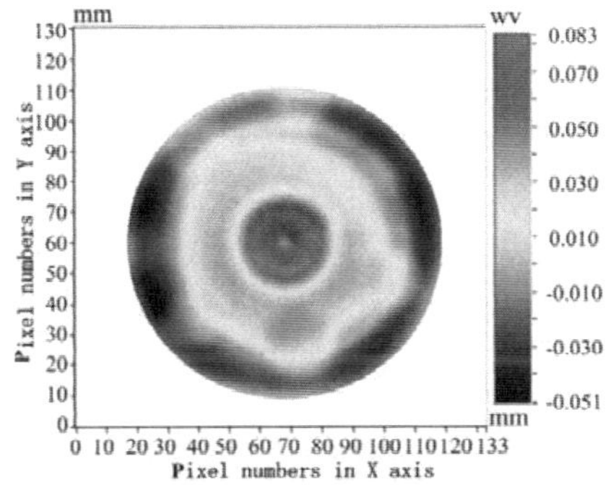

Figure. 2 The interferogram of polished surface

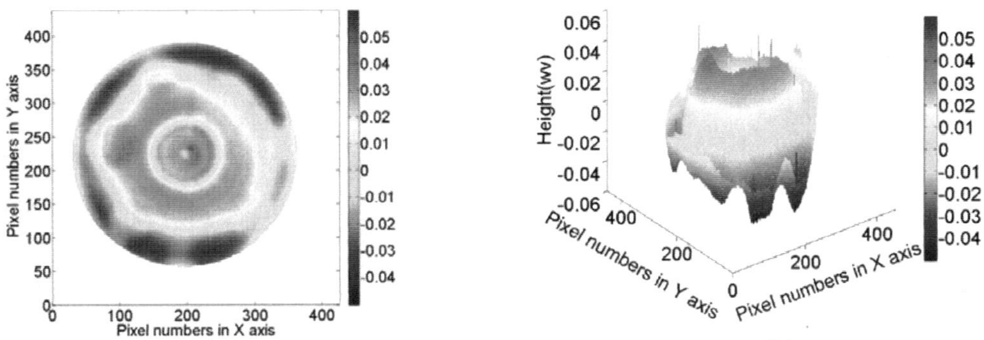

(a) (b)

Figure. 3 Original *RGB* picture of surface height in *MATLB*

According to the proposed method, the interferogram figure was split firstly into two parts. One is the interferogram picture of the whole aspheric optical surface, which is shown in Figure. 4(a). The other is the color reference column indicating the height value shown in Figure. 4(b).

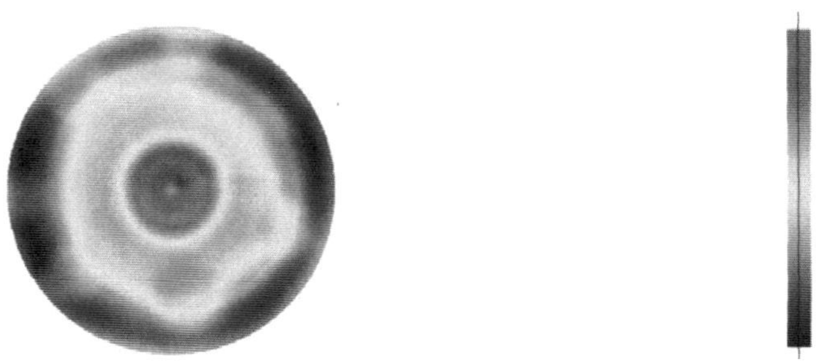

(a) Interferogram picture of the aspheric surface (b) The color reference column

Figure. 4 The trimming figure of the interferogram

Secondly, the *R*, *G*, *B* values of each pixel lying in the middle line of the reference column were fitted with polynomials. Here, in order to improve the accuracy, nine interpolation points were inserted between two neighbouring surface height point to calculate the values of *R*, *G*, *B* of each pixel. Then the reference data base that matching the surface height and the *R*, *G*, *B* value of each pixel was built, which is a 3641×1 matrix. Fig. 5 (a) shows the *R*, *G*, *B* value of middle line and Fig. 5 (b) is the fitted values. By comparing the *R*, *G*, *B* value, the residual error could be calculated and was shown in Figure. 5 (c). It can be found that the biggest error is 4.3296, which means that the fitting method is feasible.

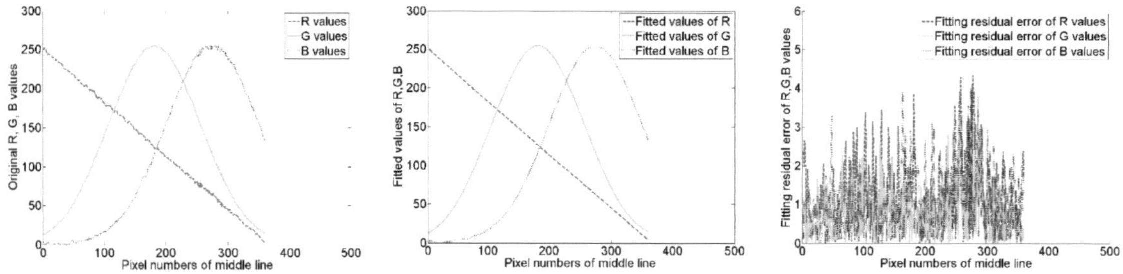

(a) Original *R*, *G*, *B* values (b) Fitted values of *R*, *G*, *B* (c) Residual error of fitting

Figure. 5 Fitting *R*, *G*, *B* values of the middle line

With the comparison method, the surface height of the whole aspheric surface height was approximated. Figure. 6 shows the approximated surface height in *MATLB*. It can be found that the original one is very similar to the approximated picture. Furthermore, the approximated deviation error of surface height was calculated as shown in Figure. 7. It is found that the deviation errors are very small compared to the surface height value. The biggest deviation error is 0.0059315 times of wave length, which is almost an order of magnitude less than the original surface height. Hence, the feasibility of proposed method in this paper has been approved and can be used in the approximated analysis of the polished aspheric optical surface.

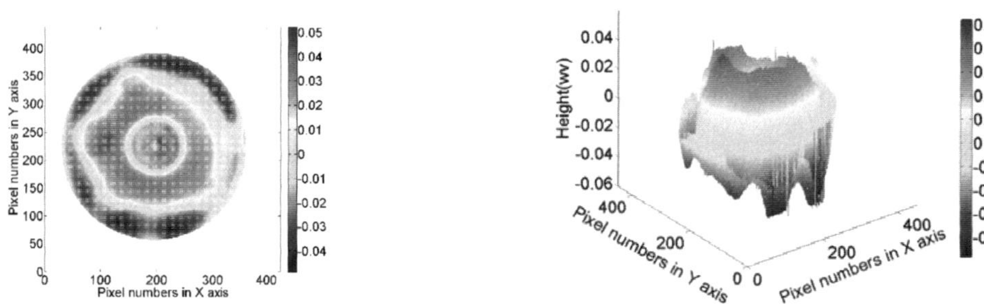

(a) (b)

Figure. 6 *RGB* picture approximated in *MATLB*

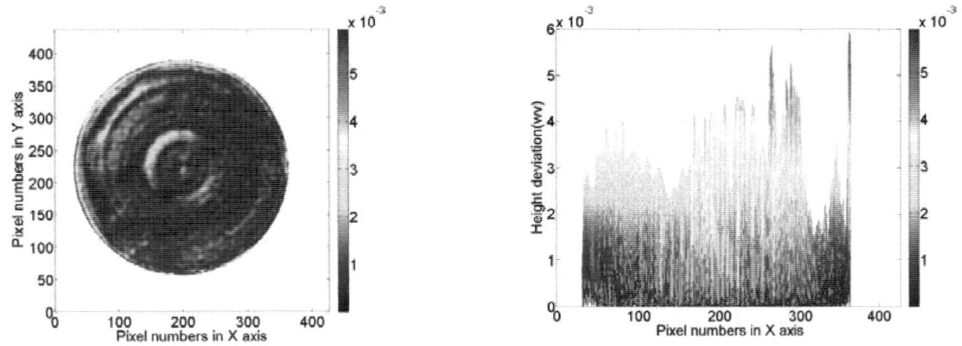

(a) (b)

Figure. 7 Deviation error between the approximated value and true value of the surface height

4. Conclusions

The analysis of aspheric optical surface always needs a simple and rapid data extraction of the surface height. In this paper, a novel method was proposed base on the *MATLB*. The interferogram picture of the aspheric surface was expressed in the *RGB* form and the ratios of red, green and blue were evaluated in each pixel. Then the different contents of RGB ratios were mapping to the surface height of the polished aspheric surface, with which the surface height can be extracted approximately. The processing procedure was done based on an aspheric surface. The result shows that the method proposed in this paper is feasible in dealing with the data extraction of surface height.

5. Acknowledgments

This work was funded by National Natural Foundation of China (No. 51675453) and Technology Projects of Shenzhen (JCYJ 20160517103720819).

6. References

[1] Yin S H, Morita S, Ohmori H, Uehara Y, lin W M, Liu Q, Maihara T, Iwamuro F, Mochida D 2025 Int. J. Mach. Tools. Manuf. **45** 1598
[2] Freeman M H, Gray C 1996 Pro. SPIE-The International Society for Optical Engineering **2775** 538
[3] Han C S, Dong S, Tang Y 2004 Key Eng. Mater. **257-258** 57
[4] Sun W, McBride J W, Hill M 2010 Precis. Eng. **34** 171
[5] Cheng H B, Wang Y W 2004 Aviation Precision Manufacturing Technolgoy, **40(4)** 8 (In Chinese).

IOP Conf. Series: Materials Science and Engineering **250** (2017) 012046 doi:10.1088/1757-899X/250/1/012046

A 3D-printed metamaterial with electromagnetically induced transmission

Zhengpeng Qin [1,2], Xiran Chen[1,2], Yawei Zheng[1,2], Guochao Wei [1,2], Luman Qin[1,2], Nangang Zhang[1,2,3],Kan Liu[1,2,3], Songzhan Li[1,2,3] and Shengxiang Wang[1,2,3*]

[1] School of Electronic and Electrical Engineering, Wuhan Textile University, Wuhan, 430020, P. R. China
[2] Advanced Micro-Nano Textile Innovation Research Center, Wuhan Textile University, Wuhan, 430020, P. R. China
[3]State Key Laboratory for Hubei New Textile Materials and Advanced Processing Technology, Wuhan Textile University, Wuhan, 430020, P. R. China
* Corresponding author: shxwang@wtu.edu.cn

Abstract: Metamaterials have recently enabled coupling electromagnetic transparency due to interference effects in coupled sub-wavelength resonators. In this work,it presents a three dimensional (3D) metamaterial designed with hollow cube whose simulation shows electromagnetically induced transparency in GHz regime with polarization insensitive to the incident electromagnetic wave due to the ultra-sharp resonance line width as a result of interaction between the constituent meta-atoms. Influences of various parameters including incident angle and frequency have been investigated numerically in detail. The simulation shows significant transmission magnitude of 99.8% at ~12GHz.

1.Introduction

The electromagnetically induced transparency (EIT) effect is a nonlinear effect which occurs in the interaction process between light and material[1].According to the wikipedia's interpret ,Electromagnetically induced transparency (EIT) is a coherent optical nonlinearity which renders a medium transparent window over a narrow spectral range within an absorption line. Extreme dispersion is also created within this transparency "window" which leads to "slow light", described below. It is in essence a quantum interference effect that permits the propagation of light through an otherwise opaque atomic medium[2]. Due to strong material dispersion caused by quantum interference, a transparency peak will emerge in the absorption curve and the group velocity of light can be rapidly slowed down[3].Therefore,the EIT effect has shown great potential in various promising applications, including slow light control, quantum information processing and sensing technology[4]. However, the realization of EIT in atomic systems is a tough task since some restrictions must be strictly fulfilled. Recently, researchers have found that the EIT-like spectrum can also be generated in some other structures which are easier to realize, such as optical resonators[5] and Fano metamaterials[6-8]. For the case of EIT, the transparency window is caused by the reduced absorption, due to the quantum destructive interference between the transitions from the two dressed states, into a common energy level. Similarly, the EIT-like effect generated by optical resonators works by means of coherent interference between the resonating modes which produces optical transparency inside the absorption window. Compared to the EIT in atomic systems, the analogue of electromagnetically induced transparency with optical resonators has many remarkable advantages such as simpler structure, smaller device size and being easier to design, which mean it has enormous potential for applications in optical sensors and optical delay lines[9]. The optical EIT-like effect has been experimentally realized in many optical structures including coupled fiber systems[10], coupled fused-silica spheres and optical parameter oscillator cavities[11]. In these structures, the line widths of the induced window can be extremely narrow. The EIT phenomenon has been mimicked by solid-state, on-chip all optical and optomechanic systems. In the past decade, MMs which are artificial electromagnetic structures provided an excellent platform to study EIT effect from microwave to visible light region. Using MMs to mimic EIT phenomenon was first proposed by Zhang et al[12]. The drawback of these devices is the bulky and incompact structure which is hardly present for integrated applications. The on-chip integrated all-optical analogue of EIT effect has been realized in photonic crystals and silicon microrings[13, 14].

In this paper, it proposes and experimentally demonstrates an electromagnetically induced transparent metamaterial, which is based on GHz regime, consisting of hollow cube array by three-dimensional printing. Amplitude manipulation in GHz regime was achieved at the resonance frequency in the transmission spectra.

2. Structure and Design

By changing the resonant frequency features of MMs, in order to realize the metamaterial structure with multiple transmission peaks, it has established resonant units with different structures in the same region. According to the resonance principle of resonant cavity, the unit cell of the planar MM in this paper is illustrated in Figure 1 (a).The purposed metamaterial is conducted with a square array(6X6) by a unit cell composed of hollow cube. The structure is made up of polytetrafluoroethylene(PTFE) covered with a layer of 0.1-mm-thick copper. Meanwhile, the structure can be fabricated by using three dimension printing technology and a subsequent burnishing process. The electric field E_{in}, magnetic field $H_{in}$and the wave vector K_{in} of the incidence are along the x, y, and z axis, respectively. As shown in the Figure 1 (b), it is a single array of 6X6, each cell is closely linked, so the periodicity P=15mm. And from the Figure 1 (b), you can see that the structure's width is w=2.5mm, L=10mm, w1=2.5mm, w2=1.25mm, h=15mm. Periodic boundary condition is used both in x and y directions. In simulations, the refractive index of the copper is assumed to be ε=2.1+0.005i. The light is normally incident onto the MMs with polarization along z direction.

The light is normally incident onto the MMs with polarization along z direction. Latter, our research shows that the peak value does not change with the angle of incidence. The simulated transmission spectra in our interested frequency range of 4GHz-14GHz are shown in Figure 2 (b) for cavity's length L=10mm.

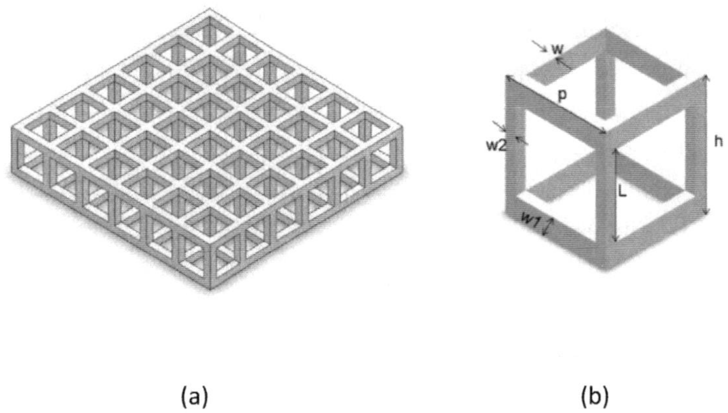

Figure 1 Schematics of the metamaterial. (a) The structure array;(b) unit cell.

3. Simulation and Discussion

To demonstrate the nature of the EIT-like effect clearly, it firstly study the transmission spectra under the various incidence with TM polarization (its magnetic field is perpendiculartothex-zplane) for the pure copper structure array and the copper covered PTFE structure array, as shown in Figure 2 (a),(b) respectively. As an example, when the incident polarization changed while the other parameters are as the mentioned above. A narrow band transmission peak centered at 11.44GHz, 11.53GHz, 11.77GHz, 11.89GHz, 11.59GHz by the degree of 0°,15°,25°,35°,45°respectively, are observed for the copper covered PTFE array in Figure 1 (b). And the pure PTFE array shows the transmission are relatively smooth curves beyond about 90% in Figure 1 (b),and have no steep transmission peak. From the simulation result, you can observe that the incident angle's change lead to the transmission peak's change, and this change is no that significant. The difference of the surface coating of the material can greatly affect the transmissivity. Thus, the proposed metamaterials present high transmission for wide incident angles at certain frequency.

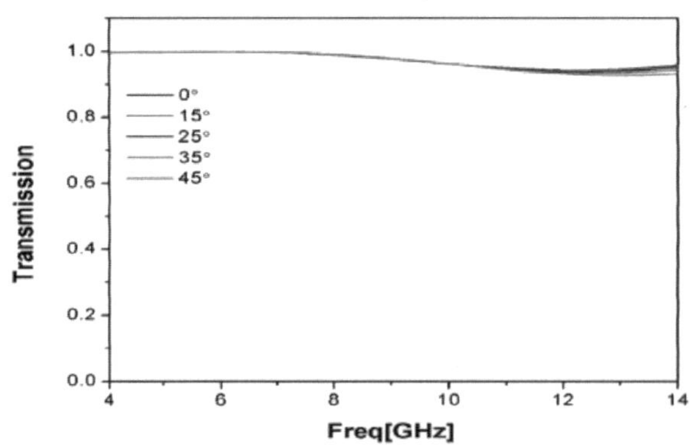

(a)

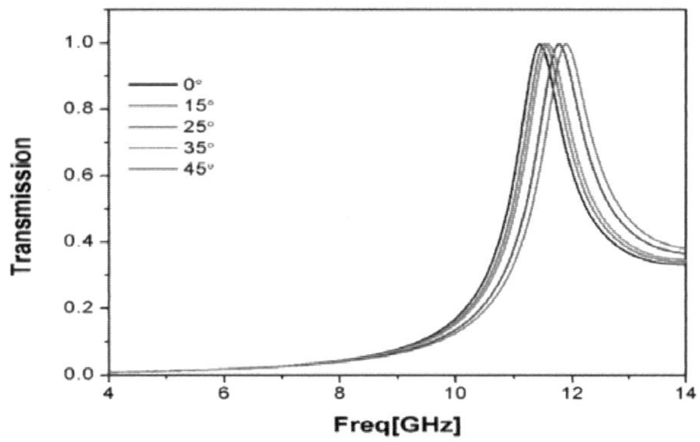

(b)

Figure 2(a) Transmission of purePTFE structure array for different incident angle;(b) Transmission ofcopper covered PTFE structure arrayfor different incident angle

4. Conclusion

In this paper, it proposed an EIT-like metamaterial structure and fabricating method, which is consisted of 3D printed PTFE with copper covered array. The electromagnetic transmission through the metamaterial was studied by changing the parameters, such as incident angles, differentmaterial, and the simulated results reveal that this metamaterial has a significant electromagnetically induced transmission for a normal incident polarized wavet in GHz regime. The peak of the EIT amplitudecould be achieved nearly 100%. The proposed metamaterial structure can be easily applied to different frequency ranges by scaling the structure sizes. The proposed novel metamaterial will easily find practical applications in active plasmonic switch,ultra-high resolution sensors and active slow-light devices.

5.Acknowledgements

This study was supported by National Natural Science Foundation of China (51302196) and State Key Laboratory for Hubei New Textile Materials and Advanced Processing Technology (ZDSYS201711).

6. References

[1] Boller K, Imamolu A and Harris S E 1991Observation of electromagnetically induced transparency*Physical Review Letters***66**2593

[2] Liu C, Dutton Z, Behroozi C H, Hau L V 2000Observation of coherent optical information storage in an atomic medium using halted light pulses*Letters to nature***409**490

[3] Safavinaeini A H, Alegre T P M, Chan J, Eichenfield M, Winger M, Lin Q, Hill J T, Chang D E and Painter O 2011Electromagnetically induced transparency and slow light with optomechanics*Nature***472**69

[4] Lukin M D and Imamoğlu A 2001Controlling photons using electromagnetically induced transparency*Nature***413**273

[5] Smith D D, Chang H, Fuller K A, Rosenberger A T and Boyd R W 2004Coupled-resonator-induced transparency*Physical Review A***69**

[6] Manjappa M, Chiam S Y, Cong L, Bettiol A A, Zhang W and Singh R 2015Tailoring the slow light behavior in terahertz metasurfaces*Applied Physics Letters***106**2075-538

[7] Cao W, Singh R, Al-Naib I A, He M, Taylor A J and Zhang W 2012Low-loss ultra-high-q dark mode plasmonic fano metamaterials*Optics Letters***37**3366

[8] Khanikaev A B, Wu C and Shvets G 2013Fano-resonant metamaterials and their applications*Nanophotonics***2**247-64

[9] Heebner J E, Wong V, Schweinsberg A, Boyd R W and Jackson D J 2004Optical transmission characteristics of fiber ring resonators*IEEE J. Quantum Electron.***40**726-30

[10] Smith D D, Lepeshkin N N, Schweinsberg A, Boyd R W and Jackson D J 2005Coupled-resonator-induced transparency in a fiber system*Optics Communications***264**163-8

[11] Di K, Xie C and Zhang J 2011Coupled-resonator-induced transparency with a squeezed vacuum*Physical Review Letters***106**153602

[12] Zhang S, Genov D A, Wang Y, Liu M and Zhang X 2008Plasmon-induced transparency in metamaterials*Physical Review Letter***101**047401

[13] Xu Q, Dong P and Lipson M 2007Breaking the delay-bandwidth limit in a photonic structure*NatPh***3**406-10

[14] Xu Q, Sandhu S, Povinelli M L, Shakya J, Fan S and Lipson M 2006Experimental realization of an on-chip all-optical analogue to electromagnetically induced transparency*Physical Review Letters***96**123901

IOP Conf. Series: Materials Science and Engineering **250** (2017) 012047 doi:10.1088/1757-899X/250/1/012047

Mechanical Behavior of Iron-bonded Grinding Wheel after ELID (Electrolytic In-process Dressing) Electrolysis Procedure by Nanoindentation

Jicai Kuai [a] and Huali Zhang [b*]

[a]School of Mechanical and Power Engineering, Henan Polytechnic Unversity, Jiao Zuo, China 454003
[b]School of Mechanical Engineering, Nantong University, Nantong, China 226019
*Corresponding author: hualiz@ntu.edu.cn

Abstract. We have evaluated the effect of oxide films on polishing behavior during ELID grinding process. The evaluation procedure used nano-indenter to analysis quantitatively the oxide film's mechanical properties on iron-bonded grinding wheel after ELID electrolysis. As a result, the oxide film's mechanical properties increased with the indent depth increasing. The plots of indent load and displacement for oxide films showed slight variations with time. The content of iron group elements in grinding wheel also has higher impact on oxide film' s properties.

1. Introduction

The friction and wear performance of oxide film on grinding wheels after ELID grinding has been gaining interest in the early 2000s, which was originated by T Kato [1]. Our early reports and results show that the friction and mechanical behaviors of oxide films by ELID grinding have profound effects on its polishing ability [2-3]. Furthermore we have explored the chemical composition and transformation procedure of oxide films and their influences on polishing ability [4]. And much is known about oxide film's temperature reduction performance [5]. Moreover, the microstructure of oxide film has shown to be core of diamond abrasive grits, surrounding with a thin film of iron oxides, and the outermost layer of α-Fe2O3 composite grits [6]. The polishing ability of oxide films is confined to the existence of α-Fe2O3 [7].

These reports were based on static state of oxide films. However, grinding with oxide film is not a static condition as expected. Oxide films are wore out and then electrolyzed to form new films under a dynamic change process by ELID grinding. These early studies have described a certain state during oxide films' dynamic change and couldn't represent actual performances of oxide films. And here we will focus on the dynamic variation rule of oxide film to obtain better understandings of its properties and effect on polishing procedure. The motivation of the present study is to evaluate the mechanical properties of oxide films on iron-bonded grinding wheels under dynamic conditions. Nano indentation tests are conducted to experiments the main features of oxide film's microstructure and its effect on hardness, stiffness and elasticity modulus of oxide films.

2. Method and Experimental Setup

Olive WC&Pharr GM developed a improved method using nano indentation experiments for determining mechanical properties of materials, especially for thin film materials. We believe it quite feasible using nanoindentation experiment for oxide film on the surface of grinding wheels. However, what needs to be clarified is that the indenting tip should keep away from abrasive grits as to obtain correct results. The indentation depth value obtained from this measurement was about 1-10μm, 1/10 of oxide film's thickness. Each indentation depth was repeated testing for 5 times. Average value was taken after indenting and represented the mechanical properties of oxide film.

In order to measure online properties of oxide films, electrolyzed abrasive wheel is necessary. In this study the grinding wheel used was W 40 iron-bonded wheel. Grinding process was followed after electrolysis using MM7120 surface grinder with ELID-5 electrolyte. Values of grinding parameters are shown in Table 1.

Table 1. Electrolysis and grinding parameters.

Supply voltage(V)	Current (A)	Pluse width (μs)	Pulse interval (μs)	Grinding speed(m/s)	Grinding depth (mm)	Feed speed (mm/min)	Content of iron group bond
60-120	10-50	1-99	1-99	15.7	0.001-0.01	150-300	30%-90%

3. Results and Discussion

3.1. Indentation Response of Oxide Films

The fact that oxide films on the wheel surface are compressible due to their porous microstructure, should help understand the indentation response of iron-bonded grinding wheels. Mechanical properties of grinding wheels were tested with a displacement of 0.001mm, 0.002mm, 0.005mm and 0.01mm by nano indenter, as shown in Figure 1. Moistened porous inside the oxide films reserved considerable number of electrolyte, while dry porous appeared hollow structure. With the indent load increasing, amounts of mirco porous were collapsed. And the resulting loading curves appear slight fluctuations during loading process (see Figure 1 a), and b) with indent displacements of 0.001mm and 0.002mm, respectively). Figure 1 c) and d) show constantly changing in loading curves with indent displacement of 0.005mm and 0.001mm. In this case, the indent tip was pierced through the oxide films and reached to the grinding wheels. The wheels were made of rather dense composite metal structure before electrolysis, with 70% iron group metal and 20% bronze, nickel, cobalt, carbon and etc. And hence no obvious fluctuations in the loading curves were shown.

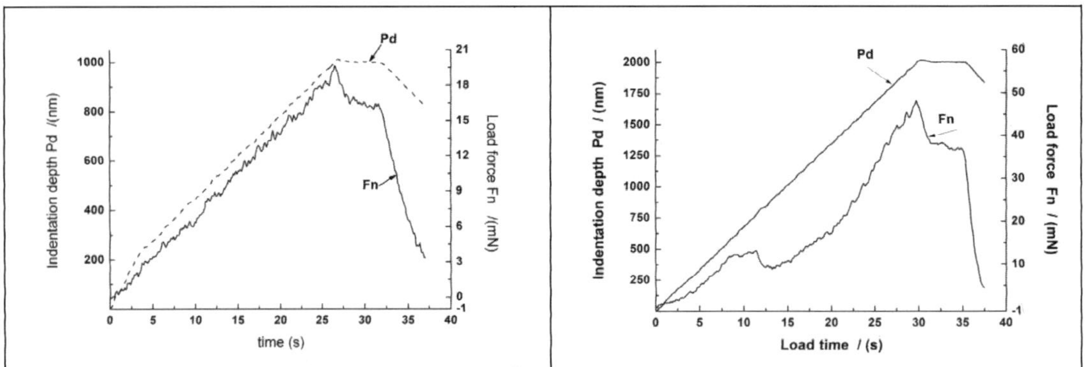

a) **0.001mm** **b) 0.002mm**

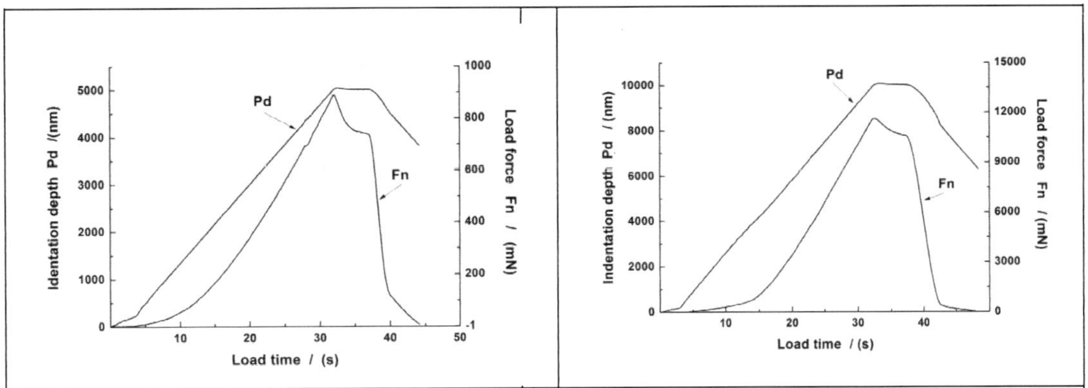

c) 0.005mm **d) 0.01mm**

Figure 1. Indentation depth as a function of load time with different displacements

Similar to applied load as a function of loading time, curves of load force verus indentation depth show same change trend as shown in Figure 2. With low indent displacement of 0.001mm and 0.002mm, the loading process was conducted inside the oxide films and the resulting loading curves presented slight fluctuations due to oxide films' microstructure (as shown in Figure 2 a) and b)), because the applied load could actually cause the micro porous in the oxide film to collapse. While with larger indent displacement of 0.005mm and 0.01mm, the oxide film's microstructure does not affect the loading curves much, and this suggests the load force was applied directed onto the wheels. In contrast to low displacements, larger displacements follow a relatively smooth curve of load force as a function of indentation depth as shown in Figure 2 c) and d).

IOP Conf. Series: Materials Science and Engineering **250** (2017) 012047 doi:10.1088/1757-899X/250/1/012047

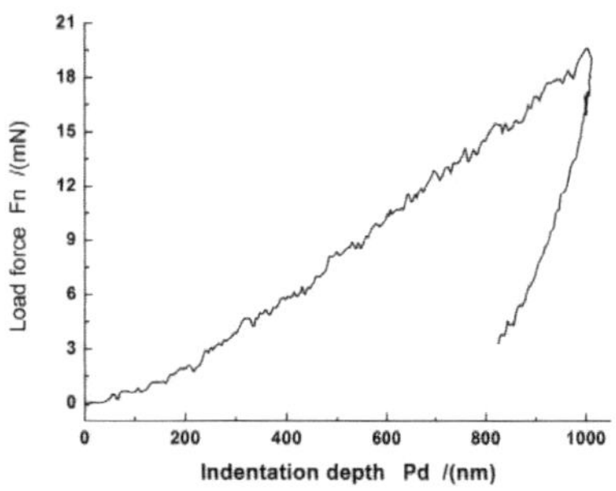

a) **0.001mm**

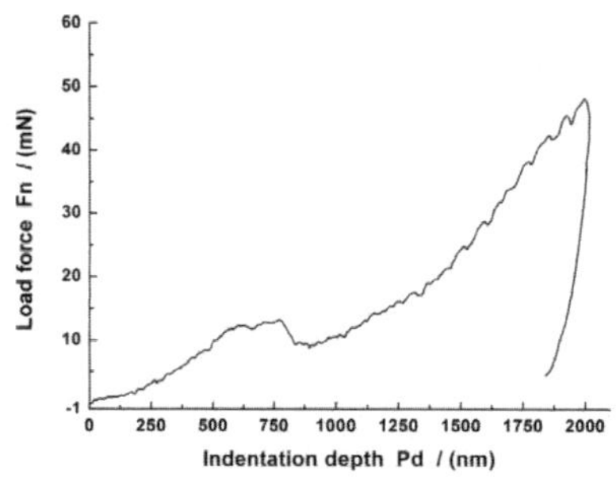

b) 0.002mm

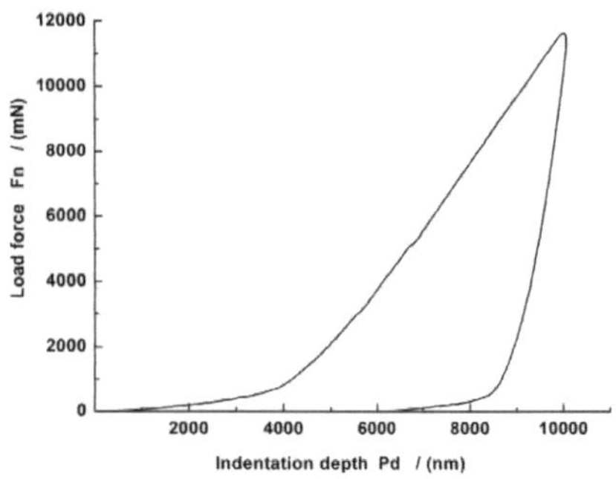

c) 0.005mm

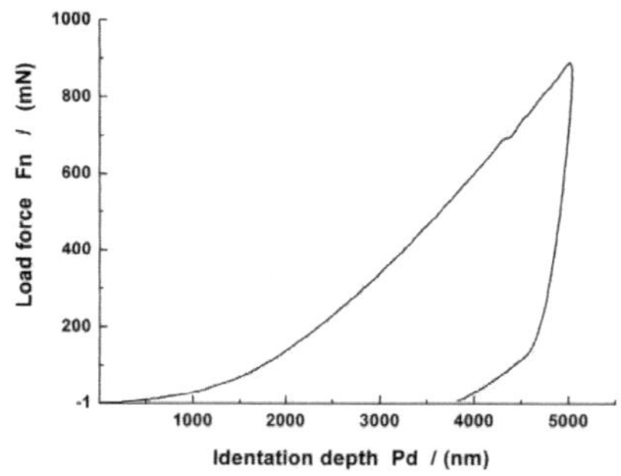

d) 0.01mm

Figure 2. The load force applied during indentation of grinding wheel with different displacements

3.2. Observation of Oxide Film Microstructure

In order to investigate the presence of porous structure, oxide film of grinding wheels was observed using SEM and the images obtained are shown in Fig. 3 a) and b). From these figures it can be seen that bump ups and downs of loose structure with a great amount of porous. Large crack can be observed in dry oxide film and easy to break off. Multi sizes of holes were found all over the oxide film, which can explain the reason for slight fluctuations in nano indent loading curves.

IOP Conf. Series: Materials Science and Engineering **250** (2017) 012047 doi:10.1088/1757-899X/250/1/012047

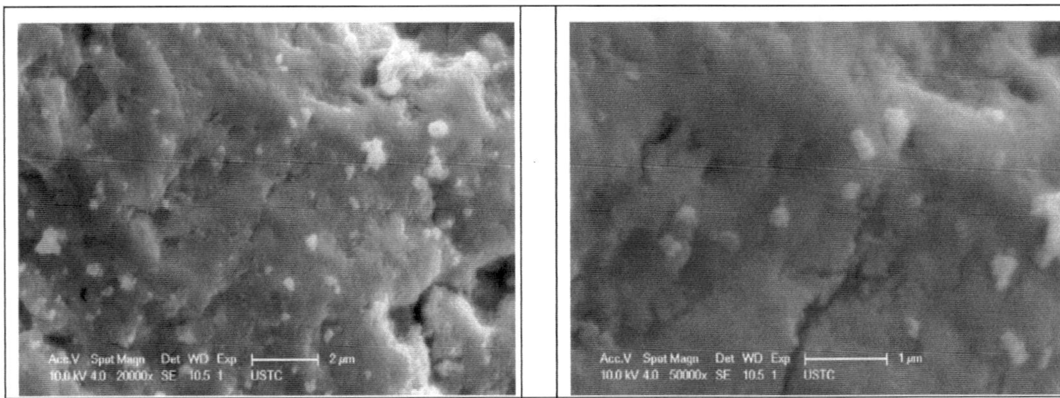

a) normal oxide film	b) dry oxide film

Figure 3. SEM images of oxide film on W 40 grinding wheel surface

3.3. Mechanical Measurements of Grinding Wheels

Experiments were carried out with indent displacement from 0.001mm to 0.01mm. Figure 4 shows the variations of stiffness, hardness and elasticity modulus of W10 grinding wheel (grain size W10, 100% concentration of iron-bond, and 90% content of iron group elements) recorded during nano indentation tests. The results obtained from the early stage (with indentation depth from 0.001mm to 0.01mm) were found to be actual oxide films' mechanical properties. The stiffness of oxide films on the grinding wheel was from 0.1117mN/nm to 11.6030mN/nm. The hardness was from 357.87 to 4833.8MPa. And the elasticity modulus of oxide films ranged from 17.92GPa to 237.1GPa. Then mechanical behavior along indentation depth down above 0.005mm, were measured for both the oxide film and the grinding wheel, in which the corresponding mechanical data are combined with the oxide film and the grinding wheel.

The mechanical properties of the grinding wheel increased with indentation depth, indicating that the oxide film on the wheel was elastic and of variable stiffness. Stiffness and elasticity modulus of oxide film increased with indentation depth. However, the hardness value clearly shows a decreasing trend with indentation depth (lower than 0.002mm) at the early stage of indenting. It is likely that some diamond grits beneath the oxide film caused local changes in hardness curve.

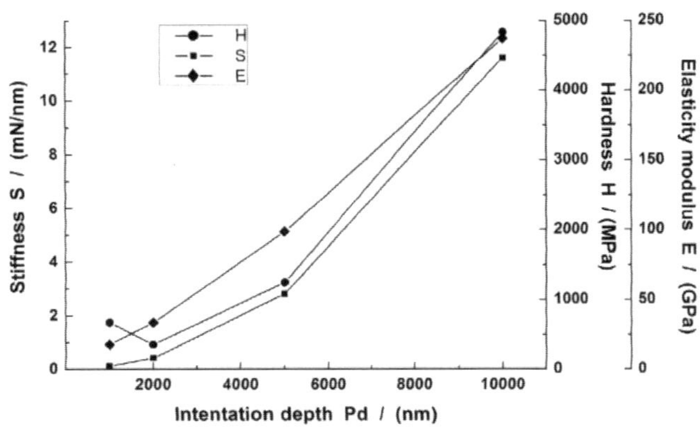

Figure 4. Variations of mechanical properties of W10 grinding wheel

3.4. Effect of Grinding Wheel Components

Nano-indentation test was repeated on iron-bonded diamond grinding wheel, which was of W40 grain size, 100% iron-bond, and 90% iron group elements. Results were displayed in Fig. 5. Stiffness value of W 40 wheel was 4-14μN/nm, hardness 210MPa, and elasticity modulus was between 30 to 80 GPa. Comparing W40 with W10 grinding wheel, interesting results were obtained that stiffness of 70% iron group elements wheel was 10 times larger than 90% iron group elements, hardness 3-5 times larger and elasticity modulus increased 3-5 times.

In addition to consisting less iron group elements, W10 grinding wheel contains almost 30% bronze bond and a small amount of nickel, cobalt and carbon. And their oxides formed thin composite film and thus changed the mechanical properties of oxide film in grinding wheel. For this thin composite film, hardness of copper oxides is higher than iron oxide. Therefore, W10 grinding wheel presents higher mechanical properties than W40 grinding wheel due to its chemical composition. However, the forming ratio of Fe2O3 reduced as a result of less iron group elements by grinding heat, and weakened the polishing ability of grinding wheels. Thus, bonding formula of grinding wheels should be given consideration to take account of oxide film's mechanical properties and its polishing ability.

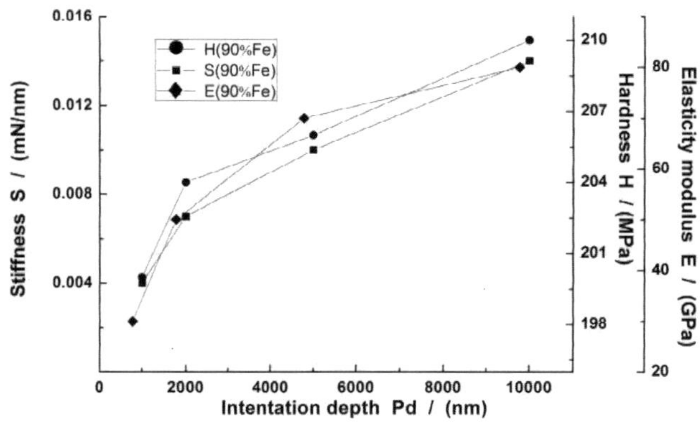

Figure 5. Mechanical properties of W40 grinding wheel (with 90% iron group elements)

4. Conclusion

From our studies conducted with nano-indenter on mechanical properties of iron-bonded diamond grinding wheel, the following conclusions may be drawn:

1) The response of the indentation process onto the oxide films, which was described in indent-loading curves, is found to be of slightly variation with loading. The porous microstructure of the oxide films was predicted to collapse during indenting and hence this could interpret the slight variation in loading curve.

2) We have used nano-indenter to study the mechanical behavior of diamond grinding wheel with 70% iron content, focusing on the effect of iron level. Level of iron group elements in iron bond grinding wheel ultimately affects the mechanical properties of the oxide films after electrolysis.

3) Mechanical measurements on the oxide films of grinding wheels show that mechanical properties of 70% content in iron group elements is higher than those of 90% content. This feature is likely related to more copper family elements in 70% iron-bonded grinding wheels, which in return reacts to form high hardness copper oxide during electrolysis process.

5.Acknowledgments

This work was supported by the National Natural Science Foundation, China (Grant No. 51475147), and by the key project of Henan Province Science and Technology Research (Grant No. 13A460341).

Also funded by the Natural Science Foundation of the Jiangsu Province, China (Grant No. BK20150406) and the technical project of Nantong, China (Grant No. MS12016015).

6.References

[1] Kato T, Itoh N, Ohmori H, Katahira K, Lin W, Hokkirigawa K 2004 *Key Engineering Materials* **257** 257
[2] Kuai J C 2013 *Advanced Materials Research* **669** 91
[3] Kuai J C, Zhang F H, Zhang Y 2010 *Nanotechnology and precision engineering* 8 447
[4] Kuai J C, Wang J W, Jiang C R 2016 *MATEC Web Conf.* **67** 1
[5] Kuai J C 2017 *Nanomaterials and Energy* **6** 1
[6] Kuai J C, Jiang C R, Wang J W 2016 *Key Engineering Materials* **709** 77
[7] Kuai J C, Ardashev D V, Zhang H L 2017 *Modern Physics Letters B* **31** 1

IOP Conf. Series: Materials Science and Engineering **250** (2017) 012048 doi:10.1088/1757-899X/250/1/012048

The Antifouling of ACLW-CAR Based on Ultrasonic Cleaner

Guohua Zhang[a], Shixuan Liu[a] and Qingliang Qin[b]

[a]Shandong Provincial Key Laboratory of Ocean Environment Monitoring Technology, Shandong Academy of Sciences Institute of Oceanographic Instrumentation, Qingdao 266001, China
[b]Qingdao University of Science&Technology, Qingdao 266001, China
E-mail: zghscu@126.com

Abstract. Equipped with ACLW-CAR, the buoy provided effective technical platform for on-site rapid monitoring of the chlorophyll and turbidity. Performance index and usage in the ocean buoy of ACLW-CAR was introduced. Ultrasonic cleaning method in seawater was developed for preventing ACLW-CAR from biofouling. Marine chlorophyll and turbidity data can serve for oceanographic research and marine resource exploitation.

1. Introduction

In situ continuous measurement for the chlorophyll and turbidity is an important part of the marine environment monitoring [1-2]. Equipped with ACLW-CAR produced by the JFE Advantech Computer, ocean buoy works automatically to observe the chlorophyll and turbidity data, and transmits these data to shore data center by Beidou system. Ultrasonic cleaner, as a new device, can prevent the ACLW-CAR from biofouling. The buoy for on-site rapid monitoring of chlorophyll and turbidity provides effective technical platform.

2. Chlorophyll and Turbidity Monitoring System

As shown in figure 1, the chlorophyll and turbidity monitoring system can be divided into sensor system, main control system, Beidou system, data center system and power system.

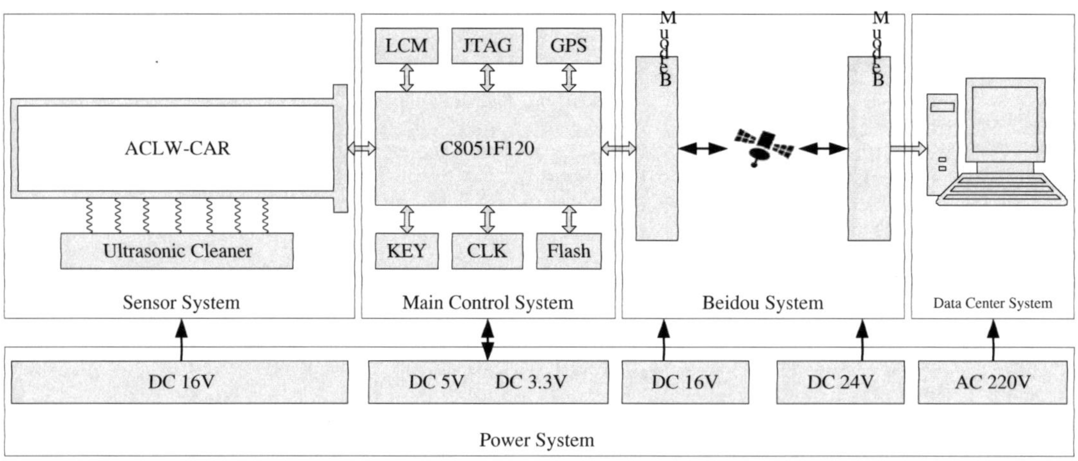

Figure 1. Block diagram of the chlorophyll and turbidity monitoring system

IOP Conf. Series: Materials Science and Engineering **250** (2017) 012048 doi:10.1088/1757-899X/250/1/012048

3. Technical Indicators of ACLW-CAR

ACLW-CAR is a chlorophyll and turbidity meter for a long term observation with a wiper that cleaned the dirt on the optical sensor. It adopts LED in the chlorophyll and the turbidity sensor luminescence part as a source of light. Especially, because the stability in the low density level is good, and the correlation is high with SS in a high density level, the turbidity sensor is suitable for an investigation not only in the sea area but also in the dam and the river. The communication protocol between sensor and PC is RS232C standard. Name of parts is shown in figure 2 [3].

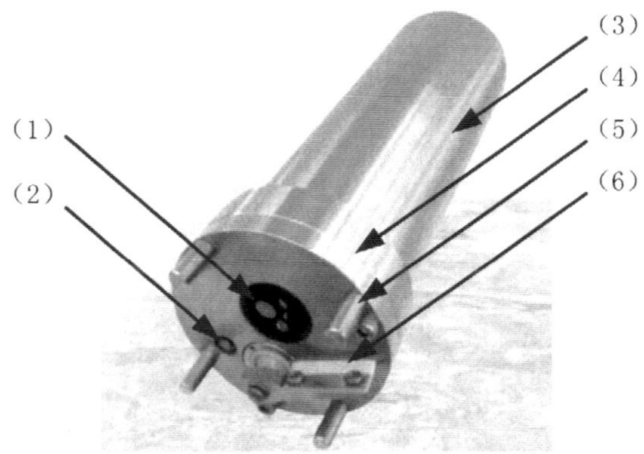

(1) Chlorophyll / turbidity sensor (2) Temperature sensor
(3) Pressure case (4) Flange (5) Wiper (6) Sensor guard

Figure 2. Name of parts

This equipment has thermistor type temperature sensor. This is very stable and biofouling can't effect the result of observation. But heavy biofouling makes the response of sensor slow down. Please clean up the sensor regularly. And be careful not to break it, this is nallow and weak parts [3].
Sensor specifications are shown in table 1[3].

Table 1. Sensor Specifications of ACLW-CAR

Parameter	Sensor type	Range	Accuracy	Resolution
Chlorophyll	Fluorescence light Scattering	0~400μg/L (Uraninite)	Linearity±1% (0~200μg/L)	0.01μg/L
Turbidity	Infrared back-scattering	0~1000FTU (Formazine)	±0.3FTU or 2%	0.03FTU
Temperature	Thermistor	-5~45°C	±0.02°C	0.001°C

Communication format of RS232 on ACLW-CAR is shown in table 2 [3].

IOP Conf. Series: Materials Science and Engineering **250** (2017) 012048 doi:10.1088/1757-899X/250/1/012048

Table 2. RS232 Communication setup

Baud rate:	38400bps
Length of character:	8 bits
Stop bit:	1 bit
Parity:	None
Character type	ASCII
A/D conversion	16-bit A/D converter
Parameters	Chlorophyll, Turbidity Tenperature
Power	DC 12V-DC24V
Power consumption	60mA (DC12V, on observation)

4. Application in Ocean Buoy

Connect wires and set up options sorrectly otherwise it can't communicate. Misinput of command line and too short interval between commands may cause wrong work. If there is not enough interval the equipment doesn't work properly. If there is not enough interval, wiper may stop moving on the way and do not move again until make maintenance. This equipment can't make proper observation during the wiper is moving[3].

The main control sends commands to ACLW-CAR by the following C statements for start wiping.

ComPutChar(COM,wipe,0);

The main control sends commands to ACLW-CAR by the following C statements for get data.

ComPutChar(COM,pval);

The main control sends commands to ACLW-CAR by the following C statements for set serial NO.

ComPutChar(COM,headR,02);

The main control sends commands to ACLW-CAR by the following C statements for get temperature constant.

ComPutChar(COM, headR,30);

The main control sends commands to ACLW-CAR by the following C statements to get chlorophyll constant.

ComPutChar(COM, headR,31);

The main control sends commands to ACLW-CAR by the following C statements for get turbidity constant.

ComPutChar(COM, headR,32);

The main control sends commands to ACLW-CAR by the following C statements for get power constant.

ComPutChar(COM, headR,33);

Figure 3 is the flow diagram of ACLW-CAR.

IOP Conf. Series: Materials Science and Engineering **250** (2017) 012048 doi:10.1088/1757-899X/250/1/012048

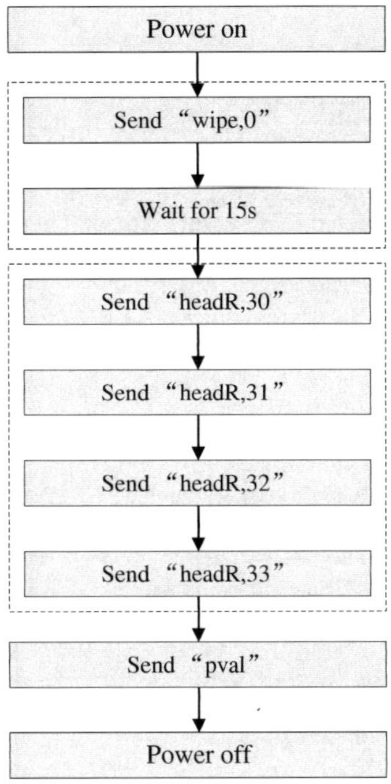

Figure 3. The flow diagram of ACLW-CAR

5. Ultrasonic cleaner for ACLW-CAR

Biofouling, the generic name given to the undesired biological adhesion to surfaces, has been a staple mark for ACLW-CAR since its very beginning. The biofouling of ACLW-CAR has resulted in the unstable or even none operation of system. A variety of approaches for inhibiting biofilm formation have been proposed. The use of toxic compounds within protective coatings has been a common practice in the past, but their adverse environmental impact is now limiting the appeal of such strategy [4].

Along with high intensity ultrasonic development, ultrasonic cleaning already has been widespread in more and more application production and life. The principle of ultrasonic cleaning is to come into being cavitation effect[5]. Figure 4 is installation instruction of ACLW-CAR and its ultrasonic cleaner. The supersonic cleaner mainly includes three parts: supersonic generator, supersonic transducer and plate sound source. Plate sound source driven by transducer creates ultrasound waves over against the ACLW-CAR. This cleaner can prevent the ACLW-CAR from biofouling and is very significant for ocean monitoring field. As can be seen from figure 5, the ACLW-CAR is clean after running for one year. It is prove that ultrasonic cleaner can prevent the ACLW-CAR from biofouling effectively.

(1) ACLW-CAR (2) Support
(3) Plate sound source (4) Supersonic transducer (5) Holder

Figure 4. Installation of ACLW-CAR and its ultrasonic cleaner

Figure 5. The effect of the ultrasonic cleaner

6. Summary

The real-time monitoring and measurement of the marine chlorophyll and turbidity is the main content for marine scientific expedition, protection of the marine environment, marine environmental assessment, the development of marine engineering, marine disaster prevention and mitigation and military oceanography are all regarded it as an important parameter. The performance parameters and the computational formulas of the ACLW-CAR were introduced in detail. Ultrasonic cleaner was developed for preventing ACLW-CAR from biofouling. Ocean current buoy for on-site rapid monitoring of current provide effective technical platform.

7. Acknowledgement

This work was financially supported by the National Natural Science Foundation of China (41406106), the Natural Science Foundation of Shandong Province (ZR2013DL007) and the Natural Science Foundation of Qingdao (14-2-4-66-jch).

8.References

IOP Conf. Series: Materials Science and Engineering **250** (2017) 012048 doi:10.1088/1757-899X/250/1/012048

[1] Yi Hu, Jian Chen. Distributions of upwelling with relation to the turbidity and chlorophyll a fluorescence of the suspended matter near Haitan Island in early summer. Marine Sciences, 2010, 34(4):58-63.
[2] Marshall T. Integrating in situ reflectance with satellite imagery to determine chlorophyll and turbidity for inland waters. Masters Abstracts International,2014,53(6).
[3] Operator's Manual of ACLW-CAR. JFE Advantech Computer. https://www.jfe-advantech.co.jp/
[4] Rodolfo E. Perez. Examination of the Effects of Localized Electric Signals on the Inhibition of Undesired Biological Adhesion. Madison: University of Wisconsin-madison (2008).
[5] Hui Wang. Research and Realization of an Ultrasonic Cleaner with Constant Power and Frequency Tracing. Baoding: Agricultural University of Hebei (2004).

Study on Material Parameters Identification of Brain Tissue Considering Uncertainty of Friction Coefficient

Fengjiao Guan[1], †Guanjun Zhang[2], Jie Liu[2], Shujing Wang[2], Xu Luo[1] and Feng Zhu[3]

[1]Science and Technology on Integrated Logistics Support Laboratory, National University of Defense Technology, Changsha, 410073, China
[2]State Key Laboratory of Advanced Design and Manufacturing for Vehicle Body, Hunan University, Changsha, 410082, China
[3]Department of Mechanical Engineering, Embry-Riddle Aeronautical University, Daytona Beach, FL, USA
†Corresponding author: zgjhuda@163.com

Abstract. Accurate material parameters are critical to construct the high biofidelity finite element (FE)models. However, it is hard to obtain the brain tissue parameters accurately because of the effects of irregular geometry and uncertain boundary conditions. Considering the complexity of material test and the uncertainty of friction coefficient, a computational inverse method for viscoelastic material parameters identification of brain tissue is presented based on the interval analysis method. Firstly, the intervals are used to quantify the friction coefficient in the boundary condition. And then the inverse problem of material parameters identification under uncertain friction coefficient is transformed into two types of deterministic inverse problem. Finally the intelligent optimization algorithm is used to solve the two types of deterministic inverse problems quickly and accurately, and the range of material parameters can be easily acquired with no need of a variety of samples. The efficiency and convergence of this method are demonstrated by the material parameters identification of thalamus. The proposed method provides a potential effective tool for building high biofidelity human finite element model in the study of traffic accident injury.

1. Introduction

The finite element (FE) method has shown its unique advantages in simulating irregular geometries, inhomogeneous media, nonlinear material properties, complex load and boundary conditions and so on. And FE simulation can capture the response that is difficult to be measured during the experiments. Therefore, many scholars study the intracranial response based on the FE model of the head, so as to better explore the damage mechanism and tissue level damage threshold. However, obtaining the accurate and reliable parameters of the brain tissue under various strain rates is helpful to improve the biofidelity of the FE model.

Based on the optimization strategy, the inverse method has increasingly become an important medium of obtaining biological tissue material parameters [1-3]. In the process of traditional biological tissue parameters identification, the positive problem model is usually built based on deterministic boundary conditions and solved by classical deterministic optimization method. However, due to the complexity of the contact surface in the actual biological tissue material experiment device, there are inevitably some uncertain factors in the modeling process of the biological tissue material parameters identification. Therefore, deterministic inverse method especially for the material parameters identification of biological soft tissue has some limitations.

Traditionally, the material mechanical properties of soft tissue can be obtained via unconstrained compression experiments. In the experiment, the soft tissue is placed in the middle of two smooth plates, assuming that the sample maintains a uniform stress / strain state during the compression process. However, if there is friction between the sample and the plate, the sample can not maintain a uniform stress / strain state when compressed. Spilker et al. [4] have shown that the friction between the specimen and the plate had a significant effect on its mechanical response, but the effect of friction on material parameters of soft tissue was not further been quantified. To reduce the impact of friction on the test results, many researchers took various measures to reduce the friction between the sample and the platen [5, 6]. However, in the actual unconstrained compression experiments, the friction between the sample and the platen cannot be completely eliminated regardless of the measures taken. Even with a well-lubricated contact, such as articular cartilage and stainless steel, the friction coefficient may exceed 0.2 [7]. Therefore, some data about compression characteristics of the soft tissue may contain errors caused by inaccurate friction.

In order to correctly estimate and evaluate the effect of friction on brain tissue material parameter identification results, this study takes the thalamus as the research object and aims at quantifying the influence of coefficient of friction on the material parameter identification based on the interval analysis method [8-12]. The possible range of material parameters can be quickly obtained via the optimization methods and interval algorithm. This will be useful to provide more accurate material parameters for the brain tissue FE model.

2. Methods

2.1 Experiment and Simulation Model Set Up

The test data came from the Wayne State University Bioengineering Centre [13]. Brain tissue samples were obtained from a 45-year-old male body (1 week after death). In the area of the thalamus (gray matter), two brain tissue samples with a size of about 15 mm by 15 mm by 8 mm were separated by surgical scalpel in the direction of the fibre. The smooth polyethylene plate was connected to the Instron Tester (Instron 1321 frame, 8500 controller) to load the sample. The unconstrained compression experiments conducted with low strain rate ($0.5s^{-1}$) and high strain rate ($35s^{-1}$) were both chosen to study the effect of friction on material parameter identification. The amount of compression of the low strain rate condition was 2.55mm, and that of high strain rate was 3.35mm.

Three-dimensional geometries of each sample were obtained using a laser scanner, and the hexahedral FE meshes were generated by ANSYS ICEM CFD. The mesh size is 0.45mm. The sensitivity analysis of the meshing shows that the convergence of the results can be sufficiently ensured by this partition density. The FE meshes of thalamus samples are shown in figure 1.

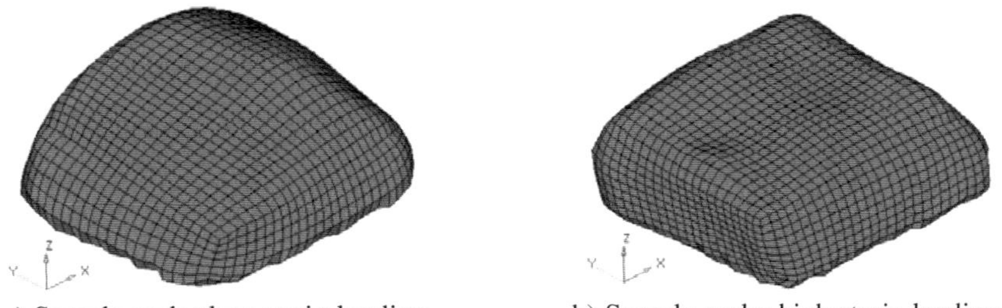

a) Sample under low strain loading b) Sample under high strain loading

Figure 1. FE meshes of thalamus samples.

There are some typical and commonly constitutive models of viscoelastic materials, such as complex constant modulus model, standard rheological model, fractional derivative model, fractional exponential model and micro-vibrator model et al[14-16]. In this paper, the characteristics of high strain rate of brain tissue samples were characterized by using the material constitutive model (fractional exponential model) which is linear and viscoelastic. The shear relaxation behavior for this kind of material is formulated through the following equation:

$$G(t)=G_i+(G_0-G_i)e^{-\beta t} \tag{1}$$

Where G_0 represents the short term modulus and G_i represents the long term shear modulus, t is time and β indicates the deacy constant that can be determinted by performaing relaxation tests. For the low strain loading tests, the viscosity of the brain tissue can be neglected and FE simulation can be performed with a higher speed than the actual experimental speed to shorten the simulation time. Thus the behavior of the sample can be estimated by a linear elastic constitutive relationship formulated through the Young's modulus E:

$$E=\frac{9KG_i}{3K+G_i} \tag{2}$$

where K denotes bulk modulus. Since K»Gi, so $E \approx 3G_i$ and the Poisson's ratio is defined:

$$v=\frac{3K-2G_i}{2(3K+G_i)}\approx 0.5 \tag{3}$$

The loading plate was moved according to the motion of impactor recorded during the test. The contac t force between plate and sample was also recorded. The initial value of the friction coefficient betwee n the sample and the test plate was defined as 0.2.

2.2 Material Parameters Deterministic Identification of Brain Tissue Based on Specimen-Specific Fe Model

Combined with thalamic sample unconstrained compression experiments, the objective function of the viscoelastic material parameters identification of thalamic can be defined as following:

$$F=\sum_{i=1}^{n}[(x_{im}^{l}-x_{ic}^{l})^2+(x_{im}^{h}-x_{ic}^{h})^2] \tag{4}$$

Where the subscript m represents the measurement data in the compression experiment, and c represents the predicted data from the simulation model, and the superscript l represents the low strain rate test, and h represents the high strain rate test, and n represents the number of points on the impact force and deformation curve obtained from the experiment and simulation. It can be seen that the fitting degree of the experimental measurement data and the simulation data can be evaluated by the objective function. The function is closer to 0, the result of material parameters identification of biological tissue is better.

According to the literature[17-20] and $E\approx 3G_i$, the ranges and initial values of linear viscoelastic material parameters (G_0, G_i, β) and elastic material parameter E are listed in table 1.

Table 1. Range of material parameters in material models

Parameter	Initial value	Lower bound	Upper bound
G_0 (kPa)	80.0	10.0	100.0
G_i (kPa)	10.0	1.0	20.0
β (s^{-1})	600	100	900
E (kPa)	30.0	3.0	60.0

Based on the FE method and sequence response surface method (FEM-SRSM)[21], the measurement data of compression experiments with two different strain rate were considered and the viscoelastic material parameters of the thalamus were investigated. The elastic material parameter E is used as a non-independent variable in the inverse process, and E is input into the three-dimensional FE model of low-strain rate compression experiment according to $E\approx 3G_i$. The result is taken as the initial value of the inverse considering uncertainty.

2.3 Material Parameters Identification of Brain Tissue Considering Uncertain Parameters
Aiming at the uncertainty of the friction coefficient, a mathematical model of the material parameter identification of the biological tissue considering the uncertain boundary condition is established. The inverse problem is transformed into an optimization problem, and the following objective function is constructed:

$$\begin{aligned} &\min \quad F(\boldsymbol{M},\boldsymbol{U}) \\ &\text{subject to } \; m_i^L \le m_i \le m_i^R \quad i=1,2,\cdots n \\ &\qquad \boldsymbol{U} \in \boldsymbol{U}^I = \left[\boldsymbol{U}^L,\boldsymbol{U}^R\right] \end{aligned} \tag{5}$$

Where F represents the error between the response curve of the experimental measurement and the response curve of the simulated prediction; $\boldsymbol{M}$ is the parameter in the constitutive equation of the material corresponding to the specific biological tissue, such as the instantaneous shear modulus G_0, the long-term shear modulus G_i, and the reduction coefficient β in constitutive equation of viscoelastic material, etc.; $\boldsymbol{U}$ is the uncertainty parameter, such as the friction coefficient f in the boundary condition.

The objective function is nonlinear and continuous. Since it is a continuous function with respect to $\boldsymbol{M}$ and $\boldsymbol{U}$, and $\boldsymbol{U}$ is an interval vector. Therefore, for any determined experimental measurement response curve, the obtained material parameters may take a range of values, rather than a constant values. Therefore, the above problems can not be solved by the traditional deterministic optimization method.

2.3.1 The Interval Model of Material Parameters Identification. Based on the interval mathematical theory [22], the interval number is defined as an ordered pair of real numbers:

$$A^I = \left[A^L, A^R\right] = \left\{A \middle| A^L \le A \le A^R, A \in R\right\} \tag{6}$$

In equation (6), the superscript I denotes the interval, the superscript L and R are the lower bound and the upper bound of the interval, respectively. When $A^L=A^R$, the interval degrades to a real number.
For the material parameters inverse problem with uncertainty of the boundary conditions, the interval vector can be used:

$$\boldsymbol{f} \in \boldsymbol{f}^I = \left[\boldsymbol{f}^L, \boldsymbol{f}^R\right], \; f_j \in f_j^I = [f_j^L, f_j^R], \; j=1,2,\cdots,n \tag{7}$$

The biological tissue material parameters identified under this boundary condition are no longer a single real number but a bounded interval corresponding to all possible uncertainty parameters of boundary condition:

$$\boldsymbol{M}(\boldsymbol{f}) \in \boldsymbol{M}^I = \left[\boldsymbol{M}^L, \boldsymbol{M}^R\right] \tag{8}$$

In equation (8), $\boldsymbol{M}^I$ is the interval of the material parameters of the biological tissue, and the upper and lower limits are the maximum and minimum values of all the material parameters that may be inversed.

$$\begin{aligned} &\boldsymbol{M}^L = \min_{f \in \Gamma} \boldsymbol{M}(\boldsymbol{f}), \quad \boldsymbol{M}^R = \max_{f \in \Gamma} \boldsymbol{M}(\boldsymbol{f}) \\ &\Gamma = \left\{f \middle| f_j^L \le f \le f_j^R, j=1,2,\cdots,n\right\} \end{aligned} \tag{9}$$

To solve the interval $\boldsymbol{M}^I$ of biological tissue material parameters, each possible value in the interval $\boldsymbol{f}^I$ of uncertainty parameter from the boundary condition can be selected in turn, and the biomechanical response and the deterministic material parameter identification method are both used in the inverse method. Finally, the minimum and maximum values for each material parameter in a series of results are searched. Thus, the whole solution process is a complicated nesting optimization problem. The FE model is called repeatedly, and the solution efficiency is not high. In order to overcome this problem, this paper introduces the interval analysis method to inverse the range of material parameters of biological tissue under uncertain boundary conditions.

2.3.2 Interval Analysis Method for Uncertain Boundary Conditions. The interval number $\boldsymbol{f}^I$ can be written as equation (10) based on the interval mathematics theory [22]:

$$\boldsymbol{f}^I = \left[\boldsymbol{f}^L, \boldsymbol{f}^R\right] = \left[\boldsymbol{f}^c - \boldsymbol{f}^w, \boldsymbol{f}^c + \boldsymbol{f}^w\right] = \boldsymbol{f}^c + [-1,1]\boldsymbol{f}^w \tag{10}$$

In equation (10), $\boldsymbol{f}^c$ and $\boldsymbol{f}^w$ represent the midpoint and radius of the interval $\boldsymbol{f}^I$. The uncertainty level γ ($\boldsymbol{f}^I$) of the interval $\boldsymbol{f}^I$ is defined as [23]:

$$\gamma(\boldsymbol{f}^I) = \frac{\boldsymbol{f}^w}{\left|\boldsymbol{f}^c\right|} \times 100\% \tag{11}$$

So the uncertainty vector $\boldsymbol{f}$ can be written as:

$$\boldsymbol{f} = \boldsymbol{f}^c + \delta \boldsymbol{f} \tag{12}$$

Where,

$$\delta \boldsymbol{f} \in [-1,1]\boldsymbol{f}^w,\ \delta f_i \in [-1,1] f_i^w,\ i = 1,2,...,q \tag{13}$$

Assuming that the uncertainty level of each parameters in $\boldsymbol{f}$ is small, the first-order Taylor expansion of the uncertain material parameter $\boldsymbol{M}(\boldsymbol{f})$ can be performed at $\boldsymbol{f}^c$

$$\boldsymbol{M}(\boldsymbol{f}) = \boldsymbol{M}(\boldsymbol{f}^c + \delta \boldsymbol{f}) \approx \boldsymbol{M}(\boldsymbol{f}^c) + \sum_{j=1}^{q} \frac{\partial \boldsymbol{M}(\boldsymbol{f}^c)}{\partial f_j} \delta f_j \tag{14}$$

The range of structural displacement response could be calculated by performing a natural interval extension on equation (14):

$$\boldsymbol{M}^I(\boldsymbol{f}^I) = \boldsymbol{M}(\boldsymbol{f}^c) + \sum_{j=1}^{q} \left| \frac{\partial \boldsymbol{M}(\boldsymbol{f}^c)}{\partial f_j} \right| [-1,1] f_j^w \tag{15}$$

Therefore, the lower and upper bounds of material parameters are respectively:

$$\begin{aligned} \boldsymbol{M}^L(\boldsymbol{f}^I) &= \boldsymbol{M}(\boldsymbol{f}^c) - \sum_{j=1}^{q} \left| \frac{\partial \boldsymbol{M}(\boldsymbol{f}^c)}{\partial f_j} \right| f_j^w \\ \boldsymbol{M}^R(\boldsymbol{f}^I) &= \boldsymbol{M}(\boldsymbol{f}^c) + \sum_{j=1}^{q} \left| \frac{\partial \boldsymbol{M}(\boldsymbol{f}^c)}{\partial f_j} \right| f_j^w \end{aligned} \tag{16}$$

Thus the material parameters identification of biological tissue with uncertain friction coefficient is transformed into two types of deterministic inverse problems. One is the material parameters identification at the midpoint of uncertain parameters $\boldsymbol{M}$ ($\boldsymbol{f}^c$) and the other is the first order partial derivative solution of material parameters with respect to the uncertain parameters $\frac{\partial \boldsymbol{M}(\boldsymbol{f}^c)}{\partial f_j}$. The material parameters at the midpoint of uncertain parameters $\boldsymbol{M}$ ($\boldsymbol{f}^c$) can be calculated by the measured biological tissue mechanics response and inversed based on material parameter identification method with the deterministic boundary. Due to uncertain boundary conditions, the upper and lower bounds of material parameters will be obtained by equation (16).

In this paper, the above-mentioned method of material parameters identification of biological tissue under the uncertainty boundary condition is applied to the inverse problem of material parameters identification of typical ultra-soft tissue, namely brain tissue.

Wu et al. [24] assume that the friction coefficient between sample and plate is 0 to 0.5 when defining the contact surface. Based on the research of friction properties between human skin and various materials [25], the friction coefficient between sample and plate is assumed between 0.1 and

0.3 under well controlled test conditions. Therefore, in the boundary condition, the friction coefficient is set as an uncertainty variable, and the median value is 0.2. The viscoelastic material parameter of the tissue sample is inversed at 10% uncertainty level (i.e. f^I = [0.18, 0.22]). The inverse problem is transformed into an optimization problem, and the following objective function is constructed:

$$\begin{aligned} &\text{Min} \quad F(G_0, G_i, \beta, f) \\ &\text{Subject to} \quad 10 \le G_0 \le 100 \\ &\qquad 1 \le G_i \le 20 \\ &\qquad 100 \le \beta \le 900 \\ &\qquad f \in f^I = [f^L, f^R] \end{aligned} \tag{17}$$

The solution flow of viscoelastic material parameters of brain tissue considering the uncertainty of friction coefficient in the material experiment is shown as figure 2. To solve the positive problem model, the FEM - SRSM is adopted, and the genetic algorithm is used as the optimization solver.

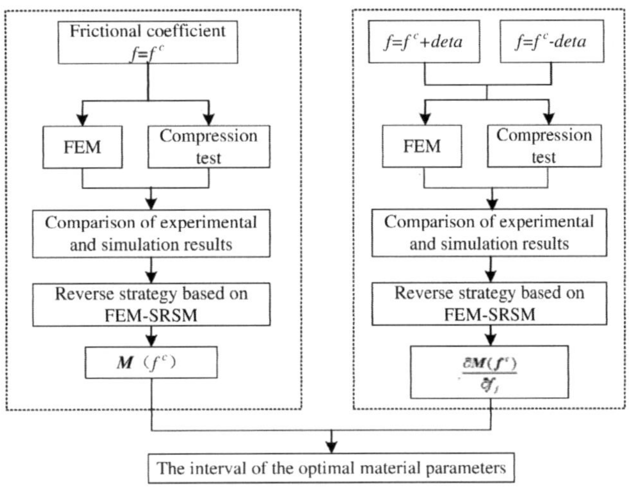

Figure 2. Flowchart of material parameters identification based on interval analysis method.

3. Results and Discussion

3.1 Deterministic Inverse Results

The material parameters were obtained by the sequence linear response surface method (shown in table 2). The FE simulation was conducted using the above material parameters. And the comparison of the force and deformation curve obtained from experiment and simulation respectively is shown in figure 3. It can be seen that the two curves fit well, which proves the feasibility of the inverse method based on the FEM-SRSM and the correctness of the material parameters obtained by this method.

Table 2. Material parameters of thalamus obtained from deterministic inverse method

material parameter		result
G_0(kPa)		11.66
G_i(kPa)		2.304
β(s^{-1})		657.7
error	Simulation	0.112%
	Prediction	0.110%

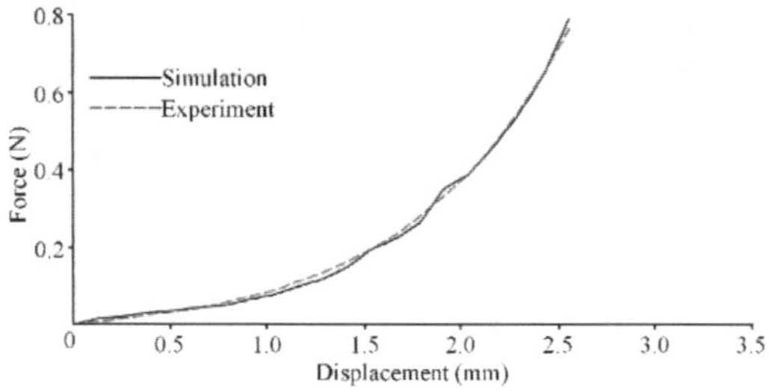

a) Compression at low speed

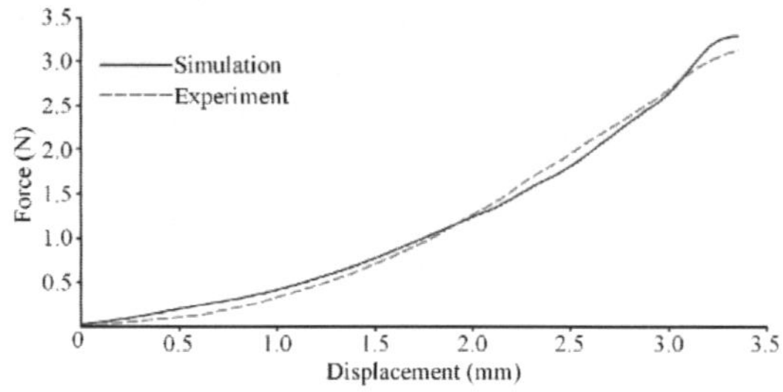

b) Compression at high speed

Figure 3. Deterministic inverse result.

3.2 Uncertain Inverse Results

Considering the uncertainty of the boundary condition (friction coefficient), the inverse results of the parameters of the thalamus based on the interval analysis method are listed in table 3. It is obvious that the uncertainty level of the parameter β is the largest, namely the friction coefficient has the greatest influence on the material parameter β. Meanwhile, the inverse result can provide an optional parameter interval for the establishment of the FE model.

Table 3. Range of material parameters of thalamus obtained from uncertain inverse method

parameter	range of material parameter
G_0(kPa)	[10.9, 12.5]
G_i(kPa)	[2.23, 2.37]
β(s^{-1})	[560.7, 754.7]

The interval analysis method is based on the approximation of first-order Taylor expansion, which means the method can be only applied to the case where the uncertainty level of the parameter is small. If the uncertainty level is large, the accurate agent model can be constructed in the material parameter space, and then the upper and lower bounds of material parameters can be acquired using the uncertainty optimization method based on the agent model.

4. Conclusion

Considering the uncertainty of friction coefficient, an identification method for parameters estimation of brain tissue based on interval analysis theory is proposed in this paper. Firstly, the intervals are used to quantify the friction coefficient in the experimental boundary condition. The interval analysis method based on first-order Taylor expansion is adopted to transform the material parameter identification problem under the uncertain boundary condition into two types of deterministic inverse problems. The intelligent optimization algorithm is used to solve the two types of deterministic inverse problems quickly and accurately, and the upper and lower bounds of the material parameters are obtained by interval operation. This method can be used to evaluate the effect of uncertainty parameters on the inverse results of biological tissue parameters. This provides an effective method for the development of human FE models with high biological fidelity and high applicability. At the same time, this method avoids the time-consuming nesting optimization problem, which greatly improves the computational efficiency.

5.Acknowledgments

This research was funded by the National Natural Science Foundation of China (11402296).

6.References

[1] Chawla A, Mukherjee S, and Karthikeyan B 2009 Characterization of human passive muscles for impact loads using genetic algorithm and inverse finite element methods *Biomech Model Mechanobiol* **8** pp 67-76

[2] Snedeker J G, Niederer P, Schmidlin F R, Farshad M, Demetropoulos C K, Lee J B and Yang K H 2005 Strain-rate dependent material properties of the porcine and human kidney capsule *Journal of Biomechanics* **38** pp 1011-1021

[3] Hu J, Klinich K D, Miller C S, Nazmi G, Pearlman M D, Schneider L W, Rupp J D 2009 Quantifying dynamic mechanical properties of human placenta tissue using optimization techniques with specimen-specific finite-element models *Journal of Biomechanics* **42** pp 2528-2534

[4] Spilker R L, Suh J K, and Mow V C 1990 Effects of Friction on the Unconfined Compressive Response of Articular Cartilage: A Finite Element Analysis *Journal of Biomechanical Engineering*, **112** pp 138-146

[5] Miller K. and Chinzei K 1997 Constitutive modelling of brain tissue: experiment and theory *J Biomech* **30** pp 1115-1121

[6] Miller-Young J E, Duncan N A, and Baroud G 2002 Material properties of the human calcaneal fat pad in compression: experiment and theory *J Biomech* **35** pp 1523-31

[7] Wang H and Ateshian G A 1997 The normal stress effect and equilibrium friction coefficient of articular cartilage under steady frictional shear *Journal of Biomechanics* **30** pp 771-776

[8] Jiang C, Han X, Guan F and Li Y 2007 An uncertain structural optimization method based on nonlinear interval number programming and interval analysis method *Engineering Structures*, **29** pp 3168-3177

[9] Liu J, Hu Y, Xu C, Jiang C and Han X 2016 Probability assessment of inverse parameters for stochastic structures using point estimation method *Reliability Engineering & System Safety* **156** pp 51-58

[10] Liu J, Han X, Jiang C, Ning H and Bai Y 2011 Dynamic load identification for uncertain structures based on interval analysis and regularization method *International Journal of Computational Methods* **8** pp 667-683

[11] Qiu Z P 2005 *Convex method based on non-probabilistic set-theory and its application* Beijing: National Defence Industry Press (In Chinese)

[12] Han X, Jiang C, and Liu G 2007 Optimization of structures with uncertain constraints based on convex model and satisfaction degree of interval *Computer Methods in Applied Mechanics and Engineering* **196** pp 4791-4800

[13] Zhu F, Jin X, Guan F, Zhang L, Mao H, Yang K H, and King A I 2010 Identifying the properties of ultra-soft materials using a new methodology of combined specimen-specific finite element model and optimization techniques *Materials and Design* **31** pp 4704-4721

[14] Rashid B, Destrade M, and Gilchrist M D 2012 Mechanical characterization of brain tissue in compression at dynamic strain rates *J Mech Behav Biomed Mater* **10 pp** 23-38

[15] Rashid B, Destrade M, and Gilchrist M D 2014 Mechanical characterization of brain tissue in tension at dynamic strain rates *J Mech Behav Biomed Mater* **33** pp 43-54

[16] Budday S, et al 2016 Mechanical characterization of human brain tissue *Acta Biomater* **48** pp 319-340

[17] Zhang L, Yang K H and King A I 2001 Comparison of brain responses between frontal and lateral impacts by finite element modelling *J Neurotrauma* **18** pp 21-30

[18] Zhou C, Khalil T B and King A I 1996 Viscoelastic response of the human brain to sagittal and lateral rotational acceleration by finite element analysis *Proceedings of the 1996 international IRCOBI conference on the biomechanics of impact* Dublin, Ireland

[19] Kang H S, Willinger R, Diaw B M and Chinn B 1997 Validation of a 3d anatomic human head model and replication of head impact in motorcycle accident by finite element modelling *Stapp Car Crash Conference* **106** 973339

[20] Al-Bsharat A, Hardy W N, Yang K H, Khalil T B, Tashman S and King A I 1999 Brain/skull relative displacement magnitude due to blunt head impact: new experimental data and model *Stapp Car Crash Conference* 99SC22

[21] Stander N, Roux W, Basudhar A, Eggleston T, Goel T and Craig K 2015 *Ls-opt user's manual* California: Livermore Software Technology Corporation

[22] Moore R E 1979 *Methods and applications of interval analysis* London: Prentice-Hall Inc

IOP Conf. Series: Materials Science and Engineering **250** (2017) 012049 doi:10.1088/1757-899X/250/1/012049

[23] Qiu Z P 1994 *Interval analysis for static response and eigenvalues problem of structures with uncertain parameters* Changchun: Jilin University of technology (In Chinese)
[24] Wu J Z, Dong R G and Schopper A W 2004 Analysis of effects of friction on the deformation behaviour of soft tissues in unconfined compression tests *J Biomech* **37** pp 147-55
[25] Zhang M and Mak A F 1999 In vivo friction properties of human skin *Prosthetics and Orthotics International* **23** pp 135-141

IOP Conf. Series: Materials Science and Engineering **250** (2017) 012050 doi:10.1088/1757-899X/250/1/012050

Preparation and Characterization of Core-Shell Al/Al_2O_3 Composites with Adjusted Structure through Fast Hydrothermal Oxidation of Aluminum

G N Ambaryan*, M S Vlaskin, E I Shkolnikov and A Z Zhuk

Joint Institute for High Temperatures of the Russian Academy of Sciences, Izhorskaya 13 Bldg 2, Moscow 125412, Russia

*Contact by email: ambaryan1991@gmail.com

Abstract. A method for preparation of core-shell Al/Al_2O_3 composites with adjusted structure was proposed. It was based on two steps: fast hydrothermal partial oxidation of aluminum in continuous reactor and subsequent thermal treatment in muffle furnace to convert aluminum hydroxide to γ-Al_2O_3. It was shown the possibility to create metal-ceramic composites with adjustable content of metallic core by temperature of hydrothermal process. During experimental study it was established that oxidation degree for aluminum powder with average size of 22 μm was changed from 40 to almost 100 % when temperature in hydrothermal reactor changed from 280 to 350 °C. The reaction time was no more than 10 min. Obtained products were studied by scanning electron microscopy, pore structure analysis and XRD analysis. Prepared core-shell Al/γ-Al_2O_3 composites can be used as catalyst materials as well as for additive technologies.

1. Introduction

The production of micro- and nano-structured metal oxides or composites with specific structure, shape and morphology have attracted a great attention in recent years due to the necessity of such materials in fabrication of ceramics, adsorbents, filters and catalyst carriers. Due to unique properties of such materials they can be used also in different areas such as optical and semiconductor industry, quantum electronics, electric-power industry, instrument engineering, etc. [1-2, 14-15].

One of simple way to produce Al/Al_2O_3 composites is aluminum oxidation in water or aqueous solutions. An interest in this process is due to the high chemical activity of highly dispersed aluminum powders in the reactions with water or water vapor. The use of pure water as an oxidant in its reaction with dispersed aluminum makes it possible to synthesize high purity hydrogen, a large amount of high temperature steam, and a variety of aluminum hydroxides [3-4]. The main element of the units for hydrothermal oxidation of aluminum by water or steam is a rector that generates also a steam-hydrogen mixture, that can be used as a working medium of conventional and future heat engines and generators [5-6]. Usually reactor represents an autoclave and in case of aluminum oxidation it operates during several hours [7-10].

One of the promising methods of advanced materials production is hydrothermal processing. The term hydrothermal technology covers the broad spectrum of different techniques [11] and commonly it implies the use of aqueous solutions at elevated temperatures (> 100 °C) and pressures (> 0.1 MPa). The success in advanced materials production under hydrothermal conditions is usually bound up with unique properties (density, viscosity, ionic product, dielectric constant etc.) of water at elevated thermodynamic parameters [11].

IOP Conf. Series: Materials Science and Engineering **250** (2017) 012050 doi:10.1088/1757-899X/250/1/012050

A way of advanced alumina or alumina-aluminum composites production is hydrothermal oxidation of aluminum. The reaction of aluminum oxidation by water, generally, can be realized in the following ways:

$$Al+3H_2O \rightarrow 0,5Al_2O_3 \cdot 3H_2O+1,5H_2+427.98 \text{ kJ/mole}, \quad (1)$$

$$Al+2H_2O \rightarrow 0,5Al_2O_3 \cdot H_2O+1,5H_2+415.24 \text{ kJ/mole}, \quad (2)$$

$$Al+1,5H_2O \rightarrow 0,5Al_2O_3+1,5H_2+407.7 \text{ kJ/mole}. \quad (3)$$

Channel (1) prevails at temperature up to 200 °C, (2) prevails in the field of 200÷400 °C and further temperature increase leads to Al_2O_3 formation (channel (3)).

Some attempts to produce advanced materials by hydrothermal oxidation of aluminum have been already carried out. Main factors that influence on both reaction kinetics and physicochemical properties of solid products of hydrothermal oxidation of aluminum are thermodynamic conditions, initial surface area of aluminum and technological conditions.

Low temperature (about 100 °C) hydrothermal oxidation of aluminum micron powder in batch reactor and subsequent thermal dehydration were proposed for the synthesis of porous Al_2O_3/Al composites (cermets) in [7]. Hydrothermal oxidation was carried out in open bucket under intensive mixing. The duration of hydrothermal treatment was from 1 to 5 hours. Although produced composites possessed high surface area (about 30-40 m^2/g) and advanced porous structure, the conversion degree of aluminum didn't exceed 60 %. In [12] it was shown that reaction can be promoted by applying the ultrasonic activation influencing also on product structure. It was shown that ultrasonic field promotes the aluminum micron particles dispersion into microcrystallites during oxidation process.

In [9] the product synthesized by hydrothermal oxidation of aluminum in autoclave at 250 °C was also used for Al_2O_3/Al composite production. Aluminum was used in the form of micron powder as well. Autoclave was heated to 250 °C during 40 min and then was maintained heated for 6.5 hours. The products of hydrothermal oxidation were dried and calcined at temperatures of 520 °C for 2 hours. After calcination the content of oxide in the sample was about 37 wt.%. Following the study of pore structure of synthesized Al_2O_3/Al composite it was established that the product has high surface area (about 60 m^2/g) and narrow pore size distribution and can be viewed as promising material in catalysis and adsorption.

In present work hydrothermal oxidation of aluminum micron powder was carried out under 240-350 °C in a flow reactor for fast up to 10-min aluminum oxidation. Main object of this study was the investigation of structural properties of Al/Al_2O_3 composites obtained.

2. Materials and Methods

Hydrothermal oxidation of aluminum powder was carried out on experimental plant designed generally for energy application [13]. Deionized water and micron powder of aluminum were used as starting reagents. Aluminum powder used in experiments was purchased from Rusal. Powder represents spherical particles with average size of about 22 μm.

The synthesis of Al/Al_2O_3 composites at different temperatures was performed during the series of model experiments. Mixture of aluminum powder with water was prepared at room temperature within mixing tank. Then small portion of mixture containing about 50 g of Al was injected by high-pressure dosage pump into preliminary heated 7.5 l reactor. After 10 minutes the mixture of oxidation product and water was withdrawn from the reactor.

Solid product produced by aluminum hydrothermal oxidation at different temperatures were separately dried and analyzed by scanning electron microscopy, X-ray analysis, analysis of specific surface, analysis of pore structure and mass spectrometry.

Surface morphology of solid products was studied on JEOL JSM-7401F scanning electron microscope (SEM). Samples were placed into the microscope on carbon substrate. The shooting was carried out at 1 kV accelerating voltage. Samples were studied at the magnification in the range of 100-200000. Phase composition of oxidation product was studied by X-ray diffraction (XRD) using Thermo ARL XTRA diffractometer using CuK_α radiation (λ=0.15418 nm). The specific surface of product was studied by low-temperature nitrogen adsorption using Sorbi 4.1 device. For thermal treatment a muffle furnace LHT 08/16 (Nabertherm) was used.

3. Results and Discussion

Table 1 shows the results of experiments of aluminum oxidation. It can be seen that by temperature adjustment we can achieve specific oxidation degree. For total oxidation of aluminum powder with the size of 22 μm temperature should be more than 350 °C. Temperature decreasing lead to the aluminum conversion decreasing (at one time of staying inside the reactor – 10 min). At 280 °C aluminum conversion after 10 min oxidation decreased to about 40 % thus producing Al/Al_2O_3 composite.

Table 1. Results of hydrothermal oxidation of aluminum powder with the size of 22 μm.

Reaction temperature, °C	Reaction pressure, MPa	Aluminum conversion degree, %
350	16.1	100
332	12.8	91
316	10.5	86
280	6.8	40

Fig. 1 shows SEM images of initial aluminum particle and solid oxidation product. It can be seen that the surface of aluminum is covered by dense aluminum hydroxide layer. Fig. 2 with larger magnification show that the size of single crystal is changed from several nanometers to hundreds nanometers.

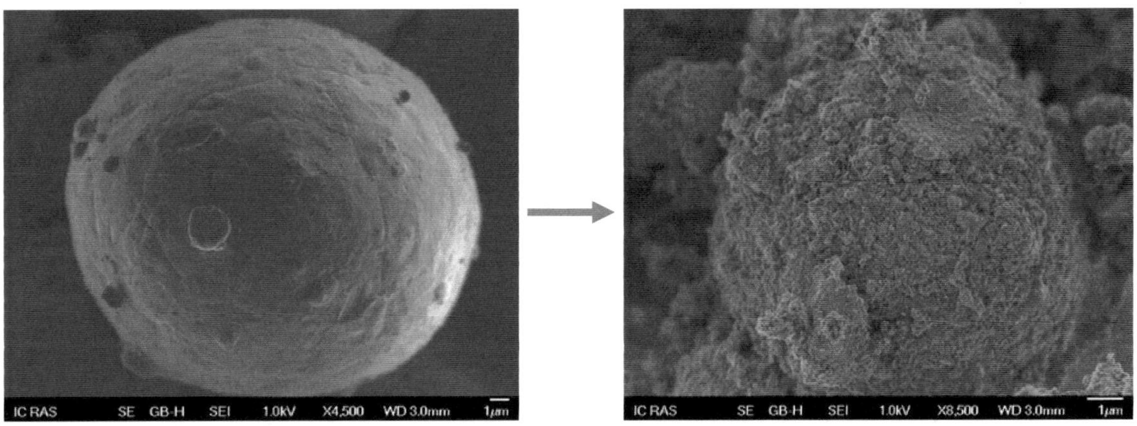

Figure 1. SEM images of initial aluminum powder and oxidation product.

X-ray analysis of oxidation product showed that the product of hydrothermal oxidation of aluminum in the range of 240-350 °C is aluminum oxyhydroxide – boehmite (γ-AlOOH). Fig. 3 shows the XRD plot for oxidation product obtained at 316 °C. All peaks on XRD plot correspond to AlOOH and just one intensive peak corresponds to non-oxidized aluminum core. The calculation of coherent scattering area for boehmite crystals produced by hydrothermal oxidation of aluminum showed that temperature increasing leads to the increasing of size of boehmite crystals.

It is important to note that although thermodynamic parameters and size of aluminum particles were the same as in previous studies with autoclave the conversion degree in case of flow reactor was turned out to be higher. For example in [7, 9] in autoclave hydrothermal conditions at 250 °C the conversion degree was less than 30 % after 6.5 hours treatment. In our work we achieved 40 % at 240 °C after 10 min oxidation. During the slow heating of autoclave the surface of aluminum in hydrothermal conditions is covered by dense layer of hydroxide. By the time when autoclave reaches maximum temperature aluminum particle has already partially oxidized and covered by oxidation product microporous layer whose thickness is several times more than the thickness of initial oxide

IOP Conf. Series: Materials Science and Engineering **250** (2017) 012050 doi:10.1088/1757-899X/250/1/012050

film covered aluminum particles. This layer is certain to complicate the diffusion of water to the surface of unreacted aluminum core and thus reducing the rate of chemical reaction. When aluminum particles come into already heated flow reactor the micron sized aluminum core is shock heated to nominal temperature by less than a second. Such shock heating expands the aluminum core and splits the oxide film on the surface of aluminum thus practically eliminating the induction period. Reaction starts immediately. Thus in flow reactor the conversion degree of aluminum is improved in comparison with autoclave method.

Figure 2. SEM image of aluminum oxidation product with larger magnification.

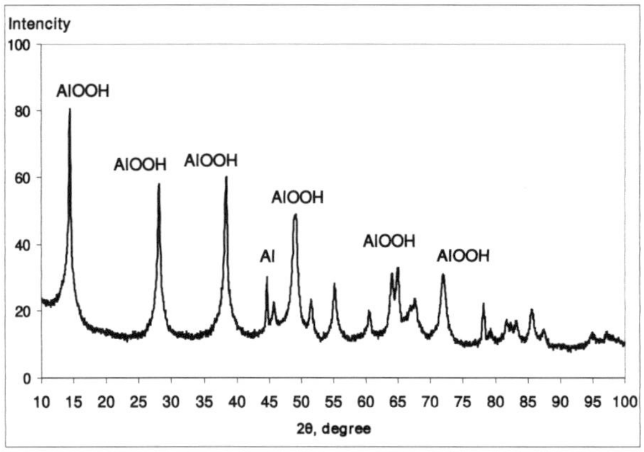

Figure 3. XRD plot for oxidation product obtained at 316 °C.

Fig. 4 shows SEM surface of Al/AlOOH composite produced at 280 °C and the product of its thermal treatment in muffle furnace at temperature of 650 °C. At this temperature the boehmite (AlOOH) was transformed into γ-Al_2O_3. It can be seen that the structure of the surface is changed. Due to thermal treatment the size of crystal becomes smaller. It can be explained by the loss of crystallized water. The surface of oxidation product at 280 °C was 58.9 m^2/g, the surface of the product after thermal treatment at 650 °C was 80.2 m^2/g.

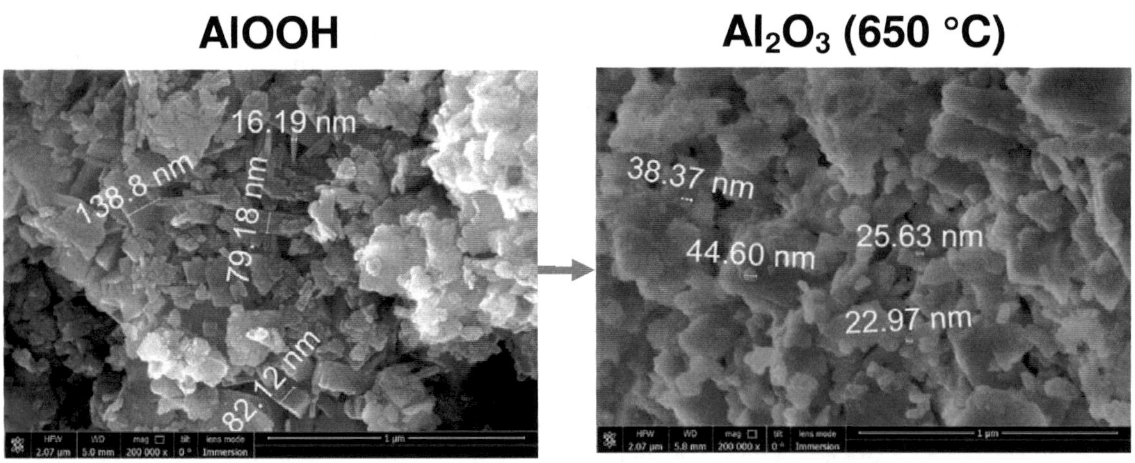

Figure 4. Surface transformation during heat treatment of Al/AlOOH producing Al/Al_2O_3.

4. Conclusion

The results of the study confirmed in general that hydrothermal oxidation of aluminum is a promising method of production of Al/Al_2O_3 or Al/AlOOH composites. It is important to note that by proposed method it is possible to adjust the size of core and shell by reaction temperature.

IOP Conf. Series: Materials Science and Engineering **250** (2017) 012050 doi:10.1088/1757-899X/250/1/012050

5. Acknowledgements
The project was implemented with the financial support of the Leading Scientific School 8406.2016.8.

6. References
[1] Zhuk A Z, Vlaskin M S, Grigorenko A V, Kislenko S A and Shkolnikov E I 2016 Synthesis of high-purity α-Al2O3 from boehmite by high temperature vacuum treatment *Journal of Ceramic Processing Research* **17** 910-8
[2] Vlaskin M S, Grigorenko A V, Zhuk A Z, Lisitsyn A V, Sheindlin A E and Shkol'nikov E I 2016 Synthesis of high-purity α-Al2O3 from boehmite obtained by hydrothermal oxidation of aluminum *High Temperature* **54** 322-9
[3] Bersh A V, Lisitsyn A V, Sorokovikov A I, Vlaskin M S, Mazalov Y A and Shkol'nikov E I 2010 Study of the processes of steam-hydrogen mixture generation in a reactor for hydrothermal aluminum oxidation for power units *High Temperature* **48** 866-73
[4] Shkolnikov E I, Shaitura N S and Vlaskin M S 2013 Structural properties of boehmite produced by hydrothermal oxidation of aluminum *The Journal of Supercritical Fluids* **73** 10-7
[5] Vlaskin M S, Shkolnikov E I, Lisicyn A V, Bersh A V and Zhuk A Z 2010 Computational and experimental investigation on thermodynamics of the reactor of aluminum oxidation in saturated wet steam *International Journal of Hydrogen Energy* **35** 1888-94
[6] Vlaskin M S, Shkol'nikov E I, Lisitsyn A V and Bersh A V 2010 Thermodynamic calculation of the parameters of a reactor for oxidizing aluminum in wet saturated steam *Thermal Engineering* **57** 794-801
[7] Tikhov S F, Fenelonov V B, Sadykov V A, Potapova Y V and Salanov A N 2000 Porous Al2O3/Al Metal Ceramics Prepared by the Oxidation of Aluminum Powder under Hydrothermal Conditions Followed by Thermal Dehydration: I. Composition and Macrocharacteristics of Composites *Kinetics and Catalysis* **41** 826-34
[8] Kim J and Lee D 2015 Synthesis and Properties of Al2O3@Al Metal–Ceramic Core–Shell Microstructures for Catalyst Applications *Topics in Catalysis* **58** 375-85
[9] Tikhov S F, Fenelonov V B, Zaikovskii V I, Potapova Y V and Sadykov V A 1999 The three-dimensional microporous structure of alumina synthesized through the aluminum hydrothermal oxidation route *Microporous and Mesoporous Materials* **33** 137-42
[10] Ambaryan G N, Vlaskin M S, Dudoladov A O, Meshkov E A, Zhuk A Z and Shkolnikov E I 2016 Hydrogen generation by oxidation of coarse aluminum in low content alkali aqueous solution under intensive mixing *International Journal of Hydrogen Energy* **41** 17216-24
[11] Byrappa K and Yoshimura M 2001 *Handbook of Hydrothermal Technology* (Norwich, NY: William Andrew Publishing)
[12] Shaytura N S, Laritchev M N, Laritcheva O O and Shkolnikov E I 2010 Study of texture of hydroxides formed by aluminum oxidation with liquid water at various activation techniques *Current Applied Physics* **10** S66-S8
[13] Vlaskin M S, Shkolnikov E I, Bersh A V, Zhuk A Z, Lisicyn A V, Sorokovikov A I and Pankina Y V 2011 An experimental aluminum-fueled power plant *Journal of Power Sources* **196** 8828-35
[14] A. Astarita, L. Carrino, M. Durante, A. Formisano, A. Langella, F. Memola Capece Minutolo, V. Paradiso and A. Squillace, Experimental study on the incremental forming of coated aluminum alloy sheets, Key Engineering Materials Vols. 622-623 (2014) pp 398-405.
[15] Astarita A., Genna S., Leone C.,Memola Capece Minutolo F., Paradiso V., Squillace, Laser Cutting of Aluminium Sheets with a Superficial Cold Spray Titanium Coating, Key Engineering Materials Vols. 611-612 (2014) pp 794-803.

IOP Conf. Series: Materials Science and Engineering **250** (2017) 012051 doi:10.1088/1757-899X/250/1/012051

Study on the Influence of Nano-SiC on the Structure and Properties of Nodular Cast Iron

Ruixiang Zhu , Yong SU and Xinyu Qin

School of Materials Science and Engineering, Hefei University of Technology, Hefei 230009, China.
Email:zhuruixiang1125@163.com

Abstract. In this paper, the effect of adding Nano-SiC on the structure and mechanical properties of nodular cast iron during inoculation and spheroidizing process is mainly studied. Three sets of samples were set up: one group is without adding Nano-SiC, and the other two groups are adding Nano-SiC in the spheroidizing process and inoculation step. The three groups of castings were sampled from the core and 1/4 respectively and treated by isothermal quenching. The influences of Nano-SiC and heat treatment process on the microstructure and mechanical properties of nodular cast iron were discussed through the observation of impact work, hardness and metallographic structure of Vivtorinox. The experimental results show that with the addition of Nano-SiC, the spheroidization rate, spheroidization degree and graphitization of the graphite in casting structure are improved. In mechanical properties, the impact toughness of the samples with adding Nano-SiC during inoculation is the best , and the hardness of nodular cast iron can be improved. After heat treatment, the impact toughness of nodular cast iron samples increased obviously, but the hardness decreased generally.

1. Introduction

Nano-SiC powder has excellent properties such as high hardness, good creep resistance, chemical resistance, good oxidation resistance, small thermal expansion coefficient and high thermal conductivity. In this paper, the microstructure, hardness and impact toughness of nodular cast iron which with adding Nano-SiC were studied , so as to improve the comprehensive mechanical properties of nodular cast iron.

2. Test and Treatment of Nodular Cast Iron

2.1. Material Preparation

There were total six casting spheres which divided into three groups, the first was a cast sphere without adding Nano-SiC, the second was the casting sphere which adds Nano-SiC in the inoculation process,the third was the casting sphere which adds Nano-SiC in the spheroidizing process. The feeding method was feeding wire, at the same time casting sphere of each group was divided into two types of Φ150 and Φ100. The casting preparation results are shown in figure 1.

Figure 1. Results of casting preparation

2.2. Wire Cutting Sample Preparation

Before testing, the castings obtained were cut into 10 * 10 * 55mm samples. According to the differences and characteristics of the casting. On three groups of nodular cast iron, respectively preparing samples from core and 1/4 cutting , 2 samples were taken at each place, there were a total of 24 samples. After sampling, all samples were grouped, as shown in table 1.

Table 1. Grouping of samples

Sample grouping		
Without nano SiC	Add Nano-SiC in the process of spheroidization	Add Nano-SiC in the process of inoculant
一,φ100 1/4	二,φ100 1/4	三,φ100 1/4
一,φ100 core	二,φ100 core	三,φ100 core
一,φ150 1/4	二,φ150 1/4	三,φ150 1/4
一,φ150 core	二,φ150 core	三,φ150 core

2.3. Heat Treatment Process

The heat treatment process in this study was treated by austempering. First, the samples was heated to 890 ~910℃ and kept for 2 hours, let primary matrix structure be completely austenitizing, and then cooled in oil of 80 ~90℃ to achieve quenching. Placed the oil quenched castings in a furnace of 240 degrees and kept warm for 3 hours, finally, taken out the samples and cooled it to room temperature. The grouping of samples for heat treatment as shown in table 2.

Table 2. Grouping of samples for heat treatment

Number	Sample	Number	Sample （heat）
1	一,φ100 1/4	13	一,φ100 1/4 heat
2	一,φ100 core	14	一,φ100 core heat
3	一,φ150 1/4	15	一,φ150 1/4 heat
4	一,φ150 core	16	一,φ150 core heat
5	二,φ100 1/4	17	二,φ100 1/4 heat
6	二,φ100 core	18	二,φ100 core heat
7	二,φ150 1/4	19	二,φ150 1/4 heat
8	二,φ150 core	20	二,φ150 core heat
9	三,φ100 1/4	21	三,φ100 1/4 heat
10	三,φ100 core	22	三,φ100 core heat
11	三,φ150 1/4	23	三,φ150 1/4 heat
12	三,φ150 core	24	三,φ150 core heat

3. Test data analysis and discussion of nodular cast iron

3.1. Results and Analysis of Impact Test

The data of impact energy measurement are shown in table 3 and figure 2.

Table 3. Impact work of sample

Number	Type	Impact energy	Number	Type	Impact energy
1	一,φ100 1/4	10.8	13	一,φ100 1/4 heat	19.2
2	一,φ100 core	4.8	14	一,φ100 core heat	8.3
3	一,φ150 1/4	3.2	15	一,φ150 1/4 heat	6.2
4	一,φ150 core	2.1	16	一,φ150 core heat	2.9
5	二,φ100 1/4	11.6	17	二,φ100 1/4 heat	20.5
6	二,φ100 core	13	18	二,φ100 core heat	22.7
7	二,φ150 1/4	3.4	19	二,φ150 1/4 heat	7.9
8	二,φ150 core	4.8	20	二,φ150 core heat	13.8
9	三,φ100 1/4	4.1	21	三,φ100 1/4 heat	7.8
10	三,φ100 core	3.7	22	三,φ100 core heat	6.6
11	三,φ150 1/4	2.9	23	三,φ150 1/4 heat	5.1
12	三,φ150 core	3.5	24	三,φ150 core heat	11.3

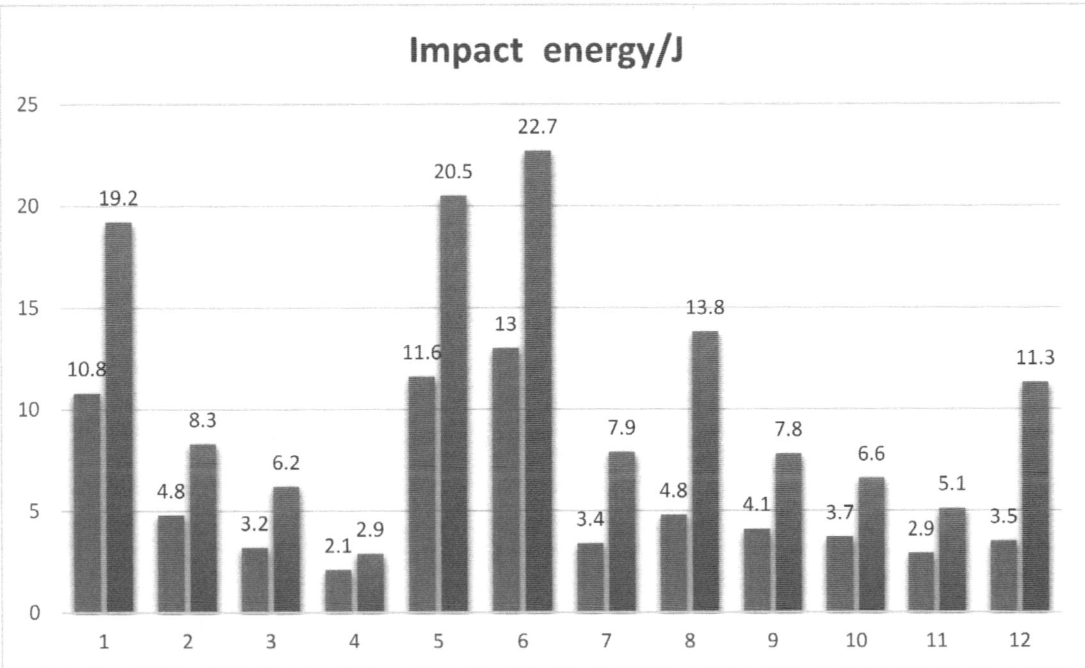

Figure 2. Impact work of sample

Blue represents the samples before heat treatment and red represents the samples after heat treatment. The following conclusions can be drawn from the chart:

(1) The impact toughness of nodular cast iron has been improved obviously after heat treatment, and the impact work of all samples is generally improved by 3~9J.

(2) The samples which without Nano-SiC and added Nano-SiC during inoculation process have higher impact toughness,but the samples that added Nano-SiC during spheroidizing process have the worst impact toughness.

(3) The impact toughness of casting is the most remarkable while adding Nano-SiC in inoculation process.And the impact toughness of the core has been improved, which is higher than the external impact toughness.

3.2. Hardness Test Results and Analysis

Measurements of microhardness are shown in table 4 and figure 3.

Table 4. Measurement data of microhardness

Number	Type	Hardness (HV)	Number	Type	Hardness (HV)
1	一,φ100 1/4	340.58	13	一,φ100 1/4 heat	256.96
2	一,φ100 core	313.58	14	一,φ100 core heat	255.06
3	一,φ150 1/4	324.56	15	一,φ150 1/4 heat	268.32
4	一,φ150 core	278.68	16	一,φ150 core heat	248.76
5	二,φ100 1/4	319.44	17	二,φ100 1/4 heat	237.72
6	二,φ100 core	348.38	18	二,φ100 core heat	238.78
7	二,φ150 1/4	359.72	19	二,φ150 1/4 heat	255.92

IOP Conf. Series: Materials Science and Engineering **250** (2017) 012051 doi:10.1088/1757-899X/250/1/012051

8	二,φ150	core	313.18	20	二,φ150 core heat	231.24
9	三,φ100	1/4	302.9	21	三,φ100 1/4 heat	249.9
10	三,φ100	core	428.42	22	三,φ100 core heat	259.14
11	三,φ150	1/4	489.16	23	三,φ150 1/4 heat	280.58
12	三,φ150	core	391.44	24	三,φ150 core heat	262.54

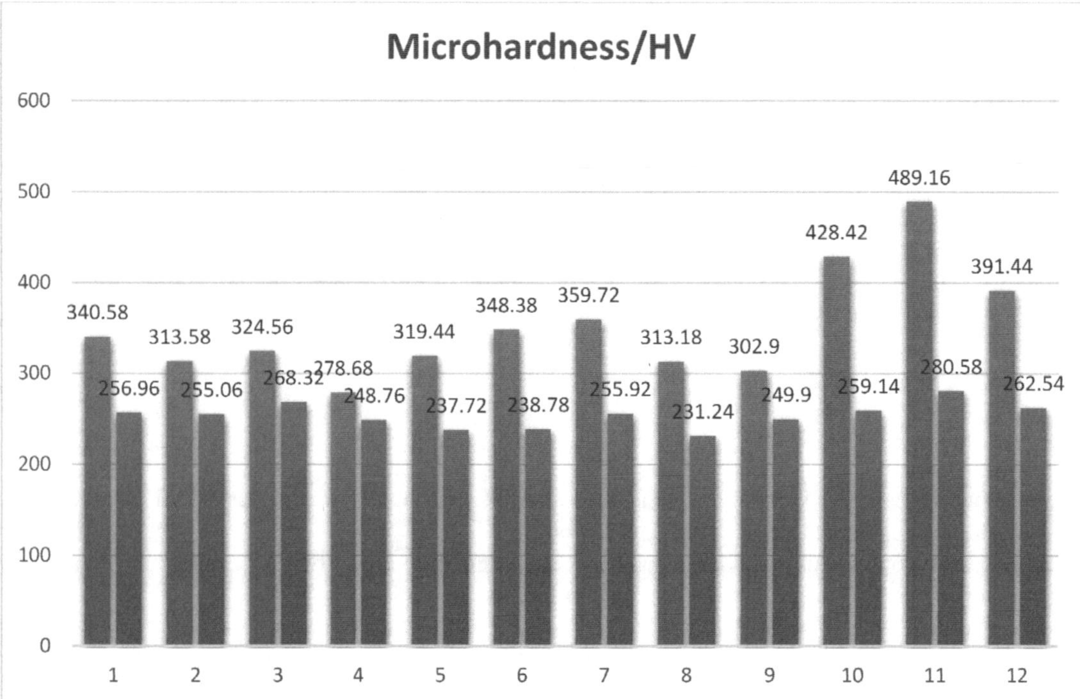

Figure 3. Histogram of measurement data of microhardness

Blue represents the samples before heat treatment and red represents the samples after heat treatment. The following conclusions can be drawn from the chart:

(1) From the experimental data, it can be seen that after low temperature tempering the hardness of the sample compared to the cast iron samples without heat treatment had a generally declined . The hardness of the cast iron samples which added with Nano-SiC in the inoculation and spheroidizing process, decreased more obviously.

(2) Before the heat treatment, the hardness of the cast iron samples which added with Nano-SiC in the inoculation and spheroidizing process compared with the samples without Nano-SiC, the hardness increased significantly.

4 . Metallographic Observation and Analysis

4.1. As Cast Microstructure Analysis

4.1.1 Graphite Morphology. Figures 4 and 5 show the graphite morphology of samples 1# and 5#. By observing , it is known that the size of the graphite ball in the 1# sample is larger and the distribution is rather disordered,the graphite spheres in the 5# sample are smaller and more evenly distributed. In contrast to "GBT 9441-2009 ductile iron metallographic examination", the size of graphite is above grade 5, and the spheroidization level is above 3, which meets the requirements of the experiment.When the nano SiC is added, the graphite spheres in the sample are more rounded, and the splitting effect on the matrix is small under the impact force, which is beneficial to the improvement of the impact toughness of the casting.

Figure 4. 1# graphite morphology

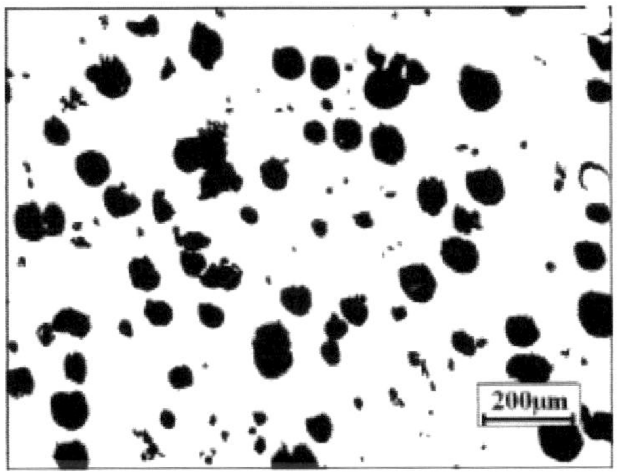

Figure 5. 5# graphite morphology

*4.1.2 Matrix Structure.*After grinding, polishing, etching, rinsing and drying the above 2 kinds of cast samples, the microstructures observed under the microscope are shown in figures 6 and 7.

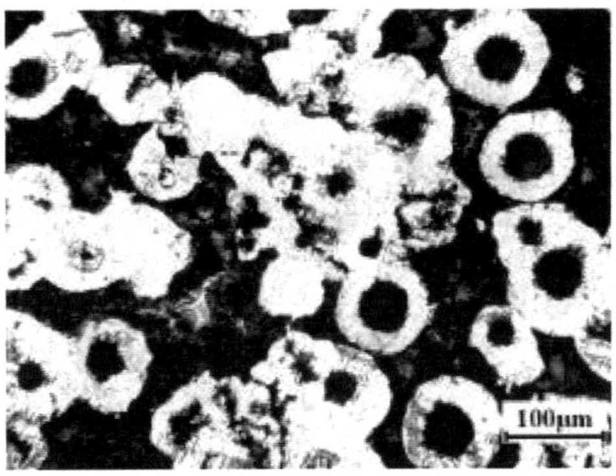

Figure 6. 1# microstructure of as cast

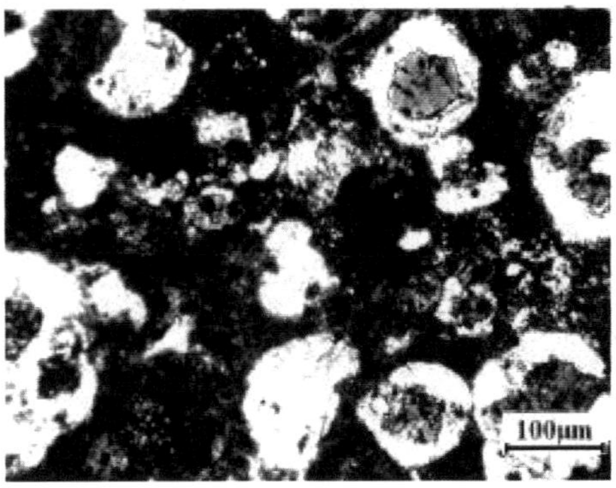

Figure 7. 5# microstructure of as cast

Because the spheroidizing agent can form graphite spheres, it leads to poor carbon content around the graphite and promotes the formation of ferrite. As shown in Figure 3.3 and 3.4, the round black is the graphite ball, and the bright area around is ferrite, in which the large black ones are mainly pearlite , a small amount of free cementite and phosphorus eutectic. According to the metallographic examination standard GB/T9441-2009 of spheroidal graphite, the metallographic analysis of two kinds of samples was carried out. After adding Nano-SiC, the content of ferrite in the as cast structure was increased, and the pearlite decreased. This is mainly because the Nano-SiC can promote the formation of ferrite and reduce the pearlite. At the same time, the content of carbide in the as cast microstructure decreases correspondingly, which is beneficial to the improvement of the impact toughness of the casting.

4.2. Observation and Analysis of Microstructure in Heat Treated State

In order to study the influence of Nano-SiC on the microstructure of austempered ductile iron, the microstructure of 13# and 17# samples was observed, as shown in figures 8 and 9.

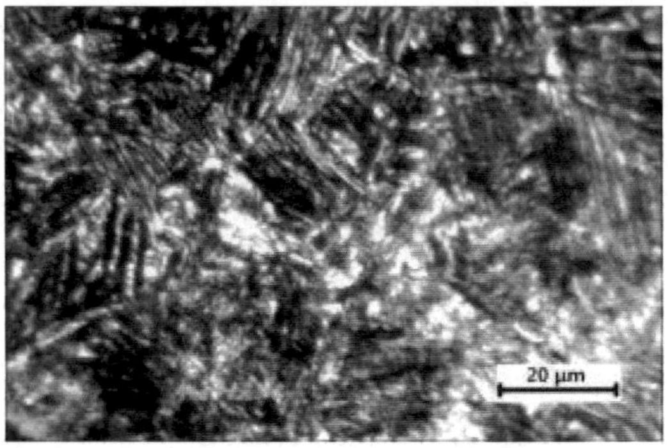

Figure 8. 13# micrograph

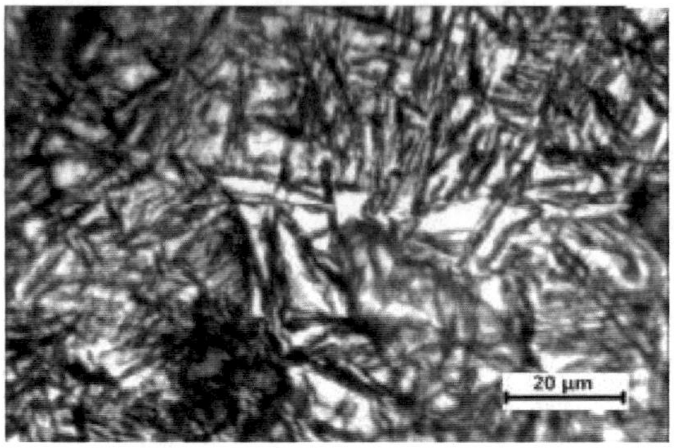

Figure 9. 17# micrograph

Figure 8 shows the microscopic morphology of the 13#, it can be finded some feathery upper bainite in the metallographic structure, the distribution of ferrite sheets in the upper bainite is relatively fine.The white bright area of retained austenite is relatively concentrated in the sample.The impact absorption work of the sample is low,this is due to the fact that cementite is distributed between the ferrite sheets in the upper bainite, which is unfavorable to the impact toughness.

Figure 9 shows the microscopic morphology of the 17# sample ,the resulting lower bainite is acicular,this is because the Nano-SiC can improve the hardenability of ductile iron,the lower bainite can be formed by adding Nano-SiC to the same heat treatment process, these acicular lower bainite are uniformly distributed in the matrix,the carbon in the lower bainite is supersaturated, and have higher dislocation density,so the dispersion of cementite in the matrix is evenly distributed, and the impact toughness is better.

When Nano-SiC was added to the inoculation process, the hardness of the samples is decreased more obviously, Because martensite is produced at 240 degrees of temperature, most austenite in the matrix is metastable austenite, and unstable austenite exists locally in the matrix. This part of austenite will change into martensite in austempering stage. At higher isothermal quenching temperature, carbon atoms can obtain more energy, the diffusion coefficient increases, and the diffusion to austenite is relatively more. In the process of martensite transformation, the total carbon content is retained, and the carbon content in the martensite is gradually decreased as the austempering temperature continues to decrease.The solid solution of carbon in martensite directly affects the hardness of martensite, eventually, make the hardness of ductile iron decreased.

5. Conclusion

1. The hardness of nodular cast iron generally decreases after heat treatment; The hardness of the cast iron samples which added Nano-SiC decreased more obviously after heat treatment; Before the heat treatment, the hardness of the cast iron samples which added Nano-SiC was significantly higher than samples that without adding Nano-SiC.

2. The impact toughness of nodular cast iron samples is obviously improved after heat treatment; When Nano-SiC was added to the spheroidizing process, the impact toughness of the samples was the worst; Samples without Nano-SiC had higher impact toughness.While adding nano SiC in the process of inoculation, the samples has the best impact toughness;

3. Nano-SiC can promote the formation of ferrite, reduce pearlite and the content of carbide in the cast structure which is beneficial to improve the impact toughness of casting.

6. References

[1] Zhao Qi, Wang Jinsheng, Li Zhaolei, et al. Study on the technology of [J]. *thermal processing of preparation and properties of low temperature and high toughness ductile iron*, 2013 (21): 43-45.

[2] Kong Lingfeng, Li Zhen Lian. Process for low temperature ductile iron production in the discussion on problems of [J]. *Heilongjiang science and technology information*, 2015 (26).

[3] Fan K L, He G Q, She M,et al. Corrigendum to "High-temperature low cycle fatigue behavior of a gray cast iron" [Mater.Charact.(2014)37–46][J].*Materials Characterization*, 2015, 99:287-287.
[4] Chen Meiling, Li Jianwei, Gao Hong. *Study on Microstructure and mechanical properties of modified nano SiC Powder Reinforced Nodular Cast Iron* [J]. materials bulletin, 2006, 20 (S1): 214-215.
[5] Wu Qiang, Guo Jingjie, Hou Li, et al. *Study on inoculation of large ductile iron castings with nano SiC core wires* [J]. Journal of Harbin Engineering University, 2016, 37 (9): 1302-1306.
[6] Yu Guangyuan, Xu Guibao, Si and C. *Impact on low temperature impact toughness of as cast Ni free cryogenic ductile iron* [J]. casting equipment and process, 2013 (1): 25-27.
[7] Xu Huijun. Analysis of influencing factors on impact toughness of spheroidal graphite cast iron [J]. *China new technology, new products,* 2009 (23): 3-3.
[8] Yang Qiyu, Chen Jun, Mao Jie. *Influence of heat treatment on impact toughness of spheroidal graphite cast iron* [J]. casting, 2008, 57 (4): 375-377.

IOP Conf. Series: Materials Science and Engineering **250** (2017) 012052 doi:10.1088/1757-899X/250/1/012052

Impact of Functional Groups in MWCNT on Surface Hydrophilicity, Mechanical and Thermal Properties of Polystyrene/CNT Composites

Yanxia Zhang[1,a] , Le Shi[1], Guangfen Li[1,b*] and Shaolei Liang[1,c]

[1]State Key Laboratory of Separation Membranes and Membrane Processes, School of Materials Science and Engineering, Tianjin Polytechnic University, Tianjin 300387, P. R. China

[a]272259193@qq.com, [b]liguangfen@tjpu.edu.cn, [c]770539199@qq.com

Abstract. We prepared nanocomposites of MWCNTs/polystyrene via a solution casting method, and DI water was used as coagulation bath. The influence of pristine and two functionalized MWCNTs on surface morphology, thermal properties and mechanical properties of MWCNT/polystyrene composites were extensively discussed. Analysis of SEM shows that the porous structures with narrow size distribution from micro- to nano-size appear in all composites, which decrease the tensile strength of the composites. Despite the glass transition temperatures (Tg) of all the composites lower compared with Tg of pure polystyrene, which increase with the loading of MWCNTs. The mechanical properties of the composite with 5 wt.% MWCNTs are enhanced greatly, and the phenomenon is especially pronounced for the addition of the MWCNT-OH.

1. Introduction

Polymer composite materials have been applied in different industrial fields due to their light weight, flexibility in process, high durability and strength[1,2]. The selection of nanofillers can significantly improve the strength, toughness, durability of the resulted composites[3]. Comparing with the conventional fillers as carbon nanofibers and spherical particles, carbon nanotubes (CNTs) are regarded as an ideal filler in preparing polymer/inorganic composites due to their outstanding electrical, thermal[4], and mechanical properties[5]. Therefore, polymers/CNTs with diverse functional and structural properties have found many applications[1,5], one of which is to produce CNTs-reinforced composite materials[6]. Nowadays, polymer materials as polystyrene[4], polymethylmethacrylate[6], and polyethylene[7] were commonly chosen as polymer matrices for producing polymers/CNTs composites. Several methods as melt mixing[8], solution casting technique[9], and in situ polymerization[10] were generally adopted for the synthesis of polymers/CNTs composites. However, the applications of these composites in different fields have been limited by poor interfacial interaction between polymer and CNTs, having the strong tendency to self-aggregate of CNTs. Two approaches as mixing process and surface treatment of pristine CNTs are generally used either for disentangling CNTs by mechanical force or for altering the surface functionality with -OH or -COOH. Surfactant such as sodium dodecyl sulfonate, sodium dodecylbenzene sulfonate, and Triton-X has been successfully applied to improve the dispersibility of CNTs in polymer matrix. The content of surfactant was also found to be a critical effect in dominating the surface morphologies caused by phase separation. Up to now, the combination effects of surfactant and pore structures on the composite properties have rarely been mentioned.

In this study, pristine and two functionalized MWCNTs on the various properties of PS based composites were investigated. The influences of type and content of MWCNTs, and presence of surfactant on the performance of composites were analyzed. The chemical component, thermal properties, as well as surface hydrophilicity of composite materials are systematically studied.

2. Experimental

0.2g polystyrene (Goodfellow, U.K.), a certain amount of pristine MWCNTs (p-MWCNT, length 10-30 μm), or hydroxylated MWCNTs (MWCNT-OH, hydroxyl content 5.58 wt.%, length 10-30 μm), or carboxylated MWCNTs (MWCNT-COOH, carboxyl content 0.49 wt.%), obtained from Beijing DK Nanotechnology Co. Ltd, 0.1 ml surfactant (Triton X-100) and 10 ml Tetrahydrofuran (THF) were mixed together and stirred until PS completely dissolved. Then the mixture was ultrasonicated for 2 h and then cast into a specific container. After 20 min, the gasket was immersed into DI water to obtain a stable membrane. The morphologies of composite surface and cross-section were studied through FESEM (Hitachis-4800, Hitachi Limited, Japan). To evaluate surface hydrophilicity of the composites, instrument of dynamic contact angle (DSA-100, Kruss, Germany) was used. Effect of the loading of carbon nanotubes on the Tg was analyzed by DSC (Perkin-Elmer, U.S.). The mechanical properties of the nanocomposite were characterized by single fiber electronic tensile strength tester (LLV-06 ED, Laizhou, China).

3. Results and Discussion

3.1 The Characterization of the Composites by DSC

Differential scanning calorimetry was performed to distinguish the effect of adding CNTs and surfactant to PS matrix on the Tg. Figure1.a depicts the values of Tg obtained for the surfactant-treated CNTs/PS composites with different loadings of p-MWCNT, MWCNT-OH, and MWCNT-COOH. The Tg of pure PS and the composites with p-MWCNT loading from 1, 5 to 12.5 wt.% were 106, 81, 95, and 97 °C, respectively. The increase of Tg with p-MWCNT loading suggests that the strong interfacial affinity between PS and p-MWCNT hinders the mobility of polymer chain segment resulting in the increase of Tg. Similar results were also found in the composites with MWCNT-OH or MWCNT-COOH. The Tg of the composites with MWCNT-OH was 66, 75, and 78 °C, whereas Tg of the composites with MWCNT-COOH different loading was 60, 66, and 72 °C. For the composites with functionalized MWCNTs, Tg values decline more than the composites with pristine MWCNTs. The highest declination of Tg was observed for the composites with MWCNT-COOH, which can be attributed to two possible effects: the length of CNTs and the type of the functional groups. As we known, the composites contained MWCNT-COOH also trend to lower thermal stability.

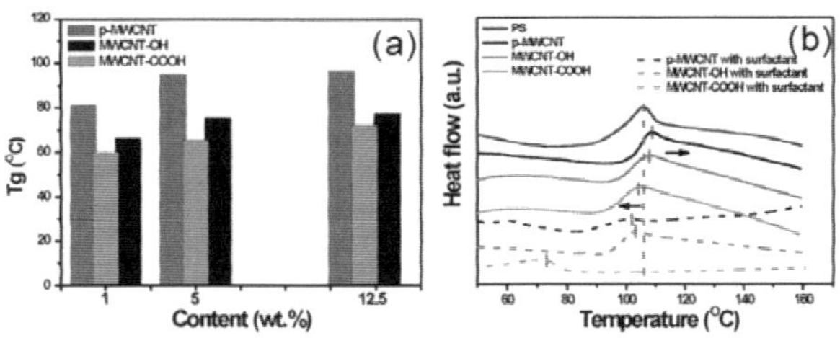

Figure1. DSC analysis of PS composites with different MWCNTs. (a) surfactant-treated composites with CNT contents from 1, 5 to 12.5 wt.%; (b) a comparison of the composites with or without surfactant.

To distinguish the effect of adding surfactant to PS matrix on the Tg. Figure 1.b displays the Tg values obtained from PS/CNTs composites with or without the surfactant. The loading of p-MWCNT, MWCNT-OH, and MWCNT-COOH was all set as 5 wt.%. Comparing with Tg for pure PS, the non surfactant contained composites with p-MWCNT and MWCNT-OH exhibit an increase in the Tg. The shift of Tg values to higher values may attribute to the interaction between polystyrene molecules and CNTs, which promotes the thermal stability. Whereas, the composite with MWCNT-COOH show a reduction in Tg, which arises from the degradation behavior of MWCNT-COOH. For the MWCNT composites with the surfactant, a decrease of Tg values was observed for all composites. As the composites contain surfactants for obtaining a better dispersion of CNTs in solution, the surfactant molecules attached on the side walls of nanotubes acts as a plasticizer within the PS matrix and reduces Tg of the composites.

3.2 Mechanical Performance of Composite Materials

Here, stress-strain curves for PS composites containing different types and contents of CNTs were tested to evaluate the mechanical properties. Comparing pure PS with composites containing 5 wt.% of p-MWCNT, MWCNT-OH, and MWCNT-COOH, the values of tensile strength are enhanced from 6.3 to 6.4MPa, 6.3 to 10.7MPa, 6.3 to 9.7MPa, respectively (Figure 2.a). Whereas the values of elongation at break are increased from 2.2 to 3.3%, 2.2 to 5.3%, and 2.2 to 4.0% (Figure 2.b). The highest tensile strength is especially found for the composite with 5wt.% loading of MWCNT-OH. The adding of p-MWCNT from 1 to 10 wt.% can not effectively improve the values of tensile strength due to CNT agglomerates. However, the fluctuation in the values of elongation at break may arise from such effects as: the addition of surfactants, the present of functional groups and the formation of agglomerates of CNTs. The present of surfactants and functional groups can improve the binding forces between CNTs and PS matrix and allows the ambient load being transferred to nanotubes when the composite is under external force. Therefore the elongation at break rises up with the adding of CNTs, which implies that CNTs are well dispersed in PS matrix. In contrary, when the loading of nanotubes exceeds a certain value, the larger agglomerates penetrate polystyrene and lower the interactions between polystyrene and CNTs, resulting in the elongation at break to decline. Our studies show that the composites with 5 wt.% CNTs tend to possess desirable mechanical properties.

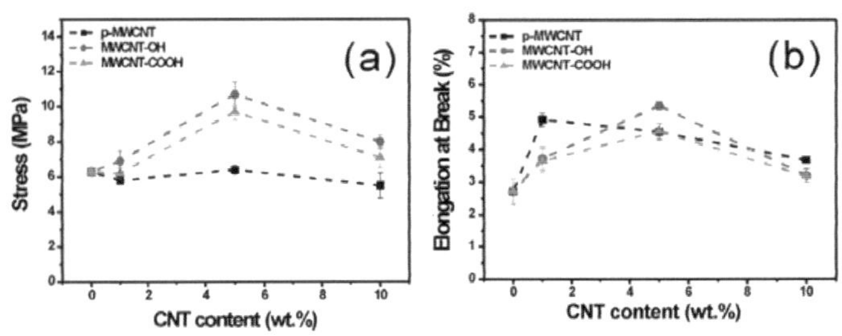

Figure2. Mechanical properties (a) and elongation at break (b) of composites with different CNTs.

3.3 Effect of Type and Content of MWCNTs on Composite Morphologies

The cross-section and surface morphologies of different surfactant-treated PS/MWCNT composites were depicted in Figure 3. The loading of p-MWCNT, MWCNT-OH, and MWCNT-COOH was varied from 1 to 10 wt.%. Microporous structures with micro-size from 6 μm to 1μm appear in the cross-section of all CNTs/PS composites. The continuous fine structures can be assigned to a morphology stabilization effect, which seems to act bridging two different material phase. Homogeneously distributed of MWCNTs can be viewed from all composite surfaces (inset of Figure 3). The white spots in the inset of Figure 3.d-f indicate the presence of PS crystals. The phenomenon of agglomerates formation can be explained by crystal nucleation during the growth of PS crystals and the clusters of CNTs acts as nucleation agent and accelerates the growth of agglomerate. Such effect is consistent with previous report [7]. And the agglomerates of the longer CNTs easily occur even for the composites with much lower CNTs loading. It is envisaged that the polymer can better penetrate in the MWCNT agglomerates, especially in the MWCNT–OH. These results also explain why the composites with 5 wt.% possess better mechanical properties, especially for the composite without porous structures (Figure 3.e).

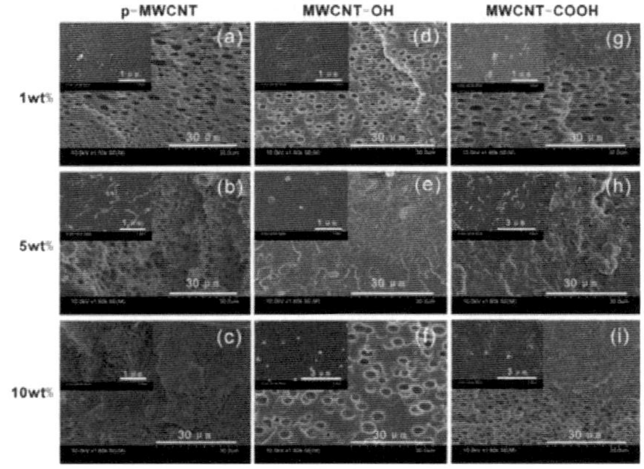

Figure3. SEM images of cross-section and surface (insets) morphologies of PS composites with different type of MWCNTs and CNT loading from 1, 5, to 10wt.%.

3.4 Surface Properties of the Nanocomposites

A better understanding of the effect of surface morphology on surface hydrophilicity was carried out by contact angle measurement to evaluate surface properties of PS composite membranes that containing different loading of p-MWCNT, MWCNT-OH, and MWCNT-COOH, as shown in Figure 4.a When the loading of CNTs is 1 wt.%, the contact angles of the composite membranes varies from 98° to 108°, which indicates that all the composite surfaces are hydrophobic regardless of the types of CNTs. With the increase of CNTs loading, the differences in contact angles of all composites diminish and almost are around 103 ± 5°. As the loading of CNTs increases to 10 wt.%, the contact angles declines especially for the composites contained MWCNT-OH and MWCNT–COOH. Likely, the amount of hydrophilic functional group as –OH or –COOH is main factor for dominating the surface hydrophilicity. The fact that the water contact angles of the composites with functionalized MWCNT decreasing with the loading of MWCNT further demonstrates that the surface hydrophilicity of the composites are correlated to the amount of functional groups in MWCNT. The effect of surfactant on the surface hydrophilicity of composite membranes was studied by plotting the contact angles of MWCNT/PS nanocomposites as a function of different MWCNT, as shown in Figure 4.b. The surfactant-treated MWCNT nanocomposite surfaces remained similar hydrophobicity as pure PS surface. The non surfactant-treated MWCNT nanocomposites exhibited relatively higher hydrophilicity than those surfactant-treated MWCNT composites. We deduce that the surface homogeneity can be one of the major effects in dominating surface hydrophilicity of their composites.

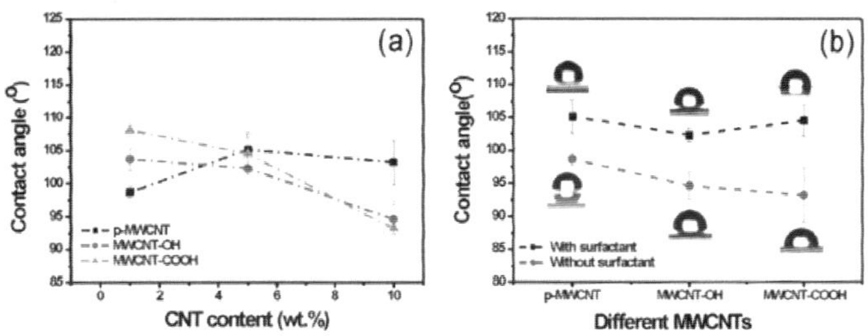

Figure4. Water contact angles of the PS composite membranes changed with fraction of three different CNTs (a) and addition of surfactant (b). The red, black and green lines represent the PS composites with p-MWCNT, MWCNT-OH, and MWCNT-COOH, respectively.

4. Conclusion

In this study, the composites were fabricated via a simple solution casting process using commercially available materials. The thermal properties, mechanical properties and the surface morphology of PS nanocomposites have been investigated with relation to several factors as the CNTs content, different functional groups, the surfactant and so on. The CNTs content was kept mainly below 12.5 wt% for obtaining the correlation between physical properties and surface morphology. It was found that the tensile strength of PS nanocomposites was promoted with increasing of CNTs content as the nanotubes loading was below 5 wt.%. The presence of surfactant and higher CNTs loading favors porous structure formation, especially for functionalized MWCNT composites. In addition, a minimum of Tg appears in MWCNT-COOH composites, indicating a prominent effect of instability of MWCNT-COOH on the thermal properties of MWCNT-COOH composites.

5. Acknowledgement

The research was supported by the Science and Technology Plans of Tianjin (No.15YFXQGX00070)

6. References

[1] Lalwani G, Henslee AM, Farshid B, Lin L, Kasper FK, Qin Y-X, et al. Two-dimensional nanostructure-reinforced biodegradable polymeric nanocomposites for bone tissue engineering. Biomacromolecules. 2013; 14: 900-9.

[2] Wu B, Li X, An D, Zhao S, Wang Y. Electro-casting aligned MWCNTs/polystyrene composite membranes for enhanced gas separation performance. Journal of Membrane Science. 2014; 462: 62-8.

[3] Hu Z, Chen G. Novel nanocomposite hydrogels consisting of layered double hydroxide with ultrahigh tensibility and hierarchical porous structure at low inorganic content. Advanced Materials. 2014; 26: 5950-6.

[4] Sun G, Chen G, Liu Z, Chen M. Preparation, crystallization, electrical conductivity and thermal stability of syndiotactic polystyrene/carbon nanotube composites. Carbon. 2010; 48: 1434 -40.

[5] Chang T-E, Kisliuk A, Rhodes SM, Brittain WJ, Sokolov AP. Conductivity and mechanical properties of well-dispersed single-wall carbon nanotube/polystyrene composite. Polymer. 2006; 47: 7740-6.

[6] Andrews R, Weisenberger MC. Carbon nanotube polymer composites. Current Opinion in SolidState and Materials Science. 2004; 8: 31-7.

[7] Potschke P, Bhattacharyya AR, Janke A. Carbon nanotube-filled polycarbonate composites produced by melt mixing and their use in blends with polyethylene. Carbon. 2004; 42: 965-9.

[8] Liebscher M, Gärtner T, Tzounis L, Micušíkc M, Pötschke P, Stamm M, et al. Influence of the MWCNT surface functionalization on the thermoelectric properties of melt-mixed polycarbonate composites. Composites Science and Technology 2014; 101 133-8.

[9] Mazov IN, Kuznetsov VL, Krasnikov DV, Rudina NA, Romanenko AI, Anikeeva OB, et al. Structure and properties of multiwall carbon nanotubes/polystyrene composites prepared via coagulation precipitation technique. Journal of Nanotechnology. 2011; 2011: 1-7.

[10] Patole AS, Patole SP, Yoo J-B, An J-H, Kim T-H. Fabrication of polystyrene/multiwalledcarbonnanotube composite films synthesized by in situ microemulsion polymerization. PolymerEngineering and Science. 2013: 1327-36.

IOP Conf. Series: Materials Science and Engineering **250** (2017) 012053 doi:10.1088/1757-899X/250/1/012053

Sensing Performances of Gaussian-shaped and Double-Side Unsymmetrical Metallic Nano-Gratings

Haiying Li and Xuesong Duan

School of sciences, Hebei University of technology, Tianjin, China
Email:lihaiying123@139.com

Abstract. A new type of sensing structure which is composed of a Gaussian-shaped and double-side unsymmetrical metallic nano-grating is presented.The reflectance spectra as functions of incident wavelengths in various refrective index conditions are analyzed by means of multiple multipole program (MMP) method. Our numerical simulation results show that the refractive index sensitivity（RIS） ~680 nm/RIU and the full-width at half-maximum (FWHM) ~3.5 nm can be obtained. The figure of merit (FOM) is over 190 RIU^{-1}. The resonant wavelength increases almost linearly with increasing of the refractive index of samples. These reflection properties not only give an insight into the physical mechanisms of the double-side unsymmetrical gratings, but also make this kind of nano-gratings more suitable to be used in the surface plasmon resonance (SPR) sensors.

1. Introduction

Free electron charges on the boundary of a metallic nanostructure can perform coherent fluctuations which are called surface plasmon resonance (SPR).For a certain frequency range of the exciting field, a complex resonant behavior can occur with strongly enhanced near-field and strong absorption of the electromagnetic waves [1-2]. SPR is highly sensitive to refractive index variation of the surrounding materials. These properties of SPR make the metallic nanostructures very popular for the application of sensing [3].

An approach with metallic nano-gratings incorporated into the SPR sensors has been proposed in recent years[4]. Use of metallic nano-gratings can enhance the refractive index sensitivity (RIS) by spectrum shifting of the peak resonant wavelength. Fortunately, focused ion beam milling (FIBM) technique provides a practical approach for fabrication of the nano-gratings [5-7]. The FIBM can realize one step direct fabrication of the nanostructures with finely controlled accuracy. However, a Gaussian profile-like shape instead of vertical sidewall of the structures in cross-section will be formed after the FIBM fabrication due to a Gaussian distribution of ion energy of the focused ion beam [8-9]. Sensing performances of metallic nano-gratings with single-side and double-side symmetrical Gaussian-shaped had been researched and reported by our group[10-11]. For obtaining higher RIS and larger FOM of the sensors, we put forth the investigation of sensing performances of the Gaussian-shaped and double-side unsymmetrical metallic nano-gratings in this paper. To our knowledge, it is first time to report the sensing performances of this type of gratings. Reflectance spectra of the sensing structure in various refractive index of the sample are calculated by multiple multipole program (MMP) method[12].

IOP Conf. Series: Materials Science and Engineering **250** (2017) 012053 doi:10.1088/1757-899X/250/1/012053

2. Modeling and Simulation Setup

There are three factors to determine the overall performance of the nano-gratings being used as an optical sensing chip. The first one is refractive index sensitivity(RIS), which is slope of the resonant wavelength over the refractive index. The second one is the full-width at half-maximum (FWHM) of the reflection curve, and The figure of merit (FOM), as the last one, is defined in reference[13] as follows:

$$\mathrm{FOM} = \frac{\mathrm{RIS(nm/RIU)}}{\mathrm{FWHM(nm)}} \tag{1}$$

The overall performance of the optical sensors is affected by the FWHM. A wide FWHM makes it difficult to distinguish minor changes in the peak value. The depth of the resonance reflectance spectra can also affect sensitivity of the sensors.

A schematic diagram of the proposed metallic nano-grating is shown in Figure1. The model adopted in our work is a thin gold film with the unsymmetrical Gaussian-shaped grooves on both surfaces. The structure in a period is described mathematically as

$$z(x) = -H_2\exp[-\left(\frac{x-x_1p}{w_2}\right)^2] - H_1\exp[-\left(\frac{x-x_2p}{w_1}\right)^2], \text{ top} \tag{2}$$

$$z(x) = -(H_1+t) + H_1\exp\left[-\left(\frac{x-x_1p}{w_1}\right)^2\right] + H_2\exp[-\left(\frac{x-x_2p}{w_2}\right)^2], \text{ bottom} \tag{3}$$

Here $\boldsymbol{H_1}$ and $\boldsymbol{H_2}$ represent the deeper groove depth and the shallower groove depth. $\boldsymbol{W_1}$ and$\boldsymbol{W_2}$ denote the deeper groove width and the shallower groove width respectively. $\boldsymbol{P}$ is the period of the metallic nano-gratings. The incomplete perforated film thickness t>0. The sample is on top surface of the gold MNG with supported glass substrate.

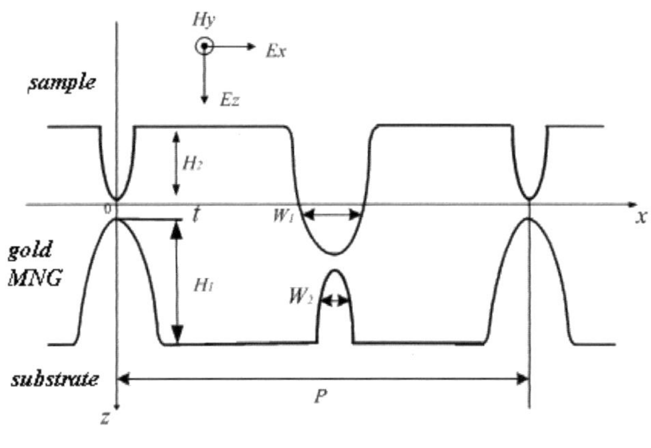

Figure1. Schematic diagram of MNG with Gaussian-shaped and double-side unsymmetrical grooves.

The reflection spectra were investigated using the MMP method[14]. The corresponding computational software is known as Max-1,which has first been proposed and developed in 1980 by Christian Hafner. MMP is a semi-analytic method for numerical field computations that has been applied to electromagnetic fields. Because of its close relations to analytic solutions, MMP is very useful and efficient when accurate and reliable solutions are desired.

3. Computational Results and Analyses

The optical constants of gold were taken from the experimental measurement [15]. Sample materials on the thin film have a refractive index ranging from 1.0 to 1.25. The substrate is SF10 glass with 1.72 refractive index. For TE polarization, surface oscillations do not exist because the electric field vector $\vec{E}$ is parallel to the interface at which E is continuous across the boundary, and thus no surface charge is induced [16]. In our computational numerical simulation, a TM-polarized with a normal incidence of light from the sample side is assumed, and the bandwidth of the incident light is set from 600 nm to 900 nm.

Influences of grating period ***P*** on reflectance and transmission spectra are investigated, which are shown in figure2(a) and figure2(b) respectively. Here the surface profile is defined by t=15nm,H_1=40nm, H_2=20nm, W_1=W_2=50nm, as ***P*** ranges from 600 nm to 800 nm.The sample is assumed to be air and its refractive index is 1.006. As can be seen from figure 2(b)， when P=600 nm, there is only one main peak at wavelength $\lambda = 638nm$. With the increasing of grating period the transmission peak becomes weaker, and its position moves to longer wavelength, which are 638nm， 729nm and 823nm respectively. From the dispersion relation of SPR, the resonant wavelength depends strongly on the structure period and the incident transverse wave vector. This can be expressed as equation (4):

$$k_{sp} = k_0 \sin\theta_0 + n \cdot \frac{2\pi}{p} \approx \left(\frac{\varepsilon_m \cdot \varepsilon_d}{\varepsilon_m + \varepsilon_d}\right)^{1/2} \cdot k_0 \quad (4)$$

Where $\boldsymbol{k_{sp}}$ and $\boldsymbol{k_0}$ represent the wave vector of surface plasmons and wave vector in free space respectively, $\boldsymbol{\theta_0}$ is the incident angle, ***P*** is grating period. The SPR wavelengthes can be calculated from equation (4), which are 626nm， 720nm and 816nmcorresponding to period of 600, 700, 800nm respectively. With the comparison of the resonant positions, the narrow transmission peaks in Fig.2(b) can be identified as SPR wavelength, and the deviations may result from the incomplete perforated film thickness ***t***. As expected, Fig. 2 shows that a peak in a transmission curve always corresponds to a minimum in the reflection curve at the same wavelength. SPR can form very deep and narrow reflection bands when grating period is 700nm. We find that the reflection curve can be deeper with a change of the groove width $\boldsymbol{W_1}$. In our following calculations, the optimized structure parameter is defined by P=700nm， t=15nm, H_1=40nm, H_2=20nm, W_1=90nm,W_2=50nm.

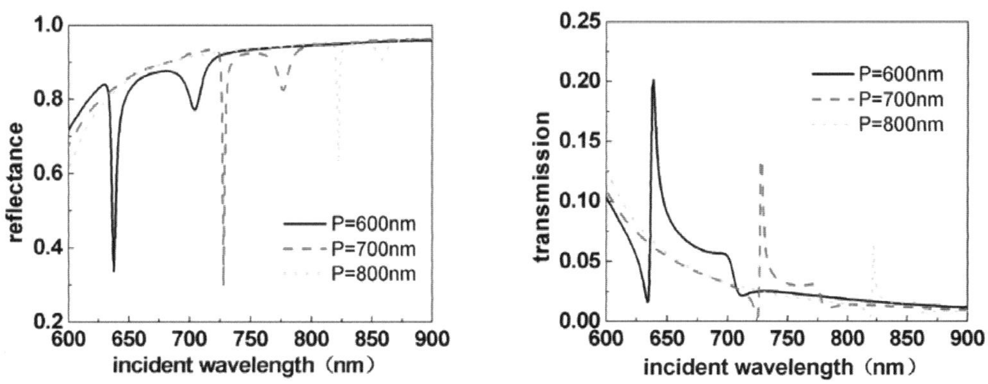

(a) (b)

Figure2. Reflectance(a) and transmission(b) spectra as a function of incident wavelength for gold MNG with different periods as marked(t=15nm, H_1=40nm, H_2=20nm, W_1=W_2=50nm).

The relationship between the reflection spectra and incident wavelength with 0.05 interval steps of refractive index of the sample was calculated, which can be seen infigure 3. With the increase of the sample refractive index , the resonant wavelength removes to longer wavelength. RIS of the structure can be obtained by the shift factor of the resonant wavelength. In order to compare the theoretical RIS and the calculated RIS, the relationship between resonant wavelength and sample refractive index is extracted from the data of figure 3, as shown in figure 4(a). The resonant wavelengths increase almost linearly with increasing of the sample refractive index. The theoretical results are obtained by equation(4). When the refractive index of sample increases from 1.0 to 1.25, the corresponding calculated resonant wavelength increases from 725.8nm to 896.8nm, and the theoretical resonant wavelength increases from 719.9nm to 896.5nm. So the theoretical RIS,i.e. the slope of the red line, is

IOP Conf. Series: Materials Science and Engineering **250** (2017) 012053 doi:10.1088/1757-899X/250/1/012053

~706.4 nm/RIU (RIU denotes refractive index unit). The calculated RIS,i.e. the slope of the black line, is ~684.0 nm/RIU, which is close to the theoretical results. FWHM of the reflection curve, which is extracted from the data of figure 3, is ~3.5 nm, as shown in figure 4 (b). The FOM is over 190RIU^{-1} while the refractive index value varies from 1.0 to 1.25.

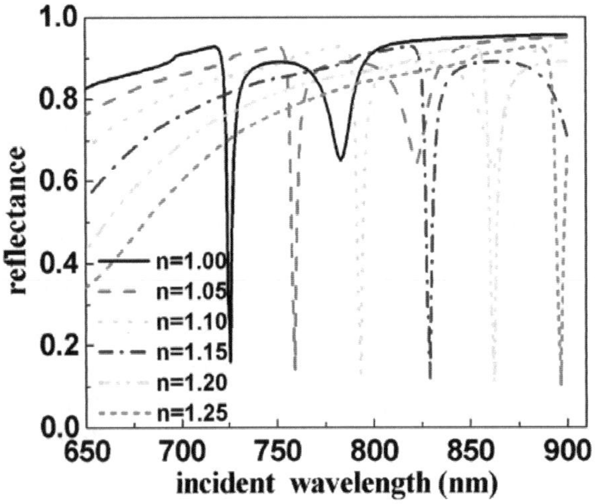

Figure3. Reflection spectra as a function of wavelength in various refractive index condition of sample (P=700nm， t=15nm, H_1=40nm, H_2=20nm, W_1=90nm,W_2=50nm) .

Previously reported value of the RIS from literature is 490 nm/RIU for the one-side gaussian-shaped metallic nano-gratings[10],420 nm/RIU for the double-side symmetrical metallic nano-gratings[11],400 nm/RIU for the gold nanohole arrays [17], 328.5 nm/RIU for the gold nanoshells [18], and 150nm/RIU for LSPR sensors [19], respectively. The reported nanoparticles-based refractive index biosensors have a FOM value less than 10 due to a wide FWHM of the extinction spectrum [20]. Recently, a silver-coated side-polished fiber based SPR sensor was reported[21]. It has a higher RIS, but due to the wider FWHM,the FOM is less than 52 RIU^{-1}. Sensing performances of the optimized Gaussian-shaped and double-Side unsymmetrical metallic nano-Gratings are superior to the performances of previously reported SPR sensors. Therefore, this type of the nano-grating based structure is more suitable to be used as a sensing chip because it can generate larger FOM.

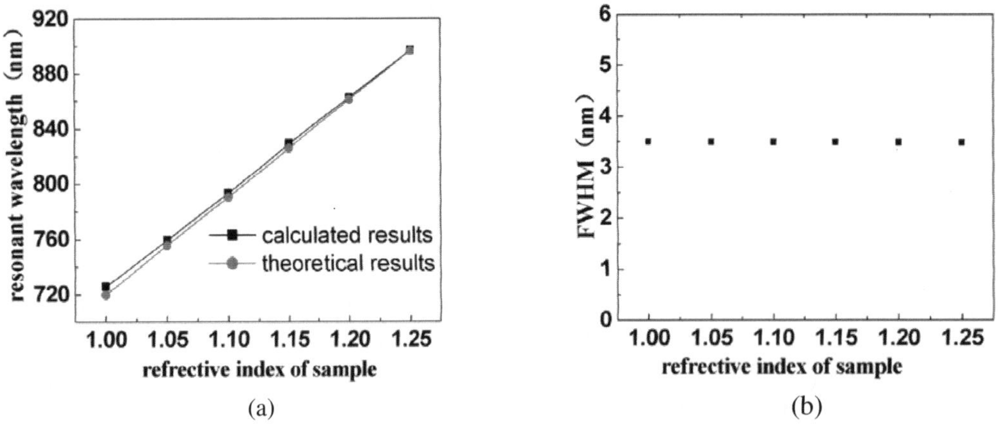

Figure4. (a) The relationship between resonant wavelength and refractive index of the sample; (b) The relationship between FWHM of reflective curve and sample refractive index which is extracted from the data of figure 3.

4. Conclusions

In summary, this paper puts forward a special type of metallic nano-Grating structure with Gaussian-shaped and double-Side unsymmetrical grooves. On the basis of the MMP method, reflection spectra as functions of wavelengths in various refractive index condition of the sample are theoretically analyzed by means of computational calculation and numerical simulation in nanofabrication point of view. For sensing application, two important factors, RIS=684nm/RIUand FOM=195RIU^{-1}, can be achieved with the optimized structure. Therefore, it is reasonable to believe that these properties make this kind of nano-gratings more suitable to be used as a SPR-based sensing chip.

5. Acknowledgments

This work was supported by science and technology planning project of Hebei province (No.15210917) and Scientific Research Foundation of Hebei University of technology.

6. References

[1] Heinz Raether,1988,Surface Plasmons on smooth and rough surfaces and on gratings,Springer, pp91-116
[2] Anatoly V. Zayats, Igor I. Smolyaninov, Alexei A. Maradudin, Nano-optics of surface plasmonpolaritons,2005,Physics Reports 408, pp131–314
[3] J. Homola, S. S. Yee, G. Gauglitz, Surface plasmon resonance sensors: review,1999, Sens. Actuators B 54, pp3-15
[4] Kyung Hun Yoon, Michael L. Shuler, Sung June Kim, Design optimization of nano-grating surface plasmon resonance sensors, 2006, Optics Express 14,pp4842-4849
[5] Yongqi Fu, NgoiKok Ann Bryan, Hybrid micro-diffractive-refractive optical element with continuous relief fabricated by focused ion beam for single-mode coupling, 2001,Applied Optics40, pp5872-5876
[6] Yongqi Fu, NgoiKok Ann Bryan, Ong Nan Shing, Experimental Study of 3D Microfabrication by Focused Ion Beam, 2000,Review of Scientific Instruments 71, pp1006-1008
[7] Yongqi Fu, etal, Diffractive optical elements with continuous relief fabricated by focused ion beam for monomode fiber coupling,2000,Optics Express 7, pp141-147
[8] Yongqi Fu, NgoiKok Ann Bryan, Wei Zhou, A quasi-direct writing of diffractive structures using focused ion beam,2004,Optics Express 12, pp1803-1809
[9] Yongqi Fu, NgoiKok Ann Bryan, Spontaneously generated sinusoidal-like structures under focused ion beam bombardment,2004,Optics Express 12,pp3707-3712
[10] Haiying Li , Xiangang Luo,Chunlei Du, Xunan Chen and Yongqi Fu,Analysis of sensing performance of Gaussian-shaped metallic nano-gratings, 2008, Journal of Nanophotonics2, p 023508
[11] Haiying Li , Zhidong Zhang,Xiangang Luo, XunanChen,Yongqi Fu, Design of gaussian-shaped and double sides flanked metallic nano-grating surface plasma resonance biosensors, 2010,IEEE Symposium on Photonics and Optoelectronics
[12] Ch. Hafner, The Multiple Multipole Program (MMP) and the Generalized Multipole Technique (GMT), 1999, In: Th. Wriedt (editor): Generalized Multipole Techniques for Electromagnetic and Light Scattering, Elsevier, Amsterdam
[13] L. J. Sherry, S.-H. Chang, G. C. Schatz, R. P. Van Duyne, B. J. Wiley, and Y. Xia, Localized surface Plasmon Resonance Spectroscopy of single silver nanocubes,2005, Nano Lett. 5, pp2034-2038
[14] Esteban Moreno and Christian Hafner, Multiple multipole method with automatic multipole setting applied to the simulation of surface plasmons in metallic nanostructures,2002, J. Opt. Soc. Am. A 19, pp101-111
[15] P. B. Johnson and R. W. Christy, Optical constants of the noble metals,1972, Phys. Rev. B 6, pp4370-4379
[16] W.-C. Tan, T.W. Preist, and J.R. Sambles, Resonant tunneling of light through thin metal films via strongly localized surface plasmons, 2000, Phys. Rev. B **62**, 11 pp 134-138

[17] G. Brolo, R. Gordon, B.Leatherm, and K. L. Kavanagh, Surface plasmon sensor based on the enhanced light transmission through arrays of nanoholes in gold films,2004, Langmuir 20, pp 4813-4815
[18] Y. G. Sun, and Y. N. Xia, Gold and silver nanoparticles: A class of chromophores with colorstunable in the range from 400 to 750 nm,2003, Analyst 128, pp 686-691
[19] M. Duval Malinsky, K. Lance Kelly, G. C. Schatz, and R. P. Van Duyne, Chain length dependence and sensing capabilities of the localized surface plasmon resonance of silver nanoparticles chemically modified with Alkanethiol self-assembled monolayers,2001, J. Am. Chem. Soc. 123, pp1471-1482
[20] N. Nath and A. Chilkoti, A colorimetric gold nanoparticle sensor to interrogate biomolecular interactions in real time on a surface, 2002, Anal. Chem. **74**, pp 504-509
[21] JingZhao, et al. Surface plasmon resonance refractive sensor based on silver-coated side-polished fiber,2016,Sensors and Actuators B 230, pp 206-211

Preparation of Si/TiO_2 Composite by the Sol-Gel Method Using As the Lithium-Ion Battery Anode

QuanweiWang [1], YueqinBan[3], HengweiZhou [1], LiliZhang[1,2], YinengHuang[1,2,*], WeiquanShao [3]and ShaouChen[1,3,*]

[1] Xinjiang Laboratory of Phase Transitions and Microstructures in Condensed Matters and College of Physical Science and Technology, Yili Normal University, Yining, 835000;
[2]National Lab of Solid State Microstructures and School of Physics, Nanjing University, Nanjing, 210093
[3]College of physics, Qingdao University, Qingdao 266071, China
The corresponding author's e-mail address:* Huang ynhuang@nju.edu.cn;
*wlxcso@qdu.edu.cn

Abstract. Si was a promising anode material for next-generation LIBs due to its extremely large capacity of 4200mAh/g ($Li_{4.4}Si$ phase). However, during repeated lithium insertion/extraction processes, the accompanied huge volume change (400%) induced the structural failure of the active material and resulted in rapid capacity fading. To overcome this problem, Si/TiO_2 composite with different mole ratio were prepared by the sol-gel method. The inclusion of TiO_2 not only worked as a stable electric conductive pathway but also buffered the volume expansion of the Si during the process of charging and discharging. The Si/TiO_2 composite with different mole ratio of 1:2, 1:3, 1:4 electrode reaches 480mAh/g, 2590mAh/g and 980mAh/g, and it delivered a charge capacity of 461mAh/g, 2510mAh/g and 891mAh/g at the first cycle, corresponding to an initial coulombic efficiency of 96%, 96% and 91%. In contrast, the cell with the pure Si nanoparticle exhibited an initial discharge/charge capacity of 48 and 33mAh/g, respectively, which was much lower than the Si/TiO_2 composite electrode.

1. Introduction

LIBs have been widely used in many types of portable devices due to their high specific energy, high battery voltage, and long cycle life. However, commercially-used graphite anode material has a low specific capacity of 372mAh/g [1].Currently, to find new anode materials with high specific capacity is an interest research direction. Therefore, alternative high capacity anode materials are strongly required. Silicon has high theoretical specific capacity (4200mAh/g) [2] and low working potential, which can be used as the lithium-ion battery anode materials in the future. However, in the process of charging and discharging, Si has large volume changes (400%) [3].This huge volume change gives rise to the collapse of electrodes and insulation of active material, eventually resulting in rapid capacity fading and poor cycle performance.

There are little volume change of the TiO_2 in the process of charging and discharging, so it can buffer volume expansion problem of the Si.In this paper, the Si/TiO_2 composite with different mole ratio were prepared by the sol-gel method [4]. It was found that the composite prepared by the different mole ratio of 1:3 has a higher energy densities and good cycle performance.

IOP Conf. Series: Materials Science and Engineering **250** (2017) 012054 doi:10.1088/1757-899X/250/1/012054

2. Material and Methods

Si powder (0.07g, 0.28g, 0.56g, 30nm in diameter, Nanostructured & Amorphous Materials, Inc) was added into the ethanol (10ml) and was mixed with the ultrasonic bath for 10min. Solution A contained the HNO3 (2ml) and the de-ionized water (5ml). Solution B prepared with the tetrabutyltitanate and the ethanol (25ml) using the magnetic stirring for 30min. The solution A and B were mixed slowly. The mixture was aged at room temperature for 24h, dried at 70℃ for 24h. After that, the mixture was grinded fully and was calcined at 500 ℃ for 3h. The composite of Si and TiO_2 with different mole ratio was prepared. For comparison, the pure Si was also prepared.

The electrode was prepared by mixing the active material, the acetylene black and polyvinylidene difluoride (PVDF) with the weight ratios of 80:10:10. The mixture was coated on the copper foil and was vacuum dried at 120 ℃ for 12h. The copper foil coated with the anode materials was cutted into electrode tabs.The CR2032 coin cell was assembled in an Ar-filled Mikrouna glove box by using the prepared electrode as anode, lithium film as the counter electrode, Celgard 2400 as separator and 1MLiPF6 in EC:DMC (1:2 in volume) as electrolyte. The electrochemical performance was investigated by carrying out galvanostatic charge–discharge experiments between cut-off potentials of 0.02 and 2.0 V at constant current densities of 42mA/g using Land cell test system (Land CT2001A, China).

3. Results and Discussion

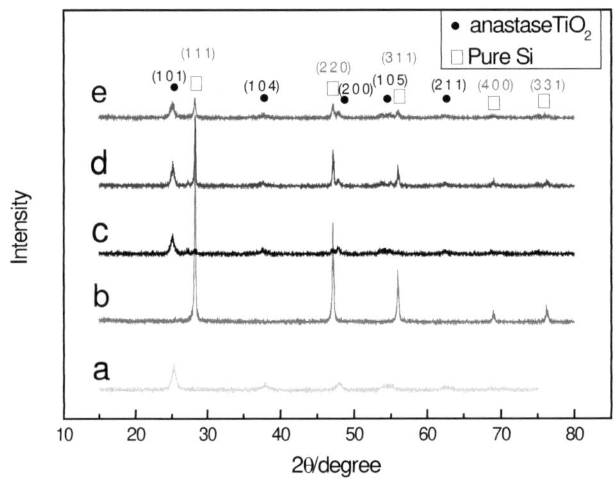

Figure 1: X-ray diffraction patterns of (a) TiO_2, (b) Si nanoparticles, the Si/TiO_2 composite with the different mole ratio of 1:2(c), 1:3(d), 1:4(e).

Figure.1 showed the XRD patterns of the sample TiO_2, pure Si and Si/TiO_2 composite with different mole ratio of 1:2, 1:3, 1:4. The TiO_2 (Fig.a) pure Si (Fig.b) presented well-defined peaks, respectively. After sintering at 500 ℃in the air, the structure of the Si/TiO_2 composite with the different mole ratio of 1:2,1:3,1:4 did not changed since no apparent changes could be observed for all XRD diffraction peaks[5] (Fig. c d and e). It show that the samples of different molar ratio of composite with the increment of TiO_2, Si and TiO_2 relative strength of peak also has the rule change. The carbon was not detected because of its amorphous structure.

IOP Conf. Series: Materials Science and Engineering **250** (2017) 012054 doi:10.1088/1757-899X/250/1/012054

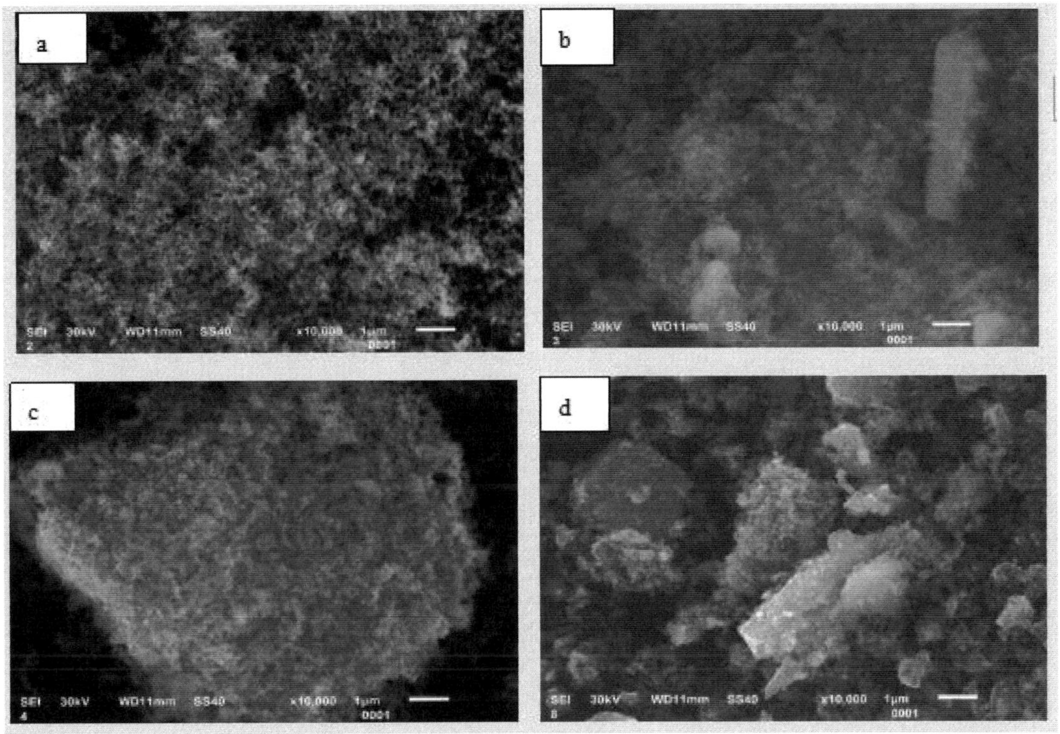

Figure 2: Images of the Pure Si(a) and the Si/TiO_2 composite with different mole ratio of 1:2(b),1:3(c),1:4(d).

Figure.2a showed the plane-view SEM images of the pure Si. From the images, the Si particles with the size of 30nm. Fig.2b shows that the Si/TiO_2 composite with the mole ratio of 1:2, a layer of the TiO_2 particles formed on the surface of the Si nanoparticles coated layer , but there is still a Si particles exposed and not fully covered live. Fig.2c shows that the Si/TiO_2 composite with the mole ratio of 1:3, there are no Si particles in the surface, the Si surface is coated the TiO_2, completely, and the TiO_2 layer is uniform. Fig.2d shows that the Si/TiO_2 composite with the mole ratio of 1:4, the Si though covered by the TiO_2 but the TiO_2 on the surface of the Si has reunion seriously [6]. The results show that the mole ratio of 1:3 is better.

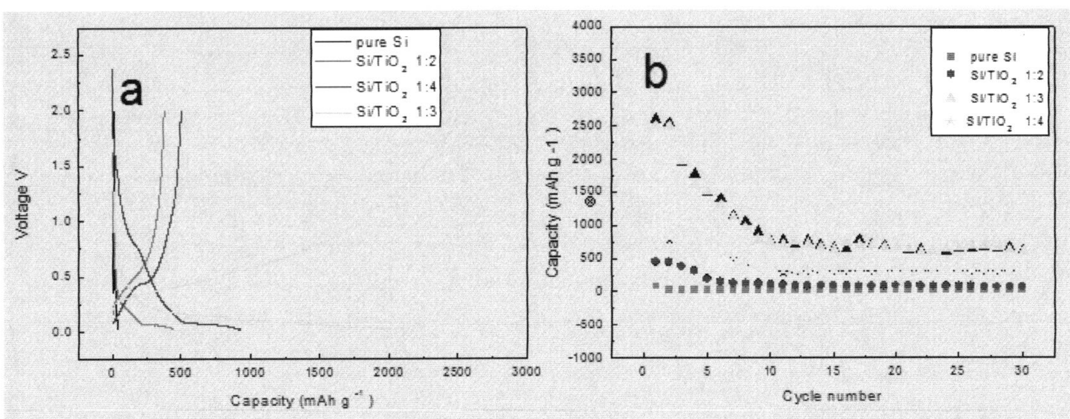

Figure 3: Voltage profiles (a) and Cycling performance (b) for the pristine Si and Si/TiO_2 composite with different mole ratio of 1:2,1:3,1:4 at 42 mA /g

Figure.3a showed the typical voltage profiles of the pure Si and the Si/TiO_2 composite with different mole ratio of 1:2, 1:3, 1:4 electrode cycled between 0.02 and 2.0 V at 42 mA/g, respectively. The first discharge capacity of the composite (1:2, 1:3, 1:4) electrode reaches 480mAh/g, 2590mAh/g and 980mAh/g and it delivered a charge capacity of 461mAh/g, 2510mAh/g and 891mAh/g at the first cycle, corresponding to an initial coulombic efficiency of 96%, 96% and 91%. The lower efficiency for the first charge/discharge could be mainly due to the larger surface area of the carbon-coated Si nanoparticles with porous nanofiber structure and formation of irreversible solid electrolyte interphase (SEI) layer[7,8] in the first discharge process. In contrast, the cell with the pure Si nanoparticle exhibited an initial discharge/charge capacity of 48 and 33mAh/g, respectively, which was much lower than the composite electrode. The results provided clear evidence that the Si/TiO_2 composite with different mole ratio of 1:3 can enhance Li storage properties.

Figure.3b displayed the cycling performances of the Si/TiO_2 composite with different mole ratio of 1:2, 1:3, 1:4 electrodes and Si nanoparticle electrode. It could be seen that the capacity of the pure Si electrode decreased to about 30mAh/g after 30 cycles. As shown in Fig. 3b, the Si/TiO_2 composite with different mole ratio of 1:2, 1:3, 1:4 anode retained the reversible discharge capacity of 198mAh/g, 350mAh/g and 780mAh/g at the 30th cycle, respectively. In all 30 cycles, Si/TiO_2 composite with different mole ratio of 1:3 had higher capacities and better capacity retentions [9] than the other mole ratio and the pure Si due to the protection of Si with TiO_2.

Fig. 4 showed the impedance spectra of the pure Si and the Si/TiO_2 composite with different mole ratio of 1:2, 1:3, 1:4 electrode with the first cycle. The electrochemical impedance spectra consist of one depressed semicircle and a line. The semicircle at high frequency can be ascribed to the charge transfer resistance, which is related to the electrochemical reaction between the particles or between the electrode and the electrolyte. The sloping line is related to lithium-ion diffusion in the active material [10, 11]. As shown in Fig. 4, the charge transfer resistance of the Si/TiO_2 composite with different mole ratio of 1:2, 1:3, 1:4 were 1200Ω, 250Ω and 1100Ω at the first cycle, whereas it increased greatly to 2070 Ω for the pure Si. The improvement [12] in the electrochemical performance for Si/TiO_2 composite with different mole ratio of 1:3 could be attributed to good electrical contact between the active materials and carbon during the lithium insertion and extraction cycles.

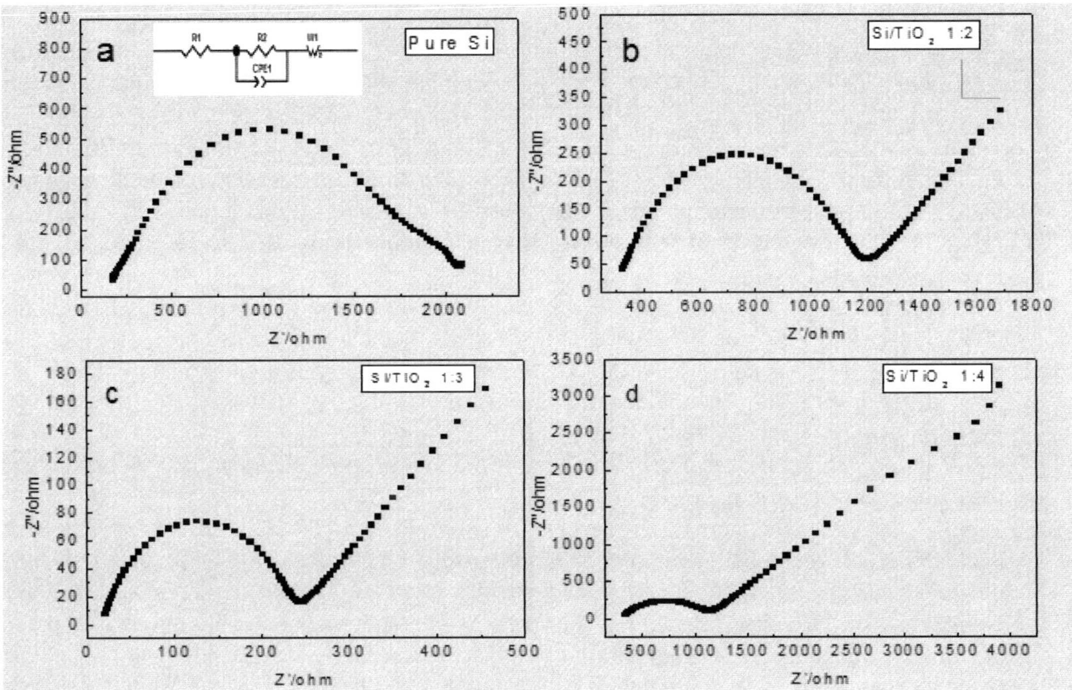

Figure 4: The impedance spectra of the pristine Si(a)and the Si/TiO_2 composite with different mole ratio of 1:2(b),1:3(c),1:4(d)

4. Conclusions

The Si/TiO_2 composite with different mole ratio of 1:2, 1:3, 1:4 electrode reaches 480mAh/g, 2590mAh/g and 980mAh/g, and it delivered a charge capacity of 461mAh/g, 2510mAh/g and 891mAh/g at the first cycle, corresponding to an initial coulombic efficiency of 96%, 96% and 91%. In contrast, the cell with the pure Si nanoparticle exhibited an initial discharge/charge capacity of 48 and 33mAh/g, respectively, which was much lower than the Si/TiO_2 composite electrode.

5. Acknowledgements

This work was supported by the project of training and construction of science and technology innovation team in the autonomous region (2014751001). This work was also supported by the Natural Science Foundation of Shandong Province (ZR2014EMQ005), the Project of Shandong Province Higher Educational Science and Technology Program (Grant No. J14LJ08).

6. References

[1] M.N. Asiah, M.H. Mamat , Zuraida Khusaimi, S. Abdullah, M. Rusop, Ahsanulhaq Qurashi, Structural and optical properties of hydrothermally synthesized mesoporous Si/TiO_2 nanowire composites. Microelectronic Engineering. 136 (2015) 31–35.

[2] Marta Buchalska, Marcin Surówka, Jani Hämäläinen, Tomi Iivonen,Markku Leskel , Wojciech Macyk, Photocatalytic activity of TiO_2 films on Si support prepared by atomic layer deposition. Catalysis Today. 252 (2015) 14–19.

[3] Supriya Gupta, Sandan Kumar Sharma, B. Venkata Manoj Kumar, Young-Wook Kim. Tribological characteristics of SiC ceramics sintered with a small amount of yttria. Ceramics International 41 (2015) 14780–14789.

[4] Shi Qian, Xuanyong Liu, Cytocompatibility of Si-incorporated TiO_2 nanopores films ,Colloids and Surfaces B: Biointerfaces. 133 (2015) 214–220.

[5] Ying Li, Guanjie Xu, Yingfang Yao, Leigang Xue, Meltem Yanilmaz, Hun Lee, Xiangwu Zhang. Coaxial electrospun Si/C–C core–shell composite nanofibers as binder-free anodes for

lithium-ion batteries. Solid State Ionics 158(2014)67–73.[6] Xiaoge Li, Ying Bai, Dong Yan, Caiyan Yu, Kai Jiang, Ning Wan, Qing Wu, Shuwei Sun, Weifeng Zhang, Preparation and electrochemical properties of TiO_2 /C/Si composites for lithium ion batteriesMaterials Letters. 126 (2014) 295–298.

[6]Yoon Hwa, Won-Sik Kim, Byeong-Chul Yu, Jae-HunKim, Seong-HyeonHong, Hun-JoonSohn. Facile synthesis of Si/TiO_2 (anatase) core–shell nanostructured anodes for rechargeable Li-ion batteries. Journal of Electroanalytical Chemistry 712(2014)202-206.

[7] Byoung-Sun Lee, Seoung-Bum Son , Kyu-Min Park , Jong-Hyun Seo, Se-Hee Lee, In-Suk Choi, Kyu-Hwan Oh, Woong-Ryeol Yu. Fabrication of Si core/C shell nanofibers and their electrochemical performances as a lithium-ion battery anode. Journal of Power Sources 206(2012)267–273.

[8] Leigang Xue, Kun Fu, Ying Li, Guanjie Xu, Yao Lu, Shu Zhang,Ozan Toprakci, Xiangwu Zhang. Si/C composite nanofibers with stable electric conductive network for use as durable lithium-ion battery anode. Nano Energy 2(2013)361–367.

[9]Magali Gauthier, David Reyter, Driss Mazouzi. From Si wafers to cheap and efficient Si electrodes for Li-ion batteries. Journal of Power Sources 256(2014)32-36.

[10] Tugrul Cetinkaya, Mehmet Uysal, Mehmet O. Guler, Hatem Akbulut, Ahmet Alp. Improvement cycleability of core–shell silicon/copper composite electrodes for Li-ion batteries by using electroless deposition of copper on silicon powders. Powder Technology 253(2014) 63–69.

[11] Congxiang Lu, Yu Fan, Hong Li, Yi Yang. Core–shell CNT–Ni–Si nanowires as a high performance anode material for lithium ion batteries. CARBON 63(2013)54–60.

[12] Lianyi Shao, Jie Shu, Kaiqiang Wu, Xiaoting Lin, Peng Li, Miao Shui. Low pressure preparation of spherical Si/C/CNT/C anode material for lithium-ion batteries. Journal of Electroanalytical Chemistry 727(2014) 8–12.

IOP Conf. Series: Materials Science and Engineering **250** (2017) 012055 doi:10.1088/1757-899X/250/1/012055

Electrochemical Performance of Zn-Al Double Layered Hydroxide for Supercapacitor Application

Yang Wen, Yu Zhao, Lu Lu, ShuoJia Zhang and Bing Xu*

Liaoning Provincial Key Laboratory of New Energy Battery, Dalian Jiaotong University, Huanghe Road 794#, Dalian, PR China

xubing@djtu.edu.cn

Abstract. Zn-Al double layered hydroxide (Zn-Al-LDH) was synthesized by a facile coprecipitation method. The morphology and microstructure of samples were characterized by X-ray diffraction (XRD) and Scanning electron microscope (SEM), respectively. The results show that the sample is uniform nanosheet with the diameter of ~400 nm and the thickness of ~80 nm. XRD characterization confirms that it is $Zn_6A_{l2}(OH)_{16}CO_3.4H_2O$ and belongs to Rhombohedral crystallographic system. Electrochemical performances of the sample were investigated by cyclic voltammery (CV) and charge/discharge. The Pure Zn-Al-LDH nanocomposites achieves a specific capacitance of 37.0 F g^{-1} at the current density of 1.0 A/g.

1. Introduction

Worldwide growth in fossil energy consumption need to develop alternative new energy of environment friendly, low cost, compatibility and efficient[1]. The rapid development of various new electronic devices and optoelectronic devices driven by modern technology also calls for more reliable, more energy-dense, longer service life and sustainable power systems, such as batteries and supercapacitors. Unlike a battery, supercapacitor causes a wide interest because of charging speed, infinite long theory of life, no temperature vulnerability, and non-toxic[2,3].

Layered double hydroxide(LDHs), a typical inorganic layered material, its chemical formula can be expressed as general formula $[M^{II}_{1-x}M^{III}_x(OH)_2]^{z+}(A^{n-})/_{z/n}.yH_2O$. M^{II} is divalent metal ions, such as Mn^{2+}, Ni^{2+}, Co^{2+}, Zn^{2+}, and Fe^{2+}. M^{III} is a trivalent metal ion, such as Al^{3+}, Fe^{3+}, Cr^{3+}, Ga^{3+}, etc. A^{n-} is an interlayer anion, which is used to balance the positive charge of the main structure of the layer of hydroxide[4,5]. Due to metal ions and the interlayer anion can be flexible modulation and does not change the main structure, LDHs structure material is widely studied in many applications, such as catalytic, adsorption, biology, and energy storage[6,7]. The high dispersion of active substances in the lattice of the layers, the easy separation and chemical modification of the single layer, provide the possibility of becoming a multi-functional electrochemical electrode material[8].

Among different kinds of LDHs materials, Ni-Co-LDH and Co-Al-LDHs show better elcectrochemical performance[4,9-10]. However, Zn-Al-LDHs are barely researched for their supercapacitor performance[11].

2. Experimental Section

2.1. Preparation of Zn-Al-LDH

Zn-Al-LDH material is prepared by coprecipitation method. 0.21 mol $Zn(NO_3)_2.6H_2O$ and 0.07 mol $Al(NO_3)_3.6H_2O$ are dissolved in 200 ml deionized water (solution I); 0.438 moles of NaOH and 0.113 mol of Na_2CO_3 are dissolved in 200 ml deionized water (solution II). Two solutions are kept in 70 °C water bath for 30 min under stirring. Solution I is dropped into solution II and react 30 min at 70 °C then aging for 20 h. The sample is washed to neutral and dry overnight[12].

2.2. X-ray Powder Diffraction (XRD)
The XRD experiment is conducted on the X'pert ProSuper ray diffractometer of PAN Analytical (Netherlands). The copper target line is the light source (λ =1.5432 nm), the graphite monochromator, the tube voltage is 40 kV, and the tube current is 100 mA.

2.3. Electrochemical Measurements.
Electrochemical performance test is performed on an Autolab electrochemical workstation (PGSTAT302 N, Ecochemie Co., the Netherlands). in 6 mol/L KOH solution. The electrochemical performance test adopts the classical three-electrode system, in which Zn-Al-LDH is used as working electrode, the Hg_2Cl_2/Hg electrode (SCE, saturated KCl) as reference electrode and Pt as the counter electrode. CV is conducted between 0-0.4 V (vs. SCE) and the constant current charging and discharge test voltage window is 0-0.35 V.

3. Results and Discussion

3.1. SEM Representation of Zn-Al-LDH.
Figure 1a represents SEM image of Zn-Al-LDH samples obtained by coprecipitation method. It can be seen that the Zn-Al-LDH sample is uniform and irregular, and the size of the slices is about 300-500 nm and the thickness is about 50-100 nm, and there is no obvious reunion phenomenon.

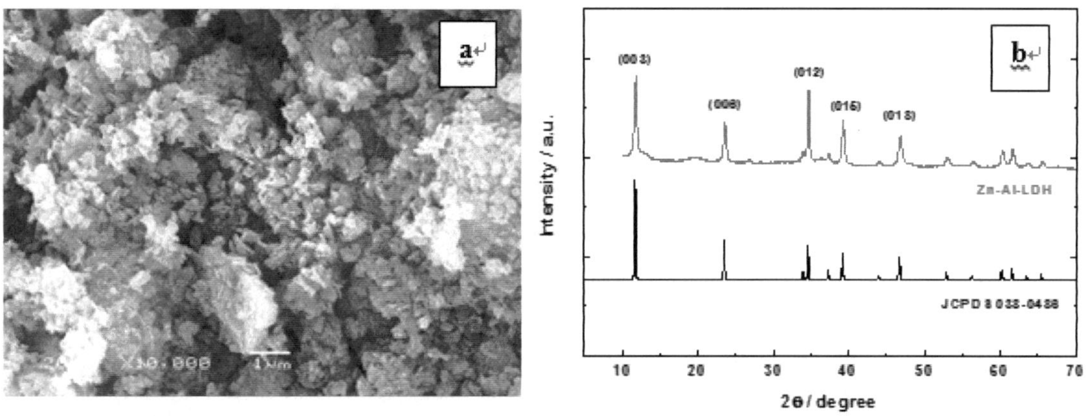

Figure 1. SEM image (a) and XRD patterns (b) for Zn-Al-LDH

3.2. XRD of Zn-Al-LDH.
XRD characterization results in shown in figure 1b. Compared with the JCPDS standard card, the peaks at 2θ=11.64, 23.39, 34.56, 39.15, and 46.64 ° corresponding to the diffraction peaks of (003), (006), (012), (015) and (018) crystal planes for hydrotalcite structure $Zn_6A_{12}(OH)_{16}CO_3.4H_2O$, trigonal system (Rhombohedral), completely consistent with the standard card (JCPDS No. 038-0486)[13]. The sample structure belongs to the space group R-3m, crystal cell parameters for a=b=3.076 Å, c= 22.800 Å,$\alpha=\beta$ = 90 °, γ = 120 °.

3.3. Cyclic Voltampere Curve of Zn-Al-LDH.
The electrochemical properties of Zn-Al-LDH electrode are tested using cyclic voltammetry (CV) technology in 6 mol/L KOH solution under different scanning rates. The voltage window is 0-0.4 V, and the result is shown in figure 2. It can be seen that the curve shows a pair of oxidation-reduction

peaks at about 0.25 V (vs SCE), which means that the electrode presents a typical pseudo-capacitance properties[14]. As the scanning rate increases, the peak value increases, and the peak position moves to high and low potential. These changes related to electrolyte solution ion diffusion ability in the LDH electrode. When lower scanning speed, time for ion diffusion of electrolyte solution is abundant, so polarization voltage is small, and usage of active material is improved.

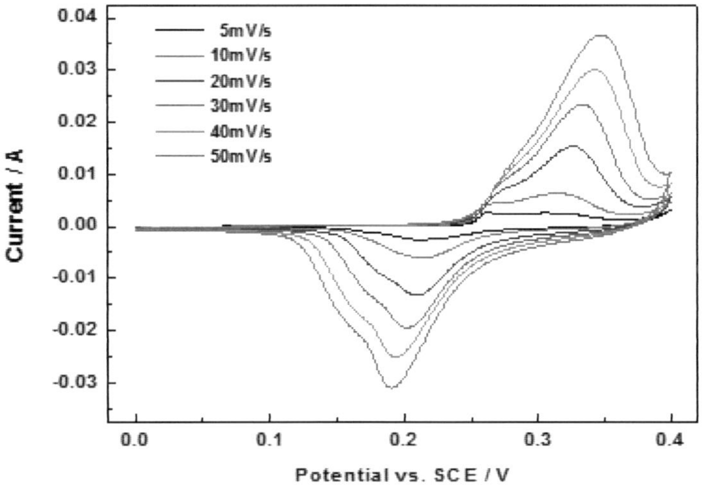

Figure 2. Cyclic voltammetry behavior of Zn-Al-LDH at different scan rates

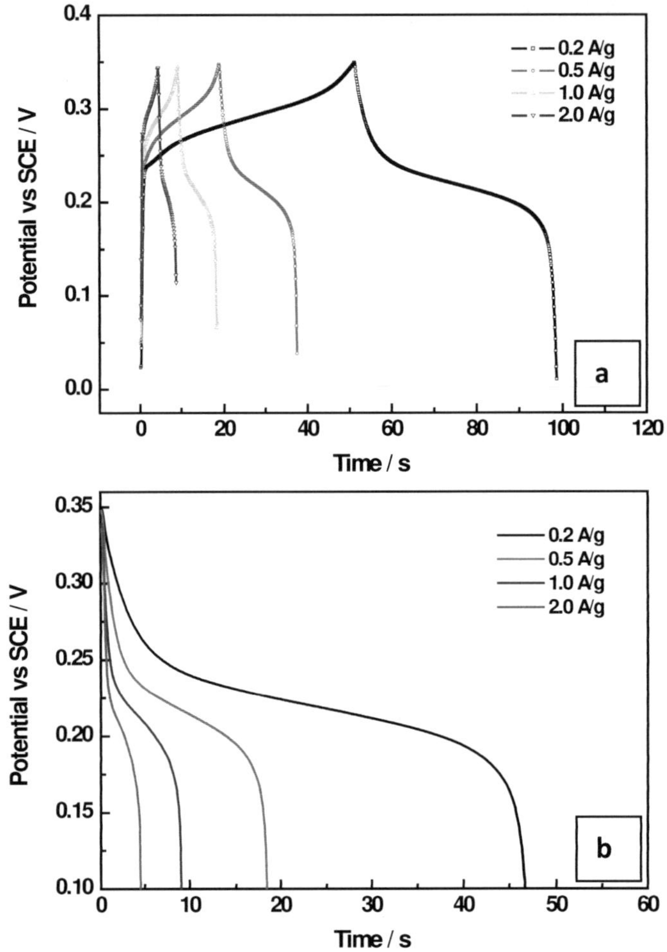

Figure 3. Galvanostatic charge-discharge curves of Zn-Al-LDH at various current densities

3.4. Charge and Discharge Curve of Zn-Al-LDH.

Figure 3 shows charge and discharge curve for Zn-Al-LDH electrode under different current density. It can be seen that the charge and discharge curve is nearly symmetrical hat (Figure 3a), which imply that Zn-Al-LDH electrochemical reaction on the surface of the electrode has a high degree of reversibility[3]. Charge and discharge platform appears near the voltage of 0.25 V. This nonlinear charge-discharge curve and charge and discharge platform is consistent with REDOX peak in CV test that further confirms the capacitance reaction on the electrode surface.

Through the discharge curve (Figure 3b), the ratio capacitance of the electrode is known to be 41.0, 39.7, 37.0, and 33.2 F/g at the current density of 0.2, 0.5, 1.0 and 2.0 A/g, respectively. The capacitance decreases slightly with the current density increasing. When the current increasing, active substances short in a period of time, adsorption of ions of active material and electrolyte interface, ion concentration to the electrolyte ion diffusion velocity cannot meet the transfer of electrode needed to charge and discharge ion number, causes the diffusion polarization between electrode and liquid interface increase.

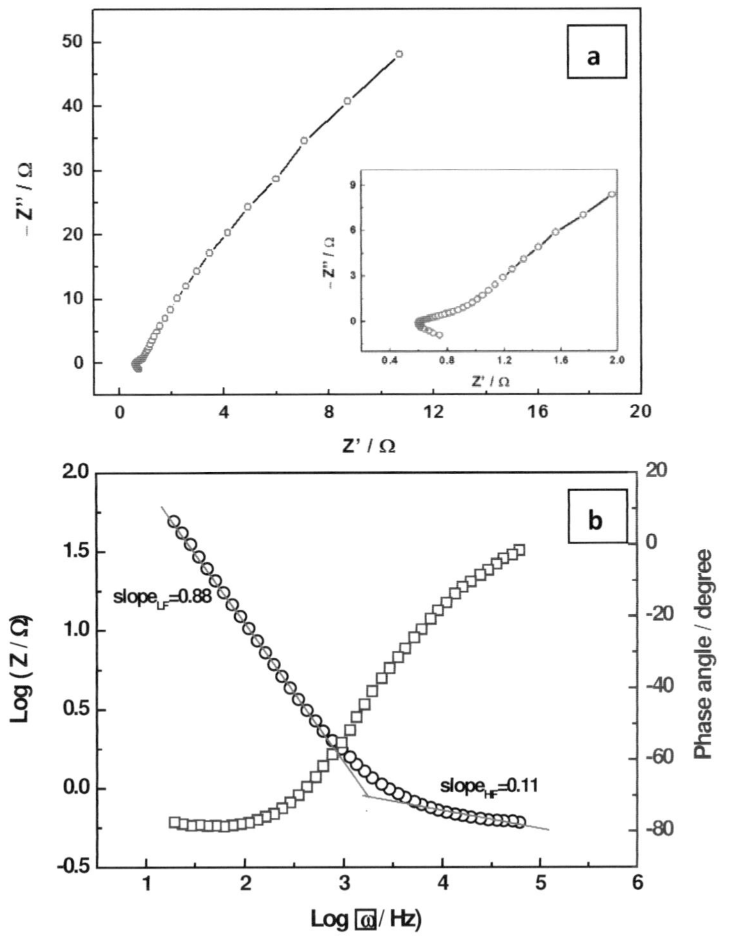

Figure 4. Nyquist (a) and Bode (b) plots of Zn-Al-LDH in 6 M KOH aqueous electrolyte

3.5. Electrochemical Communication Impedance (EIS) Test.

The ion and electron transmission capacity of Zn-Al-LDH electrode system are analyzed by electrochemical communication impedance (EIS) (Figure 4). The power output of the capacitor is closely related to the impedance of the electrode. The curve in the high frequency area forms a half circular, and in low frequency area an approximate linear form, embodies the good capacitance characteristic of electrode. High frequency area half arc curve on the x axis of the intercept the equivalent resistance of Rs is from electrochemical system (including the inherent resistance of substrate, ion resistance of the electrolyte solution and fluid collection and contact resistance of active material, etc.), semicircle diameter represents the interface between the electrodes and the electrolyte solution of the charge transfer resistance Rct. Through the Nyquist graph, the preliminary measure equivalent resistance Rs and the charge transfer resistance Rct about 0.6 and 0.6 Ω, respectively.

4. Conclusions

Zn-Al double layered hydroxide is synthesized by coprecipitation method. The sample is nanosheet-like powder and with about 80 nm thickness. The composition of Zn-Al-LDH is $Zn_6Al_2(OH)_{16}CO_3.4H_2O$ and belongs to Rhombohedral crystallographic system. Pure Zn-Al-LDH nanocomposites achieves a specific capacitance of 37.0 F g^{-1} at the current density of 1.0 A/g.

5. Acknowledgments
This research was funded by the National Natural Science Foundation of China (No. 21146007, 21406024), the National Key Research and Development Program of China (Program No. 2016YFB0101207), the Natural Science Foundation of Liaoning (No. 201602137, 201602125), and the Youth Science and Technology Star Program of Dalian (No. 2016RQ056).

6. References
[1] Scrosati B, Hassoun J, Sun, YK 2011 *Energy & Environmental Science* **4** 3287.
[2] Wang Y, Xia Y 2013 *Advanced Materials* **25** 5336.
[3] Xiao X, Han B, Chen G, Wang L, Wang Y 2017 *Scientific Reports* **7**.
[4] Nagaraju G, Raju GSR, Ko YH, Yu JS 2016 *Nanoscale* **8** 812.
[5] Lu X, Li G, Tong Y 2015 *Science China-Technological Sciences* **58** 1799.
[6] Shao M, Zhang R, Li Z, Wei M, Evans DG, Duan X 2015 *Chemical Communications* **51** 15880.
[7] Wang Z, Zhang X, Wang J, Zou L, Liu Z, Hao Z 2013 *Journal Of Colloid And Interface Science* **396** 251.
[8] Zhao J, Chen J, Xu S, Shao M, Zhang Q, Wei F, Ma J, Wei M, Evans DG, Duan X, H 2014 *Advanced Functional Materials* **24** 2938.
[9] Peng W, Li H, Song S, 2017 *Acs Applied Materials & Interfaces* **9** 5204.
[10] Li S, Cheng P, Luo J, Zhou D, Xu W, Li J, Li R, Yuan D 2017 *Nano-Micro Letters* **9**.
[11] Lonkar SP, Raquez JM, Dubois P 2015. *Nano-Micro Letters* **7** 332.
[12] Tan Y, Liu XY, Zhang L, Wang A, Li L, Pan X, Miao S, Haruta M, Wei H, Wang H, Wang F, Wang X, Zhang T 2017 *Angewandte Chemie-International Edition* **56** 2709.
[13] Liu Y, Yang Z 2016 *Rsc Advances* **6** 68584.
[14] Xi R, Wang S, Li S, Li L, Jiang C 2015 *Journal of Functional Materials* **46** 11060.

IOP Conf. Series: Materials Science and Engineering **250** (2017) 012056 doi:10.1088/1757-899X/250/1/012056

Electrochemical Micromachining with Fiber Laser Masking for 304 Stainless Steel

Xiaohai Li[1], Shuming Wang[2], Dong Wang[1] and Han Tong[1]

[1]School of Mechanical Engineering, Jiamusi University, Jiamusi City, China
[2]School of Material Science and Engineering, Jiamusi University, Jiamusi City, China
Email: li_xh@jmsu.edu.cn

Abstract. In order to fabricate micro structure, the combined machining of electrochemical micro machining (EMM) and laser masking for 304 stainless steel was studied. A device of composite machining of EMM with laser masking was developed, and the experiments of EMM with laser masking were carried out. First, by marking pattern with fiber laser on the surface of 304 stainless steel, the special masking layer can be formed. Through X ray photoelectron spectroscopy (XPS), the corrosion resistance of laser masking layer was analyzed. It is proved by XPS that the iron oxide and chromium oxide on the surface of stainless steel generates due to air oxidation when laser scanning heats. Second, the localization and precision of EMM are improved, since the marking patterns forming on the surface of stainless steel by laser masking play a protective role in the process of subsequent EMM when the appropriate parameters of EMM are selected. At last, the shape and the roughness of the machined samples were measured by SEM and optical profilometer and analyzed. The results show that the rapid fabrication of micro structures on the 304 stainless steel surface can be achieved by EMM with fiber laser masking, which has a good prospect in the field of micro machining.

1. Introduction

Electrochemical machining is a method of removing metals by ion unit, which has a broad prospect in the field of metal microfabrication. At present, in order to overcome the problem of the stray corrosion during electrochemical micromachining (EMM), many scholars at home and abroad have carried out extensive research on EMM. For example, the ultra-short pulse current was used to fabricate micro metal parts by Schuster R[1] and Zhu D and so on [2], which achieved machining accuracy of sub-micron; Fan Z J[3] superimposed magnetic field during EMM, which is conducive to improve machining accuracy, surface quality and processing efficiency; Bao H Q[4] carried out EMM experiments by using pure water, the cation-exchange membrane was used to promote hydrolysis dissociation of ultra pure water, improve the current density of electrolytic processing and make micro holes on the surface of stainless steel; Zeng Y B[5] brought forward a new method of using dry film photoresist with washed electrolyte to fabricate micro-dimple array on the surface of metal parts; Zhang C Y[6] proposed the laser-assisted electrochemical compound processing, the conductive transparent glass of indium tin oxide ITO was selected as electrolytic cathode, fiber laser scanning heating was adopted to remove the passivation film on the surface of 304 stainless steel during EMM and machine micro parts; Pajak P T[7] proposed the electrolyte jet nozzle with laser heating was used to fabricate micro parts,but the machining accuracy is low.

In this paper, in order to control the stray corrosion of EMM, the EMM with laser masking technology is adopted. Through laser surface modification technology, fiber laser masking pattern on the surface of 304 stainless steel are done by laser direct scanning, and the masking layer possessing corrosion protection can be formed. Then, the 304 stainless steel masked are used as the anode in the subsequent EMM. Due to the corrosion protection of masking layer, the localized machining can be controlled to fabricate the complex microstructures on the surface of 304 stainless steel.

2. Machining Mechanism

2.1. Principle of Making the Masking Layer

The developed device includes the unit of making the masking layer by fiber laser scanning and the unit of EMM. Two units are installed on the same three-axis feed table with resolution of 0.4 μm, which can solve the positioning problem during the EMM with laser masking, as shown in Figure 1. The pulsed fiber laser made in German IPG company is selected to form the masking layer, the galvanometer is the SCANLAB optical galvanometer made in German. The output power range of the fiber laser is from 0W to 20W, and the minimum diameter of laser spot is about 10 μm. The laser output frequency range is from 0 to 100kHz, and the output laser wavelength is 1.064μm. The laser focus can be controlled by focusing optical system of precision laser, three-coordinate precision worktable and the galvanometer, scanning along the planned path on the surface of the workpiece, direct writing and heating the patterns. The size and shape of the pattern can be strictly limited according to the set design. The patterns with complex shapes generate through laser surface modification, and patterns possess special properties as a masking layer in the subsequent EMM. Due to the small output power and ultra-short pulse output of the fiber laser with the only 100ns pulse width, the molten layer deposited by laser scanning is very thin and the thermal action region is very small, and the thickness of the masking layer formed is less than 1μm. In the process of EMM, the high-frequency pulsed power supply are used, the electrolyte is $NaNO_3$, and electrolyte circulation system is equipped for eliminating electrical corrosion products.

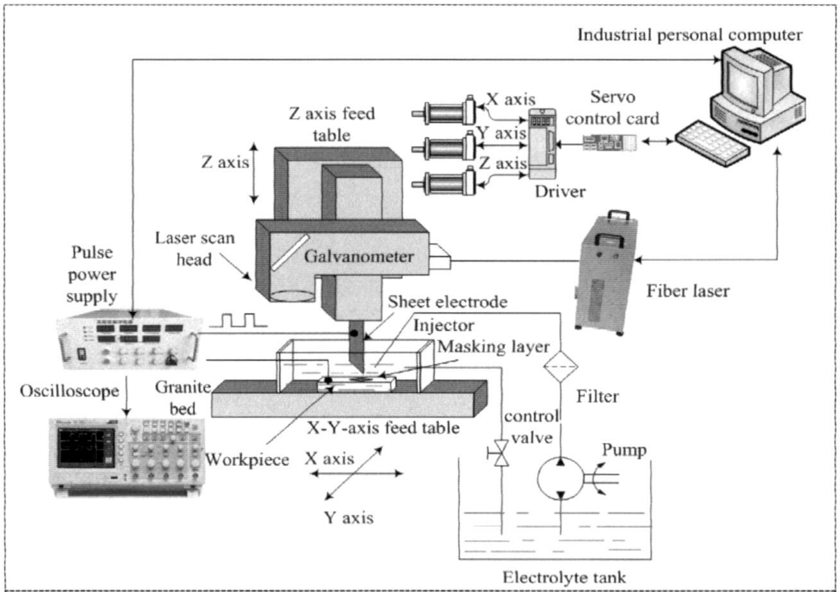

Figure 1. The developed device of EMM with fiber laser masking.

When make the laser masking layer, the scanning mode and path of the laser filling pattern are planned in advance. Since the output mode of optical fiber laser is pulsed mode , the scanning path of laser consists of a series of circles with a certain spot diameter, as shown in Figure.2. The laser scanning line width is determined by the laser spot diameter, and the laser scanning path planning is determined by the laser scanning line width and line spacing. As shown in Figure 3(a), laser surface modification are carried out first on the surface of 304 stainless steel. Fiber laser beam with high density energy begin scanning according to planned path and generate the masking layer with protective effect.Then, the EMM experiments with high frequency pulse current are done, micro sheet stainless steel metal is selected as the cathode of EMM, sheet metal of 304 stainless steel with laser masking pattern serves as anode. The distance between sheet metal electrode and the workpiece surface are kept unchanged, the cathode repeatedly scans above the anode, as shown in Figure 3(b).

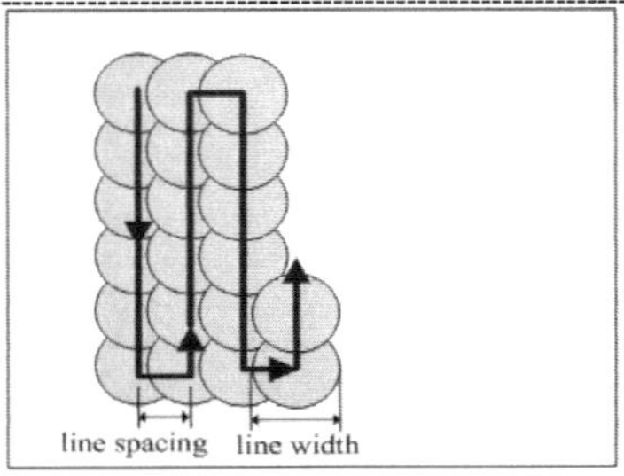

Figure 2. Laser scanning path planning as laser filling pattern.

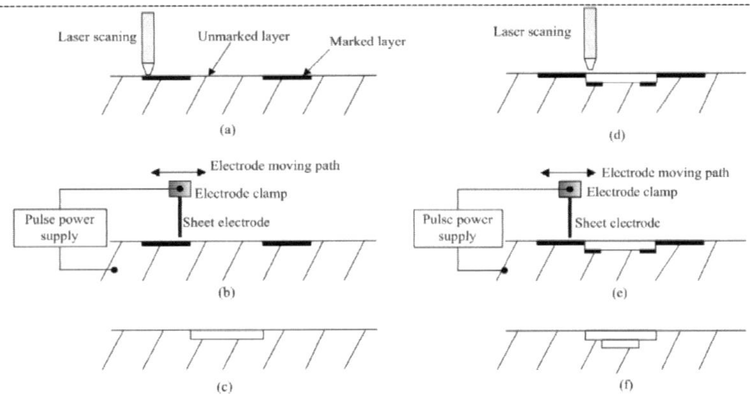

(a)laser masking;(b)electrochemical etching;(c)etching result;(d)second laser masking;(e)second electrochemical etching;(f)second etching result

Figure 3. Principle of electrochemical micromachining by laser masking.

Because the no-machining areas treated by laser masking possess the temporarily protective masking layer on 304 the stainless steel surface and the machining areas are not be treated by laser masking, the electrolytic etching metal removal rate of masking area was evidently slower than that of no-masking area on 304 the stainless steel surface. Accordingly, the machining areas can be quickly removed through electrolytic corrosion, and micro electrochemical localized processing can be realized. The two-dimensional micro cavity with certain protective masking pattern can be obtained, as Figure 3(c) shows.The depth of the cavity at one time is very shallow, and it can also be processed by many times processes of EMM with laser masking and the machining depth of cavity is improved. The repetition accuracy of laser masking pattern is less than 2 μm. If the pattern of the masking layer is different each time, a micro cavity with a multilayer structure can be machined, as Figure 3(d), (e) and (f) show.

As shown in Figure 4, under conditions of 5W laser power and 20mm/s laser scanning speed, 304 stainless steel surface with the 8K mirror precision is marked to make the surface metal melt and form the recast layer. In the process of laser masking, the metal on the stainless steel is oxidized by air. Due to rapid thermal quench, the micro cracks appear on the surface of the workpiece.The laser heating can produce the metal melting layer on the surface of stainless steel. The grains of the melting layer can be refined, which can improve the surface performance of stainless steel and also enhance the corrosion resistance on the surface of stainless steel [8].

IOP Conf. Series: Materials Science and Engineering **250** (2017) 012056 doi:10.1088/1757-899X/250/1/012056

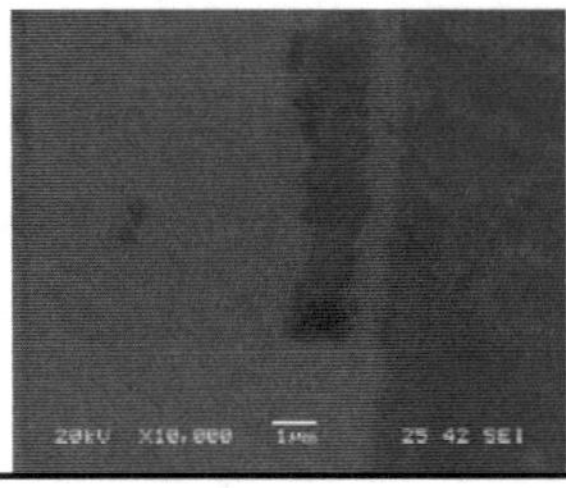

(a) The surface of stainless steel without laser masking

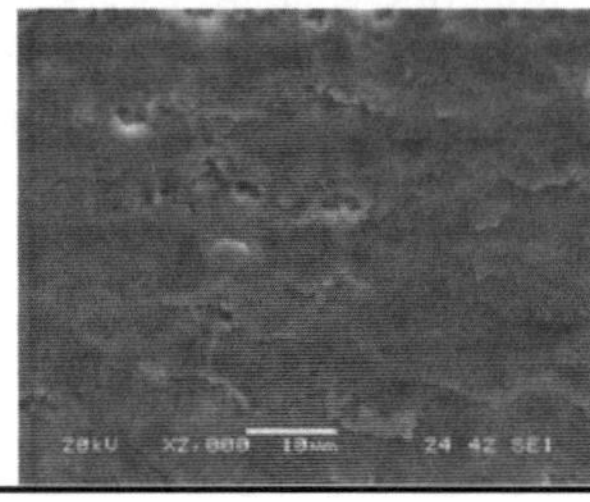

(b)The surface of stainless steel treated by laser masking

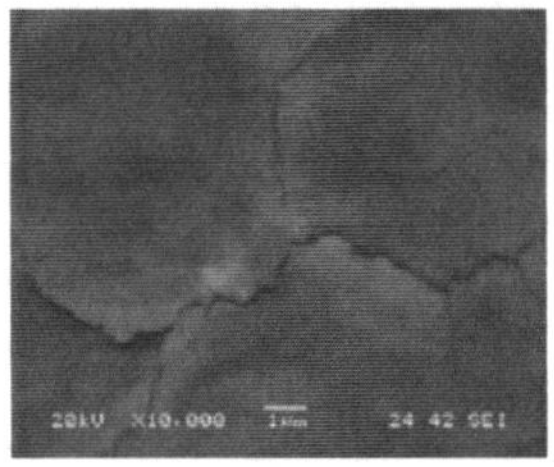

(c) Microcrack

Figure 4. SEM images of 304 stainless steel treated by laser masking

2.2. X-ray Photoelectron Spectroscopy Analysis of Laser Masking Layer

According to the processing condition of laser masking in Figure 4, the energy density of laser beam was adjusted, and the surface modification of 304 stainless steel workpiece was done by fiber laser. X-ray photoelectron spectroscopy (XPS) analysis are used to study surface characteristics and explore the mechanism of corrosion resistance enhancement of laser masking layer of 304 stainless steel. As shown in Figure 5, the XPS spectrum of the Cr composition varies with the sputtering time on the 304 stainless steel surface.

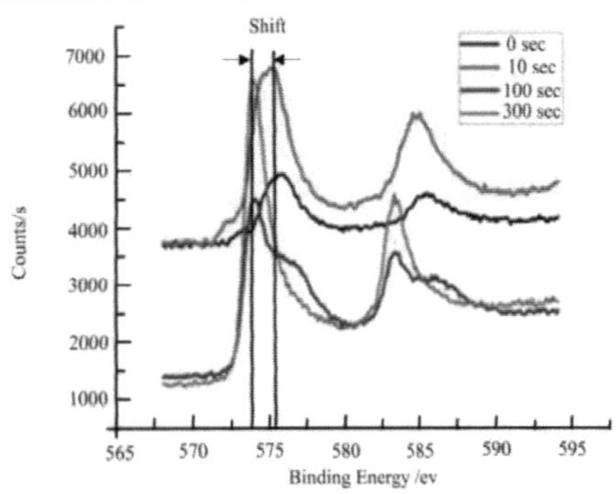

(a) stainless steel treated without laser masking

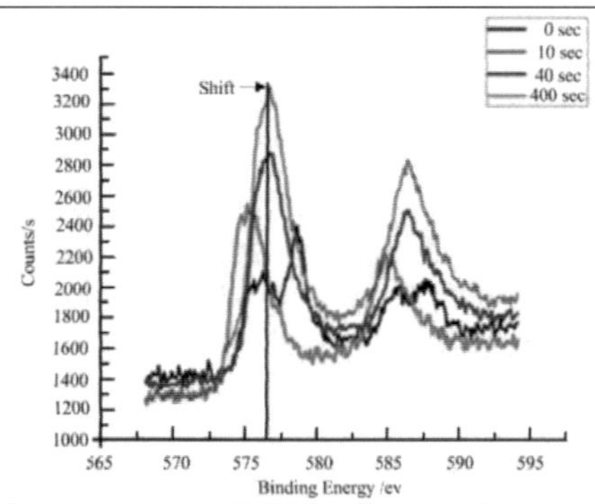

(b) stainless steel treated by laser masking

Figure 5. Cr 2p spectra with change of sputtering time

The laser masking layer test was carried out by ESCALAB 250Xi type X-ray photoelectron spectroscopy analysis.The thickness of the metal surface layer tested is related to the sputtering time, and the XPS spectrum of the inner layer on the metal surface can be obtained by using the longer sputtering time. The samples include two types: 304 stainless steel treated by laser masking and 304 stainless steel substrate without laser masking. As shown in Figure 5 (a), the peak energy of the binding energy decreases from 575.8eV (chromium oxide state) to 574.0eV (chromium metal state) as the sputtering time increases. This indicates that the Cr component on the surface of 304 stainless steel without laser masking is not modified as the sputtering time increases, which is coated with a very thin layer of chromium oxide. As shown in Figure 5 (b), as the sputtering time increases, the binding energy peak is fixed around 577.1eV, showing a chromium oxidized state of matter (Cr_2O_3, CrO_3). This indicates that there is a thick layer of chromium oxide layer (Cr_2O_3, CrO_3) on the 304 stainless steel treated by laser masking.

Figure 6 shows the XPS energy spectrum of Fe on the surface of 304 stainless steel with the change of sputtering time. With the increase of sputtering time, the binding energy peak decreases from 710.1eV (for iron state) reduced to 707.1eV (iron metal state) on the surface of 304 stainless steel substrate without laser masking , as shown in Figure 6 (a), which indicates that Fe valence does not change and there is a very thin layer of iron oxide layer on the surface of 304 stainless steel surface without laser masking. But as shown in Figure 6 (b), with the increase of sputtering time, the binding energy peak is fixed at 710.0eV(corresponding to the ferric oxide state), which indicates that the surface of 304 stainless steel treated by laser masking is covered with a thick layer of iron oxides (FeO and Fe_2O_3) .

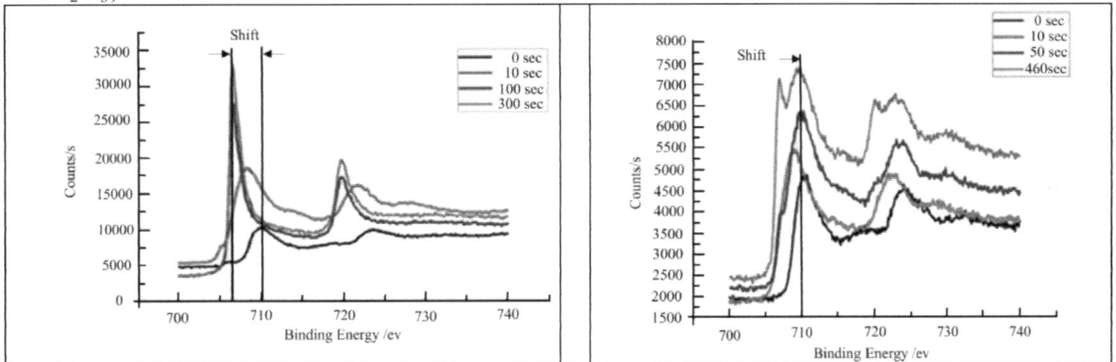

(a) stainless steel without laser masking

(b) stainless steel treated by laser masking

Figure 6. Fe 2p spectra with change of sputtering time

According to Faraday's law, electrochemical removal rate v_a

$$v_\alpha = \eta \omega I \quad (1)$$

where η is the machining current efficiency, I is current density, ω is electrochemical equivalent. In Figure 7, the surface of 304 stainless steel treated by laser masking is coated with Cr, Fe and other oxides, so the resistance R_1 of masking layer is greater than the resistance R_2 of no-masking layer.Then, the current density I_1 between the electrode and the masking layer is much smaller than the current density I_2 between tool electrode and no-masking layer current density, and the material dissolved removal rate of no-masking region is much greater than that of masking region. Therefore, the oxide layer produced by the laser masking plays a major protective role in EMM and can realize the localized machining.

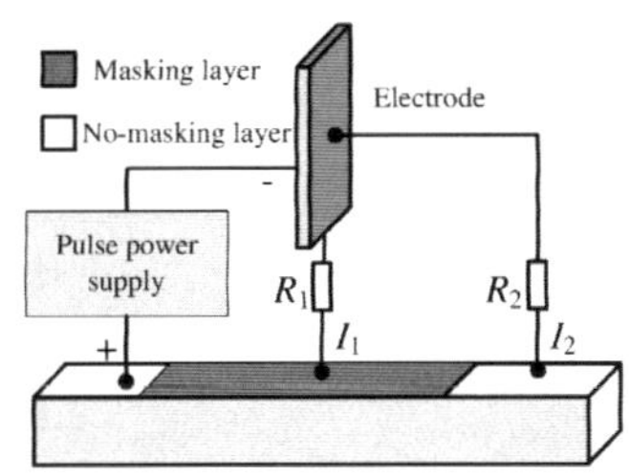

Figure 7. Localized processing mechanism of electrochemical machining with laser masking

IOP Conf. Series: Materials Science and Engineering **250** (2017) 012056 doi:10.1088/1757-899X/250/1/012056

3. Experiments of EMM with Laser Masking

The experimental researches on EMM with laser masking were carried out on the developed device of EMM with fiber laser masking. First, the surface modification of 304 stainless steel workpiece was treated by laser masking with suitable laser parameters, as shown in table 1. The samples of 304 stainless steel treated by laser masking are fabricated by EMM in the electrolytic tank, the simple sheet electrode of stainless steel serves as the cathode electrode, 304 stainless steel treated by laser masking is anode, and the CNC system makes the tool electrode scan reciprocally above the stainless steel workpiece during EMM and strictly controls the unchanged distance between the electrodes in the process of EMM. The EMM parameters are shown in Table 2, the electrolyte is 1.8mol/L sodium nitrate solution, the electrolyte is powered by micro diaphragm pump to flow with high-speed through small nozzle jet, the electrolyte flow rate is 320ml/min,which makes electrolytic products fast wash out and makes electrolytic processing smoothly.

Table 1. Laser processing parameters

Laser machining parameter	Value
Average laser power /W	5
Frequency /kHz	80
Laser scaning speed / （mm/s）	20
Beam diameter /μm	50
Line spacing /μm	10

Table 2. Electrochemical processing parameters

EMM parameter	Value
Pulse width /μs	50
Pulse interval /μs	750
Machining voltage /V	9.5
$NaNO_3$ electrolyte concentrations /(mol/L)	1.80
Interelectrode gap /μm	220
Sheet electrode scaning speed / （μm/s）	200

The cavity depth of one time EMM with laser masking is very small, and the depth of the machined cavity is remarkably improved by many times repeated processes of EMM with laser masking. Figure 8 is the samples fabricated through EMM with laser masking three times, and the cavity depth reach about 23μm, the width of the groove is about 120μm in Fig.8(b). In the process of EMM with laser masking, the scanning speed of the laser masking is 20mm/s, and many samples can simultaneously be fabricated.

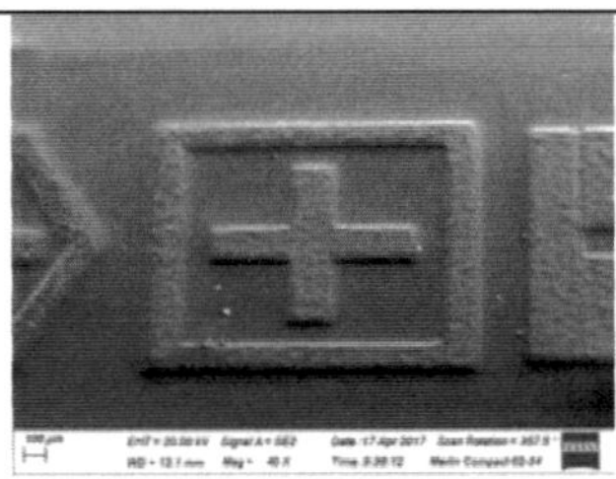

(a)Raised cross cavity

(b)Recessed cross cavity

Figure 8. Typical structural sample

Figure 9 is a morphology of a cavity of figure 8 (a) measured by an optical profilometer, including the height difference, roughness, and sectional contour of the cavity. From Figure 9, the cavity side wall fabricated is very steep, micro cavity surface fabricated is very smooth, and the surface roughness Sa value is 0.263μm. The stainless steel surface formed by electrolytic corrosion at no-masking area is as shown in Figure10. Because the electrode scanning speed of 200 μm/s during EMM and it is unnecessary for the micro die-sinking electrode to be fabricated, so the efficiency of fabricating each cavity by EMM with laser masking is very high. The technology has a good prospect in respect of fabricating micro channels, micro mold cavities, etc.

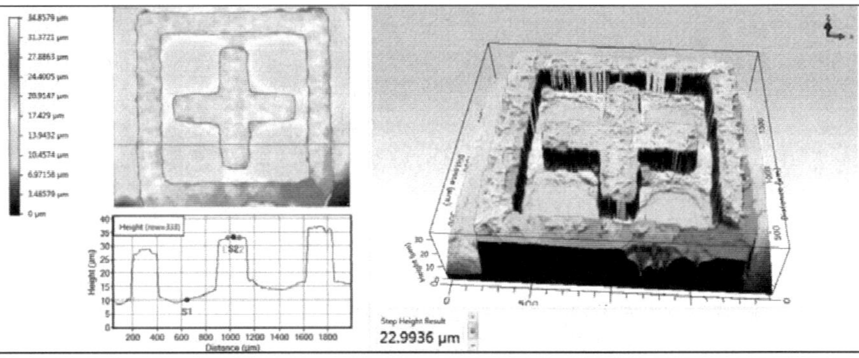

Figure 9. Measurement of microstructure by optical profilometer

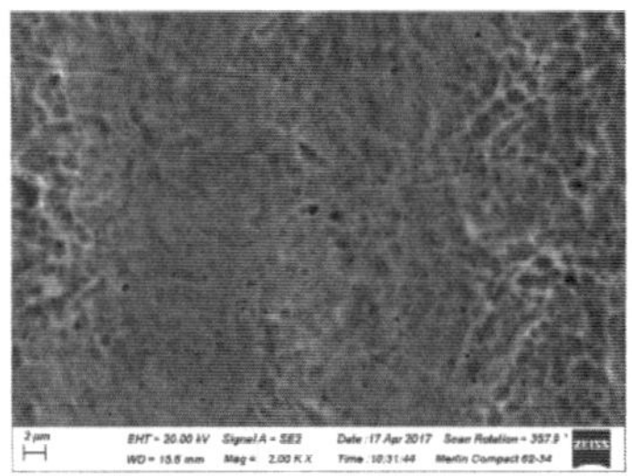

Figure 10. Surface morphology after electrochemical machining

4. Conclusion

The surface modification of 304 stainless steel surface was done by using fiber laser masking on the surface of 304 stainless steel to produce Fe and Cr oxide masking layer with corrosion resistant performance. A device of EMM with laser masking was developed to realize micro machining by combining laser masking with EMM. The experiments indicate that EMM with laser masking technology can rapidly fabricate the micro size cavities, which has a good prospect in micro machining .

5. Acknowledgements

The project is supported by National Natural Science Foundation of China(51375208) and Scientific and Technological Innovation Team of Jiamusi University (Cxtdpy-2016-04).

6.References

[1] Kock M, Kirchner V, Schuster R 2003 *Electrochimica Acta* Vol. **48** pp3213-3219
[2] Yu Q, Zhu D, Zeng Y B 2012 *Acta Aeronautica ET Astronautica Sinica* Vol. **33** pp1-8
[3] Fan Z J, Wang G G, Tang L 2010 *Journal of Mechanical Engineering* Vol.**46**(1)pp194-198
[4] Bao H Q, Xue J W, Cao X M 2009 *Journal of Harbin Institute of Technology* Vol.**5** pp234-237
[5] Zeng Y B, Cai W W, Li H S 2014 *Electromachining & Mould* (5) pp27-41.
[6] Zhang C Y, Qin C L, Feng Q Y 2015 *Mechanical science and technology* Vol. **34** (7) pp1031-1034
[7] Pajak P T, Desilva A K, Harrison D K 2006 *Precision Engineering*,Vol.**30**(3)pp288-298.
[8] Styer P, Valette S, Forest B 2006 *Surface Engineering* Vol. **22**(3) pp167-172.

IOP Conf. Series: Materials Science and Engineering **250** (2017) 012057 doi:10.1088/1757-899X/250/1/012057

Effect of Discharge Rate on Positive Active Material of Lead Carbon Battery for Energy Storage

Kailun Chen[1,2], Hao Liu[2,*], Chen Hu[2], Fei Gao[2], Kai Yang[2] and Hao Wang[1]

[1]Beijing University of Technology, No.100, Pingleyuan, Beijing, China
[2]China Electric Power Research Institute, No.15, Xiaoying Rd(E), Qinghe, Beijing, China
Email: 13811586984@163.com

Abstract. Lead carbon battery has been widespread concern with its excellent performance of charge and discharge under High Rate Part State of Charge (HRPSoC) as well as its cycle performance. In this paper, the cycling performance of lead carbon battery for energy storage was tested by different discharge rate. The effects of different discharge rate on the composition and morphology of positive active materials in the cycle was studied by XRD and SEM. The effect of different discharge rate on the ohmic impedance of lead carbon battery was studied by testing Electrochemical Impedance Spectroscopy with different capacity retention rates. The results show that with the increase of the discharge rate, the content of PbO_2 in the positive active material increases, the active substance utilization and the particle size of PbO_2 crystal declines, and the ohmic impedance of the battery decreases.

1. Introduction

Long life, low cost and high security are the basic requirements of a power system for energy storage batteries. Lead-acid batteries are widely used as storage batteries currently because of their low cost, mature technology and high security. But their short cycle life[1] cannot meet the needs of the large-scale energy storage technology in the new energy field for long life batteries[2].

A lead carbon battery has some carbon material introduced into its negative electrode based on lead-acid batteries, and thus significantly improves its performance of rapid charge and discharge in the high rate part state of charge (HRPSoC) as well as its cycle performance. The negative carbon material can significantly improve the negative electrode sulfation of the lead carbon battery in the HRPSoC mode[3-4]. With the progress of the research, the performance of the negative electrode is becoming better, the bottleneck of battery life transferred to the positive electrode, which fails mainly due to active material softening and directly affects the battery life[5].

In order to further analyze the performance of the lead carbon battery in the field of energy storage, this paper probes into the cycle performance of the lead carbon battery tested at different discharge rates and analyzes the effect of different cycle conditions on the positive active materials.

2. Experimental Method

2.1. Experimental Battery

Tailor-made 2V/15Ah lead carbon batteries are used to cycle under different conditions for cycle performance test, electrochemical impedance spectroscopy (EIS) test, X-ray diffraction (XRD) test and scanning electron microscope (SEM) test.

2.2. Battery Charge and Discharge Cycle Test

The Digtron and Maccor battery testers are used to test the capacity and cycle performance of lead carbon batteries. A research is made through high-temperature accelerated experiments on the effect of different discharge rates on the positive active materials and its mechanism. A total of four batteries are employed, with one of them as reference not cycled. The reference battery is used to compare with the cycled batteries in composition and morphology. The C_4 capacity is calibrated and the electrochemical impedance spectroscopy is tested after each cycle. Table 1 shows the initial C_4 capacity and cycle conditions of each battery:

Table 1. The working conditions of batteries

Experimental Condition	#1	#2	#3	#4
C_4 capacity	14.06	14.01	13.96	13.94
Temperature	\	40°C	40°C	40°C
Charge rate	\	0.2C	0.2C	0.2C
Discharge rate	\	0.5C	1C	2C
Depth of charge and discharge	\	30% SoC - 80% SoC		

2.3. EIS Test

PGSTAT 302N workstation made by Metrohm is used to test the electrochemical impedance spectroscopy of batteries. The batteries are set at 80% SoC, with the test amplitude at 3mV and the frequency range between 0.1Hz and 100kHz.

2.4. XRD and SEM Tests of Positive Plates

The batteries to be dismantled for positive active materials are set at 80% SoC (base on the latest C_4 capacity calibration after cycle). D8 Advance X-ray diffractometer made by Bruker is used for XRD tests, with the test range between 20° and 80° and the scan speed at 0.2°/s. S4800 field emission scanning electron microscope made by Hitachi is used for SEM tests

3. Results and Discussion

The C_4 capacity is calibrated at 25 °C after cycle and compared with its value before the cycle, and the capacity retention rate of the battery is thus obtained.

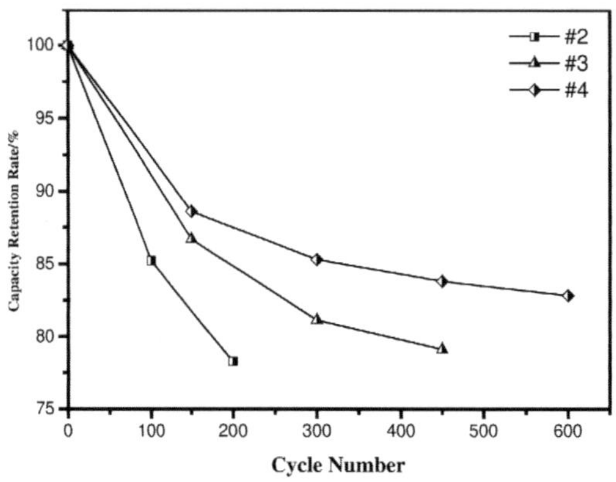

Figure 1. The cycle number vs. capacity retention rate

Figure 1 shows the relations of cycle times with capacity retention rates of #2, #3 and #4 batteries at different discharge rates. The capacity retention rates of #2, #3 and #4 batteries drop to 78.29% (after 200 cycle times), 79.12% (after 450 cycle times) and 82.83% (after 600 cycle times) respectively. These three batteries almost have the same state of health at the end of the cycles.

3.1. Macroscopic Morphology of Positive Plates

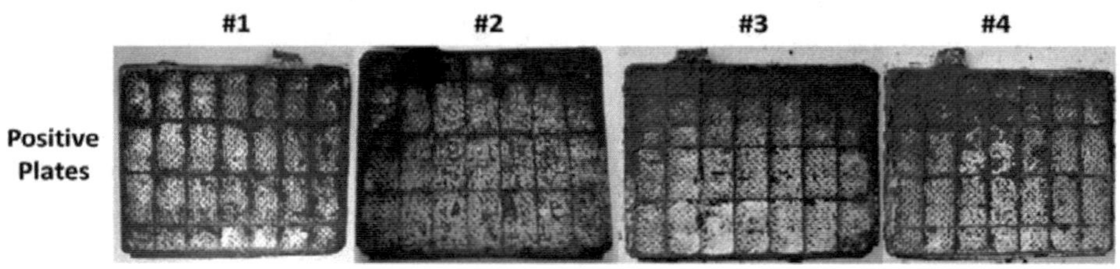

Figure 2. The macroscopic features of positive plates and separator before and after cycles

Figure 2 shows the positive plates and diaphragms before and after cycles. The positive plate of #1 battery which is not cycled is compared with those of #2, #3 and #4 batteries which are cycled under different working conditions in macro-surface state, micro-surface morphology and composition. The black solid matter on the positive plates is PbO_2, and the white solid matter is $PbSO_4$ which is dense and hard. The black residue on the diaphragms is softening PbO_2 powder.

3.2. Effect of Discharge Rate on Positive Active Material Composition

In figure 3, (a) shows the XRD patterns of the positive active materials of #1, #2, #3 and #4 batteries respectively. Since the dismantled batteries are set at 80% SoC, the four batteries contain both β-PbO_2 and $PbSO_4$. #1 battery is not cycled and therefore should theoretically contain some α-PbO_2[6], but the XRD test does not show any α-PbO_2. The reason is that most of the original α-PbO_2 is consumed in the process of activation and calibration. α-PbO_2 can only be produced under weak base conditions and no α-PbO_2 is further produced in the battery. As a result, there is no α-PbO_2 on the positive plate of #1 battery. In figure 3, (b) shows contents of the positive active materials calculated according to the K-value method. The β-PbO_2 content in #1 battery is greater than those in #2 and #3 batteries, probably because the reversible capacity of #1 battery has not been yet recovered before cycles and the utilization of the active materials is lower at 80% SoC.

The capacity retention rates of #2, #3 and #4 batteries are similar to each other, and their β-PbO_2 contents directly reflects the utilization of active materials at 80% SoC. It can be seen from the data in the figures that the β-PbO_2 contents in the active materials increase along with the discharge rates after cycles. That is, the battery with high discharge current needs more active materials in the event of the same charge capacity. It means that high current discharge reduces the utilization of active materials. On the other hand, the $PbSO_4$ contents in the positive active materials decrease along with the increase in discharge current.

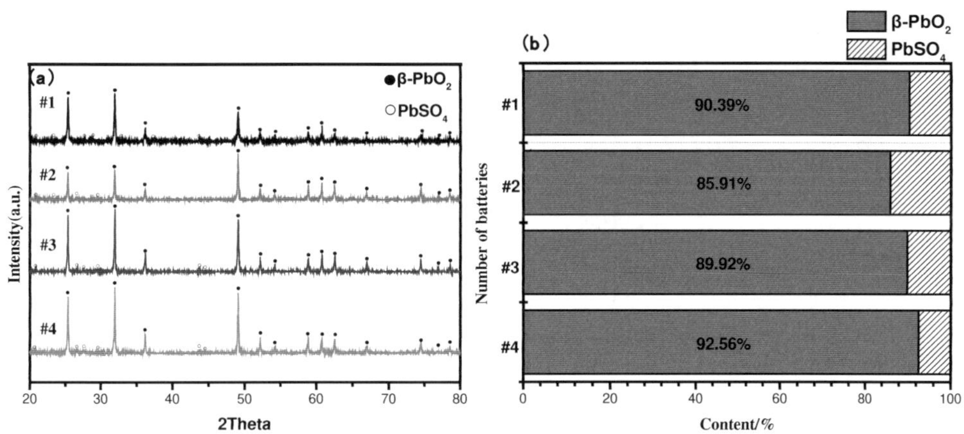

Figure 3. (a) The XRD patterns of positive active materials of #1-#4 Batteries; (b) β-PbO_2 content on positive plates of #1-#4 batteries

3.3. Effect of Discharge Rate on the Morphology of Positive Active Materials

In figure 4, (a), (b), (c) and (d) show the microscopic morphology patterns (magnified by 2000) of the positive active materials of #1-#4 batteries respectively. It's (a) shows that the initial positive active materials mainly consisting of homogeneously distributed fine β-PbO_2 particles and a small amount of $PbSO_4$ particles. (b), (c) and (d) in figure 4 show the positive active materials after cycles, with the particles significantly bigger.

The particles of positive active materials become bigger after cycles. The positive active materials of the three batteries after cycles have the similar particle adhesion, which is less than that of the initial positive active materials. Therefore, the charge and discharge cycles of the batteries can gradually change the β-PbO_2 in the active materials from small dense particles to large loose ones.

It is known from figure 4 that (b), (c) and (d) that the β-PbO_2 particles in the active materials become smaller with the increase in the discharge rate. A high discharge rate causes more electrochemical activity points on the plate in the process of discharge, and thus produces more $PbSO_4$ particles per unit time. However, the $PbSO_4$ particles are smaller due to the short discharge time. Small $PbSO_4$ particles could change into small β-PbO_2 particles in the process of charge at the same rate. It is more

IOP Conf. Series: Materials Science and Engineering **250** (2017) 012057 doi:10.1088/1757-899X/250/1/012057

likely for the particles to come off from the plate due to the loose structure of the small particles.

Figure 4. (a), (b), (c) and (d) The SEM patterns of positive active materials of #1-#4 batteries

3.4. Effect of Discharge Rate on Electrochemical Impedance Spectroscopy of Battery

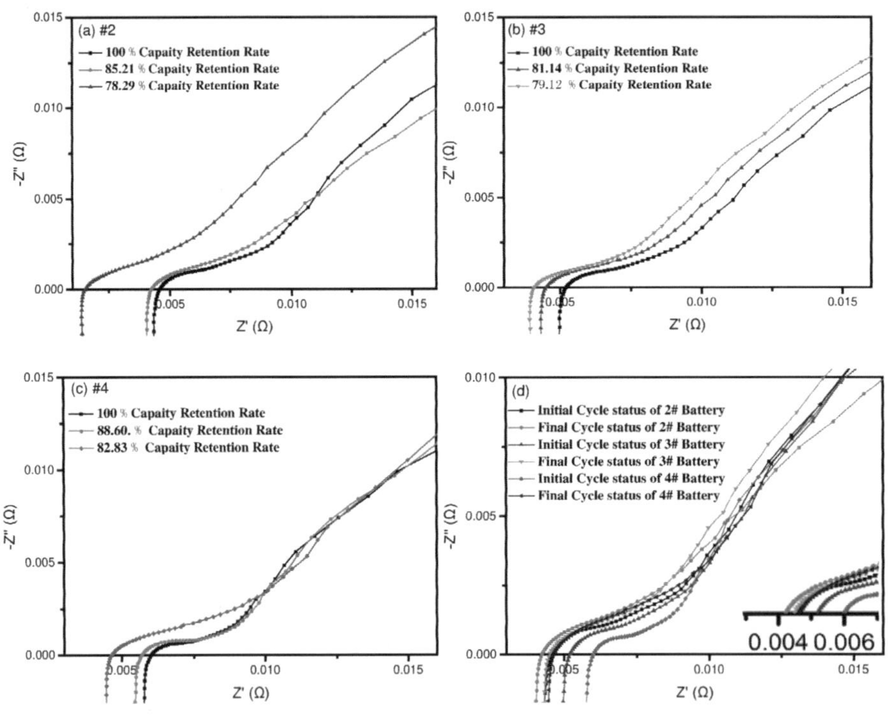

Figure 5. (a), (b) and (c) The EIS patterns of #2-#4 batteries at 80% SoC
(d)The EIS pattern of #2-#4 batteries at 80% SoC before and after Cycles

In figure 5, (a), (b) and (c) show the EIS patterns of #2, #3 and #4 batteries at 80% SoC at different capacity retention rates. It can be seen from figure 5 that the batteries have the higher ohmic impedance before cycles. The ohmic impedance decreases gradually with the decrease in capacity retention rate, because the positive plates are covered with $PbSO_4$ at 80% SoC before cycles and the batteries have a higher ohmic impedance. The irreversible dense $PbSO_4$ on the positive plates is gradually converted into more dispersed reversible $PbSO_4$. As a result, the active materials are more conductive, with electrons more easily reaching inside the active materials, so the ohmic impedance decreases. (d) shows that the ohmic impedance of #2, #3 and #4 batteries increasingly decreases with the decrease in capacity retention rate. It means that a higher discharge rate can decrease the ohmic impedance greatly when the batteries have the same capacity retention rate. It is evidenced that a higher discharge rate could decrease the $PbSO_4$ content on the positive plate, and thus reduce the plate resistance.

4. Conclusions

The electrochemical performance of lead carbon batteries and the composition and morphology of the active materials are obtained by means of testing the electrochemical impedance spectroscopy of lead carbon batteries at different discharge rates during cycles and the XRD and SEM of the positive active materials after cycles. The results show that:

1) When lead carbon batteries are used at higher discharge rate, the discharge rate directly affects the size of PbO_2 particles in the positive active material. The higher the discharge rate, the smaller the PbO_2 particles.

2) The discharge rate directly affects the utilization rate of PbO_2 during cycles. The utilization rate of PbO_2 decreases with the increase in discharge rate.

3) The $PbSO_4$ near the tab is converted into the active material PbO_2 during cycles, and the internal resistance of battery decreases. The higher discharge rate, the lower ohmic impedance.

5. References

[1] Sawai K, Funato T, Watanabe M, et al. Development of Additives in Negative Active-material to Suppress Sulfation During High-Pate Partial-State-of-Charge Operation of Lead–acid Batteries[J]. *Journal of Power Sources*, 2006, 158(2):1084-1090.

[2] Qiangqiang Liao, Lin Qiu, Xiaodong Lou, et al. Performance of Lead Carbon Battery and Its Application in Electric Energy Storage [J]. *Power Construction*, 2014(11):117-121.

[3] Lam L T, LoueyR, Haigh N P, et al. VRLA ultrabattery for high-rate-partial-state-of-charge operation[J]. *Journal of Power Sources*, 2007, 174(1):16-29.

[4] Schaeck S,Karspeck T,Ott C,et al. A field operational test on valve-regulated lead-acid absorbent-glass-mat batteries in micro-hybrid electric vehicles.Part I. Results based on kernel density estimation[J]. *Journal of Power Sources*,2011,196(5):2924-2932.

[5] Mian Zhao, Yan Zhang, Binjuan Wei, et al. Characterization and Electrochemical Properties of Softening Positive Plate Materials for Lead Acid Batteries[J]. *Battery*, 2017,54(2).

[6] Duyou Wang. Study on Softening and Shedding of Cathode Active Material[J].*Electric Bicycle*,2014(11):40-44

Corrosion Behavior of L80Steel in Different Temperature and Sulfur Content

Zhichao Qiu, Chunming Xiong, Ran Yi and Zhengrong Ye

PetroChina Research Institute of Petroleum Exploration & Development,
P.O.Box 910, No.20 Xueyuan Road, Haidian District, Beijing 100083 P.R.China
E-mail: qiuzhichao@petrochina.com.cn

Abstract. To understand the corrosion behavior of L80 steel in different temperature and sulfur content, the experiment which simulated the downhole corrosive environment was conducted. From the experiment result, when other factors were constant, the lowest corrosion rate was appeared when the temperature was 90° C. The influence of sulfur was complex. When temperature was low, the corrosion rate was decreased with the increase of sulfur content and the experimental result was opposite when temperature was high.

1. Introduction

During the development of oil and gas reservoirs, there often exist acidic elements, such as CO_2gas, H_2S gas and sulfur element which may corrode the pipes. In recent years, as more and more high temperature and high acidic carbonate reservoirs have been, or will be, developed, it is increasingly imperative to address the pipe corrosions.L80 steel is the most common used steel in oil and gas fields. So it is meaningful to understand the corrosion behavior of L80 steel in different corrosive environment, especially in different temperature and sulfur content. According to the literatures presented before[1-8], most of the research focus on the influence of the partial pressure of sour gas, such as H_2S and CO_2. However in the production of oil field, sulfur which is the corrosive element was found in the content of oil. In this paper, the experiment which simulated the downhole corrosive environment was conducted to test the corrosion behavior of L80 steel.

2. Experiment Preparation

The sample of L80 was used for corrosion test. The chemical composition of L80 steel was shown in table 1.The size of the sample for corrosion test was 50×10×3mm.

Table 1.The chemical composition of L80 steel

Steel	C	Mn	Mo	Cr	Al	Ni	Cu	P	S	Si
L80	~0.22	~1.00	/	~1.2	/	~0.50	~0.25	~0.20	~0.010	~1.00

The medium for corrosion test was shown in table 2. All of medium was prepared by the de-ionized water and AR regent. The sour gas of H_2S and CO_2 was inserted into the autoclaveand the content of gas was adjusting by the gas flow valve.

IOP Conf. Series: Materials Science and Engineering **250** (2017) 012058 doi:10.1088/1757-899X/250/1/012058

Table 2.The medium composition for corrosion test

No	T/°C	P_{H2S}/ MPa	P_{CO2}/ MPa	Cl^-/ g.L^{-1}	S/%	H_2O/%
1	60	0.1	0.5	0.13	2	30
2	90	0.1	0.5	0.13	2	30
3	150	0.1	0.5	0.13	2	30
4	60	0.1	0.5	0.13	4	30
5	90	0.1	0.5	0.13	4	30
6	150	0.1	0.5	0.13	4	30
7	60	0.1	0.5	0.13	2	70
8	90	0.1	0.5	0.13	2	70
9	150	0.1	0.5	0.13	2	70
10	60	0.1	0.5	0.13	4	70
11	90	0.1	0.5	0.13	4	70
12	150	0.1	0.5	0.13	4	70

The sample treatment: First, the sample with size of 50×l0×3mm was cut from the block of the raw materials of L80 steel, polished by sand paper with grid degree of 300#-1000# and cleaned by alcohol and acetone, and dried for half of an hour in dryer. Finally the mass of the sample was weight out by scale.

Experiment processing:①All of chemicals were added into the autoclave.②The support hold three parallel samples were put into the autoclave, and the autoclave was covered tightly.③The heater and stirrer were open to mix the medium and make the test temperature up to the need.④The down-pressure value and flow counter were open, and the gas of H_2S and CO_2 were inserted into the autoclave. The input value of the autoclave and down-pressure value of the vessel were shut down when the pressure of H_2S and CO_2 was up to the design value. ⑤The corrosion test was started. The temperature fluctuation range was controlled to ±1°Cand the test time was continued to 720h during the test. ⑥After the experiment was finished. Firstly, the power of heater was shut off, and the sample was taken out from the autoclave after the autoclave was cooled to room temperature. Secondly, the corrosion production on the sample surface was cleaned according to the standard of "GB/16545-1996 corrosive product on the corrosive sample of metal and alloy", and the sample was cleaned and got rode of water using alcohol and acetone. Finally the mass of sample was weight and recorded after the sample cleaned was kept in the dryer for at least 24h.

3. Experimental Result and Discussion

The corrosion rate of L80 steel sample in corrosive medium was shown in table 3. The table 3 indicated that the lowest corrosion rate of 0.6833 mm·a^{-1} of the sample was arrived in the medium that was composed of Temperature of 90°C, S of 4% and water of 70%, and the highest corrosion rate of 2.0835 mm.a^{-1} of the sample was arrived in the medium that was composed of Temperature of 150°C, S of 4% and water of 30%.

Table 3.The corrosion rate of L80 steel

No	T/°C	P_{H2S}/ MPa	P_{CO2}/ MPa	Cl^-/ g.L^{-1}	S/%	H_2O/%	V/mm.a^{-1}
1	60	0.1	0.5	0.13	2	30	1.5568
2	90	0.1	0.5	0.13	2	30	1.0627
3	150	0.1	0.5	0.13	2	30	1.3941
4	60	0.1	0.5	0.13	4	30	1.4670
5	90	0.1	0.5	0.13	4	30	0.7794
6	150	0.1	0.5	0.13	4	30	2.0835
7	60	0.1	0.5	0.13	2	70	1.1466
8	90	0.1	0.5	0.13	2	70	0.9401
9	150	0.1	0.5	0.13	2	70	0.9807
10	60	0.1	0.5	0.13	4	70	1.0967
11	90	0.1	0.5	0.13	4	70	0.6833
12	150	0.1	0.5	0.13	4	70	1.0032

The effect of temperature, sulfur contenton the corrosion rate of the sample was shown as following:

1) Temperature: From the table 3 and figure 1, the lowest corrosion rate was achieved when the temperature is 90°C. It was seen that the corrosion rate of L80 steel decreased with the increasing of temperature firstly when the content of sulfur and water was constant. Then as the increase of temperature, the corrosion rate increased.

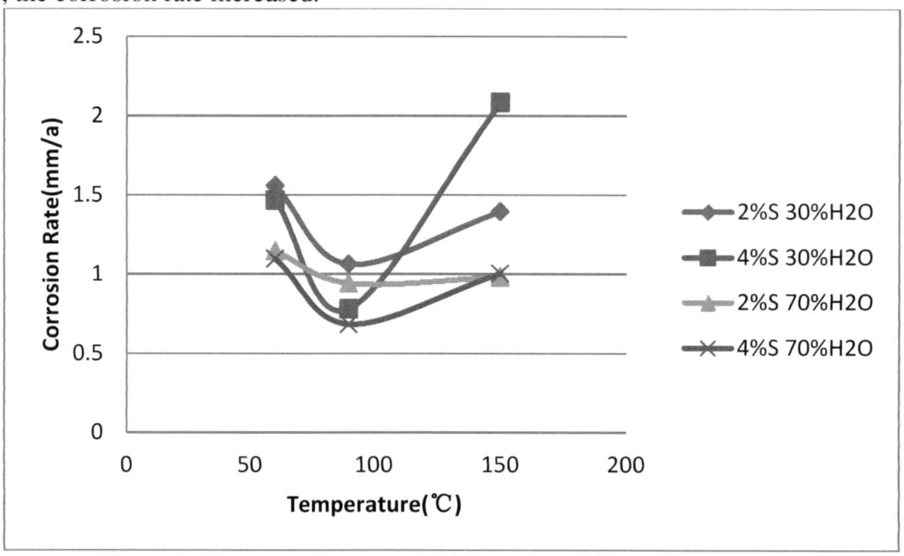

Figure 1. The corrosion rate of L80 steel in different temperature

2）Sulfur: When sulfur content increased from 2% to 4%, and the content of temperature and water was constant, if the temperature of corrosion medium was lower than 90°C(60°Cand 90°C), the corrosion rate of L80 steel would decrease, and the corrosion would be inhibited. However if the temperature of corrosion medium was up to 150°C, the corrosion rate of L80 steel would increase with the increase of sulfur content.At the lower temperature the sulfur maybe only deposited on the surface of the steel. Although the deposition of sulfur could induce the severe localized corrosion, the general corrosion was mild. So the corrosion rate decreased with the increasing of sulfur content. Nevertheless at high temperature the sulfur would react with the corrosion product and the corrosion product would be destroyed. The protection and stabilization of corrosion film was destroyed. So the corrosion rate was increased with the increase of sulfur content. From the experimental result, the contribution of sulfur to the L80 steel can be summarized in two aspects as following: on the one hand, the deposition of sulfur could induce the severe localized corrosion; On the other hand, the sulfur could react with the corrosion product and corrosion film.

4. Conclusions

From the result of corrosion experiment, when other corrosive factors, such as water cut, sulfur content and the partial pressure of CO_2 and H_2S gas were constant, the lowest corrosion rate was appeared when the temperature was 90°C. The influence of sulfur content was complex. When temperature was low(60°Cand 90°C), the corrosion rate was decreased with the increase of sulfur content. When temperature was high(150°C), the corrosion rate was increased with the increase of sulfur content.

5.Acknowledgement

This research was financially supported by the National Science and Technology Major Project of the Ministry of Science and Technology of China (2016ZX05016-004) and (2016ZX05015-002).

6.References

[1] Michel Bonis, *Weight loss corrosion with H_2S: from facts to leading parameters and mechanisms*, NACE, 2009, 09564

[2] TapanChakrabarty, Richard J.Smith, *Identification of SSC (Sulfide Stress Cracking)-Susceptible Wells and Risk Prediction*, SPE, 2012, 160489-MS

[3] Craig B.D, Thompson R.S, *An Expert Advisor Program for SSC and NACE MR-01-75*, SPE, 1991, 21741-MS

[4] Masakatsu Ueda, Akio Ikeda, *Effect of Microstructure and Cr Content in Steel on CO_2 corrosion,* Corrosion, 1996, 13

[5] Bruce Craig, *Metallurgical Limits to Increasing Tubular Strength While Maintaining SSC Resistance,* SPE, 2005, 97580-MS

[6] Raul B.Rebak,, Teresa E. Perez, etal, *Effect of Carbon Dioxide and Hydrogen Sulfide on the Localized Corrosion of Carbon Steels and Corrosion Resistant Alloys,* NACE, 2017, 8933

[7] Li Hui, Kwei Meng Yap, Sridhar Srinivasan, *Enhanced Corrosion Prediction Model for Multiphase Oil & Gas Production Systems,* NACE, 2017,9419

[8] G.Verri, K.S.Sorbie, M.A.Singleton, et al, *Iron Sulphide Scale Management in High H_2S and CO_2 Carbonate Reservoirs,* SPE, 2016, 179871-MS

IOP Conf. Series: Materials Science and Engineering **250** (2017) 012059 doi:10.1088/1757-899X/250/1/012059

Williams Element with Generalized Degrees of Freedom for Fracture Analysis of Multiple-Cracked Beam

Hua Xu[1,a], **Quyang Wei**[1,b] **and Lufeng Yang**[1,2,c]

[1]Key Laboratory of Disaster Prevention and Structural Safety of China Ministry of Education, Guangxi Key Laboratory of Disaster Prevention and Engineering Safety, School of Civil Engineering and Architecture, Guangxi University, 530004, Nanning, China
[2]Department of Housing and Urban-Rural Development, Guangxi Zhuang Autonomous Region, 530028, Nanning, China
E-mail: [a]xuhua@gxu.edu.cn(corresponding author), [b]weiquyang@163.com, [c]lfyang@gxu.edu.cn

Abstract. In this paper, the method of finite element with generalized degrees of freedom (FEDOFs) is used to calculate the stress intensity factor (SIF) of multiple cracked beam and analysed the effect of minor cracks on the main crack SIF in different cases. Williams element is insensitive to the size of singular region. So that calculation efficiency is highly improved. Examples analysis validates that the SIF near the crack tip can be obtained directly though FEDOFs. And the result is well consistent with ANSYS solution and has a satisfied accuracy.

1. Introduction

Beam system is widely used in architectural engineering, bridge engineering and underground engineering. Structure is easily cracked cracks during application. And most of the cracks are multiple cracks. These cracks will reduce structural stiffness and integrity, which causes a serious threat on the structure. This paper is applied to fracture analysis of multiple cracked beam using FEDOFs compared with ANSYS.

The safety analysis of structure with cracks usually needs to determine the stress field or the displacement field near the crack tip. Due to SIF could be efficiently manifested as the degree of crack tip stress - strain field, crack tip SIF becomes an important research object on structural linear elastic fracture analysis. Researchers apply different kinds of methods to calculate the value of crack tip SIF on a single crack beam, such as mathematical and mechanical theory [1], the boundary collocation method [2], and the singularity element method [3]. However, the SIF study of multiple crack beam is rare no matter at home or abroad. Rohde [4] combined analytical theory with geometry factors proposed an efficient algorithm for calculating the crack tip SIF containing multiple boundary cracked beam. Ooi [5] integrated SBFEM(scaled boundary finite element method) with finite element, and used SBFEM in the extended area to analyze the structure displacement changes on three sided-cracks and double edge cracks beam under different loads and the cracks propagation with loading. Qingyuan Wang [6] applied the singularity element to analyze the three-point bending beam that contains three cracks. In additional, the influence of the change of minor crack length and its position modification to the main crack's crack tip SIF as well as mid-span deflection was discussed. Obviously, the research of multiple cracks fracture analysis still lack of achievement and remains to be furthered.

Like industry standards such as UML activity diagrams, Business Process Model and Notation and EPCs, Petri nets [7] offer a graphical notation for stepwise processes that include choice, iteration, and concurrent execution. Unlike these standards, Petri nets have an exact mathematical definition of their execution semantics, with a well-developed mathematical theory for process analysis. Williams generalized parameters element (W element) has been well applied [8-10] in fracture analysis that relates to the center-cracked plate and the three-point bending beam with higher precision and efficiency. This thesis will establish W unit calculating format for fracture analysis of multiple cracked beam. At the same time, using singularity element as contrast to verify the correctness of the results, and providing possible engineering advice.

2. Williams Element and SIF

2.1. *Williams Element*

Assuming n strips of vertical cracks are on the low edge of a simply supported beam, now taking 3 strips as example and discretizing them using finite element mesh, as shown in figure 1.

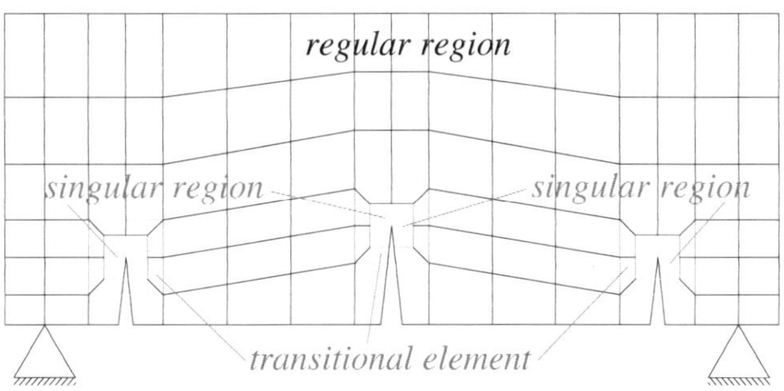

Figure 1. Discretized mesh with 3 edge cracks.

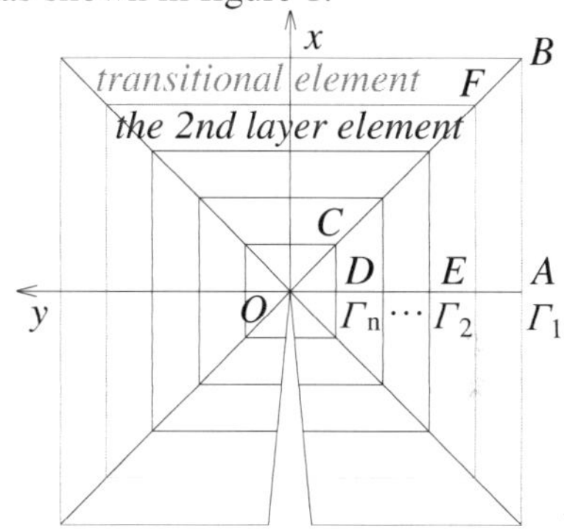

Figure 2. Discretized elements around the crack tip.

The interior of the red region represents the ranges of crack tip singular region, and out of it is regular region. It can be learned from reference [11], the size of the crack tip singular region almost has no effect on the result of SIF, and every sub-element crack tip region selects the same size of the singular region for easily to discretize finite element mesh. The outermost of the singular region called sub-element ABFE, own the same edge with adjacent regular region element, as a result, the element one part in the regular region, and another in the singular region, termed as transitional element, which is between the blue line and the red line in figure 1 and figure 2.Crack tip triangle element sub-element OCD is in the crack tip damaged region, when ignore its element stiffness contribution to the global stiffness, termed trapezoidal element as Williams element (W element), or Williams macro-element.

2.2. *Sub-Element SIF at the Crack Tip*

In this thesis, Williams series is selected as the arbitrary sub-element global displacement interpolation function. Thus its displacement field can be rewritten in matrix form, and intercept first m terms of the series:

$$\{w\}=[H(r,\theta)]\{a\} \tag{1}$$

where, $[H(r,\theta)]=\begin{bmatrix} 1 & 0 & r^{1/2}h_{11}^{x}(\theta) & r^{1/2}h_{12}^{x}(\theta) & \cdots & r^{m/2}h_{m1}^{x}(\theta) & r^{m/2}h_{m2}^{x}(\theta) \\ 0 & 1 & r^{1/2}h_{11}^{y}(\theta) & r^{1/2}h_{12}^{y}(\theta) & \cdots & r^{m/2}h_{m1}^{y}(\theta) & r^{m/2}h_{m2}^{y}(\theta) \end{bmatrix}$; $\{w\}=\{u,v\}^{\mathrm{T}}$;

$\{a\}=\{u_0 \quad v_0 \quad a_1 \quad b_1 \quad \cdots \quad a_i \quad b_i \quad \cdots \quad a_m \quad b_m\}^{\mathrm{T}}$. And has

$$h_{i1}^{x}(\theta)=\frac{1}{2G}[(\kappa+\frac{i}{2}+(-1)^{i})\cos\frac{i}{2}\theta-\frac{i}{2}\cos(\frac{i}{2}-2)\theta];h_{i2}^{x}(\theta)=\frac{-1}{2G}[(\kappa+\frac{i}{2}-(-1)^{i})\sin\frac{i}{2}\theta-\frac{i}{2}\sin(\frac{i}{2}-2)\theta]$$

$$h_{i1}^{y}(\theta)=\frac{1}{2G}[(\kappa-\frac{i}{2}-(-1)^{i})\sin\frac{i}{2}\theta+\frac{i}{2}\sin(\frac{i}{2}-2)\theta];h_{i2}^{y}(\theta)=\frac{1}{2G}[(\kappa-\frac{i}{2}+(-1)^{i})\cos\frac{i}{2}\theta+\frac{i}{2}\cos(\frac{i}{2}-2)\theta]$$

in which G is the shear modulus; κ is defined as: take 3-4μ in plane strain, take $(3-\mu)/(1+\mu)$ in plane stress , μ is the Poisson ratio; a_i、b_i is the undetermined coefficients, namely the general parameters in equation (1), they can be defined from external loading and boundary condition; (r, θ) is the polar coordinate taking sub-element crack tip as origin.

Every stress components can be obtained according to Williams series displacement expression, and it can be known from the linear elastic fracture mechanics: SIF at the crack tip can be obtained from corresponding stress components. In order to avoid morbid equations, post-processing shear modulus G, so when $r \to 0$ and $\theta=0$, one has:

$$K_{\mathrm{I}} = \sqrt{2\pi} G a_1 \ , \ K_{\mathrm{II}} = \sqrt{2\pi} G b_1 \tag{2}$$

Only needs obtaining corresponding column vector $\{a\}_i$ to each sub-element crack tip, and extracting general parameters a_1 and b_1 of each column vector, corresponding crack tip SIF can be obtained.

3. The Global Governing Equation

Solving global governing equation of domain. Discretizing the whole solution domain to n crack tip singular region and corresponding transition region and peripheral regular region, each crack tip singular region and transition region are discretized to multiple Williams element. It can learned from the constitution of Williams element that its element stiffness equation can be divided into blocks as follow:

$$\begin{bmatrix} [K]_{\mathrm{bb}}^{\mathrm{e}} & [K']_{\mathrm{bs}}^{\mathrm{e}} \\ [K']_{\mathrm{sb}}^{\mathrm{e}} & [K']_{\mathrm{ss}}^{\mathrm{e}} + [K'] \end{bmatrix} \begin{Bmatrix} \{w\}_{\mathrm{b}}^{\mathrm{e}} \\ \{a\} \end{Bmatrix} = \begin{Bmatrix} \{f\}_{\mathrm{b}}^{\mathrm{e}} \\ \{f\}_{\mathrm{s}}^{\mathrm{e}} + \{f'\} \end{Bmatrix} \tag{3}$$

where $[K']_{\mathrm{sb}}^{\mathrm{e}} = [T']^{\mathrm{T}} [K]_{\mathrm{sb}}^{\mathrm{e}} = [K']_{\mathrm{bs}}^{\mathrm{eT}}$; $[K']_{\mathrm{ss}}^{\mathrm{e}} = [T']^{\mathrm{T}} [K]_{\mathrm{ss}}^{\mathrm{e}} [T']$; $[K'] = \sum_{k=2}^{n} [K]^{(k)}$, in which b is the common boundary nodes of regular region and crack tip singular region , subscript s is the singular region.

Integrating the whole Williams element of n crack tip and the whole element stiffness of regular region to obtain global stiffness, and without considering the condition of loading in crack tip singular region, then solve global governing equation of domain:

$$\begin{bmatrix} [K]_{\mathrm{rr}} & [K]_{\mathrm{rb}} & [0] & [0] & [0] & [0] \\ [K]_{\mathrm{br}} & [K]_{\mathrm{bb}} & [K']_{\mathrm{bs_1}} & [K']_{\mathrm{bs_2}} & \cdots & [K']_{\mathrm{bs_n}} \\ [0] & [K']_{\mathrm{s_1 b}} & [K']_{\mathrm{s_1 s_1}} & [0] & [0] & [0] \\ [0] & [K']_{\mathrm{s_2 b}} & [0] & [K']_{\mathrm{s_2 s_2}} & [0] & [0] \\ [0] & \vdots & [0] & [0] & \ddots & [0] \\ [0] & [K']_{\mathrm{s_n b}} & [0] & [0] & [0] & [K']_{\mathrm{s_n s_n}} \end{bmatrix} \begin{Bmatrix} \{w\}_{\mathrm{r}} \\ \{w\}_{\mathrm{b}} \\ \{a\}_1 \\ \{a\}_2 \\ \vdots \\ \{a\}_{\mathrm{n}} \end{Bmatrix} = \begin{Bmatrix} \{f\}_{\mathrm{r}} \\ \{f\}_{\mathrm{b}} \\ \{0\} \\ \{0\} \\ \vdots \\ \{0\} \end{Bmatrix} \tag{4}$$

where subscript r is the regular region. According to the boundary conditions, solving equation (4) then generalized parameters $\{a\}_1$、$\{a\}_2$,..., $\{a\}_n$ of each crack tip region can be obtain, extracting general parameters a_1 and b_1 of each column vector, substituting it into equation (2) so SIF of each crack tip can be obtain.

4. Numerical Example

A main crack of length c_1 and two minor cracks of length c_2 exist in a three-point bending beam with span length l=4000 mm, section size $b \times h$=250 mm×600 mm, elastic modulus $E=3.0\times10^4$ MPa and Poisson ratio μ=0.2. A concentrated load P=2.5 kN is concentrate in mid-span, as shown in figure 3. Calculating the changing law of main crack and two minor cracks SIF when the minor cracks' length and position have changed by Williams elements.

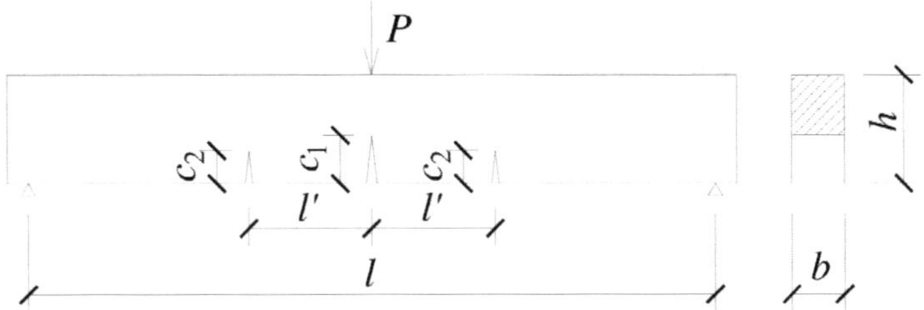

Figure 3. Three-point bending beam with three edge cracks.

4.1. Preserve Minor Crack Length Constant with Position Changed

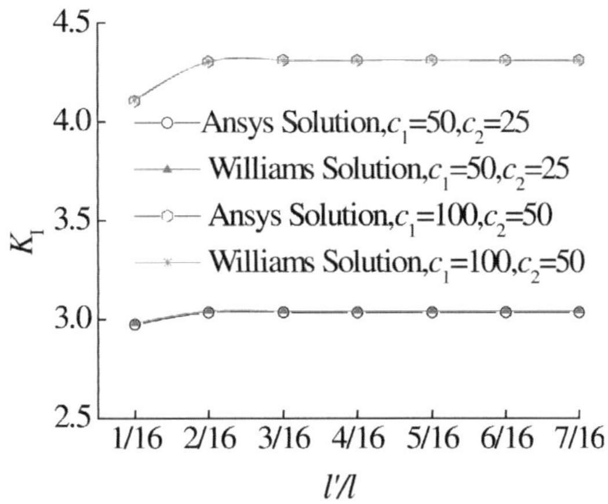

Figure 4. Main crack K_{I} versus minor cracks' position changed.

Figure 4 shows that when l'/l smaller than 1/8(that is near the mid-span), there is a decreasing trend of main crack K_{I}; when l'/l is larger than 1/8, the change of the position of the minor crack has no effect on the K_{I} of the main crack.

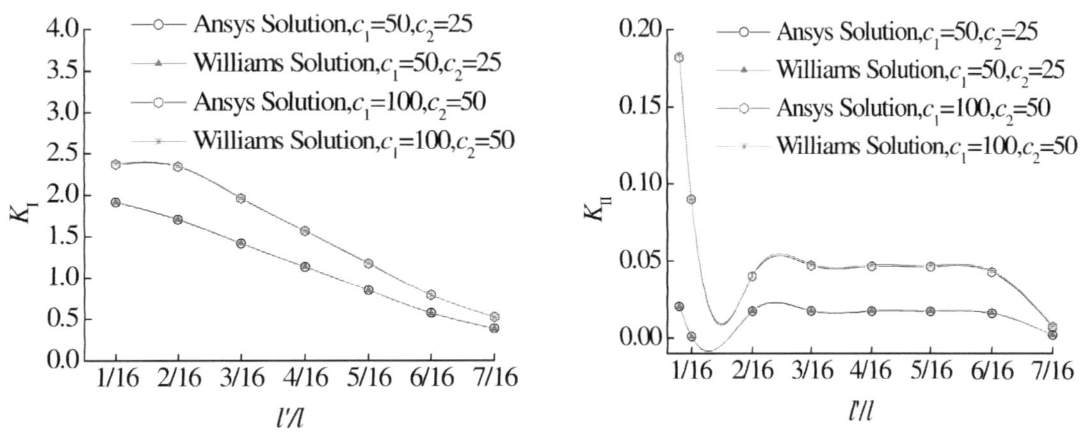

Figure 5. Minor crack K_{I} (left) and K_{II} (right) versus minor cracks' position changed.

Figure 5 shows that with the minor crack furthering from the mid-span, the moment of the section in which the crack position is decreased, thus K_{I} decreases, but near the mid-span, the decreasing trend relatively small. And minor crack K_{II} decreased first and then increased. When the minor crack is very close to the symmetry axis, K_{II} has its maximum value because of the shear mutation in mid-span.

When *l'*/*l* smaller than 1/8, minor crack leaves the stress concentration area, K_{II} decreases sharply and then has a certain recover, so it comes to a minimum value. When *l'*/*l* is larger than 1/8, the shear remains constant, so the K_{II} also tends to be constant. When the minor crack is near the endpoint, the moment gradually decreases to zero, thus K_{II} decreases to zero.

It can be inferred from the example: When the minor crack relatively close to the main crack of mid-span, the changes of position have a great influence on the main crack tip SIF. And the minor crack tip SIF is relative large, which shows that minor crack and main crack absorb energy together, thus main crack tip K_I is smaller, reducing the risk of structural failure, proving the number of crack don't represent the risk of failure. When minor crack relatively far from main crack of mid-span, effect of change in position of the main crack is not obvious, or even negligible. And the minor crack tip of crack K_I decreases gradually, K_{II} tends to be constant for a while and decreases to zero when the minor crack is near the endpoint. It shows that with minor crack position changes, K_I is affected by the moment and K_{II} is affected by moment and shear.

4.2. Preserve Minor Crack Position with Length Changed (Value C_1=100 Mm; L′ =500 Mm, 1000 Mm, 1500 Mm)

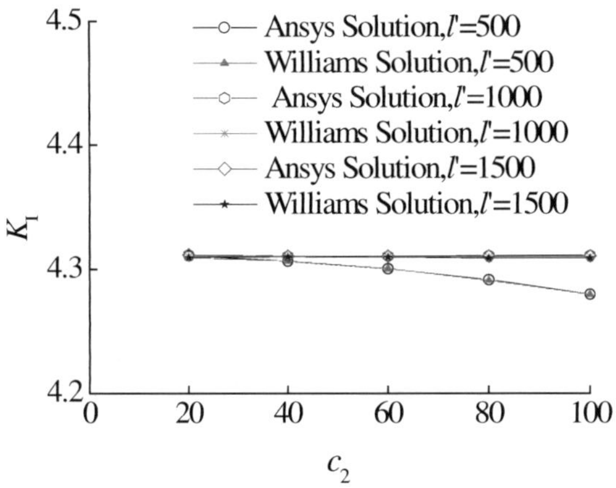

Figure 6. Main crack K_I versus minor cracks' length changed.

Figure 6 shows that results of W unit and the singular element calculation are approximately equal, whose relative error is around 1%, and both the changing trends are almost the same. When *l* '= 1000 mm or 1500 mm, growth with minor crack not be affected the main crack K_I. When *l* '= 500 mm, the main crack K_I tended to decreases with minor crack extends.

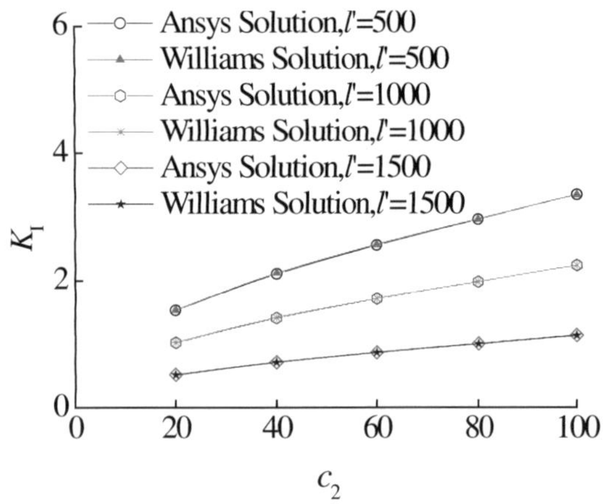

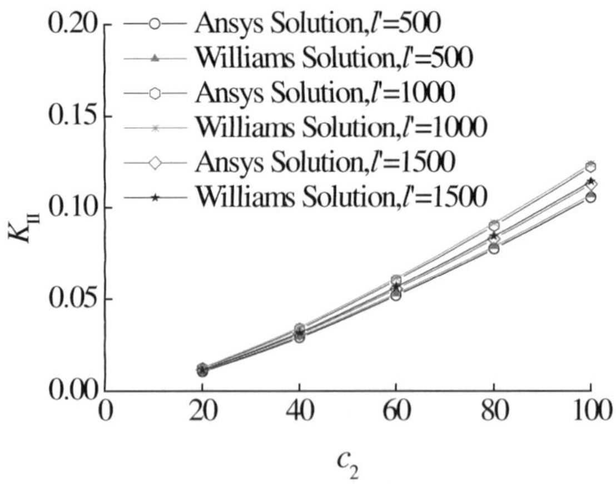

Figure 7. Minor crack K_I (left) and K_{II} (right) versus minor cracks' length changed.

Figure 7 shows that the stress concentration becomes more obvious with minor crack extend. Among them, when the minor crack is more near the main crack, the stress concentration of K_{I} is more obvious, but the distance changes are insensitive to K_{II}.

From this example, there comes the following conclusion. The changes of the minor crack length has little influence on the main crack tip SIF, the main crack tip of crack K_{I} will be affected only when length of the minor crack and the main crack are relative equal. The minor crack tip SIF becomes much larger with minor crack extend.

5. Conclusion

This thesis has established the generalized parameter Williams more crack beam element analysis model, analyzed three cracks in three point bending beam crack tip SIF, which shows that results of W unit and the ANSYS finite element solution are identical. This method can obtain SIF directly and has a satisfied accuracy, thus calculation efficiency can be highly improved. W unit shows the strong vitality and can be used widely in the future.

Possible engineering advice is provided according to the results. Crack on the same side of beam in actual project needs major consideration main crack, take main reinforcement measures for main crack. When the minor crack length is short, its impact could be neglected .When length of minor crack and main crack are almost equal, the interaction between main crack and minor crack should be considered.

6. Acknowledgment

Financial supports for this study provided by National Natural Science Foundation (No. 51478125), Foundation of Guangxi Natural Science (No. 2013GXNSFBA019257), Systematic Project of Guangxi Key Laboratory of Disaster Prevention and Structural Safety(No. 2013ZDX05; 2016ZDX06), Foundation of Guangxi Zhuang Autonomous Region-education (No. KY2016YB039) are gratefully acknowledged.

7. References

[1] FU P, JOHNSON S M, SETTGAST R R, Carrigan C R. Generalized displacement correlation method for estimating stress intensity factors [J]. Engineering Fracture Mechanics, 2012, 88:90-107.

[2] SRAWLEY J E. Wide range stress intensity factor expressions for ASTM E 399 standard fracture toughness specimens [J]. International Journal of Fracture, 1976, 12(3): 475-476.

[3] DaChun Guan. Finite element analysis of the stress intensity factor of the crack in the three point tending beam [J]. Journal of Beijing University of Technology, 1994, 20(4): 19-24.(In Chinese)

[4] ROHDE L, KIENZLER R, HERRMANN G. On a new method for calculating stress-intensity factors of multiple edge cracks [J]. Philosophical Magazine, 2005, 85(33-35): 4231-4244.

[5] OOI E, YANG Z. Modelling multiple cohesive crack propagation using a finite element–scaled boundary finite element coupled method [J]. Engineering analysis with boundary elements, 2009, 33(7): 915-929.

[6] QinYuan Wang, YongJie Liu, XiangGuo Zeng, Jun Ding. Finite element numerical analysis of fracture parameters of concrete under multi crack interaction [J]. Sichuan construction science research, 2007, 32(6): 73-77. (In Chinese)

[7] C.A.Petri. Kommunikation mit Automaten [D]. Bonn, West Germany University of Bonn, 1962.

[8] LuFeng Yang, Hua Xu, Li Peng, Ran Li. Williams general parameters of the fracture problem analysis unit [J]. Chinese Journal of Computational Mechanics, 2009, 26(1): 33-39. (In Chinese)

[9] LuFeng Yang, Hua Xu, Ran Li, Li Peng. Generalized parametric finite element method for calculating stress intensity factor [J]. Engineering Mechanics, 2009, 26(3): 48-54. (In Chinese)

[10] LuFeng Yang, Hua Xu, ZhengPing She, Li Peng. Williams element analysis I—II mixed mode crack stress intensity factor [J]. Journal of Ship Mechanics, 2014, 18(1): 115-123. (In Chinese)

[11] LuFeng Yang, Hua Xu, MeiLan Wang. Triangular Williams element for the analysis of stress intensity factor [J]. Chinese Journal of Applied Mechanics, 2011, 28(6): 570-575. (In Chinese)

IOP Conf. Series: Materials Science and Engineering **250** (2017) 012060 doi:10.1088/1757-899X/250/1/012060

Prediction on the Enhancement of the Impact Sound Insulation to a Floating Floor with Resilient Interlayer

Xianfeng Huang[1, 2, 3], **Yao Meng** [1, 2, 3] **and Riming Huang** [1, 2, 3]

[1] College of Civil Engineering and Architecture, Guangxi University, Nanning 530004.

[2] Guangxi Key Laboratory of Disaster Prevention and Engineering Safety, Guangxi University, Nanning 530004.

[3] Key Laboratory of Disaster Prevention and Structural Safety of Ministry of Education, Guangxi University, Nanning 530004.

[a]Email: seulee@163.coml, [b]Email: 1148500367@qq.com, [c]Email: 1729753937@qq.com.

Abstract. This paper describes a theoretical method for predicting the improvement of the impact sound insulation to a floating floor with the resilient interlayer. Statistical energy analysis (SEA) model, which is skilful in calculating the floor impact sound, is set up for calculating the reduction in impact sound pressure level in downstairs room. The sound transmission paths which include direct path and flanking paths are analyzed to find the dominant one; the factors that affect impact sound reduction for a floating floor are explored. Then, the impact sound level in downstairs room is determined and comparisons between predicted and measured data are conducted. It is indicated that for the impact sound transmission across a floating floor, the flanking path impact sound level contribute tiny influence on overall sound level in downstairs room, and a floating floor with low stiffness interlayer exhibits favorable sound insulation on direct path. The SEA approach applies to the floating floors with resilient interlayers, which are experimentally verified, provides a guidance in sound insulation design.

1. Introduction

The floating floor is usually adopted in a building to achieve higher impact sound insulation, since the resilient interlayer in the floating floor can reduce the sound transmission effectively. The commonly used resilient materials include rubber mat, mineral wool and foam etc. The performance of the resilient material against impact sound is related to the dynamic stiffness (stiffness coefficient) of the interlayer. Kim[1] studied 51 groups of resilient interlayers to investigate the relationship between the impact sound insulation and its dynamic stiffness, it was found that the dynamic stiffness reduced with the increase of resilient interlayer thickness, the impact sound insulation improvement increased as dynamic stiffness reduced, and it is also related to frequency. Based on ISO140-8 international regulations, Schiavi[2] analyzed the comparison between the measured data and the predicted data and indicated that the dynamic stiffness was positively correlated with the preloading of the floating floor. P Mees.[3] studied the topic of installing resilient interlayer between wall and floor, and analyzed the sound attenuation pattern of the structural sound excited on a wall and transmitted to a floor across a resilient interlayer. Based on the EN-12345-2, Craik[4] analyzed the direct and flanking sound transmission from the upstairs room to downstairs rooms which separated by a floating floor, and

IOP Conf. Series: Materials Science and Engineering **250** (2017) 012060 doi:10.1088/1757-899X/250/1/012060

used SEA method to predict the impact sound pressure level in a receiving room, and investigated to optimized the stiffness material type and thickness of the interlayer in a floating floor structure with the enhancement of impact sound insulation. All above researches provide inspirations to the study in this paper to explore the patterns which the enhancement of impact sound insulation varies with the properties of a resilient interlayer.

2. SEA Prediction Model

A floating floor refers to a concrete slab floor with a resilient interlayer to isolate impact sound, and then a face-layer is set on the top of the resilient interlayer, as shown in Figure 1. In accordance with the requirements of the code, standard tapping machine was used to strike a floating floor, then the impact sound pressure levels were measured in room 5 underneath the floor. Sound energy is generated by tapping machine on the floor 2 to cause the impact sound transmission through two paths: 2-3-4-5 and 2-3-4-6-5, one is direct transmission path, the other are flanking transmission paths which include 4 paths due to four side walls[4].

Impact sound transmission across a floor involves a direct transmission path, which the impact sound transmitted from an up-stair room into the down-stair room directly; and four flanking transmission paths, which impact sound passed from floor through each of the side walls, and to the down-stair rooms finally, which is only considered the first-order flanking sound transmission. Sound transmission would be encountered in the process of T junction, and the transmission loss will be occurred, this is sound coupling within a T junction. two paths are considered: direct path: Floor 2 - resilient interlayer 3 - concrete slab 4 - room 5; flanking path: Floor 2 - resilient interlayer 3 - concrete slab 4 - wall 6- room 5.

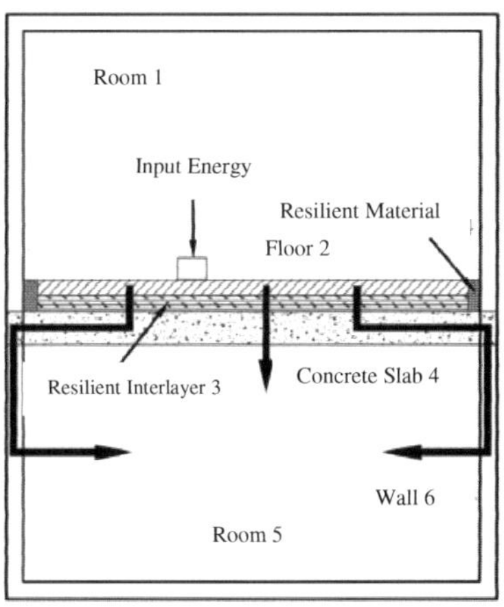

Figure 1. The floating floor system

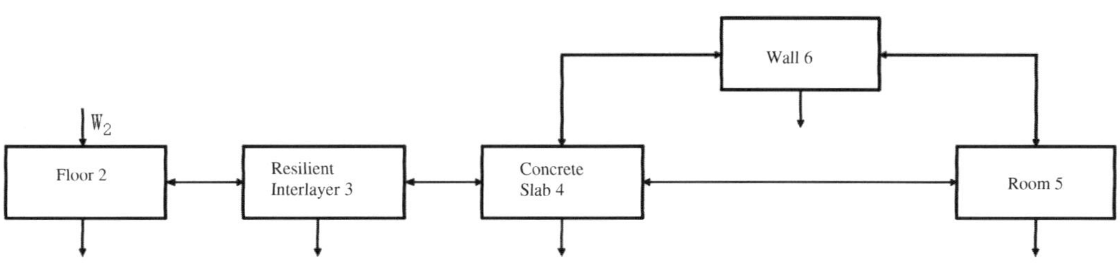

Figure 2. The SEA model of a floating floor

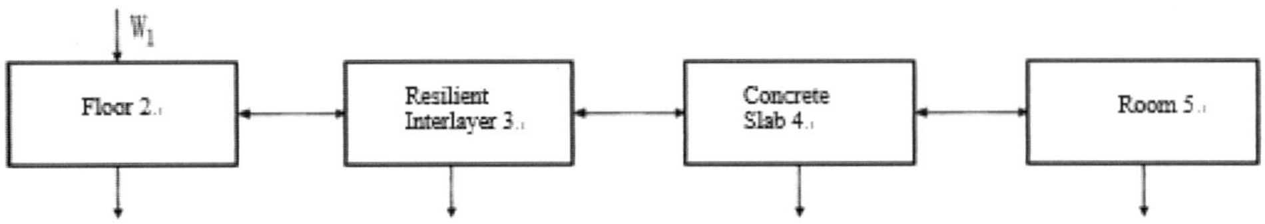

Figure 3. The direct path

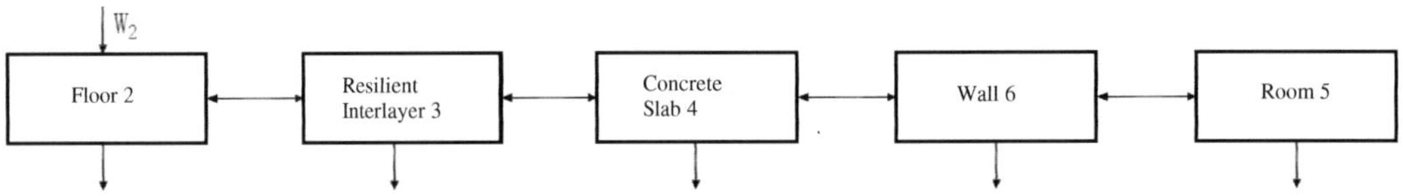

Figure 4. The flanking path

The above model ignored the sound energy inverse flows from the resilient interlayer, and since subsystems 3, 4, 6, and 5 have very few input sound energy, the sound energy returned from the subsystem and the sound energy input subsystems 3, 4, 6, and 5 could be ignore, therefore, the sound energy balance equation of the model can be written as:

$$\begin{bmatrix} -\eta_2 & & & & \\ \eta_{23} & -\eta_3 & & & \\ & \eta_{34} & -\eta_4 & & \\ & & \eta_{46} & -\eta_6 & \\ & & & \eta_{65} & -\eta_5 \end{bmatrix} \begin{bmatrix} E_2 \\ E_3 \\ E_4 \\ E_6 \\ E_5 \end{bmatrix} = \begin{bmatrix} -W_2/\omega \\ \\ \\ \\ \\ \end{bmatrix} \tag{1}$$

Where η_i is the internal loss factor, η_{ij} is the coupling loss factor , E_i is the energy of each subsystem and W_2 is the energy input subsystems 2.

Based on equation (1), the impact sound insulation can be obtained, the impact sound level on the direct path is:

$$L_{p,direct} = 10\log(\frac{p^2}{p_0^2}) = 10\log\frac{W_2\eta_{23}\eta_{34}\eta_{45}\rho_0 c_0^2}{\omega\eta_2\eta_3\eta_4\eta_5 V p_0^2}$$
$$= L_{w,2} + 10\log(\frac{\eta_{23}\eta_{34}\eta_{45}}{\eta_2\eta_3\eta_4}) - 10\log(\frac{V}{T}) + 14 \tag{2}$$

The impact sound level on the flanking path is:

$$L_{p,flank} = 10\log(\frac{p^2}{p_0^2}) = 10\log\frac{W_2\eta_{23}\eta_{34}\eta_{46}\eta_{65}\rho_0 c_0^2}{\omega\eta_2\eta_3\eta_4\eta_6\eta_5 V p_0^2} \tag{3}$$
$$= L_{w,2} + 10\log(\frac{\eta_{23}\eta_{34}\eta_{46}\eta_{65}}{\eta_2\eta_3\eta_4\eta_6\eta_5}) - 10\log\left(\frac{V}{T}\right) + 14$$

Where ρ_0 is the density of the air, c_0 is the velocity of the air, p_0 is the reference sound pressure.

IOP Conf. Series: Materials Science and Engineering **250** (2017) 012060 doi:10.1088/1757-899X/250/1/012060

The total impact sound level in room 5 is:

$$L_p = L_{p,direct} + 10\lg[1 + 10^{-\frac{L_{p,direct} - L_{p,flanking}}{10}}] \quad (4)$$

where V and T are the volume of room 5 and the reverberation time in room 5 respectively.

3. Material Parameters

The above theoretical model should be compared with the existing data to verify the reliabiliy of the theory. So, three kinds of floating floor with different interlayer are list in Table 1 and 2. And the physical parameters of the floor slab and face-layer are shown in Table 1. The resilient interlayer was closely connected with the floor, Mineral wool, Fibrous resilient and Polystyrene foam are selected as a resilient interlayer, whose physical properties are shown in the Table 2.

Table 1. Properties of the face-layer and reinforced concrete slab.

Name	Density ρ (kg/m^3)	Thickness d(m)	Young modulus E(N/m^2)
Timber floor	542	0.022	4.0x10^9
Reinforced concrete slab	2200	0.120	2.8x10^{10}

Table 2. Properties of the resilient layers.

Resilient Interlayer	Thickness (mm)	Stiffness coefficient (KN/m^2)	Damping
Mineral wool	15	86	0.12
Polystyrene foam	3	165	0.05
Fibrous resilient	35	77	0.17

As seen from Table 2, polystyrene foam has the highest stiffness and fibrous resilient has lowest stiffness which exhibit favorable resilient among three kinds of resilient interlayer. If these resilient materials are adopted as the interlayer, the coupling loss factor between the two slabs should be determined.

4. Results and Discussion

The measured and predicted data were compared as shown in figure 5-8. The floating floor contains a 140mm concrete floor, a resilient interlayer and a 10mm oak wood face layer. And these measurement are conducted in according with Standard ISO140-8.

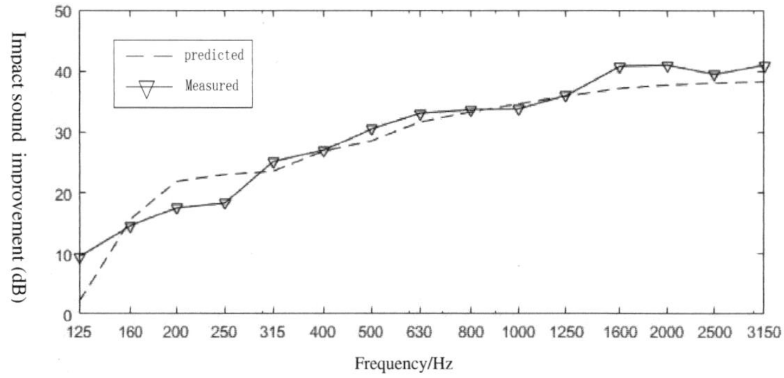

Figure 5. Measured and predicted impact sound level enhancement of a floating floor with 15 mm mineral wool interlayer (K=86KN/m^2)

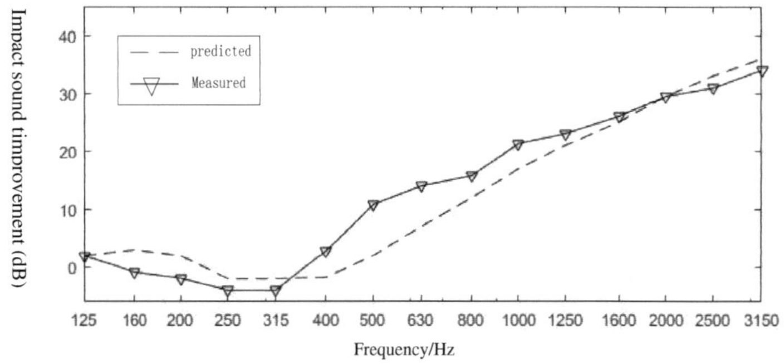

Figure 6. Measured and predicted impact sound level enhancement of a floating floor with 4mm Polystyrene foam interlayer (K=77KN/m^2)

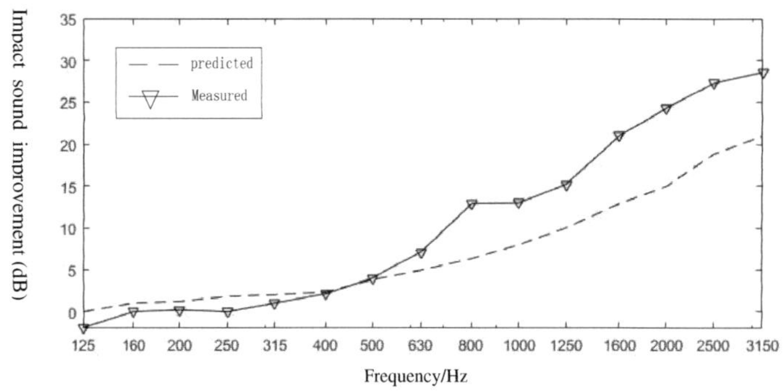

Figure 7. Measured and predicted impact sound level enhancement of a floating floor with 4mm Fibrous resilient interlayer (K=165KN/m^2)

IOP Conf. Series: Materials Science and Engineering **250** (2017) 012060 doi:10.1088/1757-899X/250/1/012060

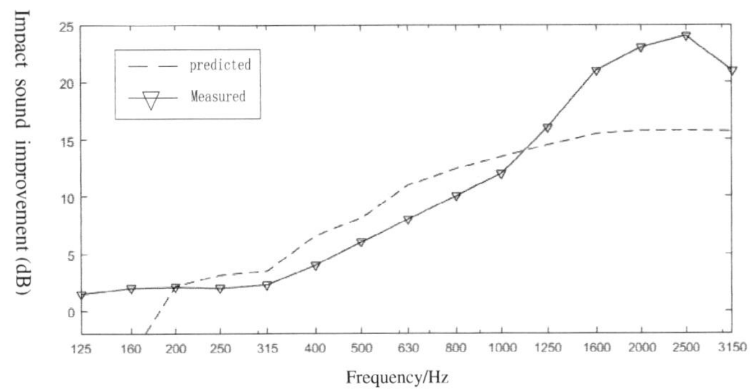

Figure 8. Measured and predicted impact sound level of a concrete floor with no interlayer

As shown in Fig5-8: the impact sound transmission is attenuated due to the effect of resilient interlayer within a floating floor, and it's insulation varied with properties of the interlayer material. Suitable interlayer material and it's thickness are chosen to obtain favorable performance of sound insulation. Polystyrene foam, fibrous resilient have higher impact sound insulation at middle and high frequency range, and not same at low frequencies. For instance, a 4mm polystyrene foam interlayer don't have high impact sound reduction below 315Hz, but it have obvious impact sound reduction at middle and high frequency range, especially above 1000 Hz, and it's average sound reduction index reaches above 28dB.

There are four flanking transmission paths through the building members, this paper selected one typical flanking path to analyze impact sound pressure level and compare it's values between direct path and flaking paths. And the floating floor composed of resilient interlayer material was adopted to calculate the impact sound pressure level: with fibrous resilient, polystyrene foam and without interlayer. These results are shown in figure 9-11.

By the comparison of three group of data, within the whole 1/3 octave frequency band range, the first group of floor structure, with wooden face-layer and without resilient interlayer, has higher impact sound insulation below 500Hz than that at other frequency range. Meanwhile, The dominant sound transmission is still the direct one, since the SPL on the direct path into the downstairs room is 10 dB more than those on any flanking path, and the two sound pressure level superposition, the total sound pressure is increased by only 0.1dB, therefore, there was no obvious influence of the flanking transmissions of the structure without resilient interlayer on overall sound pressure level in downstairs room.

After the comparison of the floating floor with three kinds of resilient materials of interlayer, It can be found that a floating floor with higher stiffness interlayer exhibits poor sound insulation on direct path. Conversely, favorable sound insulation can be obtained. But, the resilient materials with low stiffness barely meet the requirements of strength.

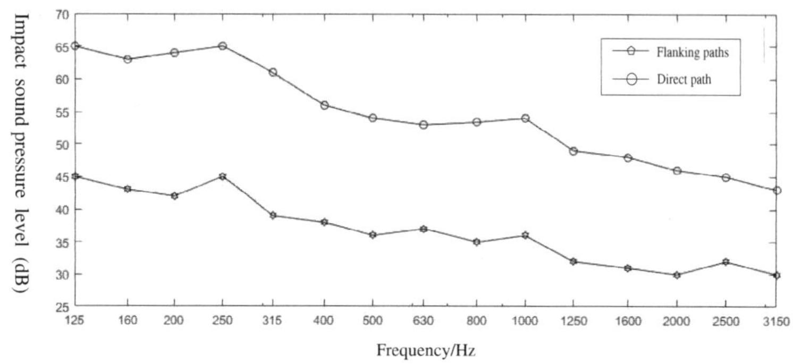

Figure 9. Impact sound level of two paths across a floating floor with 20 mm fibrous resilient interlayer($K=165KN/m^2$)

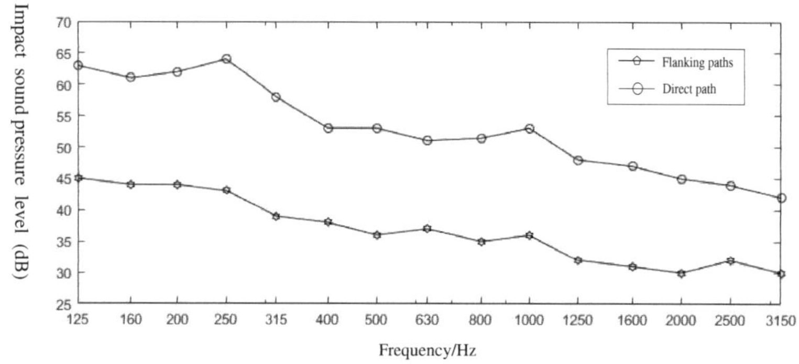

Figure 10. Impact sound level of two paths across a floating floor with 20 mm polystyrene foam interlayer($K=77KN/m^2$)

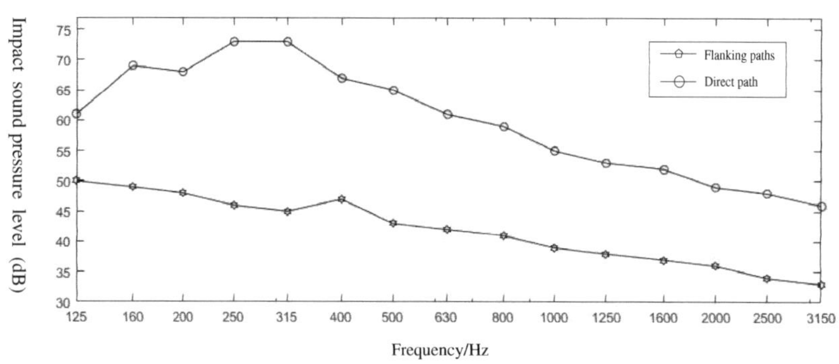

Figure 11. Impact sound level of two paths across a concrete floor without interlayer

5. Conclusions

The SEA model for impact sound transmission across a floating floor is established to determine two kinds of impact sound transmission ---direct path and flanking path, and equilibrium equations of a floating floors is gained by considering the performance of bending wave, which solves the problem of structural acoustic transmission through a resilient interlayer. And according to this SEA mode, the prediction method of the impact sound pressure level of floating floor structure is derived. The accuracy of the calculation can be verified by comparing with the measured data. The method is employed to predict the sound insulation of the floating floor accurately and provide a guidance for the sound insulation design.

The comparative analysis of the impact sound pressure level come from direct path and flanking paths to downstairs rooms is carried out, it is revealed that the sound insulation of floating floor is mainly affected by the direct path, the difference of impact sound pressure level between direct path and flanking paths is almost 20dB, when the resilient interlayer with different stiffness coefficients is used, and the stiffness coefficient decreased gradually.

6. Acknowledgement

This research is supported by the National Natural Science Foundation of China (Project No. 51568003), the Natural Science Foundation of Guangxi Province, China (Project Nos. 2015GXNSFAA139254 and 2014GXNSFAA118017).

7. References

[1] Kim K-W, Jeong G-C, Yang K-S, et al. Correlation between dynamic stiffness of resilient materials and heavyweight impact sound reduction level. *Building and Environment*, 2009, **44(8)**: 1589-1600.
[2] Schiavi A, Belli A P, Russo F. Estimation of acoustical performance of floating floors from dynamic stiffness of resilient layers. *Khirurgiia*, 2005, **12(2)**: 58-62.
[3] Mees P, Vermeir O. Structure-borne sound transmission at resilientally connected plates. *Journal of Sound & Vibration*, 1993, **166(1)**: 55-76.
[4] Craik R. Statistical energy analysis of sound transmission through buildings. *Gower Press*, UK, 1996.
[5] C H Zhu. Study on impact sound insulation using floating floor. Guangxi University, 2015.

IOP Conf. Series: Materials Science and Engineering **250** (2017) 012061 doi:10.1088/1757-899X/250/1/012061

Experimental Study of Fatigue Damage Strength of Concrete Lining under Dry-Wet Cycles

Yan Zhang[1], Jun Dai[1], Xuanrong Zheng[1], Xinyan Zhang[1] and Ning Li[2]

[1] Institute of Geotechnical Engineering, Xi'an University of Science and Technology, Xi'an, Shaanxi 710054, China

[2] Institute of Rock and Soil Mechanics, Xian University of Technology, Xi'an, 710048, China

ylozy@126.com

Abstract. The laboratory mortar test is used to simulate the actual force state of concrete lining in a dry-wet cycling environment of a high-temperature tunnel. The influence characteristics of temperature and the number of cycles on the lining concrete strength and strain are also explored. The test results suggest that the number of cycles has a great impact on the deterioration of the mechanical properties of the cement-based materials. During the early period, the higher the temperature, the more serious the loss of strength. As the number of cycle increases, the strength damage caused by dry-wet cycles gradually increases. Mortar test shows the fibbers have a good fatigue resistance. The quantitative results presented in this paper can be used as reference in similar projects.

1. Introduction

As part of a tunnel structure, lining concrete is in direct contact with the surrounding rock, and its fatigue damage strength property in a dry-wet environment is the key connection to effective stability of the supporting structure. In respect to the gradual occurrence of high temperature and thermal damage of a tunnel and the dry-wet condition, serious strength damage of the lining concrete might occur; in particular, under the action of cycling, the amount of lining concrete damage will increase continuously, which may lead to the loss of the supporting effect because of excessive strength damage. In engineering designs, if the lining fatigue effect is not taken into account, the safety and durability of a high-temperature tunnel will be greatly affected during the operational period, which may shorten the design lifespan of the tunnel or even result in the emergence of a safety incident.

In recent years many researches on the concrete lining of a high temperature tunnel are mainly focused on the distribution of the temperature field and stress field. A study was performed on the temperature distribution pattern of the rock surrounding the tunnel and the stress characteristics under an external high geothermal condition [1]. This study provides the stress characteristics of the lining structure influenced by external temperature. At the same time, some scholars carried out a study on damage degradation characteristics of different type of rocks [2-6], such as silty mudstone (2005), shale (2004), diabase (2014), etc., during a freeze-thaw cycle. The damage deterioration mechanism and mechanical properties of sandstone in different chemical solutions were studied after a freeze-thaw cycle, and it was found that the damage degree of sandstone gradually increases as the number of freeze-thaw cycle increases [7]. Han TL et al.[8]studied the damage characteristics of different rocks during a freeze-thaw cycle, and the results showed various differences in damage with a change in the number of freeze-thaw cycles and in temperature. In the studies above, although the influence of environmental factors, such as temperature and cycles, on the rock and concrete was considered, a systemic discussion is still lacking for the fatigue damage characteristics of a lining concrete that is subjected to a dry-wet cycling action in high temperature field [9]. Obviously, more research is deserved to dispose the effect of environmental factors on the fatigue of concrete.

According to the high geothermal and high rock temperature environment formed by high temperature diversion tunnel in Xinjiang, China, and 0 °C of ice and snow melt water will pass the diversion tunnel during the operation period, the stress of the tunnel lining is very complicated. That is, the lining is in a wet state during operation while in a dry state during the maintenance period. So the lining concrete is subjected to alternating cycles of dry and wet.

Accordingly, the objective of this paper is to develop a fatigue damage law for concrete subjected to dry and wet cycles at temperatures. The fatigue damage strength of lining concrete in a dry-wet environment of a tunnel is simulated through the laboratory test. And the temperature is selected as the major parameter affecting the fatigue property of specimens because the mechanical performance and characteristics of concrete are significantly changed with temperature. In addition, the effect of cycles on fatigue is also considered. The objectives of this study are to investigate an influence of environmental factors such as temperature, dry and wet conditions on mechanical performance of mortar and concrete, and to evaluate the resistance to fatigue performance considering the temperature effect.

2. Program for Dry - Wet Cycles Test

2.1. Condition of Dry-Wet Cycles Test
The key to the dry-wet cycles test was the simulation of the dry-wet environment where the concrete lining was located. In this paper, the small specimens were selected to carry out the cycle test, and four kinds of curing conditions were chosen to simulate four kinds of dry and wet environment. To evaluate the temperature's effect on the mechanical properties, four different temperatures (25° C, 45° C, 60° C, and 80° C) were considered. The samples were molded at 25℃ as a comparative object The samples at 25°C could be molded in a room temperature environment. After being stripped, the samples were maintained in a curing room for 28 d.

For the samples in dry-wet cycles test at 45°C, 60°C, and 80°C, the samples were maintained at a given temperature for 28 d, and then they were placed in 25°C water to cool for 10 min after being removed from the temperature control box, see Figs. 1 and 3. The samples were naturally air-dried for 30 min after cooling, see Fig. 2; then, they were placed into a temperature control box for 1 hour, cooled in water for 10 min, and finally naturally air-dried for 30 min. Repetition of this process to simulate the dry-wet cycles effect on the samples. The water immersion test was performed according to the national standard GB 2575-89 "Glass fiber reinforced plastic water-resistance test method".

Figure1. Specimens immersed in water **Figure2.** Dried specimens after soaking **Figure3.** Schematic view
of the cooling water temperature

2.2. Specimen Preparation
In order to show the effect of dry-wet cycles on cement-based materials clearly, this experiment replaced the concrete samples with cement mortar samples for production and maintenance. The

sample was a prism of 40 mm × 40 mm × 160 mm. Raw materials of the sample were as follows: the cement was Qinling general silicate cement of Shanxi (PO42.5R), the sand was a top-grade medium sand with a fineness modulus of μf = 2.9 originally extracted from the Bahe River of Xi'an, and the mixing water was tap water. The three selected polymer fibers were polyester, polypropylene, and polyacrylonitrile. The fiber content of the sample was 0.9 kg/m3. The fiber property parameter and form were both shown in Table 1.

Table 1. The performance parameters of fibers

Type	**Length /mm**	**Diameter /μm**	**Density /(g/cm^3)**	**Melting point /℃**	**Tensile strength /MPa**
polyester	8	20	1.38	>249	>900
polypropylene	14	24	0.91	165	>400
polyacrylonitrile	12	12	1.18	260	>500

The mass ratio of sand/cement was set to 1.0. The specific material mix ratio of the sample was that cement: water: sand = 1:0.52:3.

2.3. *Test Procedures*

The dry-wet cycles test had four levels of cycles: 0 was the baseline, and the other levels were 10, 20, and 30 cycles. The specimens were classified as pure mortar (PM), polyester fiber mortar (PEM), polypropylene fiber mortar (PPM), and polyacrylonitrile fiber mortar (PAM). There were three samples for each combination of temperature and number of cycles, which results in a total of 192 specimens. The flexible strength of specimens was tested after the completion of the dry-wet cycles at the corresponding temperature. The maximum deviation between each specimen flexible strength test value and the average value was controlled within 10%.

3. Results and Discussion

3.1. *Strength*

The experimental results were listed in Tables 2 and 3. The flexible strength distribution patterns were shown in Figs. 4 and 5. Compared with the strength under 0 cycle, the strength influenced by the dry-wet cycles was lower than that of 0 cycle. This strongly suggests that the sample fatigue effect was produced by the cycles, and its anti-fracture toughness gradually decreased.

Table 2. Strength results of specimens under dry-wet cycles /(MPa)

Type	Curing temperature /℃	the number of cycles 0	10	20	30
PM	25	4.13	3.87	3.64	3.44
	45	3.91	3.61	3.34	3.13
	60	3.61	3.28	3.01	2.8
	80	3.39	3.03	2.73	2.45
PPM	25	4.34	4.11	3.88	3.66
	45	4.31	4.03	3.71	3.46
	60	4.26	3.91	3.61	3.39
	80	3.96	3.62	3.33	3.11
PAM	25	4.32	4.03	3.79	3.57
	45	4.24	3.97	3.72	3.43
	60	4.13	3.87	3.59	3.44
	80	3.91	3.59	3.29	3.16
PEM	25	4.19	3.96	3.73	3.61
	45	4.11	3.88	3.69	3.33
	60	3.98	3.69	3.33	3.14
	80	3.82	3.54	3.32	3.09

Table 3. Strain results of specimens under dry-wet cycles

Type	Curing temperature /℃	the number of cycles 0		10		20		30	
		ε	$\varepsilon-\varepsilon_0$	ε	$\varepsilon-\varepsilon_0$	ε	$\varepsilon-\varepsilon_0$	ε	$\varepsilon-\varepsilon_0$
PM	25	0.363	0	0.388	0.025	0.429	0.066	0.478	0.115
	80	0.443	0.08	0.668	0.305	0.77	0.407	0.908	0.545
PAM	25	0.349	0	0.378	0.029	0.453	0.104	0.477	0.128
	80	0.421	0.072	0.525	0.176	0.643	0.294	0.727	0.378

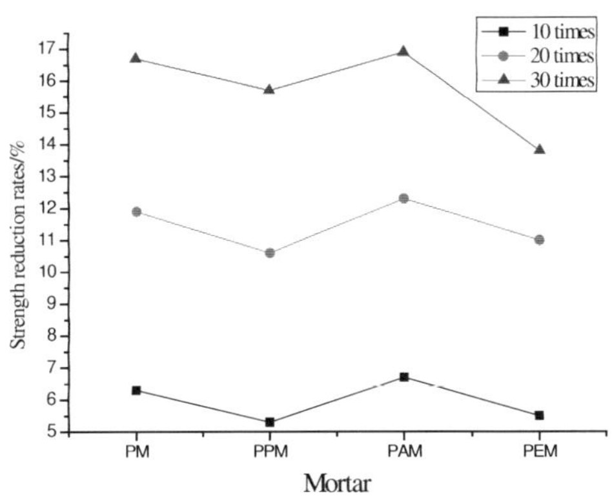

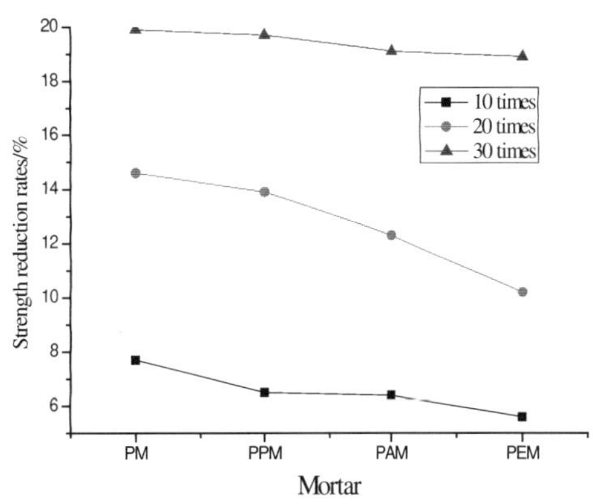

(a) 25℃ (b) 45℃

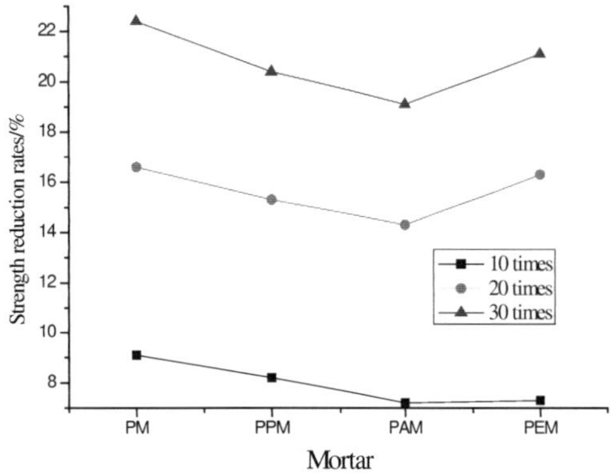

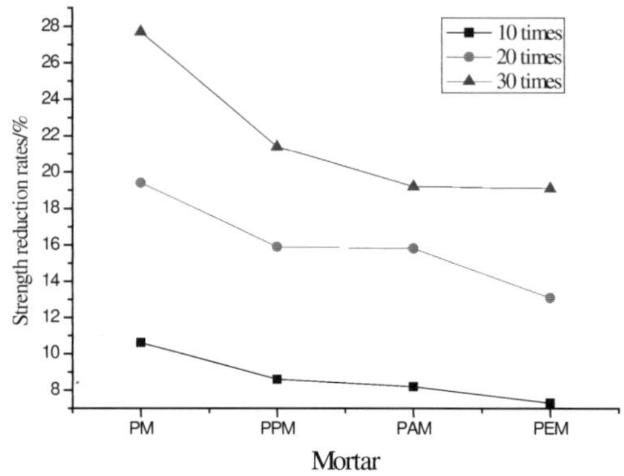

(c) 60℃ (d) 80℃

Figure 4. Relationship between the mortar strength reduction rate and the number of cycles at different temperatures

The flexible strengths of the pure mortar samples and the fiber mortar samples were shown in Table 2 for the experimental conditions of 0 cycle, 10 cycles, 20 cycles and 30 cycles.

Take samples at 60°C for example, the lowest strength reduction rates of four types of mortar after 10 cycles were, in order of large to small, polypropylene fiber mortar, polyacrylonitrile fiber mortar, polyester fiber mortar, and pure mortar. The lowest reduction rate was 5.3%, and the highest was 6.7%.

The lowest strength reduction rates of four types of mortar after 20 and 30 cycles were, in order of large to small, pure mortar, polyester fiber mortar, polypropylene fiber mortar, and polyacrylonitrile fiber mortar. The lowest strength reduction rates after 30 cycles was 20%, and the highest was 22%. And the degree of the strength reduction was even more severe at 80°C under dry-wet cycles.

It could be seen from Table 2 and Fig. 4 that (1) the sample strength decreased as the temperature increases for the same number of cycles, and (2) the sample strength decreased as the number of cycles increased at the same temperature.

In addition, the addition of fibers to the mortar could improve the loss of strength at the same temperature and cycles. For example, the strength of the pure mortar samples decreased 27% after 30 cycles at 80°C. But the strength of fiber mortar samples only decreased about 20% under the same test conditions.

3.2. Strain

Take pure mortar and polyacrylonitrile fiber mortar samples as an example. It could be observed in Table 3 that the strain of the pure mortar and polyacrylonitrile fiber mortar increased as the number of cycles and temperature increased. Correspondingly, the strain variation of the mortar specimens increased as the temperature and the number of cycles increased.

Compared with the strain at 0 cycle and 25°C, the strain of the mortar sample at 10 cycles increased by 0.025 at 25°C and 0.305 at 80°C, respectively. The strain of the mortar sample at 20 cycles increased by 0.066 at 25°C and 0.407 at 80°C. And the strain of the mortar sample at 30 cycles increased by 0.115 at 25°C and 0.545 at 80°C.

Based on the cumulative fatigue damage theory, a certain amount of fatigue damage would occur under any cyclic stress, and the severity of the damage corresponds to the number of cycles of a given stress amplitude. The cumulative fatigue damage was also affected by the number of non-cyclic loadings for the same stress amplitude.

It could be seen that the strain and strain variation of the mortar specimens increased with the increasing of temperature and cycle times. The mortar specimens showed the corresponding fatigue property. The results were consistent with the experimental results of Huang, Atkinson and Bassuoni[10-12] et al.

4. Conclusions

(1) For the same number of cycles, the mortar strength tends to decreased as the curing temperature increases. For a given temperature, the mortar strength decreased as the number of cycles increased.

(2) The strains of the mortar samples increased as the number of cycle increased, as well as when the temperature increased. Correspondingly, the amount of change in the mortar specimen strain also increased as the temperature and number of cycles increased. The incorporation of fibers facilitates improved strength and strain.

5. References.

[1] Zhang XJ 2005 *Chin. J Rock Mech Eng.* **24** pp 1086–1089.
[2] Xu GM and Liu QS 2005 *Chin. J Rock Mech Eng.* **24** pp 3076–3082.
[3] Zhu LP, Whalleyz WB and Wang JC 1997 *J Glaciol Geocryol.***19** pp 302–310.
[4] Zhang HM and Yang GS 2011 *Eng Mech.* **28** pp 161–164.
[5] He GL, Zhang L and Wu G 2004 *Rock Soil* **Supp. 2** pp 52–56.
[6] Luo X，Jiang N，Zuo C，Dai Z and Yan S 2014 *Rock Mech Rock Eng.* **47** pp 1997–2004.
[7] Li N, Chen W, Zhang P and Swoboda G 2001 *Int J Rock Mech Min.* **38** pp 1071–1079.
[8] Han T, Shi J and Chen Y 2016 *J water conservancy.* **47** pp 644–655.
[9] Zhang Z, Zhou H and Zhao J 2004 *J Hohai University (Natural Sciences).* **32** pp 674–679.
[10] Huang W 1997 *Cement Concrete Res.* **27** pp 395-406.
[11] Atkinson A, Haxby A and Hearne J A. 1988 *The chemistry and expansion of limestone-portland cement mortars exposed to sulphate-containing solutions* (NIREX rep. NSS/R127). Oxfordshire: Nirex, Harwell.
[12] Bassuoni M T and Nehdi M L 2009 *Cement Concrete Res.***39** pp 206–226.

IOP Conf. Series: Materials Science and Engineering **250** (2017) 012062 doi:10.1088/1757-899X/250/1/012062

Creep Behavior of UFG CP Ti at Room Temperature

Lei Luo, Xicheng Zhao, XiaoYan Liu and XirongYang

College of Metallurgical Engineering, Xi'an University of Architecture and Technology, Xi'an 710055, China.
Email: luolei0301@126.com

Abstract. Ultra-fine grained commercial purity titanium (UFG CP Ti) is processed by Composite refining process (Equal channel angular pressing (ECAP), cold rolling and rotary swaging) at room temperature. The grain size is refined from 19 μm to 180 nm, and the ultimate tensile strength increase to 870 MPa. Creep tests were carried out on Ultra-fine grained commercial purity titanium with the stresses of 640, 660, 680, 700, 720, 740, 760 MPa at room temperature. Steady state creep rate and stress exponent n at various stresses were calculated for Ultra-fine grained commercial purity titanium, and creep deformation mechanism was also investigated. (With the rise of stress, the steady creep rate increases while the creep time decrease). The steady state creep rate reached maximum 1.416×10^{-6} s^{-1} (under) stress of 760 MPa. The stress exponent is 17.3 when the stress was 640 ~ 700MPa, while the stress exponent is 55.7 when the stress was 700 ~ 760MPa, UFG CP Ti shows good creep property at room temperature. The creep deformation mechanism of UFG CP Ti is the dislocation creep.

1. Introduction

Titanium and titanium alloys are widely used in aerospace, shipbuilding, automobile manufacturing, chemical industry, biomedicine, etc. because of the good general mechanical performance, such as low density, high specific strength, good corrosion resistance and excellent biocompatibility[1]. The application in biomedicine as implant is restricted due to its low strength, poor abrasive resistance and other disadvantages of commercial purity titanium. However, it is an effective approach to improve mechanical property of commercial purity titanium and expand its application by acquiring ultra-fine microstructure [2].Equal Channel Angular Pressing, abbreviated for, ECAP, is one of the main techniques for preparing bulk UFG CP Ti to improve comprehensive mechanical properties of materials via the method of refining grain [3]. Nevertheless, the microstructure of CP Ti after ECAP deformation is inhomogeneous, and the strength is less than the strength of TC4 titanium alloy. The study on further cold working to improve the strength of commercial purity titanium after ECAP has been conducted in order to improve mechanical property of commercial purity titanium after ECAP[4]. In this study, composite refining process of ECAP+ cold rolling + rotary swaging is applied to prepare UFG CP Ti at room temperature. UFG CP Ti has overall mechanical properties (room-temperature tensile property, fatigue property, thermal stability and the like) at room temperature, UFG CP Ti as biomedical materials does not only demand good room-temperature tensile property, fatigue property and fracture property, but also good creep property.

Creep usually refers to the rheological phenomenon that strain increases with prolongation time under constant stress less than yield strength under constant temperature. Creep property is affected by stress, time, structure property and other factors [5]. Creep performance of UFG CP Ti is one of the main bases to evaluate comprehensive mechanical properties. As biomedical material, the creep property of UFG CP Ti has a decisive effect for the service life and safety of material, UFG CP Ti is required to possess very small creep plastic strain at room temperature [6]. This experiment is focused on the creep property of UFG CP

Ti under different levels of stress at room temperature, and the creep mechanism of UFG CP Ti at room temperature is analyzed.

2. Experimental Material and Method

Experimental material is the UFG CP Ti prepared via complex refinement (ECAP+ cold rolling + rotary swaging) at room temperature, and its chemical component (mass fraction, %) is shown in table 1, tensile strength is 870 MPa.

Table 1 Chemical composition of the UFG CP Ti (*wt*%)

Fe	C	N	H	O	Ti
<0.20	<0.014	<0.03	<0.0015	<0.18	Bal.

TEM texture image and SADP pattern of UFG CP Ti are shown in Fig.1. It can be observed that, after severe plastic deformation, amount of dislocation aggregates in microstructure and a mass of high density dislocation tangles to cellular structure whose boundary is indistinct. Subgrain forms in internal microstructure with the misoritation between adjacent dislocation cells increasing[7]. In selected-area diffraction pattern, the diffraction spot distributes approximately like a ring, indicated that the grain of CP Ti gets refined remarkable after complex refinement process. The average grain size is about 180 nm.

Figure 1. TEM image and the corresponding SAED pattern of UFG CP Ti

Creep tests are carried out on UFG CP Ti on a RD-50 creep testing machine at room temperature under different creep stresses of 640, 660, 680, 700, 720, 740 and 760 MPa respectively. The gauge length is 25 mm and gauge diameter is Φ5 mm. The creep curve contructed according to test results is analysed, and the steady creep rate is calculated with the slope of the curve and the gauge length.

JEM-200CX transmission electron microscopy is adopted to observe microscopic structure of UFG CP Ti after creep test. 0.5 mm sample is cut up from the creep sample, coarsely grinded to 0.1 mm, and then, mechanically grinded to 40μm. MTP-1A automatic twin-jet electropolisher is adopted for electrolytic polishing to obtain transmission sample, the electrolyte is methyl alcohol: n-butyl alcohol: perchloric acid=12:7:1, voltage is 30V, power supply is 50mA and temperature is about -30℃.

3. Results and Discussion

3.1 CreepCurves of UFG CP Ti

The creep curves of UFG CP Ti under different stress contructed according to experimental data is shown in Fig. 2.

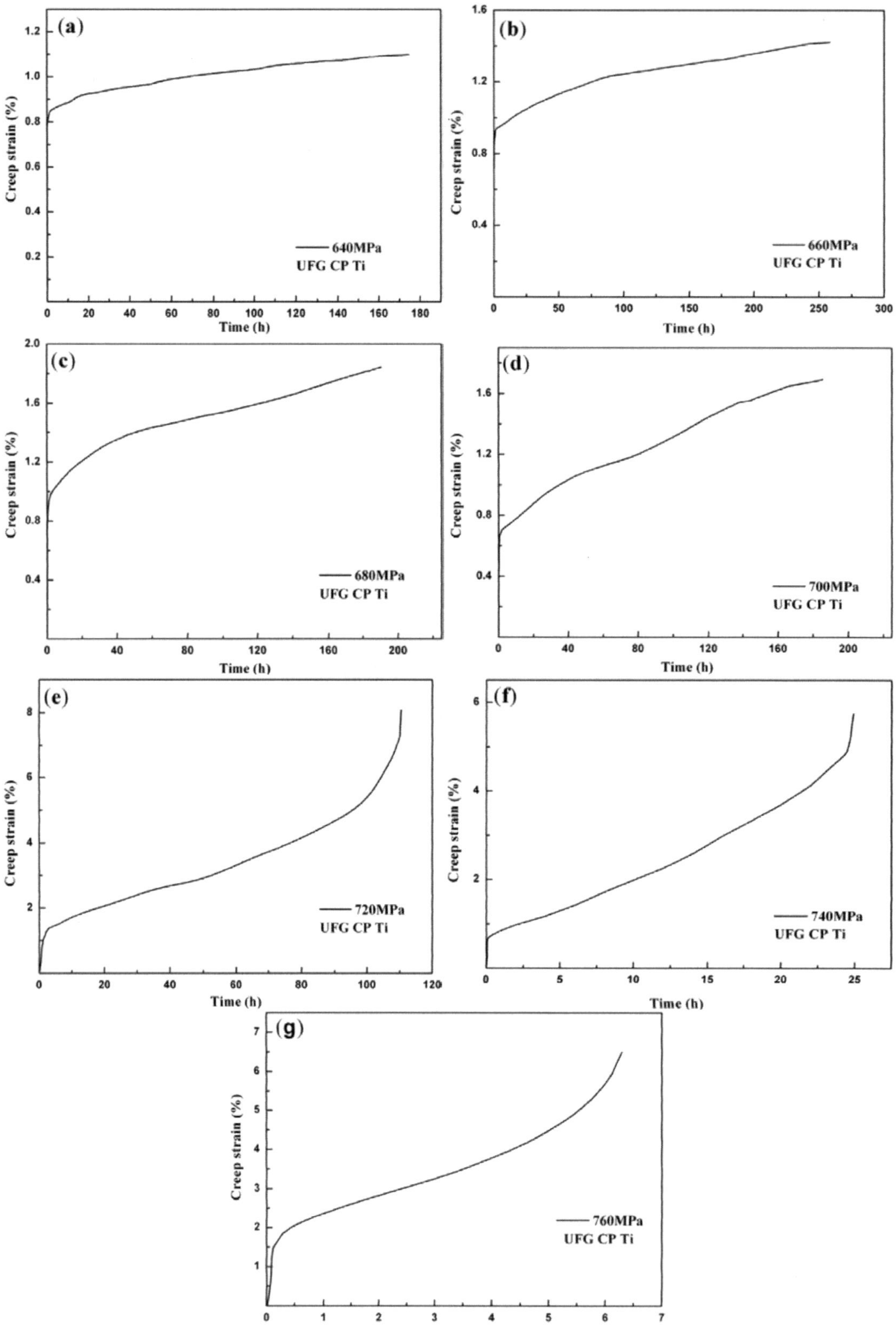

Figure 2. Creep curves of UFG CP Ti
(a)640MPa; (b)660MPa; (c)680MPa; (d)700MPa; (e)720MPa; (f)740MPa; (g)760MPa

It can be seen from Fig.2 that when creep stress from 640MPa to 700MPa, creep curve of UFG CP Ti only presents stages I and II of typical creep curve, while the creep stress is 720MPa, 740MPa and 760MPa, creep curve of UFG CP Ti would present 3 stages I, II and III of the typical creep curve. Stage I is primary creep, the creep rate of primary creep stage is large; with the prolongation of time, it decreases gradually, and then, to a constant value. Stage II is steady state creep, the creep rate in such stage nearly keeps same, and the graph of creep stress and time is a straight line whose slope is steady state creep rate. Stage III is accelerated creep, with the prolongation of time, creep rate increases gradually till samples rupture [8]. Under low stress level, the creep is slow; the steady-state creep stage is big and the accelerated creep even does not happen, as shown in Fig.2 (a), (b), (c), (d), which contributes to the slowly intergranular dislocations sliping and climbing. With the creep stress level increasing, the steady-state creep stage is shortening and the creep enters the accelerated stage quickly, as shown in Fig.2 (e), (f), (g). Under the high stress level, the intergranular dislocations motion occurs due to the experiment load [9]. The dislocation movement is the dominated creep mechanism of UFG CP Ti at room temperature.

The room temperature creep presents α-type creep under the condition of the test creep stress far blow the tensile strength of UFG CP Ti. The transient creep rate, namely, is high at the beginning, and it decreases gradually with the time, even to a certain constant range, as shown in Fig.2 (a). Thus, the room temperature creep of UFG CP Ti behaves saturation phenomena of creep under high stress level [10]. The creep value of UFG CP Ti at room temperature increases with the creep stress increasing, as shown in Fig. 2 (b), (c), (d), (e), (f) and (g). Therefore, room temperature creep is sensitive to creep stress, namely, room temperature creep value increases much more under high stress level when creep stress increment is same.

3.2 Steady Creep Rate of UFG CP Ti

According to creep mechanics theory, steady state creep rate is the minimum creep rate, which is one of the important indicators to characterize creep property of materials. Depanding on Fig.2, the steady state creep rate can be calculated out according to the slope of steady-state creep and the gauge length of sample.

Table 2 Steady state creep rate of UFG CP Ti at room temperature

Creep stress /MPa	Total testing time /h	Experimental result	Steady state creep rate/s^{-1}
640	174.5	not rupture	3.194×10^{-9}
660	258.4	not rupture	4.806×10^{-9}
680	190.6	not rupture	8.722×10^{-9}
700	185.4	not rupture	1.472×10^{-8}
720	110.4	creep rupture	1.023×10^{-7}
740	24.9	creep rupture	4.763×10^{-7}
760	6.4	creep rupture	1.416×10^{-6}

As shown in table 2, the steady state creep rate increases with the creep stress increasing at room temperature and the steady-state creep stage is shorten. When Creep stress increases from 640 MPa to 700 MPa, UFG CP Ti creep sample fail to rupture, and creep curve only represents stages I and II (as shown in Fig.2 (a), (b), (c), (d)). While creep stress increases to 720 MPa, 740 MPa and 760 MPa, UFG CP Ti creep sample suffers creep rupture failure, and creep curve shows 3 complete stages (as shown in Fig.2 (e), (f), (g)), And with the increase of creep stress, the time of suffering creep rupture in UFG CP Ti sample shortens, creep rapidly transits to stage III from stage I. From Fig.2, under the creep stress of 760 MPa, the maximum steady state creep rate is 1.416×10^{-6} s^{-1}, while the steady state creep rate of commercial purity titanium on the codition of 150 ℃/240 MPa, is 4.339×10^{-6} s^{-1} [11]. Hence, room temperature creep resistance of UFG CP Ti is higher and creep property is good.

3.3 CreepStress Exponent of UFG CP Ti

The steady state creep rate of most metals or alloys can be represented as follows [12]:

$$\dot{\varepsilon} = A\sigma^n \exp(-\frac{Q_{\mathrm{App}}}{RT}) \quad (1)$$

In the formula, $\dot{\varepsilon}$ is steady state creep rate, A is material constant, σ is Apparent Creep Activation Energy, Q_{App} is nominal creep activation energy, R is universal gas constant, ($R=8.314J\cdot mol^{-1}\cdot K^{-1}$), T is absolute temperature, unit is K, n is creep stress exponent.

When the temperature of creep process is constant, the expression of stress exponent n is as follows by equation (1) differential:

$$n = (\partial \ln \dot{\varepsilon} / \partial \ln \sigma)_T \quad (2)$$

According to the data of table 2, the graph of $\ln \dot{\varepsilon}$ - $\ln \sigma$ is shown in Fig.3. The slope of $\ln \dot{\varepsilon}$ - $\ln \sigma$ relation curve is namely the creep stress exponent n. Through calculation, when creep occurs in UFG CP Ti at room temperature, creep stress is constant., which is 640 ~ 700 MPa, the creep stress exponent n of UFG CP Ti is 17.3; when creep stress is 700 ~ 760MPa, the creep stress exponent is 55.7.

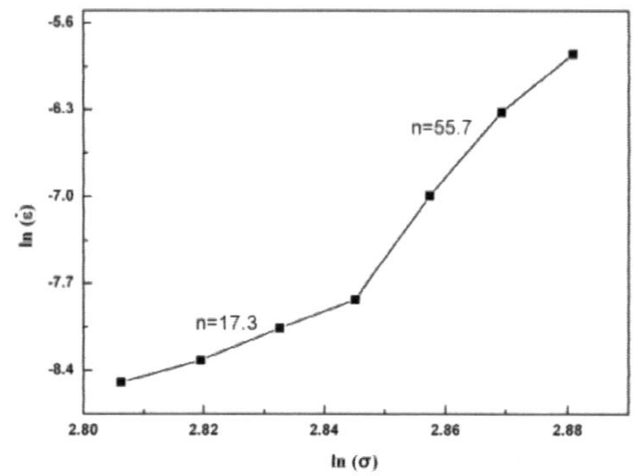

Figure 3. The plot of $\ln \dot{\varepsilon}$ against $\ln \sigma$ for UFG CP Ti

In this experiment, UFG CP Ti has higher creep stress exponent n (17.3 ~ 55.7), which shows that UFG CP Ti has better creep resistance at room temperature. Combined with literature[13], under 500 ~ 600 °C, when stress is 97 ~ 472 MPa, the creep stress exponent n of Ti-6Al-4V alloy is approximate 5.2 ~ 11.3, consistenting with the rules of the experimental result. In this study, the creep stress exponent n of UFG CP Ti is bigger than the high-temperature creep stress exponent n of Ti-6Al-4V alloy, which means that room temperature creep property of UFG CP Ti is good.

3.4 Microscopic Structure of UFG CP Ti

Comparing the TEM images of UFG CP Ti after creep test with the before, as shown in Fig.4, it can be investgated that, after composite refining deformation, amount of cellular substructures forms in the microstructure of CP Ti and the twin crystal develops into subgrain by self-intersection. At the same time, severe plastic deformation causes greatly dislocation accumulation in the microstructure; high-density dislocation accumulates into cell wall; and dislocations tangle into clusters, the boundary of cellular structure is distinct. With the increase of cell dislocation density, misorientation between adjacent dislocation cells gradually increases to develop into subgrain .

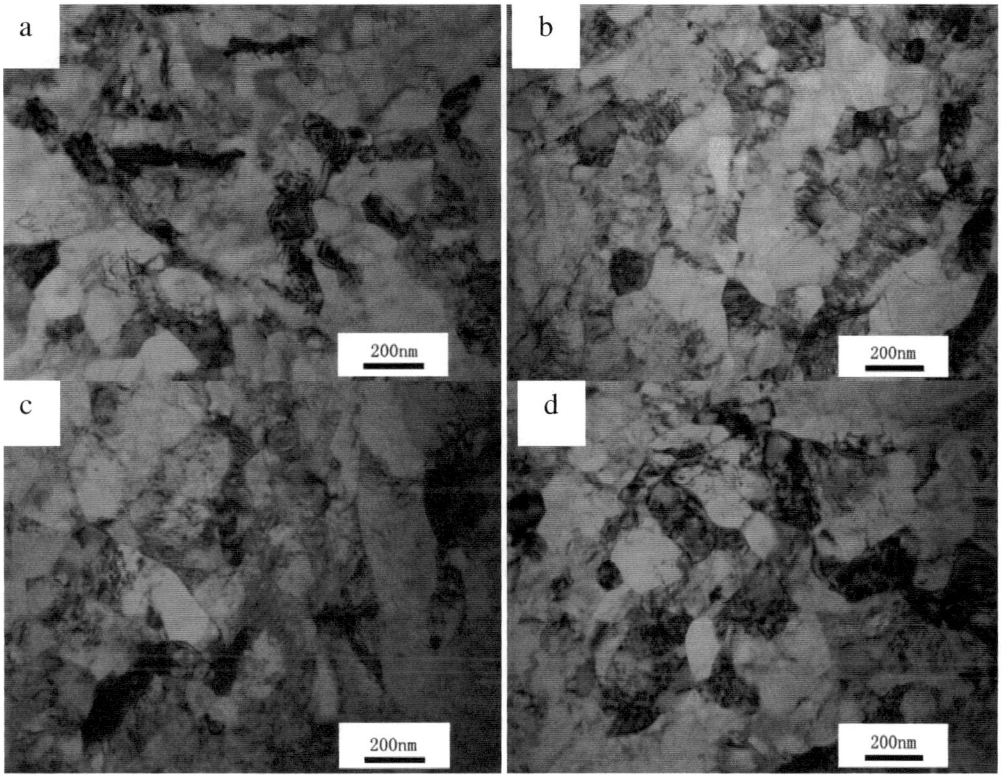

Figure 4. TEM micrographs of UFG CP Ti
(a) before creep; (b) 660 MPa; (c) 700 MPa; (d)740 MPa

Micromechanism of creep is closely related to the change of internal material microstructure and the behavior of dislocation configuration. Creep deformation mainly includes two mechanisms, one is dislocation deformation and the other is diffusion creep deformation. Under biggsh stress and low temperature (< 0.5Tm), the creep mechanism is mainly the dislocation movement, while under higher temperature (0.6 ~ 0.7Tm) and relatively less stress, it is mainly diffusion creep deformation. At room temperature, external load causeing stress applied on materials, intergranular dislocation is enable to move, resulting in plastic deformation, and new dislocation forms during the process of plastic deformation, causing multiplication. Dislocation moves on the glide plane, avoiding from accumulating on grain boundary, so as to make deformation on-going[14].

It can be seen from Fig.4 that, under applied stress, at the beginning of creep, intergranular dislocation could move freely and multiplicate constantly. Therefore, the deformation rate of creep in stage I is very high. With creep proceeding, dislocation pileup happens on the grain boundary, the original dislocation with free distribution is transformed into dislocation cell, as shown in Fig.4 (b), (c), (d). Screw dislocation leaves blocking area via dislocation gliding and climbing, at this time, dislocation cell wall is the small angle grain boundary resulted by dislocation gliding and climbing. The dislocation distribution of intracellular sliding and cell wall climbing forms a long-range inner gravitational field. Under the action of external force, dislocation moves on glide plane along with the gliding direction, till to surface and then disappearing. Therefore, the equilibrium state of the original inner gravitational field is broken, and dislocation source starts moving and generates new mobile dislocation. The dislocation desity is very high in the microstructure of UFG CP Ti and dislocations movement results in the creep. Therefore, the creep mechanism of UFG CP Ti room temperature creep is dislocation creep, its dominate deformation is dislocation glide, and at the same time, the creep property of material is also affected by the interaction between UFG CP Ti microstructure, such as, grain size, the number of grain boundary, and dislocation.

4. Conclusions

1) Under the low stress level, the creep behave saturation phenomenon. With the increase of creep stress, creeps value of UFG CP Ti room temperature creep increases gradually.

2) At room temperature, when creep stress of UFG CP Ti is 760 MPa, steady state creep rate is 1.416×10^{-6} s^{-1}, and creep property is good.

3) When creep stress is 640 ~ 700 MPa, creep stress exponent n of UFG CP Ti is 17.3, and when creep stress is 700 ~ 760 MPa, the creep stress exponent n is 55.7.

4) The creep mechanism of UFG CP Ti during room temperature creep is dislocation creep, and it's dominate creep deformation is dislocation glide.

5. References

[1] N. S. Weston, F. Derguti, and A. Tudball,"Spark plasma sintering of commercial and evelopment titanium alloy powders," Journal of Materials Science, vol. 50, pp. 4860-4878, 2015.

[2] S. Hariprasad, M. Ashfaq, and T. Arunnellaiappan, "Role of electrolyte additives on in-vitro corrosion behavior of DC plasma electrolytic oxidization coatings formed on Cp-Ti," Surface and Coatings Technology, vol. 3, pp.16-44, 2016.

[3] Christopher S. Meredith and Akhtar S. Khan,"The microstructural evolution and thermo-mechanical behavior of UFG Ti processed via equal channel angular pressing," Journal of Materials Processing Technology, vol. 219, pp. 257-270, 2015.

[4] P. Rodriguez-Calvillo and J.M. Cabrera,"Microstructure and mechanical properties of a commercially pure Ti processed by warm equal channel angular pressing,"Materials Science and Engineering: A, vol.625, pp. 311-320, 2015.

[5] D.V. Gunderov, A.V. Polyakov, and I.P. Semenova, "Evolution of microstructure, macrotexture and mechanical properties of commercially pure Ti during ECAP-conform processing and drawing,"Materials Science and Engineering: A, vol. 562, pp. 128-136, 2013.

[6] P. SudharshanPhani and W.C. Oliver. "A direct comparison of high temperature nanoindentation creep and uniaxial creep measurements for commercial purity aluminum," ActaMaterialia, vol. 111, pp.31-38, 2016.

[7] D. Andrés, R. Lacalle, and J.A. Álvarez. "Creep property evaluation of light alloys by means of the Small Punch test: Creep master curves," Materials & Design, vol. 96, pp. 122-130, 2016.

[8] W. Pachla, M. Kulczyk, and S. Przybysz, "Effect of severe plastic deformation realized by hydrostatic extrusion and rotary swaging on the properties of CP Ti grade 2," Journal of Materials Processing Technology, vol.221, pp. 255-268, 2015.

[9] Qiao Dai, Chang-Yu Zhou, and Jian Peng, "Room-temperature creep behavior on crack tip of commercially pure titanium," Materials & Design, vol. 85, pp. 618-625, 2015.

[10] B. Barkia, V. Doquet, and J.P. Couzinié, "Room-temperature creep and stress relaxation in commercial purity titanium–Influence of the oxygen and hydrogen contents on incubation phenomena and aging-induced rejuvenation of the creep potential," Materials Science and Engineering: A, vol. 624, pp. 78-89, 2015.

[11] Ş. Ţălu, Micro and nanoscale characterization of three dimensional surfaces. Basics and applications. Napoca Star Publishing House, Cluj-Napoca, Romania, 2015.

[12] Peng Jian, Zhou Changyu, and Dai Qiao, "The temperature and stress dependent primary creep of CP-Ti at low and intermediate temperature,"Materials Science and Engineering: A, vol. 611, pp. 123-135, 2014.

[13] D.A.P. Reis, C. Moura Neto, and C.R.M. Silva, "Effect of coating on the creep behavior of the Ti–6Al–4V alloy," Materials Science and Engineering: A, vol. 486, pp. 421-426, 2008.

[14] Yi-jie Zhao, Yong Zheng, and Wei Zhou, "Effect of carbon addition on the densification behavior, microstructure evolution and mechanical properties of Ti(C, N)-based cermets,"Materials Characterization, vol. 106, pp. 266-272, 2015.

IOP Conf. Series: Materials Science and Engineering **250** (2017) 012063 doi:10.1088/1757-899X/250/1/012063

Reliability analysis of crankshaft for high-speed punch based on Monte-Carlo method

Xinzhou Zhang, Lan Chen[1], ShuyuanGan, Yuan Wang, Naifei Ren and Jian Zhou

School of Mechanical Engineering, Jiangsu University, 301 Xuefu Road, Zhenjiang, 212013, China.
Email: chenlan5321@126.com

Abstract. The crankshaft, which has fateful consequence to the performance and reliability, is a key component of high-speed punch. In manufacturing process, the high reliability value of crankshaft contributes to the increasing of the reliability of the whole punch system. The study builds a reliability analysis model of the crankshaft for punching process under the premise of regarding design parameters as random variables. Monte-Carlo method is employed to make reliability analysis for crankshaft based on ANSYS. The numerical results present that the failure probability of strength and stiffness for crankshaft satisfied the use requirement. The impact of every input variable on reliability are obtained from the sensitivity analysis of status function factors. The design variable D4 and PRXY have the greatest impact on the strength of crankshaft and the design variable YOUNG and L3 have the greatest impact on the stiffness of crankshaft. And the result matches up with the response surface graph. The reliability analysis result provides some useful information for the improvement of the reliability of crankshaft for high-speed punch.

1. Introduction

High-speed punch is an efficient, high-precision and high-automation punching machine. It has been widely used in aeronautics and astronautics, automobile, traffic and transportation, metallurgical chemical industries and so on [1-2]. The crankshaft is a key part of main transmission system for high-speed punch and its reliability has great impact on the reliability of high-speed punch.

Modeling and simulation of the crankshaft is obviously not new, as the crankshaft is a very popular mechanism used in many different types of machinery. Zhou [3] constructed a 2-D stress strength interference model, considering the disparity of the load asymmetric factor, to make reliability analysis for diesel engine crankshaft and the analysis result was consistent with the statistical data of actual failure cases. Ye [4-5] analyzed the fatigue reliability of crankshaft for reciprocating pump using the radius vector method and he developed a practical object-oriented application software to make fatigue reliability analysis for improving the design of reciprocating pump. Wang [6] built a quarter finite element model of the entire assembly to make reliability analysis and he got the temperature distribution, thermal deformation and thermal stress distribution of the piston. Taraza [7] developed an algorithm capable to identify faulty cylinders based on the phases of the lowest three harmonic orders of the measured speed. Sun [8] made the fatigue reliability analysis of crankshaft using stochastic response surface method with implicit performance function and some examples were employed to demonstrate the advantages of the proposed method. Wang [9] presented a sensitivity-based algorithm about tool change time and the reliability of the machining process to improve the reliability of both tools and the overall process. Singh [10] provides an optimal model in predicting the reliability for an automobile crankshaft through the time dependent failure analysis based on three parameter Weibull distribution.

As stated previously, there is few study on reliability analysis of crankshaft for high-speed punch. This work introduces the reliability analysis theory including the basic assumptions of reliability and the reliability calculation method. Then, the reliability analysis model of crankshaft is built in ANSYS and the reliability analysis based on Monte-Carlo method is carried out.

2. Reliability analysis theory

2.1. Basic assumptions of reliability

Reliability is defined as: the probability that a device will perform its intended function during a specified period of time under stated conditions. The stress-strength interference model is usually employed to make reliability analysis for mechanical components. Stress in the model is typically the factors that has an impact on functionality, such as stress, torque, temperature, vibration, corrosion, and so on. Strength in the model is typically the resistance ability of external loads. The influence in strength is corrosion, temperature, materials performance, quality of machining and external loads. So the definition of reliability is described as: the probability of the strength is greater than the stress.

Some basic assumptions of reliability should be made to analyze the reliability for mechanical products [11]:

- The strength of mechanical components is nonnegative random variable or random process.
- The stress is nonnegative random variable or random process.
- When the strength is greater than the stress, the part is believed to be reliable. Otherwise, the part is believed to be of failure.
- For the stress-strength interference model, the part is of failure only because of the stress.
- The formula for calculating the strength and stress still applies and all quantitative symbols are treated as random variable or random process.

In the stress-strength interference model, the stress and strength are all regarded as random variable or random process, which could reflect the exact thing that the stress and strength are of objective uncertainty, discreteness and time-varying uncertainty. For this reason, the probabilistic method must be used to calculate the reliability of mechanical components.

2.2. Reliability calculation method

The reliability could be calculated in the case that the probability density function or joint probability density function of random variable X is known. That may not always be the case, and the distribution probability is difficult to be determined in the operating condition. There are some approximation methods to calculate the reliability, such as Moment method, Monte-Carlo method, Response Surface method, and so on.

*2.2.1. Moment method.*The moment method mainly includes the first-order second-moment method and the higher order moment method. And the first method is widely used in the actual projects. The basic idea is that Taylor expand the nonlinearity performance function according to the first two moments, and remain the once migration, and then approximately calculate the mean value & standard deviation of performance function to get the reliability.

*2.2.2. Monte-Carlo method.*The basis theory of Monte-Carlo method is large numbers law in probability theory. Monte-Carlo method is one of the most intuitive, accurate and the most efficient for high non-liner problem structural reliability analysis approach. The reliability can be got from statistical analysis of a large random sample.

*2.2.3. Response surface method.*Response surface method is employed to solve the reliability problem of complex mechanical system. The basic idea of this method is to suppose a simple analytic expression between limit state variable and basic variable firstly, then to determine the unknown parameters using interpolation method, and to get the certain analytic expression finally. The reliability calculation is simplified and the calculation precision could be improved using response surface method. And it can be calculated using finite element analysis program.

3. Reliability model of crankshaft

3.1. Structure of crankshaft

The structure sketch of crankshaft is shown as Figure 1 [12-13]. This type of crankshaft is used for 800KN high-speed punch made by a company. The working conditions of the punch are as follows: Nominal Pressure is 800 KN, Nominal Pressure Stroke is 2 mm, Slider Stroke is 30 mm, number of strokes is 600 SPM (Strokes per Minute), the thickness of punched part is 2 mm, the material of crankshaft is 40Cr.

The crankshaft transfer drive torque from the flywheel, which is installed in location 4 with clutch. The four groups of support bearings, which are installed in location 1, 3, 5, 8, form a revolute pair with the punch frame. The two groups of support bearings, which are installed in location 6, 7, form a revolute pair with main connecting rod. One group of support bearing, which is installed in location 2, form a revolute pair with balancing connecting rod.

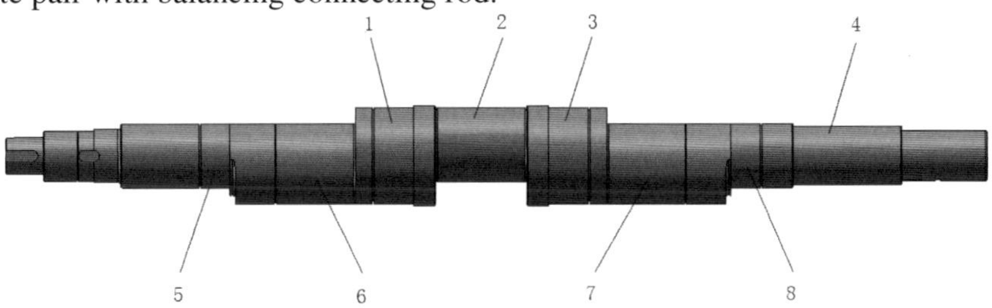

Figure1. Structure sketch of crankshaft

1、3、5、8—Installation locations of support bearing

2—Installation location of connecting rod bearing of Balancing slider

3.2. Design variables of crankshaft

Design variables of crankshaft fall into three types: material characteristics parameters, load parameters and geometry characteristic parameters. The name of parameter, the distribution type, the variation coefficient and the mean are shown in Table 1. The data is from reference [12-13].

The output variables of crankshaft are *SUBS* and *SUBU*.

$$SUBS = YIES - maxstr \begin{cases} >0 & \text{reliable state} \\ =0 & \text{limit state} \\ <0 & \text{failure state} \end{cases} \quad (1)$$

$$SUBU = U_{max} - maxu \begin{cases} >0 & \text{reliable state} \\ =0 & \text{limit state} \\ <0 & \text{failure state} \end{cases} \quad (2)$$

Where U_{max} is the maximum allowable displacement of crankshaft in vertical direction, *maxu* is the maximum allowable displacement of crankshaft in vertical direction.

4. Results and discussion

4.1. Histogram of status function

The histogram of design variable *D*1 for reliability analysis is shown in Figure 2. As can be seen from the chart, the histogram of design variable approaches the normal distribution probability function curve. The histogram generates a smooth curve without larger gap and the sampling number of simulation is bigger enough to keep the calculating accuracy.

4.2. Graph of response surface

The response surface graph of output status function is shown in Figure 3- Figure 4. Figure 3 shows the impacts of input variable *PRXY* and *D*4 on output variable *SUBS*. *SUBS* decreases with the increase of the *D*4, and increases with the increase of *PRXY*. The optimal combination of input variables *PRXY* and *D*4 could be found to keep an appropriate *SUBS*.

Table 1. The statistical table of input random variable

Parameter type	Parameter name	Distribution type	Mean	Variation coefficient	Standard deviation
Material characteristics parameters	Elastic modulus *YOUNG*/GPa	GAUS	206	0.05	10.3
	Poisson ratio *PRXY*	GAUS	0.28	0.05	0.014
	Yield strength *YIES*/MPa	GAUS	345	0.05	17.25
	Density ρ/kg/m^3	GAUS	7700	0.05	385
	Gravitational acceleration *g*	GAUS	9.8	0.05	0.49
Load paramet ers	Load of neck journal of main connecting rod *F*1/N	GAUS	64897.95	0.07	4542.9
	Load of neck journal of balancing connecting rod *F*2/N	GAUS	141180	0.07	9882.6
Geometry characteristic parameters	Diameter of constrained neck journal*D*1/m	GAUS	0.118	0.025	0.00295
	Diameter of neck journal of main connecting rod *D*2/m	GAUS	0.15	0.025	0.00375
	Diameter of constrained neck journal *D*3/m	GAUS	0.182	0.025	0.00455
	Diameter of neck journal of balancing connecting rod *D*4/m	GAUS	0.14	0.025	0.0035
	Length of constrained neck journal *L*1/m	GAUS	0.055	0.025	0.001375
	Length of neck journal of main connecting rod *L*2/m	GAUS	0.126	0.025	0.00315
	Length of constrained neck journal *L*3/m	GAUS	0.074	0.025	0.00185
	Length of neck journal of balancing connecting rod *L*4/m	GAUS	0.1	0.025	0.0025

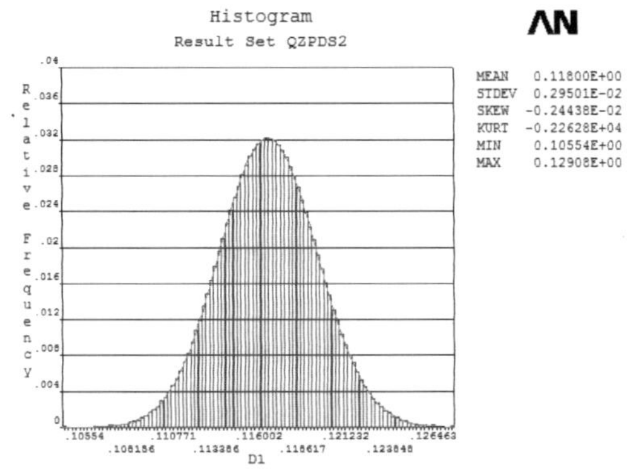

Figure2. Histogram of design variable *D*1

Figure 4 shows the impacts of input variable *L*3 and *YOUNG* on output variable *SUBU*. SUBU decreases with the increase of the *L*3, and decreases with the increase of *YOUNG*. The optimal combination of input variables *L*3 and *YOUNG* could be found to keep an appropriate *SUBU*. The response surface graph can present the impact of multiple variables on output variable and it provides guidance on crankshaft design.

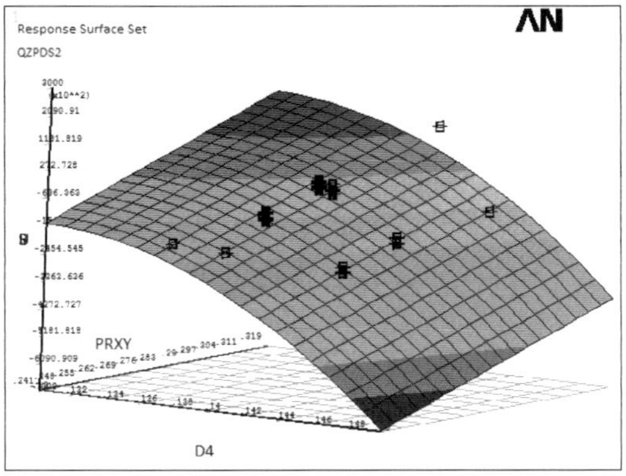

Figure3. Response surface graph of*SUBS-PRXY-D*4

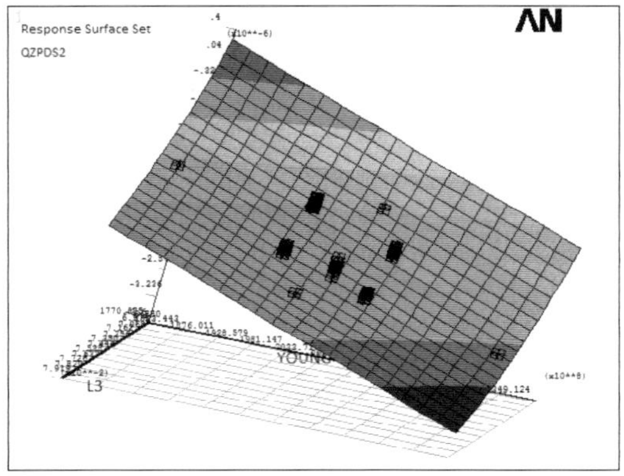

Figure4. Response surface graph of*SUBU-L*3-*YOUNG*

4.3. Cumulative distribution function ofoutputstatusfunction

Thecumulative distribution function of output status function isshowninFigure 5-Figure 6. The value of any point in the distribution function equals to the probability value of data appeared in this point. As shown in Figure 5, the failure probability of strength for crankshaft is probability value of when *SUBS*<0. And the probability value of *SUBS* is 0.94863. As shown in Figure 6, the failure probability of stiffness for crankshaft is probability value of when *SUBU*<0. And the probability value of *SUBU* is 0.99953.

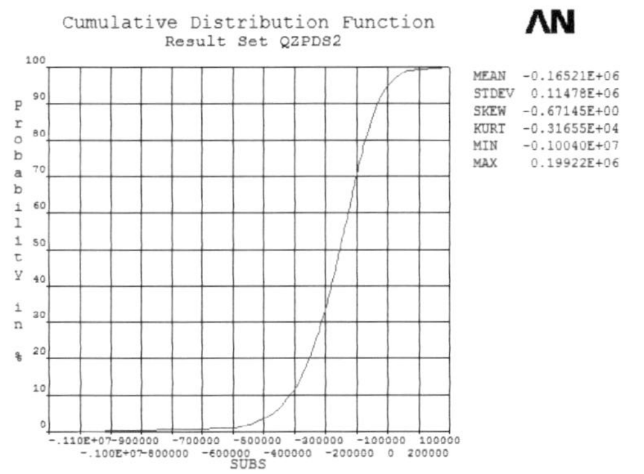

Figure5. Cumulative distribution function of *SUBS*

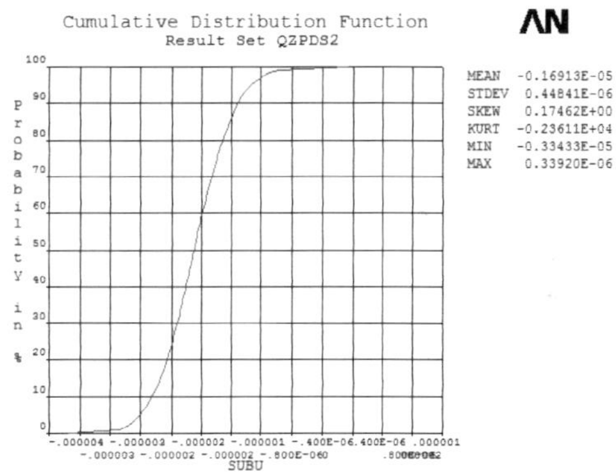

Figure6. Cumulative distribution function of *SUBU*

4.4. Sensitivity analysis of statusfunction factors

The probability sensitivity reflects the impact of every input variable on reliability. The input variable couldbe controlled inthedesignandmanufacturingprocesstogetbetterproduct. The sensitivity graph of *SUBS* and*SUBU* isshownin Figure 7-Figure 8. As shown in Figure 7, the input variable *D*4 and*PRXY* have the greatest impact on *SUBS*. *D*4 isbelow zero graduation line, which means that *SUBS* decreases with the increase of *D*4, and *PRXY* is above zero graduation line, which means that *SUBS* increases with the increase of *PRXY*. And it matched up with the response surface graph. As shown in Figure 8, the input variable *YOUNG* and*L*3 have the greatest impact on *SUBU*. They are all below zero graduation line and *SUBU* decreases with the increase of *YOUNG* and*L*3.

5. Conclusion

This paper studies the reliability ofcrankshaft forhigh-speed punch.Thereliability analysis theoryisintroduced, and the basic assumptions of reliability and the reliability calculation method are analyzed.Monte-Carlo method is employedtomakereliabilityanalysisforcrankshaftbasedon ANSYS. The numerical results present thatthe failure probability of strength for crankshaft is 0.94863andthe failure probability of stiffness for crankshaft is 0.99953. The impact of every input variable on reliabilityareobtainedfromthesensitivity analysis of status function factors. The input variable*D*4 and*PRXY*have the greatest impact on*SUBS*andthe input variable*YOUNG* and*L*3 have the greatest impact on*SUBU*. Thesensitivity analysisresultmatches up with the response surface graph.

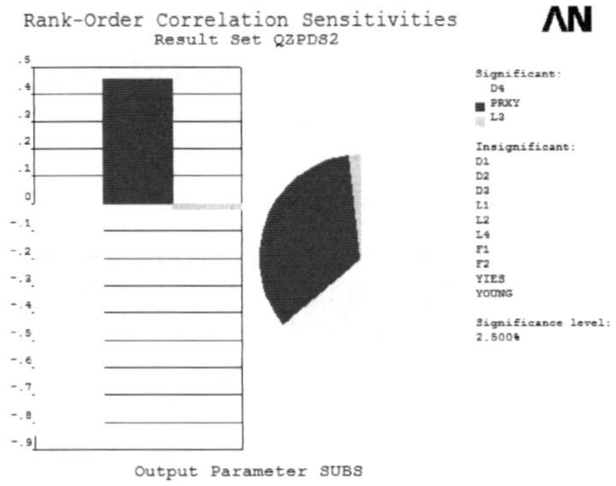

Figure7. Sensitivitygraph of*SUBS*

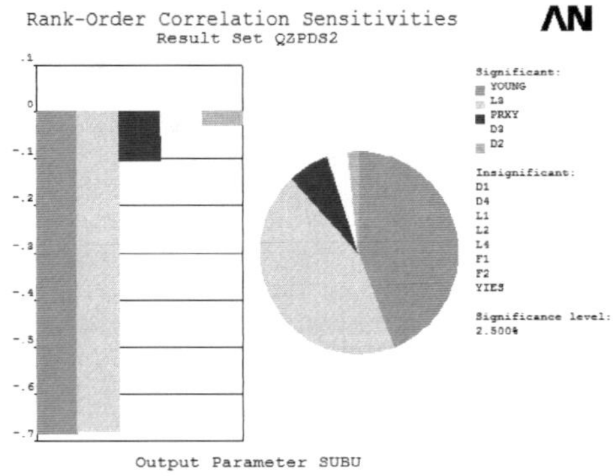

Figure8. Sensitivitygraph of*SUBU*

6. Acknowledgments
This work is financially supported by Fundamental Re-search Project of the National Natural Science Foundation of China(Grant No. 71601086), the National Natural Science Foundation of Jiangsu Province(Grant No. BK20160485), China Postdoctoral Science Foundation(Grant No. 2015M581748), the Natural Science in Colleges of Jiangsu Province of China (Grant No. 15KJB460005 & 16KJB140002), Jiangsu Planned Projects for Postdoctoral Research Funds (Grant No. 1501102C), Advanced Talent Foundation of Jiangsu University of China (Grant No. 14JDG138 & 15JDG033).

7. References
[1] ChenL, Zhang X Z, Gan S Y and others 2016*MATEC Web of Conferences***61** 04003
[2] Breitling J, Wallace DandAltanT1996 *Journal of Materials Processing Technology***59** 18-23
[3] Zhou X, Yu XLandLi Y. 2006*Transactions of the Chinese Society of Agricultural Machinery* **37**(10) 149-152, 131
[4] YeX Y, JiangX P, XuJQand others 2008*Chinese Journal of Mechanical Engineering***44**(10) 272-276
[5] YeX Y, JiangX P, XuJQand others 2008*Transactions of the Chinese Society of Agricultural Machinery***39**(6) 94-97,72
[6] Wang Y X, Qi X L and Hu K 2009 *Proceedings of the Third International Conference on Mechanical Engineering and Mechanics***1-2** 1905-1909
[7] Taraza D, Henein N A and Bryzik W 2001 *Journal of Engineering for Gas Turbines and Power-Transactions of the ASME* **123**(2) 428-432
[8] Sun Y F, Qiu H B, Gao Land others 2010*Proceedings of the ASME International Mechanical Engineering Congress and Exposition***13** 263-268
[9] Wang X G, Wang B Y, Lv CMand others2017 *The International Journal of Advanced Manufacturing Technology***89** 1535-1544
[10] Singh S S K, Abdullah Sand Mohamed NAN 2015*Maintenance and Reliability***17**(3) 408-415
[11] Wang XG 2009 *Some Problems Research on Time-Dependent Reliability-Based Robust Optimization Design of Mechanical Components*(Shenyang: Northeastern University)
[12] Zhang X Z, Chen L, Gan S Y and others 2017*International Journal of Science***4**(7) 190-195
[13] Zhang X Z, Chen L, Gan S Y and others 2017*International Journal of Science***4**(7) 203-209

IOP Conf. Series: Materials Science and Engineering **250** (2017) 012064 doi:10.1088/1757-899X/250/1/012064

Investigations on Reliability Growth for High-Speed Punch Based on Whole Life Cycle

Xinzhou Zhang, Lan Chen*, Naifei Ren, Jian Zhou, Shuyuan Gan and Yuan Wang

School of Mechanical Engineering, Jiangsu University, 301 Xuefu Road, Zhenjiang, 212013, China.
Email: chenlan5321@126.com

Abstract. This paper presents a study on the reliability growth analysis of high-speed punch. The main contribution of the paper is the development of an efficient approach to analyze the reliability growth and to make reliability growth management of high-speed punch. The reliability growth methods for high-speed punch in different stages are presented. The approach is applied to improve the reliability growth of high-speed punch in whole life cycle, and in the meantime to provide some useful information for the improvement of the design, manufacturing and using of high-speed punch.

1. Introduction

High-speed punch is an efficient, high-precision and high-automation punching machine. It has been widely used in aeronautics and astronautics, automobile, traffic and transportation, metallurgical chemical industries and so on [1].

Weiss [2] proposed the first reliability growth model in 1956 when he studied the guided missile system. And the model is considered to be the prototype of reliability growth technology. Duane [3] proposed a model, which is called "Duane Model", in 1964 and it has extensive applications in many fields. Department of defense in USA firstly took reliability growth technology into reliability working plan in national military standard [4]. Crow [5] gave an explanation of probability theory to Duane Model and proposed an Army materiel system analysis activity model based on Non-homogeneous Poisson process. In 1981, USA awarded a military handbook that included 17 reliability growth models and it was firstly used in military, then extended to civilian product [6]. And it was also used by many countries. The international electrotechnical commission set an international standard in 1989, which stipulates the requirements and rules of making the reliability growth plan [7]. Then, reliability growth problem has been paid attentions widespreadly in the world and then came so many papers, reports, standards and handbooks [8-16].

As stated previously, there is few study on reliability growth of high-speed punch. This work proposed a reliability growth method based on whole life cycle for high-speed punch.

2. Analysis of Reliability Growth

There will be a gradual process for a new product from design to manufacturing. The process usually can be broadly divided into three phases, as shown in Figure 1.

Reliability growth of the development process: the new product has a low reliability in the development process. The engineer uses design experience of traditional similar products for the design of new product development. The reliability comprehensive experiment is used to find the failure of new product. Then, analyze the reason of failures to correct the design and to eliminate the weak link. Finally, the reliability of new product will grow to prescribed level.

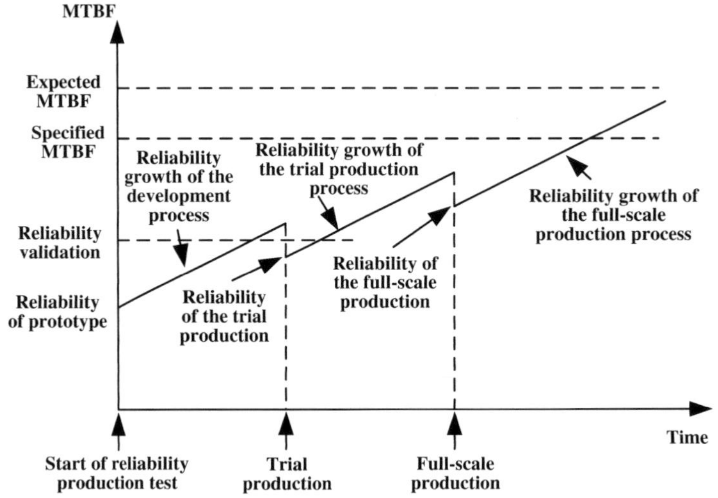

Figure 1. The reliability growth

Reliability of the trial production: in the beginning of this stage, the reliability will be influenced by the equipment, the technology and the worker's proficiency. With the corrected design and worker's training, the reliability would grow.

Reliability of the full-scale production: the product has been in the full-scale production process. There will be some problems in design and manufacturing and it will still influence the reliability. The reliability finally will grow after some corrective actions.

3. Reliability Growth Management of High-Speed Punch

The high-speed punch is a complex system that involves numerous fields, like mechanical structure design, manufacturing and assembly technology, electric and electronic technology, hydraulic and pneumatic transmission technology, computer automatic control technology and sensing technology. If any of the components goes bad, the punch will fail. Whether the main punch can work normally will mainly depend on the coordination and cooperation of all components.

The reliability growth management of high-speed punch is as follows:

- Rules and regulations of reliability growth management. To keep an efficacious management on reliability growth, the policy, regulation, rule of the management need to be made for institutionalization, normalization and standardization of information management.
- Reliability information analysis. There need to analyze practical requirements of information unions and users at all levels and to investigate the accuracy of collected information. The information collecting should be organized orderly to improve the actual benefit of information management.
- The closed-loop reliability information management. In the reliability growth process, resources need to be allocated properly by informational needs and then supply feedback information. The information management will be carried out according to this principle to form closed-loop management system.
- The fault report, analysis and correcting system. All failure information in reliability growth test will be collected by the fault report, analysis and correcting system. The failure information will be analyzed timely and the corrective action will also be recorded.

4. Reliability Growth Analysis of High-Speed Punch

4.1. Reliability Growth Programming of High-Speed Punch

According to the characteristics of high-speed punch, the reliability growth programming of high-speed punch is as shown in Figure 2.

Planning approaches and procedures are shown as follows:

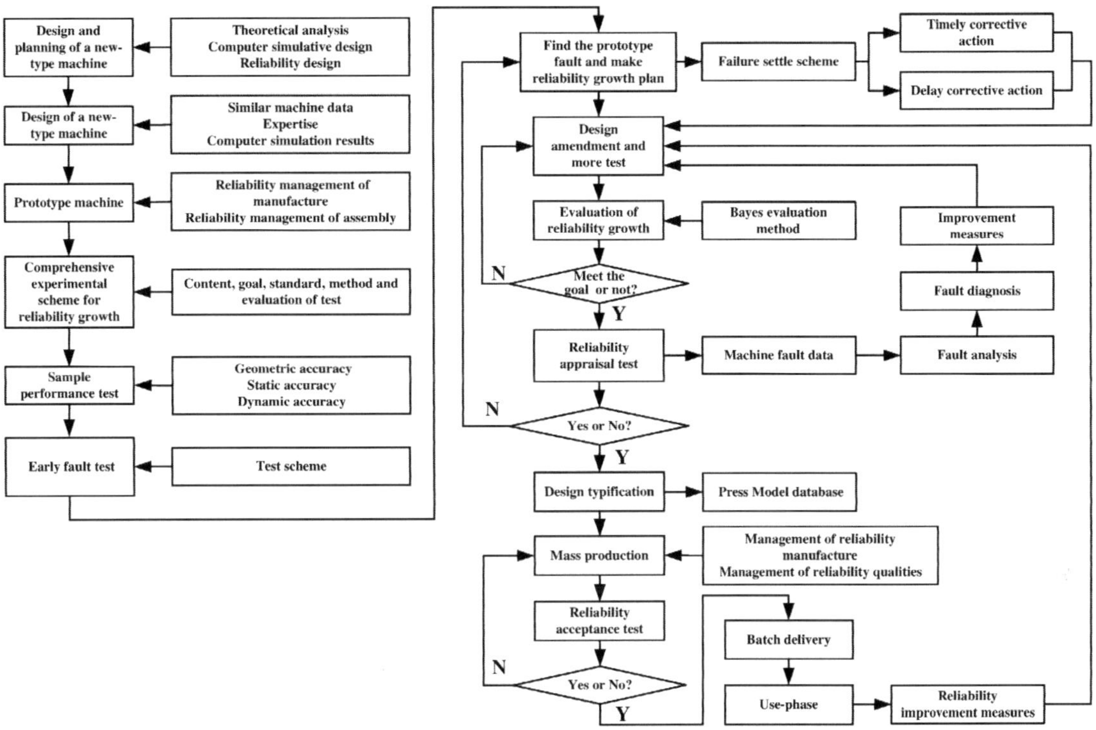

Figure 2. Planning procedure of reliability growth for high-speed punch

- (1) The planning procedure of the new type high-speed punch should firstly be made after the theoretical analysis, computer simulation analysis and reliability design.
- (2) To make prototype design of new type high-speed punch with reference to product data, expertise and computer simulation results of similar product in the same category.
- (3) To product the prototype punch based on reliability manufacturing and management.
- (4) To make comprehensive experimental scheme, which includes the experiment objective, content, standard, method and evaluation, for reliability growth for the new type prototype punch.
- (5) To carry out sample performance test, which includes geometric accuracy test, static accuracy test and dynamic accuracy test.
- (6) To make early fault test scheme according to the relevant norms and standards for high-speed punch and then carry out the test.
- (7) To find the prototype fault and to make reliability growth plan.
- (8) To make failure settle scheme: take timely corrective action to failures which can be corrected timely and take delay corrective action for others.
- (9) To make design amendment in accordance with the corrective action and then carry out more test.
- (10) To make evaluation of reliability growth based on a small sampling of Bayes method for new type high-speed punch.
- (11) To analyze the evaluation result: if the result achieves the target, then go to next step, otherwise go back to last step.

- (12) To count the failure of reliability appraisal test into fault database. The analysis and fault diagnosis will be made and then draw the relevant improvement measures.
- (13) To make design typification if the reliability appraisal test pass and type it into Press Model database, otherwise go back to step (7).
- (14) To make mass production with reliability manufacture and reliability qualities management.
- (15) To carry out reliability acceptance test: if it passes the test, to deliver into users, or go back to step (14) for more improvement measures.
- (16) To take relevant improvement measures for reliability growth in accordance with the failure in using stage.

By the steps, to achieve the significant reliability growth.

4.2. Reliability Growth Analysis of High-Speed Punch

Based on the foregoing analysis, the reliability growth of high-speed punch is divided into 3 stages: reliability growth of design phase, reliability growth of manufacturing phase and reliability growth of using phase.

*4.2.1. Reliability Growth of Design Phase.*The reliability design requires of high-speed punch are made firstly based on planning requirements. Then, make reliability distribution of components and make reliability analysis and reliability optimization of components with high failure frequency. Finally, make reliability prediction of high-speed punch and if the result does not meet the requirement, modify the design. The process of reliability growth in design phase of high-speed punch is shown in Figure 3.

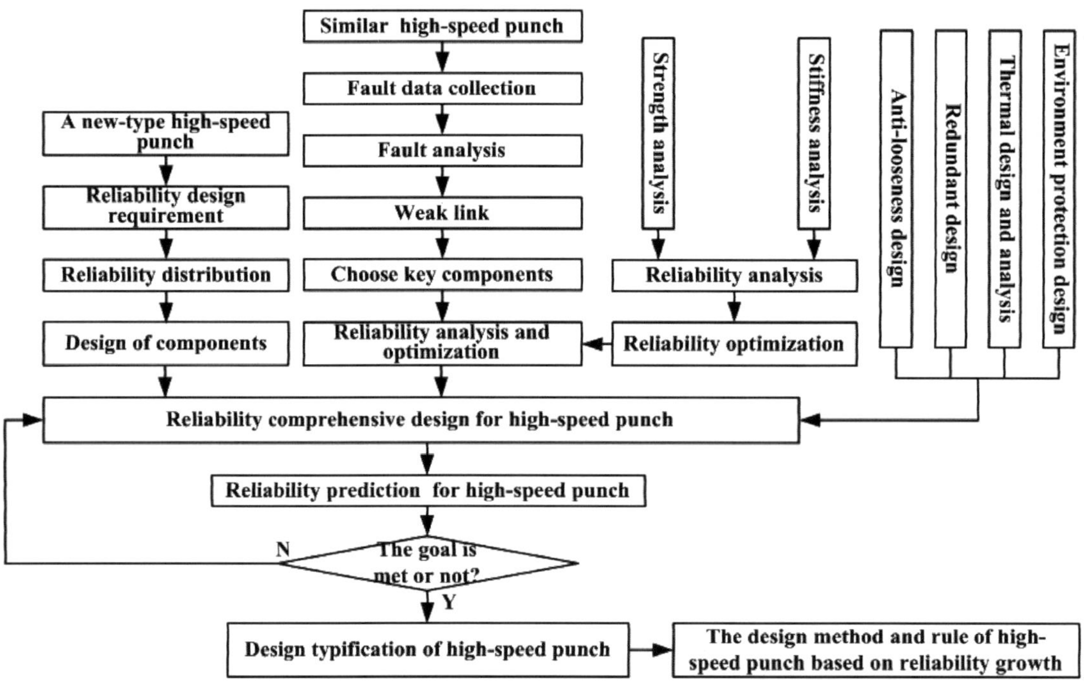

Figure 3. The process of reliability growth in design phase of high-speed punch

4.2.2. Reliability Growth of Manufacturing Phase. The manufacturing process and quality control during manufacturing greatly affect system reliability. The reliability will grow by removing systematic defects in manufacturing. The reliability of complex system usually decreases from design and development to production and the quality control technology is more and more important for reliability growth.

To improve the reliability, the early malfunction and weak links of high-speed punch should be found and improved design should be made. The test of early malfunction of high-speed punch is as shown in Figure 4.

The early malfunction test is to find potential early malfunction of high-speed punch, to check the stability of the punch. So the weak links can be found and some relevant maintenances and corrections can be made for accumulating design experience. The reliability growth process of high-speed punch is as shown in Figure 5. While the testing is under way, the systematic failure should be analyzed as soon as possible and some relevant corrections should be made to the prototype punch.

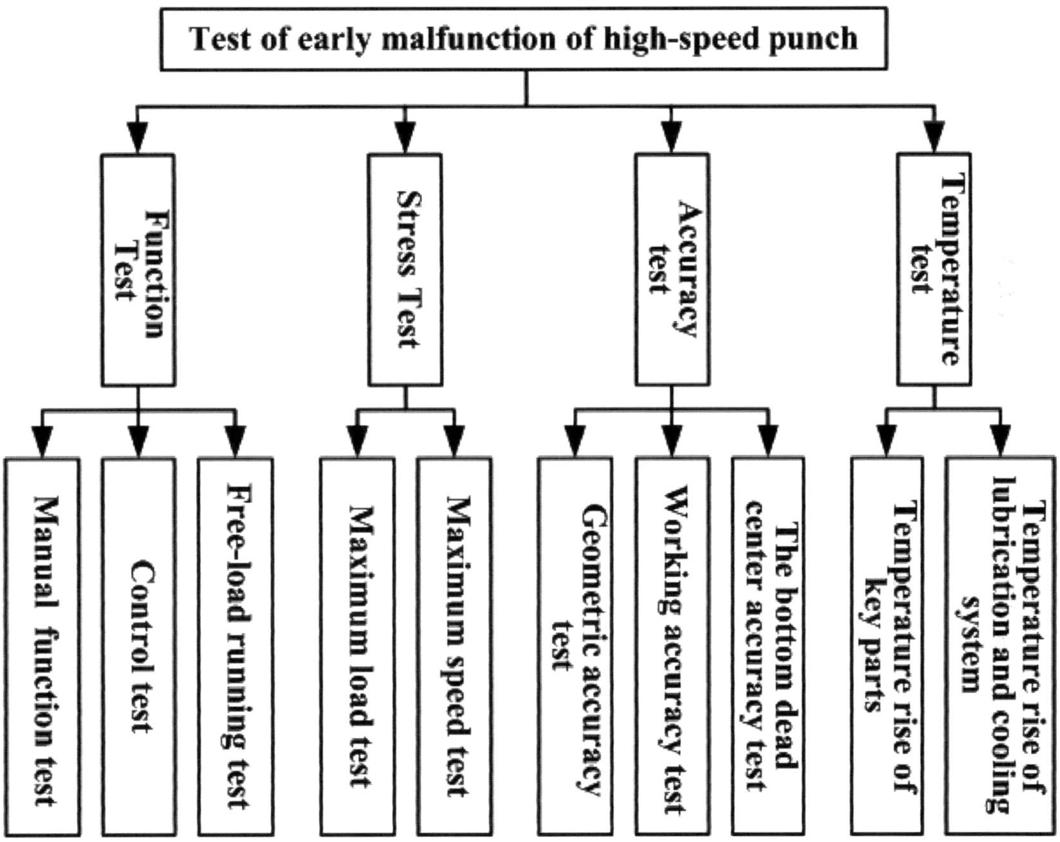

Figure 4. The test of early malfunction of high-speed punch

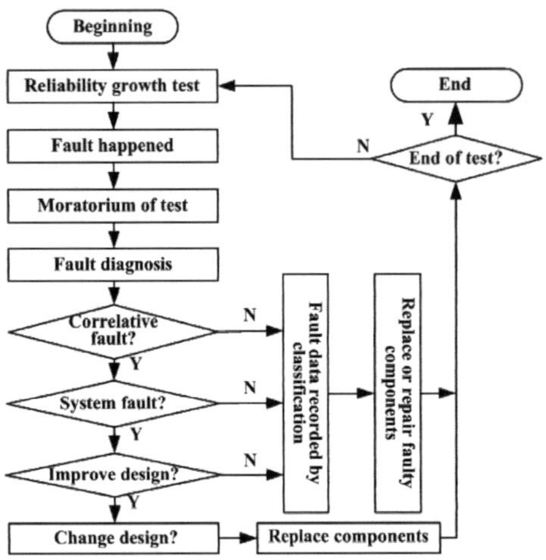

Figure 5. The reliability growth process of high-speed punch

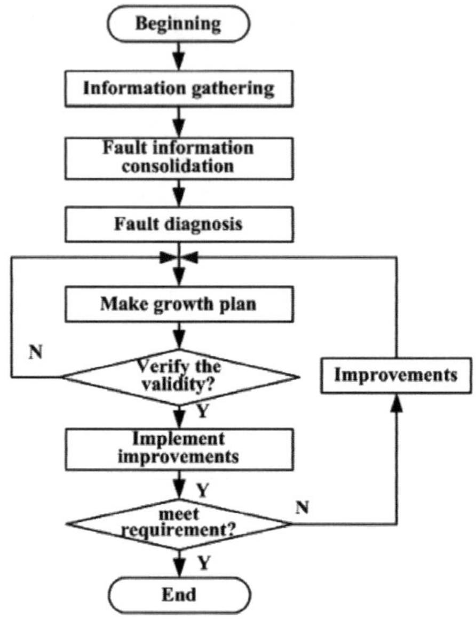

Figure 6. The process of reliability growth in using phase of high-speed punch

4.2.3. Reliability Growth of Using Phase. The performance and reliability of any product is far away from perfect from design, to manufacturing, to using primitives. There is always quality feedback and quality improving in the manufacturing and using phase and the reliability is growing all the time. In fact, it is the reliability growth.

The process of reliability growth in using phase of high-speed punch is as shown in Figure 6.

- To make record of utilization information of high-speed punch, such as on-off time, operating conditions, failure situations and maintenance status.
- To make fault information consolidation of high-speed punch.
- To make statistical analysis of fault information and to find weak links of high-speed punch.
- To make reliability growth plan based on fault analysis.
- To verify the validity of reliability growth plan: if it works, to implement improvements; or to make reliability growth plan again.
- To examine whether the punch could satisfy the requirement: if it works, it is the end; or to make reliability growth plan again.

5. Summary

The reliability growth analysis was analyzed based on the process from design to manufacturing. The reliability growth management of high-speed punch was carried out by 4 ways. The reliability growth programming of high-speed punch was made by 16 steps. The reliability growth of high-speed punch was divided into 3 stages: design phase, manufacturing phase and using phase, laid the foundation for the reliability growth of high-speed punch.

6.Acknowledgments

This work is financially supported by Fundamental Re-search Project of the National Natural Science Foundation of China(Grant No. 71601086), the National Natural Science Foundation of Jiangsu Province(Grant No. BK20160485), China Postdoctoral Science Foundation(Grant No. 2015M581748), the Natural Science in Colleges of Jiangsu Province of China (Grant No. 15KJB460005 & 16KJB140002), Jiangsu Planned Projects for Postdoctoral Research Funds (Grant No. 1501102C), Advanced Talent Foundation of Jiangsu University of China (Grant No. 14JDG138 & 15JDG033).

7.References

IOP Conf. Series: Materials Science and Engineering **250** (2017) 012064 doi:10.1088/1757-899X/250/1/012064

[1] ChenL, Zhang X Z, Gan S Y and others 2016 MATEC *Web of Conferences***61** 04003
[2] Weiss H K1956 *Operations Research***4**(5) 532-545
[3] Duane J T1964*IEEE Transaction on Aerospace***2**563-566
[4] MIL-STD-785A 1969*Reliability program for system and equipment development and production*U.S. Department of Defense
[5] Crow L H1974*Aberdeen Proving Ground, MD: AMSAA*
[6] MIL-HDBK-1891981*Reliability growth management* U.S. Department of Defense
[7] IEC610141989*Program for reliability growth* International Electrotechnical Commission
[8] Jelinski Zand Moranda P B1972*Proceedings of the statistical methods for the evaluation of computer system performance***1972** 465-484
[9] MIL-STD-1635(EC)1978*Reliability growth testing*U.S. Department of Defense
[10] Crow L H1982*Technometrics***24**(1)67-72
[11] Crow L H and Basu A P1988*Proceedings annual reliability and maintainability symposium***1988** 476-483
[12] Zhou Y Q and Weng Z X1992*The reliability growth technology*Beijing: Science Press
[13] IEC61164 2004 *Reliability growth—statistical test and estimation methods*International Electrotechnical Commission
[14] Gaver D P and Jacobs P A 2014 *Reliability Engineering & System Safety***130** 27-32
[15] Byuna J E, Nohb H.M and Song J 2017 *Reliability Engineering & System Safety***165** 410-421
[16] Mobin M, Li Z J and Komaki G A 2017 *IEEE Transactions on Reliability***66**(1) 99-110

IOP Conf. Series: Materials Science and Engineering **250** (2017) 012065 doi:10.1088/1757-899X/250/1/012065

Reliability optimization of crankshaft for high-speed punch based on genetic algorithm method

Xinzhou Zhang, Lan Chen[1], Naifei Ren, ShuyuanGan, Yuan Wang and Jian Zhou

School of Mechanical Engineering, Jiangsu University, 301 Xuefu Road, Zhenjiang, 212013, China.
Email: chenlan5321@126.com

Abstract. The crankshaft is one of the most important components of high-speed punch and the design and optimization of crankshaft has fateful consequence to the performance and reliability.The high reliability value of crankshaft contributes to the increasing of the reliability of the whole punch systemin the manufacturing and using process. This study builds a reliability optimization model of the crankshaft for punching process under constraints of the strength,stiffnessand geometry.Genetic algorithm method is employed to make reliability optimization for crankshaft based on ANSYSand MATLAB. The strength and stiffness reliabilityare both set as 0.997, namely, the minimum reliability of crankshaft is 0.997. Theoptimum design parameters based on reliability are obtainedunder the abovecondition. The obtained results demonstrate that genetic algorithm method can be effectively employed to investigate the reliability optimization of crankshaft for high-speed punch.

1. Introduction

High-speed punch is an efficient, high-precision and high-automation punching machine. It has been widely used in aeronautics and astronautics, automobile, traffic and transportation, metallurgical chemical industries and so on [1-2]. The crankshaft is a key part of main transmission system for high-speed punch and its reliability has great impact on the reliability of high-speed punch.

Design and optimization of the crankshaft is obviously not new, as the crankshaft is a very popular mechanism used in many different types of machinery. Sun [3]made multi-objective optimization design of crankshaft-bearing based on the crankshaft-bearing system, which reducedthe frictional power loss of crankshaft-bearingsand increasedthe crankshaft safety factor.Renzi [4] built an integrated design environmentbased on genetic algorithm, whichis used for the preliminary design and optimization of aeronautical piston engine components, andthe test results show that it could reduce 20% weight of the crankshaft.Liu [5] programed a multi-island genetic algorithm combined with multidisciplinary cooptimization method, which is used to carry out the multi-objective optimization of crankshaft structure. Sun [6] carried outthe optimum design of crankshaft bearings considering crankshaft strength. Song [7] carried out a multi-objective optimization considering the mutually exclusive changes of local machining allowances of the final forged product.Kim [8] proposed a hybrid method based on micro-genetic algorithm and regression-based sequential approximate optimizer, the result shows the radial stiffness and axial stiffness ofthe main shafthave been greatly improved.Additionally, there are more references for thedesign optimization of crankshaft [9-10].

As stated previously, there is few study on reliability optimization of crankshaft for high-speed punch. This work introduces the reliability optimizationmodel. Then, the reliability optimization model of crankshaft is built in ANSYSand MATLAB and the reliability optimization based on genetic algorithm method is carried out.

2. Reliability optimizationmodel

2.1. Fundamental of reliabilityoptimization

Reliability based design optimization is defined as:

$$\begin{cases} \min: f\left(d,\mu_x,\mu_p\right) \\ \text{s.t.} \quad P\left\{g_i\left(d,\text{X},\text{P}\right)\le 0\right\}\ge R_i^T \quad i=1,2,\cdots,n_g \end{cases} \tag{1}$$

Where $d$is the certain design variable,μ_x is the mean value of the uncertain design variable,μ_p is the mean value of the uncertain state parameters, $n_g$is the number of reliability constraints, $g_i$isthe limit state function, R_i^T is the reliabilityvalueneeded to be reached.

Thereliabilityanalysisistocalculatetheprobability when the limit state function is less than zero by the effect of input random variable, as shown asEq. (2).

$$R_i=P\left\{g_i\left(\mathbf{x}\right)\le 0\right\}=\int_{g_i(\mathbf{x})\le 0} f_x\left(\mathbf{x}\right)d\mathbf{x} \tag{2}$$

Where $X=\{x_1, x_2, \dots,x_n\}$ is the random design variable,$f_x(X)$is the joint probability density function, $g_i(X)\leq 0$ is the integral area.

2.2. Reliability optimizationmodelin narrow sense

For illustrative purposes, take the relationship between the reliabilityandcostofcomponentasan example.

2.2.1. The cost of component is given (or volume, weight, performance).

Maximizethereliabilityvalueunderthe constraint ofcost, the objectivefunctionisthereliabilityofthecomponent, themodelisshownas:

$$\begin{cases} \max: R(X) \\ \text{s.t.} \quad f(X)\le C_0 \quad g_u(X)\le 0(u=1,2,\dots,\ m) \end{cases} \tag{3}$$

Where $R(X)$ is the requiredreliability, Xisthebasisrandomvariable, C_0isthepre-concerted costofcomponent,$f(X)$ and$g_u(X)$ arethefunctionofdesignvariable (cost, performanceorparameter).

2.2.2. The reliability of component is given.

Minimizethecost (or volume, weight)ofcomponentormaximizetheperformanceandthereliabilityistheconstraint, the objectivefunctionisthecostofcomponent, the modelisshownas:

$$\begin{cases} \max: f(X) \\ \text{s.t.} \quad R(X)\le R_0 \quad g_u(X)\ge 0(u=1,2,\dots,\ m) \end{cases} \tag{4}$$

Where$R(X)$ is the actual reliabilityofthecomponent, R_0isthepre-concerted reliabilityofthecomponent.

2.3. Reliability optimizationmodelin broad sense

Ifconsideringthereliabilityorprobabilityofalltheconstraint, the probabilistic reliability optimization model should be built, which is called the reliability optimizationmodelin broad sense for mechanicalcomponent. Thereareseveralmodelsfordifferentpurposes.

The dynamic reliability model is shown as:

$$\begin{cases} \min: E\left\{\sum_{i=1}^{m} f_i(X)\right\},(i=1,2,\cdots,m) \\ \text{s.t.} \quad \prod_{j=1}^{n} R_j \ge R^0,(j=1,2,\cdots,n) \quad a_u\le X_u\le b_u,(u=1,2,\cdots,p) \quad g_l(X)=0,(l=1,2,\cdots,q) \end{cases} \tag{5}$$

Where $f_i(X)$ is the i-thobjectivefunction, R_j is the reliabilityofthe$j$-thsystem, R^0isthelimit reliabilityvalue, X_u is the$u$-th design variable, a_u and b_u are theupper and lower boundary values of the$u$-th design variable X, $g_l(X)$ is the constraint of the l-th design variable X.

3. Reliability optimizationmodel of crankshaft

3.1. Theobjectivefunction of crankshaft

Thereliability optimization model of crankshaftisshownas:

$$\begin{cases} \min F(x) \\ \text{s.t.}\ \ R_0^i\text{-}P\left(g_i(x,X)>0\right)\le 0,(i=1,2,\cdots,m) \quad x_k^1\le x_k\le x_k^u,(k=1,2,\cdots,n) \quad x=\mu(X) \end{cases} \tag{6}$$

Where$F$is the objectivefunction,x is the vector of the design variable,μ is the mean value of therandom variable$X$,$R_0^i$is the objective reliability of the$i$-thprobabilityconstraint, $P(g_i(x,X)>0)$is the probability when $g_i(x,X)>0$, $m$is the number of probabilityconstraint, x^1_k and x^u_k is the upper and lower boundary values of the k-thdesign variable, $n$is the number of the design variable.

Set F as the volume of the crankshaft, it is shown as:

$$F(x)=\frac{\pi}{4}\left(0.324D_1^2+0.47D_2^2+0.3D_3^2+0.001312D_3+0.17D_4^2+0.005687\right) \tag{7}$$

Where$D_1$is the diameter of constrained neck journal,D_2 is the diameter of neck journal of main connecting rod, $D_3$is the diameter of constrained neck journal, $D_4$is the diameter of neck journal of balancing connecting rod.

3.2. Design variables of crankshaft

Therearealotofparameters whichcanhaveeffectontheperformanceandreliabilityforthecrankshaft. According to the result of the reference [11-12], the design variableD_1, D_2, D_3, D_4, L_1 (Length of constrained neck journal), L_2 (Length of neck journal of main connecting rod), L_3 (Length of constrained neck journal) and L_4 (Length of neck journal of balancing connecting rod)have an impact on the reliability ofthecrankshaft.Therefore, these design variables are selected to the optimization variables, it is shown as:

$$x=\left(x_1,x_2,x_3,x_4,x_5,x_6,x_7,x_8\right)^T=\left(D_1,D_2,D_3,D_4,L_1,L_2,L_3,L_4\right)^T \tag{8}$$

3.3. Constraints of crankshaft

3.3.1. Strength constraint.

*3.3.1. Strength constraint.*According to thestress-strength interference theory,the equation of stateis describedin the ultimatestress state:

$$SUBS=YIES-maxstr>0 \tag{9}$$

Where$YIES$ is the yield strength, $maxstr$ is the maximumstressof crankshaft.

The reliability constraint of the strength is shown as:

$$g_1(X)=P(SUBS>0)-R_0^1\ge 0 \tag{10}$$

Where $P(SUBS>0)$ is the reliability of strength for crankshaft, $R_0{}^1$is the minimumallowed valueof the reliability of strength for crankshaft.

*3.3.2. Stiffness constraint.*To ensure the crankshaft to work properly, the stiffness should be considered in the design process.Take the allowedvalueof thestiffness: U_{max}=0.05mm. The equation of stateforthestiffness is shown as:

$$SUBU=U_{max}-maxu>0 \tag{11}$$

Where$SUBU$ is the stiffness reliability of the crankshaft, $maxu$ is the maximum allowable displacement of crankshaft in vertical direction.

The reliability constraint of the stiffness is shown as:

$$g_2(X)=P(SUBU>0)-R_0^2\ge 0 \tag{12}$$

Where $P(SUBU>0)$ is the reliability of stiffness for crankshaft, $R_0{}^2$is the minimumallowed valueof the reliability of stiffness for crankshaft.

*3.3.3. Geometrical constraint.*To ensure the crankshaft to be suited for complex assembly requirements, the equation of stateforthegeometry is shown as:

$$\begin{cases} g_3(X)=0.13-x_1=0.13-D_1\geq 0 & g_4(X)=x_1-0.11=D_1\text{-}0.11\geq 0 \\ g_5(X)=0.16-x_2=0.16-D_2\geq 0 & g_6(X)=x_2-0.14=D_2\text{-}0.14\geq 0 \\ g_7(X)=0.19-x_3=0.19-D_3\geq 0 & g_8(X)=x_3-0.17=D_3\text{-}0.17\geq 0 \\ g_9(X)=0.15-x_4=0.15-D_4\geq 0 & g_{10}(X)=x_4-0.13=D_4\text{-}0.13\geq 0 \\ g_{11}(X)=0.065-x_5=0.065-L_1\geq 0 & g_{12}(X)=x_5-0.045=L_1\text{-}0.045\geq 0 \\ g_{13}(X)=0.136-x_6=0.136-L_2\geq 0 & g_{14}(X)=x_6-0.116=L_2\text{-}0.116\geq 0 \\ g_{15}(X)=0.084-x_7=0.084-L_3\geq 0 & g_{16}(X)=x_7-0.064=L_3\text{-}0.064\geq 0 \\ g_{17}(X)=0.11-x_8=0.11-L_4\geq 0 & g_{18}(X)=x_8-0.09=L_4\text{-}0.09\geq 0 \end{cases} \quad (13)$$

Considering that there are two failure modes: strength failure and stiffnessfailure,thereliability interval estimationof the crankshaft is shown as:

$$\prod_{i=1}^{m} R_i \leq R_S \leq R_{\min} \leq R_i \quad (14)$$

Where $m$is the number of the failure, set m=2, $R_i$is the reliability of each failure mode, $R_{\min}$is the minimumvalue of R_i, R_S is the system reliability.
Taken together,the reliability optimization model of crankshaftis shown as:

$$\begin{cases} \min \quad F(x)=\frac{\pi}{4}\left(0.324x_1^2+0.47x_2^2+0.3x_3^2+0.001312x_3+0.17x_4^2+0.005687\right) \\ \text{s.t.} \quad g_1(X)=P(SUBS>0)-R_0^1\geq 0 \qquad g_2(X)=P(SUBU>0)-R_0^2\geq 0 \\ \qquad g_3(X)=0.13-x_1\geq 0 \qquad g_4(X)=x_1-0.11\geq 0 \\ \qquad g_5(X)=0.16-x_2\geq 0 \qquad g_6(X)=x_2-0.14\geq 0 \\ \qquad g_7(X)=0.19-x_3\geq 0 \qquad g_8(X)=x_3-0.17\geq 0 \\ \qquad g_9(X)=0.15-x_4\geq 0 \qquad g_{10}(X)=x_4-0.13\geq 0 \\ \qquad g_{11}(X)=0.065-x_5\geq 0 \qquad g_{12}(X)=x_5-0.045\geq 0 \\ \qquad g_{13}(X)=0.136-x_6\geq 0 \qquad g_{14}(X)=x_6-0.116\geq 0 \\ \qquad g_{15}(X)=0.084-x_7\geq 0 \qquad g_{16}(X)=x_7-0.064\geq 0 \\ \qquad g_{17}(X)=0.11-x_8\geq 0 \qquad g_{18}(X)=x_8-0.09\geq 0 \end{cases} \quad (15)$$

3.4. Optimization algorithm
The genetic algorithm (GA) is employed to make reliability optimization for crankshaft.GAis a metaheuristic inspired by the process of natural selectionandit is commonly used to generate high-quality solutions to optimization and search problems by relying on bio-inspired operators such as mutation, crossover and selection.The genetic algorithmwhich is based on reliabilityis written in MATLAB and ANSYS is employed to makestrength and stiffness reliability analysis for crankshaft.The basic flowchartis shown as Figure 1.
Planning approaches and procedures of thegenetic algorithmare shown as follows:

- According to the reliability optimization model of the crankshaft, determinethe population sizeof genetic algorithm.
- According to theinitial design value, determinethe ranges ofthedesignvariable.
- Initialize the population in binary random way,decode thedesignvariable, obtain theinitial design variable value.
- Write theinitial design variable valueto a file.
- Invokethe ANSYS, read theinitial design variable value, importthem in reliability calculation file, calculate thestrength and stiffness reliability of crankshaft.

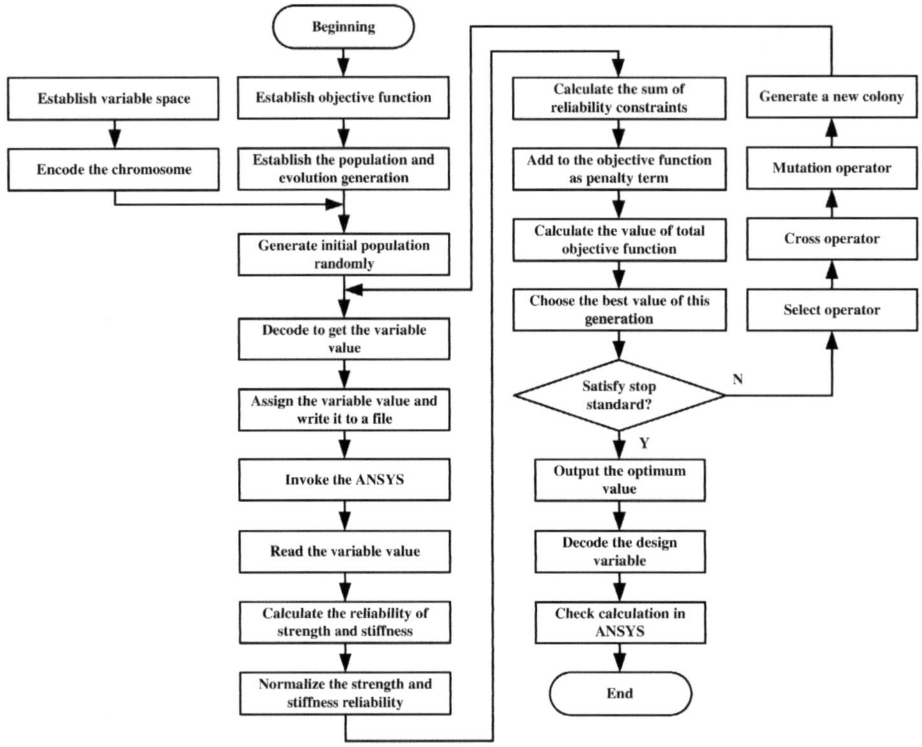

Figure1. The basic flowchart of genetic algorithm

- The reliability analysis result is normalizedwith a base ofthe minimum allowed reliability value.
- Calculatethe summation of reliabilityconstraint forall individualsof each generation.
- Addthe reliability constraint to the objective function as penalty term.
- Calculate the value of the total objective function, selecttheindividual with the minimum value as the optimization value of this generation.
- Make a judgementof that whetherthe evolution generations are satisfied.
- If the evolution generations are satisfied, then output the optimization value, decode the design variable, invoke the ANSYS to check the reliability result, finally exit the program.
- If the evolution generations arenot satisfied, select, cross and mutateoperators, produce a new generation, returnto the third step.

4. Results and discussion

The convergence process of genetic algorithmfor reliability optimization of crankshaft is shown as Figure 2. As seen in the chart,after the 4th generation,the change ofthe value of the total objective functionpresents a fairly stable statuswith two small fluctuations.

The reliability optimization result is shown as Table 1.

Table 1. The reliability optimization result of crankshaft

Design variable	*D*1	*D*2	*D*3	*D*4	*L*1	*L*2	*L*3	*L*4
Initial design value/mm	0.12	0.15	0.18	0.14	0.055	0.126	0.074	0.1
Optimization value /mm	0.11	0.155	0.195	0.14	0.055	0.125	0.095	0.105

The result shows that the design variable *D*4and *L*1do not change, the design variable *D*1 gets smaller, the rest ofthe design variableget bigger. Thestrength and stiffness reliability value of the initial design for crankshaft are 0.94863 and 0.99953, the reliability optimization analysis both set the strength and stiffness reliability valueas 0.997. For this reason, most design variable value become bigger and the total reliability of the optimizedcrankshaftis higher than that of the initial crankshaft.

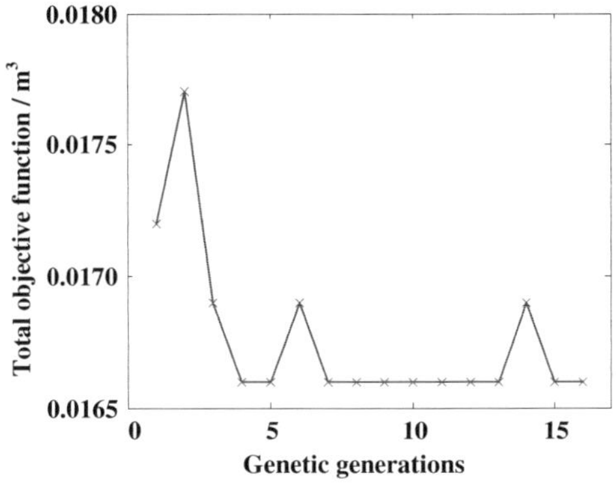

Figure2. The convergence process of genetic algorithm

5. Conclusion

This paper studies the reliabilityoptimization ofcrankshaft forhigh-speed punch.Thereliability optimizationmodel (including the model in narrow senseand in broad sense)isintroduced.Thereliability optimization model of crankshaftis built, which considers the strength,stiffnessand geometry constraints.Genetic algorithm method is employedtomakereliabilityoptimizationforcrankshaftbasedon ANSYSand MATLAB. Thereliability optimization results present thatmost design variable value become bigger that of the initial value under the condition of the strength and stiffness reliabilityare both set as 0.997. And the total reliability of the optimizedcrankshaftis higher than that of the initial crankshaft.The obtained results demonstrate that genetic algorithm method can be effectively employed to investigate the reliability optimization of crankshaft for high-speed punch.

6. Acknowledgments

This work is financially supported by Fundamental Re-search Project of the National Natural Science Foundation of China(Grant No. 71601086), the National Natural Science Foundation of Jiangsu Province(Grant No. BK20160485), China Postdoctoral Science Foundation(Grant No. 2015M581748), the Natural Science in Colleges of Jiangsu Province of China (Grant No. 15KJB460005 & 16KJB140002), Jiangsu Planned Projects for Postdoctoral Research Funds (Grant No. 1501102C), Advanced Talent Foundation of Jiangsu University of China (Grant No. 14JDG138 & 15JDG033).

7. References

[1] ChenL, Zhang X Z, Gan S Y and others 2016*MATEC Web of Conferences***61** 04003
[2] Breitling J, Wallace DandAltanT1996 *Journal of Materials Processing Technology***59** 18-23
[3] SunJ, ShuL, Song X H and others 2016 *Industrial Lubrication and Tribology***68**(1) 86-91
[4] Renzi, C 2016*International Journal optimization of Advanced Manufacturing Technology***86**(10)3365-3381
[5] Liu J, Yu GY, Li Yand others 2016*Mathematical Problems in Engineering***2016**9596089
[6] Sun J, Fu Y, Deng Mand others 2009*Proceedings of the Institution of Mechanical Engineers Part J-Journal of Engineering Tribology***223**919-924
[7] Song M C, VantyneChester JandCho, J R 2017*Journal of Manufacturing Science and Engineering-Transactions of the ASME***139**(9) 091014
[8] Kim SW,Kang K,Yoon Kand others 2016*Mechanism and Machine Theory***104**287-302
[9] Wang SJ, Xu ZJ2012*Information-An International Interdisciplinary Journal***15**(12B)5593-5604
[10] Cui Z Q, Zhang T, Wang XHand others2012*Applied Mechanics and Materials***128-129**1426-1429
[11] Zhang X Z, Chen L, Gan S Y and others 2017*International Journal of Science***4**(7) 190-195
[12] Zhang X Z, Chen L, Gan S Y and others 2017*International Journal of Science***4**(7) 203-209

IOP Conf. Series: Materials Science and Engineering **250** (2017) 012066 doi:10.1088/1757-899X/250/1/012066

Product Conceptual Design Based on Agent Federation

Minghai Yuan, Song Wang and Jinting Huang

College of Mechanical and Electrical Engineering, Hohai University, Changzhou 213022, China.
Email: ymhai@126.com

Abstract. A cooperative Agent Federation was used in the complicated products conceptual design after studying the characteristics of the conceptual design. Each Agent Federation designed products cooperatively under the control of the chief Agent Federation. Thus, the mapping of 'function- behaviour -structure' was achieved, and the whole process of the conceptual design was accomplished. The functions, communications, cooperation and the theories about running of the key Agent Federation were discussed, and the frame of the conceptual design based on cooperative Agent Federation and work group was established in this paper.

1. Introduction

Conceptual design is the most creative section of the whole design process. Compared with other phases, this phase occupies so less resource that belongs to the low cost part but which will decide the most costs of the product's whole life cycle [1]. It is not easy to get a reasonable scheme for complex product conceptual design that just rely on a single designer, so the conceptual design is the process of group cooperation [2-3]. Due to multi-agent having the advantages of obvious stability, high efficiency, dynamic adaptability and easy realization [4], it has been applied in many fields, such as collaborative design, manufacturing, communication, medical treatment, management, etc. Wenxian Tang from Shanghai University [5], Qianfan Zhang from Hua zhong University of Science and Technology [6], Gek Woo Tan [7] and other researchers have researched cooperation design technology [8] and framework of multi-agent [9-10]. Facing the conceptual design of complex product, the above-mentioned pure distributed Agent structure will confront with complicated control, communication complex, difficult to coordination and other disadvantages.

According to the characteristics of complex product conceptual design and the structure of the pure distributed agent, this paper proposes agent federation-based conceptual design on the basis of previous studies, and it will overcome the drawback of pure distributed Agent structure and solve the coordination problem in conceptual design.

2. Conceptual Design

2.1. The Definition and Characteristic of Conceptual Design

Conceptual design has many definitions from different angles, two present authoritative definitions as: in Engineering Design, G Pahl and W.Beitz [11]defined it as "After the mission determined, through abstraction propose function structure and seek appropriate action principle and their assemblies, etc., then determine the basic solving approach, last get the solving scheme, all these designing working is called conceptual Design." M. J.Frence [12]defined conceptual design in Conceptual Design for Engineers as "Conceptual Design is a design phase that need consider the content of the design problem and put forward a variety of solutions."

IOP Conf. Series: Materials Science and Engineering **250** (2017) 012066 doi:10.1088/1757-899X/250/1/012066

Conceptual design has many characteristics, such as innovation, diversity, hierarchy, disability and so on [13-14].

Innovation: Innovation is the "soul" of conceptual design, include modify and innovate function structure and other things.

Diversity: The diversity in conceptual design refers to the designing path and result.

Hierarchy: The process of conceptual design is divided into function layer, behavior layer and structure layer, each layer also has hierarchy.

Disability: This characteristic mainly reflects in the acquired product information is qualitative, inaccuracy, inconclusive and incomplete.

2.2. The Process of Conceptual Design

Found and satisfied the requirements are the starting point and ending result of product design. The task of conceptual design is to satisfy the market demand, and the creative and conception are also coming from the understanding of market demand [15-17]. So the first step of conceptual design is to investigate and analyse the market demand and establish demand assemble. Based on previous knowledge and experience, analyse and abstract the demand, then get the exact product global function. Decomposition the global function into exhaustive function unit, make the function unit mapping to behaviour. As for using different structures will achieve the same behaviour movement, according to some regular, combining those structures to get conceptual product family which can achieve global function. Finally evaluate the conceptual product family and get the optimal conceptual product scheme. The layer structure and process of product conceptual design as shown in Figure 1.

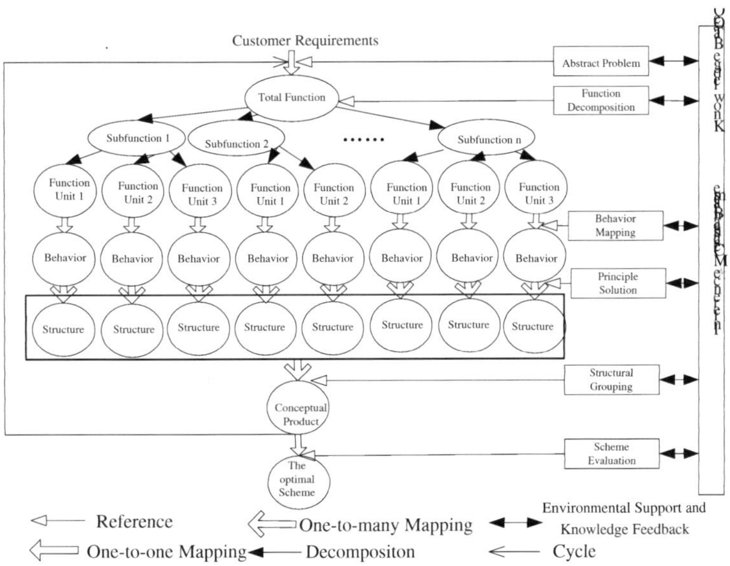

Figure 1. The layer structure and process of product conceptual design

3. The Conceptual Design Frame of Agent Federation

The Agent Federation includes many sub-agents and mediators, mediator harmonize the behaviour of Agent in Agent Federation and communicate and cooperate with other Agent. According to the process of conceptual design mentioned in chapter 1.2, the system adopts five Agent Federation and blackboard structures. The five Agent Federations are: global control Agent Federation, function analysis Agent Federation, substructure design Agent Federation, product design Agent Federation and domain service Agent Federation.

The function of global control Agent Federation is to distribute tasks and resources, cooperate with each Agent Federation behavior, and interact with information between systems and outside and so on.

IOP Conf. Series: Materials Science and Engineering **250** (2017) 012066 doi:10.1088/1757-899X/250/1/012066

The function of function analysis Agent Federation is abstract the total function and decompose it into function unit.

Substructure design Agent Federation is the core of the system; the process "function- behavior-structure" can be completed. According to the conceptual design process, the substructure design Agent Federation mainly including: principle of function unit analysis Agent, behavior map Agent, principle solve Agent.

The function of product design Agent Federation is to form the conceptual product and get the optimal scheme, mainly including: structure composes Agent, scheme evaluates Agent, and scheme saves Agent.

The function of domain service Agent Federation is to provide a variety of tools, knowledge and examples to support the designing smoothly.

The blackboard is system for each Agent Federation communicating to share storage. There four regions: public region and every exclusive region, public region decomposed to information release area, conflict coordination scheme area, cooperative scheme area and problem area, exclusive region on the basis of team is divided to several conceptual spaces.

If there are multiple products simultaneously concept design, each Agent Federation of the product design will compose to Teamwork, the Agent Federation served as collaboration. The System frame is shown in Figure2, the communication pattern of Teamwork and blackboard is shown in Figure 3,and the structure of blackboard is shown in Figure 5.

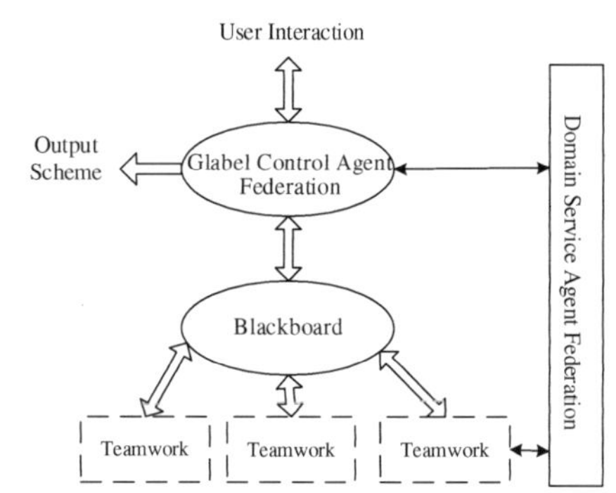

Figure 2. The frame of agent federation conceptual design system

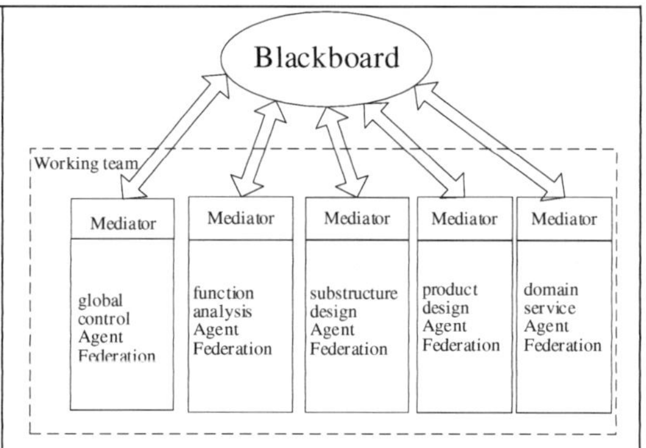

Figure 3. The communication pattern of each working team and blackboard

4. The Agent Federation Cooperation in Conceptual Design

4.1. The Agent Federation Communication

The Agent Federation communication has two communications: one communication is in federation interior, another communication is between federations, they all adopt hybrid communication. One communication method is adopted by "Private line", that means only by calling and called process can see the communication information (Figure 4 is the private line-based federal communication pattern). Another method is blackboard-based communication which divided into public region and exclusive region, according to working nature, the exclusive region will be divided into several conceptual regions, each Agent Federation in teamwork only pay attention to the regions that related to itself (Figure 5 is the structure of blackboard).

All the Agent in federation communicate with mediator through the mediator coordinate the behavior between internal Agent, the mediator stand the federation communicate and coordinate with outside. The structure of mediator communication interface system in this paper is as shown in Figure 6, the information processing module make evaluation and analysis for received information, coordinate internal Agent or release the result.

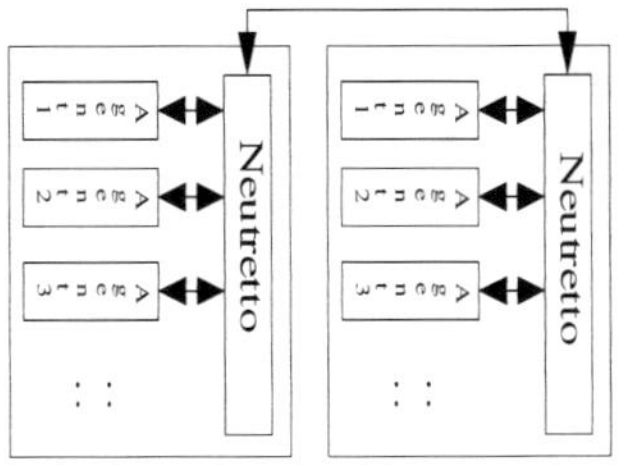

Figure 4. The private line-based federal communication pattern

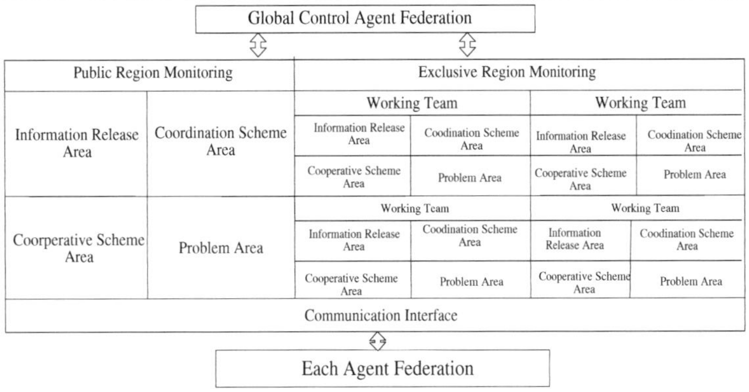

Figure 5. The internal structure of blackboard

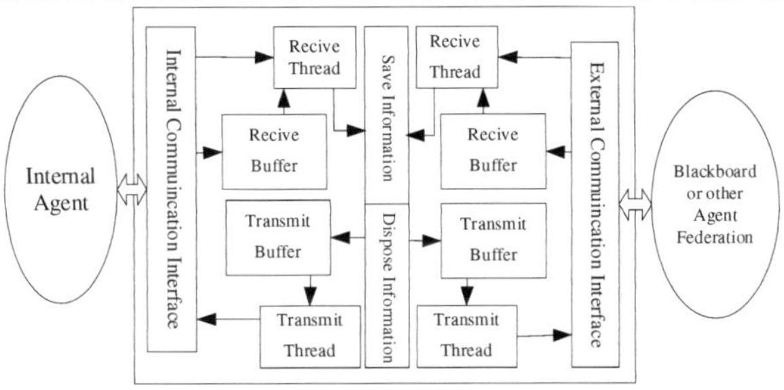

Figure 6. The structure of mediator communication interface

4.2. The Agent Federation cooperation

When make conceptual design for complex product, it is impossible for any single Agent or Agent Federation to accomplish, it need the cooperation between Agents. The constructed Agent Federation system (shown in figure 2, 3) can classify the cooperation in conceptual design into cooperation between teams, cooperation between internal Agent Federations and cooperation between Agent Federation's Agents. Cooperation and competition are two behaviors in the process of cooperation. Competition occurs mainly between isomorphism Agent Federation, such as substructure design between Federations is interested in the same task, and then they will compete. Otherwise, heterogeneous Agents have different functions, so there is no competition between them, such as there is no competition between substructure design Agent Federation and functional analysis Agent Federation.

The process of cooperation includes three steps: (1) Determine goals; (2) Seek for cooperation company and confirm the cooperation scheme; (3) Achieve the goals.

The goal of cooperation includes two parts: Mission object, namely the work to be done; and function object, namely the constraints and requirements in the process of cooperation. The general task object is to get the optimal scheme of product design, specific Agent Federation has specific submission objectives. The function object includes the ability the Agent Federation design, resources conditions, constraints in designing requirements and so on. Global-control Agent Federation will realize the general task aim and bring about the subtask aim determine.

After determining the design aim, the next step is to seek cooperation partner and determine the cooperation scheme. The process of seeking cooperation partner is equal to the process of determining the cooperation scheme and distributing the task. Distribute the task in this way need to make effectiveness evaluation for the biding Agent Federation, and choose the most effective Agent Federation or teamwork to cooperate. We can assume $U(t_i)$ is the effect of finished task t_i, so the formula $U(t_i)$ can be shown as following:

$$U(t_i) = \alpha T_i + \beta M_i + \gamma C_i + \lambda B_i + \nu O_i \tag{1}$$

T_i stand the time for finishing task t_i, M_i stand the quality of finishing task t_i, C_i stand the cost of finishing task t_i, namely occupy resources, B_i stand busy idle state, O_i stand conflict situation happened between t_i and dealing t_i. α, β, γ, λ, ν is the weights of T_i, M_i, C_i, B_i, O_i. Global control Agent Federation through blackboard biding, task distribution, supervise and cooperate the behavior of Agent Federation.

After determined the cooperation scheme, under the supervision and coordinate of global control Agent Federation, each designing Agent Federation will mutual response, fulfil the promise, and meet the corresponding constraint in collaboration process, and achieve the goal together.

5. The Operation Mechanism of Agent Federation Conceptual Design

System operation mechanism is as show in Figure 7. It uses man-machine interaction of the global control Agent Federation to input new problem, function analysis Agent Federation and make function abstract and decomposition, the global control Agent Federation according the function unit to invite. Evaluate each bidding substructure designing Agent Federation, distribute tasks to the winning federal and sign the contract. Substructure design Agent Federal enforce the contract, when the structures of federal designing encounter conflicts, the solving scheme from global control Agent Federation will be feedback to each Federation until the conflict solved. After finishing the substructure designing, structure composite to Agent and compose twice in product designing Agent Federation. By Agent product design evaluation in the federal Agent to evaluate product solutions, when the product can't meet the designing constraints, it will conflict.Transfer the conflict to blackboard, the global control Agent Federation will propose solutions, according to the solutions, substructure design Agent will modify the design of the structure or decompose the function, allocate the task and redesign.Repeat these steps until the scheme meet the customer's requirement, and then save the optimal scheme. Finally, the global control Agent Federation will output the result and form to a file or drawings.

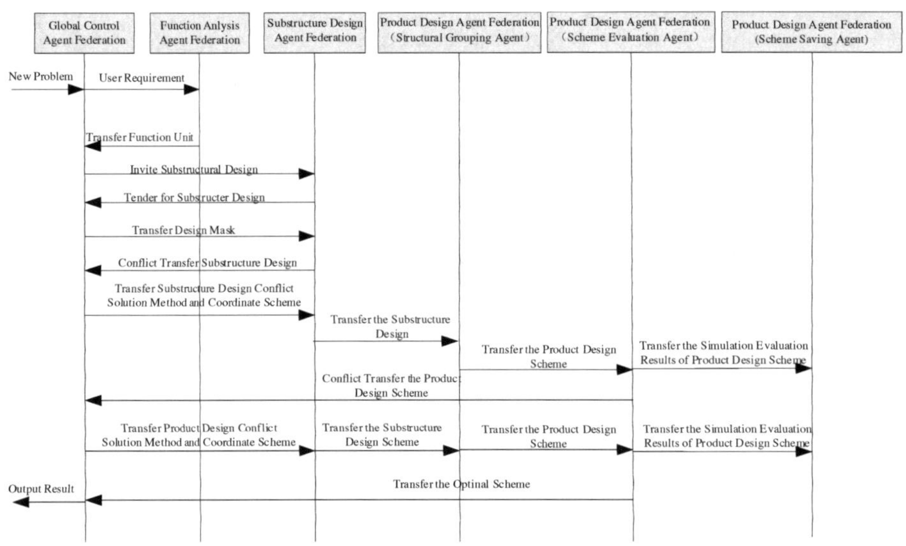

Figure 7. The operation mechanism of agent federation conceptual design

6. Conclusion

With the development of Internet technology, artificial intelligence and other technology, multi-agent has been widely applied in many fields. Based on deep research conceptual design, this paper proposes a conceptual design frame of Agent Federation cooperation. Each agent in the form of distribution coordinates to finish the conceptual design of product that makes the whole system operability, flexibility and extensibility. Agent Federation-based product conceptual design will reduce the complexity of pure distributed multi-agent communication, practice have been proven that is a good way to solve the problem of complex products' conceptual design, and it is easy to come true.

7. Acknowledgments

This work was supported by Postgraduate Research & Practice Innovation Program of Jiangsu Province under Grant number KYCX17_0532, Sany enterprise special under Grant number XZX/14B001-05 and Guangxi Key Laboratory for manufacturing system and advanced manufacturing technology under Grant number 13-051-09S09.

8. References

[1] Ullman D., "The Mechanical Design Process", New York: McGraw- Hill, 1997

[2] Hong-Zhong H,Yu L,Yanfeng L,Lihua X,Zhonglai Wa. New evaluation methods for conceptual design selection using computational intelligence techniques[J]. Journal of Mechanical Science and Technology, pp .273,2013

[3] Cristina Renzi,Francesco Leali. A Multicriteria Decision - Making Application to the Conceptual Design of Mechanical Components[J]. J. Multi - Crit. Decis. Anal., pp.233-234,2016

[4] Kuhlmann T., Lamping R. and Massow C., "Agent-based production management", *Journal of Material Processing*, Vol.**9**, no.2, pp.23-27, 1998

[5] Wenxian T., Minglun F. and Limin L., "design strategies oriented to knowledge driven collaborative of products' life cycle", *Journal of Shanghai Jiaotong University*, Vol.**38**, no.6, pp. 1017-21, 2011

[6] Qinfan Z. and Zigang Z., "Research on the designing system of multi-agent coordinate product", *Journal of Wuhan University of Technology*, Vol.**26**, no.5, pp.100-102, 2012

[7] Tan G., Hayes C. and Shaw M. "An intelligent-Agent framework for concurrent product design and planning", *IEEE Transaction on Engineering Management*, Vol.**43**, no.3, pp.297-306, 2012

[8] Zhili Tang,An Zhang. Human-Machine-Cooperation Design Methodology for Civil Aircraft Cockpit[A]. Information Engineering Research Institute, USA.Proceedings of 2013 2nd International Symposium on Computer,Communication,Control and Automation(3CA 2013)[C].Information Engineering Research Institute, USA,5,2013

[9] Wenle Z,Jianchang L. Ultra-fast consensus of discrete-time multi-agent systems under a unified framework[J]. International Journal of Control, pp.886,2015

[10] Haitao Z,Dunbing T,Tengxiao H,Chao X. An Agent Based Intelligent Distributed Control Paradigm for Manufacturing Systems[J]. IFAC PapersOnLine, pp.4912,2016

[11] Paul G,Beitz W.Engineering Dseign.London:The Design Council,1984

[12] Frence M J.Conceptual Design for Engineerings(2nd Edition).London: The Design Council,1985

[13] Aurich J.C., Fuchs C. and Wagenknecht, C., "Life cycle oriented design of technical Product-Service Systems", *Journal of Cleaner Production*, Vol.**14**, pp.1480-94, 2006

[14] Becker J., Beverungen D.F. and Knackstedt R., "The challenge of conceptual modeling for product-service systems: Status-quo and perspectives for reference models and modeling languages", *Information Systems and e-Business Management*, Vol.**8**, no.1, pp.33-66, 2012

[15] Chang X., Sahin A. and Terpenny J., "An ontology-based support for product conceptual design", *Robotics and Computer-Integrated Manufacturing*, Vol.**24**, no.6, pp.755-62, 2008

[16] Kitamura Y. and Mizoguchi R., "Ontology-based systematization of functional knowledge", *Journal of Engineering Design*, Vol.**15**, no.4, pp.327-51, 2004

[17] Berners-Lee T., Hendler J. and Lassila O., "The Semantic Web", Sci. Amer., Vol.**284**, no.5, pp.34–43, 2012

IOP Conf. Series: Materials Science and Engineering **250** (2017) 012067 doi:10.1088/1757-899X/250/1/012067

Design and Implementation of a Quadruped Bionic Robot Based on Virtual Prototype Technology

Li Wang

Zhuhai College of Jilin University, Zhuhai,519041,China
Email: wangli00019@163.com

Abstract. Design out a quadruped bionic robot with nine degrees of freedom. Conduct virtual assembly and trotting gait simulation on the robot by using NX software. Present the angular velocity and angular displacement curves of the diagonal two legs' hip joints and knee joints, thus to instruct the practical assemble and control of the robot. The fact that the movement effect of the physical model is consistent with the simulation verifies the validity and practicability of virtual assembly and motion simulation. both.

1. Introduction

The wheel type robot and the tracked robot are able to exhibit good athleticism on flat ground [1], but they are difficult to walk on complex irregular terrain, and even can't get across at all. In the process of biological evolution, creatures have formed structural characteristics and material properties adapt to their living environment[2]. The bionic creeping robot realizes its motion with the crawling mechanism similar to biology. Because its foothold is discrete, it can choose the optimum supporting point on the ground it can reach. This makes the robot have better adhesion with the contact surface and better obstacle climbing ability. This kind of robot has very wide application potential in the field of military reconnaissance and the cleaning of civil high-rise building external wall and etc[3].Research on foot-type walking robot has become one of the hottest spots of robotics now[4].In this paper, a quadruped bionic robot with nine degrees of freedom is designed. First, UG NX software is used to build a virtual model. Then the trotting gait imitating quadruped insect in the nature is adopted to establish the virtual simulation. Finally, an experiment with a physical robot is carried out. The realize process of the quadruped bionic robot is introduced from the following aspects, such as body structure, gait planning, virtual prototype simulation analysis and results, experimental result and its analysis and so on.

2. Body Structure

Each leg of the quadruped bionic robot has two degrees of freedom. They are the swing of the hip joint and the knee joint. With these degrees of freedom, the robot can walk straight and turn a corner. There are a total of 9 degrees of freedom including a swing degrees of freedom of the head. Figure 1 is the structure sketch of the quadruped bionic robot's legs. All of the robot's legs, head and body in the physical model are plastic components of PC material. This kind of material not only has a certain strength, but also can reduce the quality of the robot. Each degrees of freedom of the robot is controlled by a MG995 steering gear with a quality of 55g, the maximum torque of which is 1.2NM. The quadruped bionic robot is directly drived by installing the steering gear directly in the robot joint. This kind of drive model can reduce the intermediate transmission links and the structure is compact.

IOP Conf. Series: Materials Science and Engineering **250** (2017) 012067 doi:10.1088/1757-899X/250/1/012067

The MultiFLEX 1.0 control card made in Beijing Bochuang company is adopted as the hardware control system of the robot.

The Virtual Prototype can replace the physical model, it also can Shorten the product design cycle and reduce the design cost. In this paper, build a quadruped bionic robot simulation model by using UG NX software, as shown in figure 2.

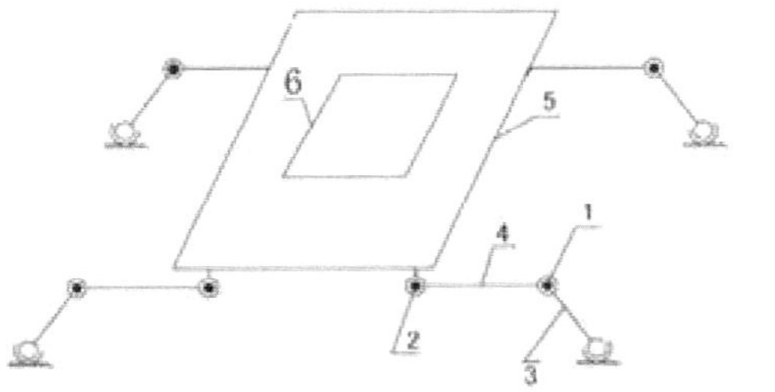

1.knee joint 2.hip joint 3.crus 4.thigh 5.body 6.control system

Figure 1: structure sketch of the quadruped bionic robot's legs

Figure 2: virtual prototype of the quadruped bionic robot

3. Gait Planning

The stability of the foot-type robot is mainly determined by the gait[5].The plan and control of the gait is critical to coordinated motion between the legs. Gait refers to the leg and leg placing sequence of the walking system. The swing- support sequence is used to control the robot's walking in the working circumstance. Robot with uniform velocity walking changes periodically, so it is called gait period. The moving distance of the walking robot body's barycenter in one gait period is defined as a step length. In the nature, the gait of the quadruped animal can be summarized as crawling, trotting, unilateral trotting, double foot jumping, jogging, race and four foot jumping and etc. In all of the basic gaits, crawling is the most common static gait for the crawl of general mammalian animal; while trotting, unilateral trotting and double foot jumping are belong to dynamic gait with two legs swing together. Among them, trotting gait is a gait with diagonal two legs swing together.

Trotting gait was adopted in this paper for the quadruped bionic robot. That is to say, four legs are at the standing phase in the initial state, as shown in figure 3(a); then the hind leg1 and fore leg 2 leap forward. During this process, thigh swings slightly around the hip joint 01, crus swings outwards around the knee joint 02, and the other two legs keep standing phase, as shown in figure 3(b); when the hind leg1 and fore leg 2 have leaped and balanced, the fore leg 1 and hind leg2 begin to leap forward, as shown in figure 3(c). The robot moves forward with four legs repeating the action.

4. Virtual Prototype Simulation Analysis and Results

Although the sketch map of the gait can clearly reflect whether each leg is at the support phase or at the free phase, it is much more complex to realize the gait transition and control the movements of the legs. Despite the theoretical analysis in the country has achieved certain development and has lay a foundation for robot, its operability remains to be verified and improved. So, by using the UG NX software, build a virtual prototype, establish simulation scheme, shown in Fig.4, simulate each leg joint motion, check the movement interference, simulate the trotting gait. The results conclude that the robot structure is feasible and the motion is stable. This can provide theoretical basis for the following design and control of the physical model.

IOP Conf. Series: Materials Science and Engineering **250** (2017) 012067 doi:10.1088/1757-899X/250/1/012067

5. Experimental Result and Its Analysis

According to the virtual prototype of UG NX, implement harmonious control of a physical quadruped bionic robot, as shown in figure 6. The movement screenshots of the experimental process are shown in Figure 11.

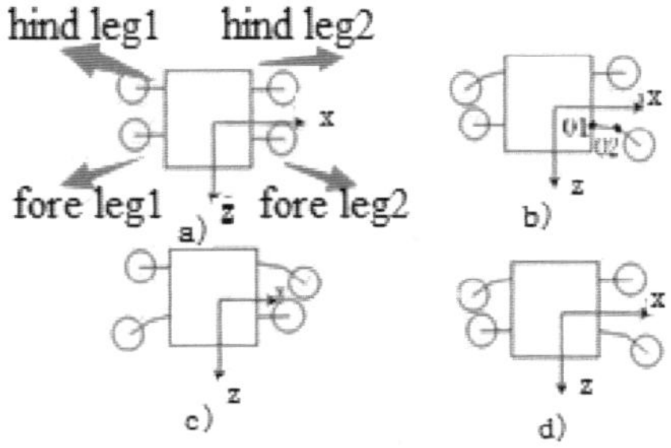

Figure 3: the sketch map of the quadruped bionic robot's trotting gait

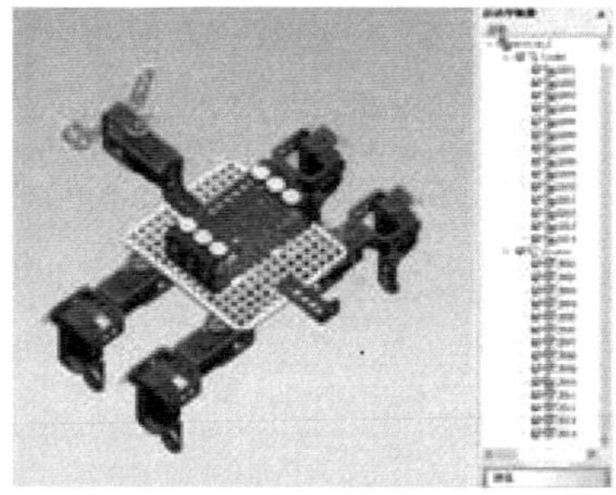

Figure 4: simulation connecting rod and joints

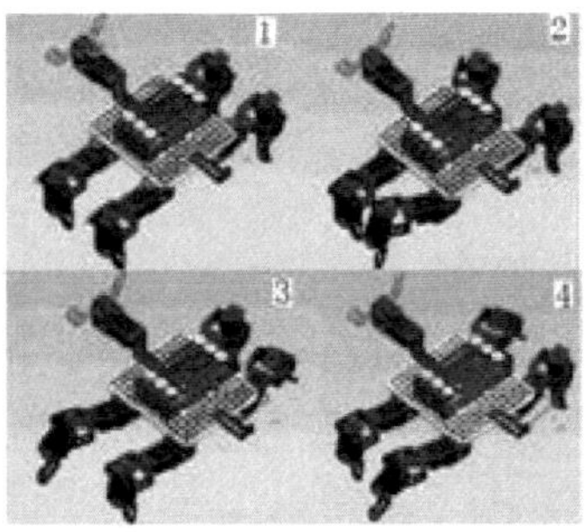

Figure 5 Motion simulation screenshot

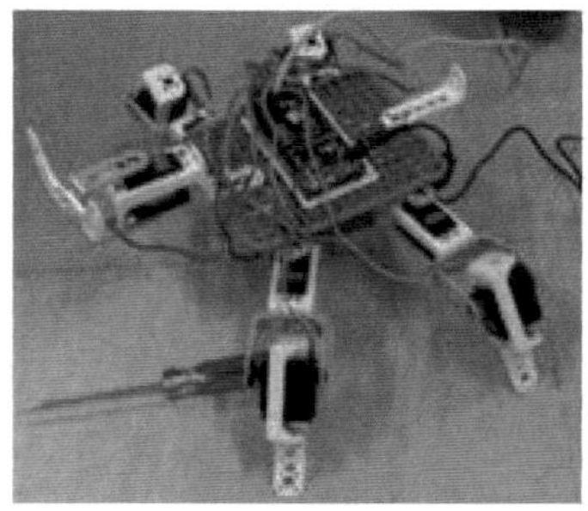

Figure 6 the photo of the quadruped bionic robot

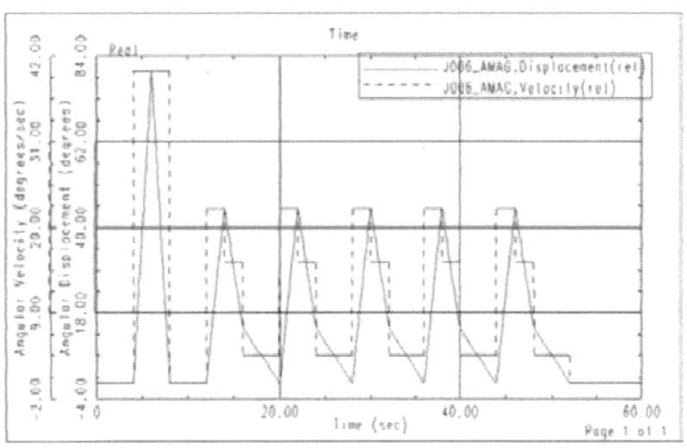

Figure 7 The angular velocity and angular displacement curve of left fore leg's knee joint

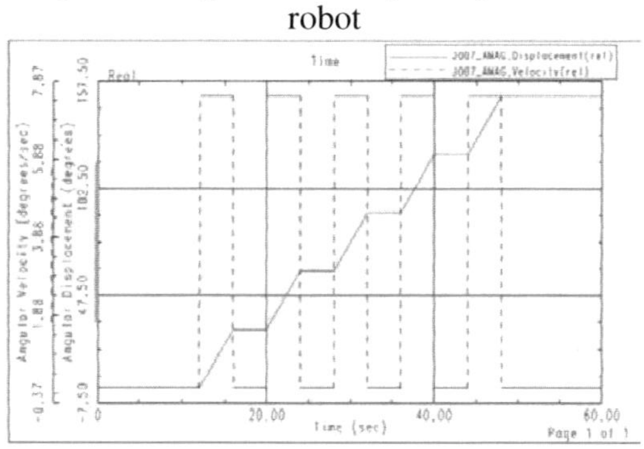

Figure 8 The angular velocity and angular displacement curve of left fore leg's hip joint

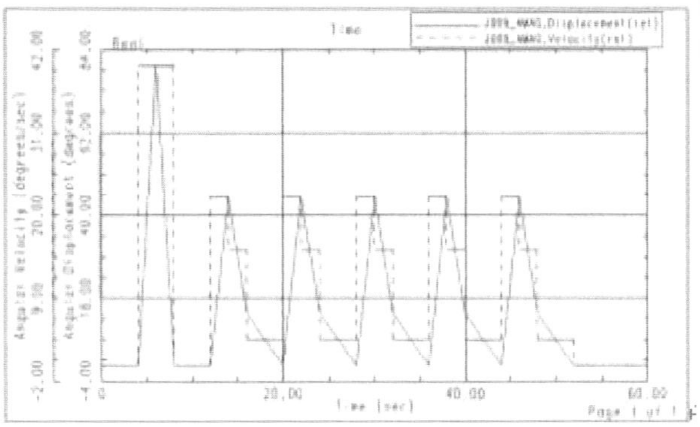

Figure 9 The angular velocity and angular displacement curve of right hind leg's knee joint

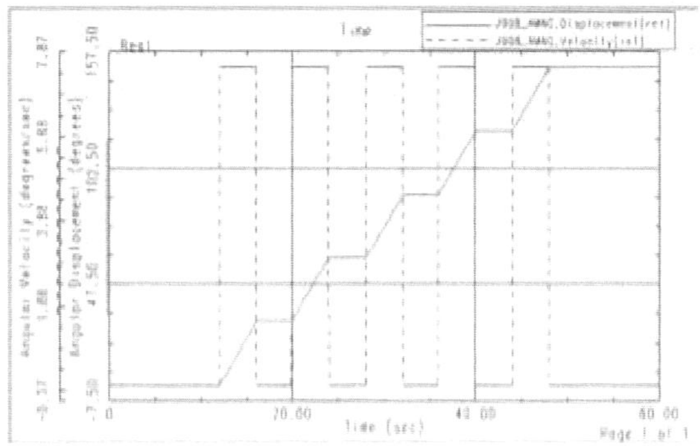

Figure 10. The angular velocity and angular displacement curve of right hind leg's hip joint

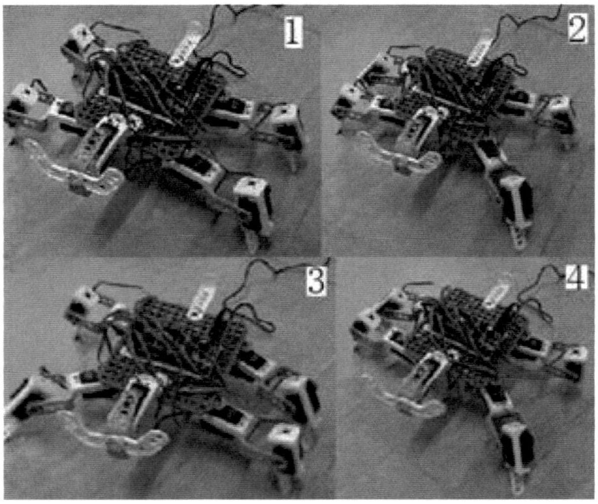

Figure 11 experimental screenshot

In the experiment, because of the small contact area between the foot end and the ground and small friction, the quadruped bionic robot has a slippery motion. So its actual move velocity is slower than the simulation. But on the whole, the robot crawls according to the trotting gait simulated in the virtual prototype. It moves stabled and basically agrees with the simulation result.

6. Conclusions

In this paper, design a quadruped bionic robot adapt to walking on complex irregular terrain. First build a virtual prototype , then simulate and finally do physical experiment. The virtual prototype technology's instruction to productive practice and its superiority have been fully reflected. The results of the simulation indicate that the quadruped bionic robot can walk smoothly with the trotting gait. The results provide theoretical basis for the following design and experiment of the physical prototyping. The physical experiment movement agrees well with the simulation results . The virtual assemble and simulation are verified to be effective and practical.

7. References

[1] Yin XL, Zhang H, Zhao J, Liu YB and Zhao LH 2013(**10**) Gait control of hexapod robot based on local rules, *J. HrbCU* (Natural Sciences Edition) pp 565-70

[2] Lu YX 2004(**1**) Significance and progress of bionics, *J. Bionic Eng* pp 1-3

[3] Xiao Y, Wang JP, Jiang XK, Li XF and Liu ZG 2006(**6**) Design and Implementation of a Robot that is Imitative of Spider with Eight Legs, *J. Mechatronics* pp 79-82

[4] Zhang ZhY, Liu YF, Feng M and Yang YF 2007(**10**) The new hexapod robot design and simulate based on suppositional model machine technology, *J. Equip Manu Tec* pp 35-43

[5] Ma DX, Wang YH and Yue L, 2008 *Rev.* **37** (3) Gait Simulation and Implementation of a New Quadruped Robot, *J. Mech Manu and Res* pp 21-4

IOP Conf. Series: Materials Science and Engineering **250** (2017) 012068 doi:10.1088/1757-899X/250/1/012068

Parametric Design on Internal Gear of Cycloid Gear Pump with NX10.0

Enguang Zhang [1, a] **Min Chen**[1] **and Baoqing Deng**[1,*]

[1]Zhuhai College of Jilin University, Zhuhai, 519041, China

[a] teamcenter@163.com * corresponding author's e-mail address: dengbq@jlu.edu.cn

Abstract. The parametric equation of the theoretical contour curve of cycloid gear was established in this paper. Using the NX10.0 software, parametric design of cycloidal gear profile curve was carried out, and the radius of curvature with the contour curve of cycloid gear was analyzed. The consistency between the simulation results and the theoretical design results was verified by assembly analysis, and the parametric drive of the cycloidal gear was verified simultaneously.

1. Introduction

The cycloidal gear pump is an internal gear pump. Its outer rotor tooth-form is a circular arc, and the inner rotor tooth-form is an equidistant curve of curtate epicycloid, so it is also called as internal meshing cycloidal gear pump. The two tooth lines are a pair of conjugate curves. The tooth-form of cycloid gear is composed of a cycloid curve. The so-called cycloid, that is, a rolling circle, along the basic circle outside rolling, the trajectory of a point on the rolling circle is called the outer cycloid, while the trajectory of a point that rolling along the basic circle inside is called the hypocycloid. Cycloidal gear pump has the advantages of compact structure, small size, light weight, less parts, speed up to 10000r/mim, smooth movement, low noise, high volumetric efficiency, small force on the shaft and so on. While its disadvantages are large flow pulsation, complex rotor manufacturing process and etc. It is widely used in the hydraulic systems such as aviation, spaceflight, automobile, ship, chemical machinery and so on. For example, the cycloidal pumps are widely used in the fuel system and lubricating system of engines as oil supply ump and oil return pump[1].

It is difficult to design the inner rotor in designing the cycloidal gear pump. That is because the inner rotor tooth profile is continuous closed equidistant curve of curtate epicycloid, which is complex and involves many geometric parameters. Meanwhile, the related geometry parameter values also change along with different cycloidal gear pump models. The profile curve of the inner rotor tooth-form of the cycloidal gear pump should be redesigned every time the gear is updated in traditional design method, which will take a lot of time and effort. So it is necessary to introduce parametric modeling in the design of the inner rotor[2]. With the rapid development of computer aided design and analysis technology, the CAD method of the contour curve of cycloidal gear is gradually enriched, and more and more manufacturing information software can realize the design of cycloidal gear contour curve and the motion simulation of internal gear pump. At present, many researchers[3-6] use UG, PRO/E and SolidWorks to do parametric design. This paper uses NX10.0 for design and simulation. NX10.0 is a powerful 3D parametric software which integrates CAD/CAE/CAM as a whole, and its function covers the whole life cycle of the product. This paper introduced the parametric design method of cycloidal gear profile curve on the platform of NX10.0, analyzed the radius of curvature with the contour curve of cycloid gear, proved the consistency between the simulation results and the theoretical design results by assembly analysis, and verified the parametric drive of the cycloidal gear simultaneously. The internal and external gears of the gear pump involved in this paper are shown in figure 1.

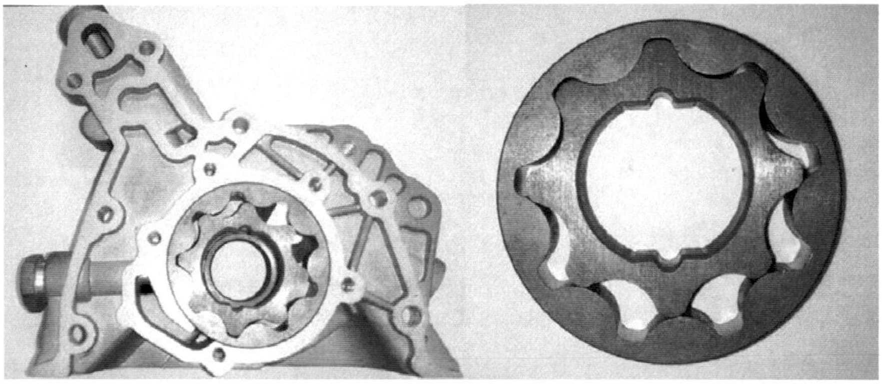

Figure 1. Trochoid Pump

2. Parameter Determination

The profile curve of the inner rotor of the cycloidal gear pump is related to the base circle that forms the curtate epicycloid, spheronization radius r2, number of inner rotor teeth Z1, number of external rotor teeth Z2 (Z2=Z1+1), center distance e , formation coefficient K and the radius factor H of the tooth circle. Once these parameters are determined, other geometries relevant to the tooth profile can be computed. The key parameters of the internal and external rotors of the gear pump are shown in figure 2.

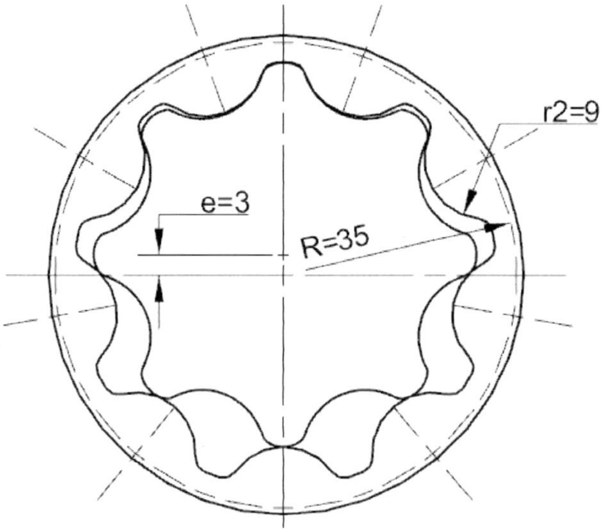

Figure 2. Internal and external rotor parameters of cycloidal gear pump

3. Build Models Using NX10.0

The key of Internal Gear of Cycloid Gear Pump design lies in the design of its contour, but the parametric design of the contour depends on "expression" and "law curve" of NX10.0 modeling module. Expressions created in NX10.0 are shown as follows. The contour curve of the cycloid gear generated with the "law curve" is shown in figure 3. After changing the number of teeth, a new gear contour is generated, as shown in Figure 4, which indicates that parameterized driver is valid. Figure 5 shows the analysis about the radius of curvature of the original gear profile. Since the outer rotor is an arc gear, use sketch to draw the contour curve, and then stretch to create the internal and external rotors separately. After that, using the assembly module of NX10.0 to assemble and perform interference analysis about the inner and outer rotors. If there is no interference, the rationality of the design is preliminarily proved, and then motion simulation can be used to further verify the rationality of the motion state. No more narrative here. The assembly model is shown in Figure 6.

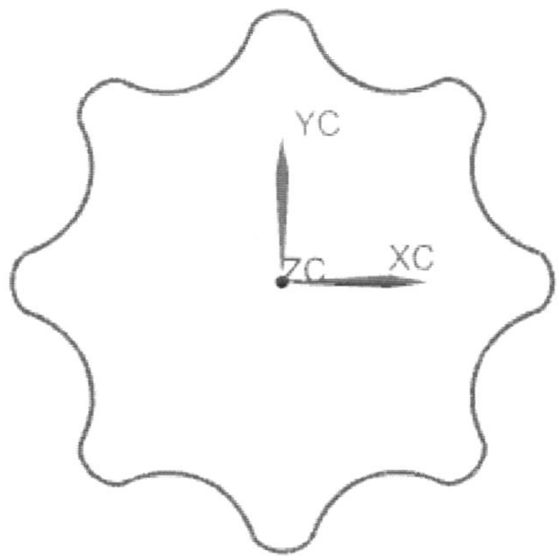

Figure 3. contour curve of cycloidal gear

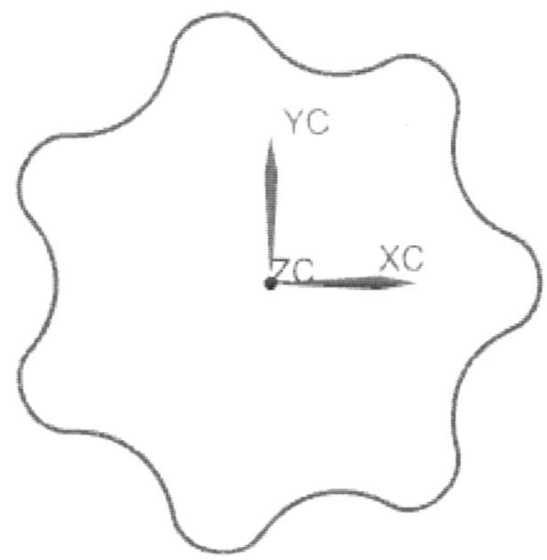

Figure 4. updated contour curve of cycloidal gear

The formula formats under NX software are as follows:

R=35

a=360*t*z2

b=sqrt(e^2+R^2*(1-i)^2+2*R*e*(1-i)*cos(i*a))

e=3

h=r2/(e*z2)

i=z1/z2

k=R/(e*z2)

r2=9

t=1

x1=R*cos((1-i)*a)+e*cos(a)

x2=r2*(R*(1-i)*cos((1-i)*a)+e*cos(a))/b

xt=x1-x2

y1=R*sin((1-i)*a)+e*sin(a)

IOP Conf. Series: Materials Science and Engineering **250** (2017) 012068 doi:10.1088/1757-899X/250/1/012068

y2=r2*(R*(1-i)*sin((1-i)*a)+e*sin(a))/b

yt=y1-y2

z1=8

z2=z1+1

zt=0

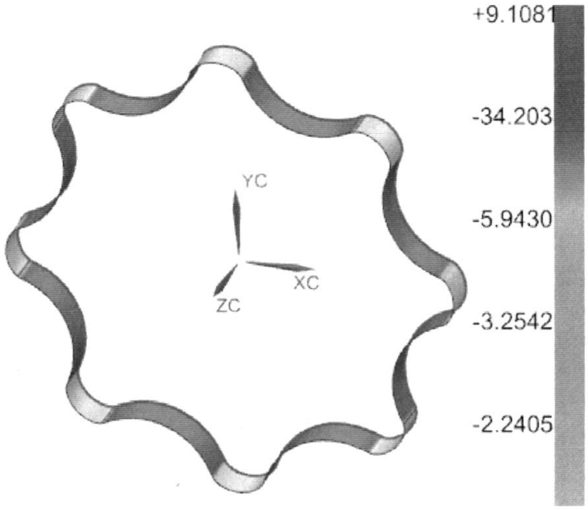

Figure 5. the minimum radius of the curtate epicycloid

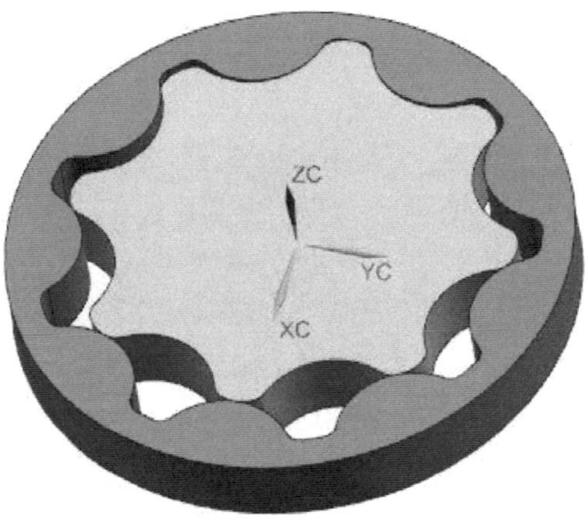

Figure 6.the assembly impression drawing of the internal and external rotors

4. Conclusions

From the above examples, it can be seen that parametric design of Cycloid Gear based on NX 10.0 can fully ensure the design accuracy. The parameterization of the key variables greatly improves designers' speed in modifying models, shortens the design cycle, and increases the reusability of models. Besides, The assembly module of NX10.0 can provide interference analysis and ensure the rationality of the design

5. References

[1] LIU Jie, LI Hua-cong. Study on Parametric Modeling Method of Cycloid Gear Pump Rotor Based on UG [J]. Science Technology and Engineering,2012,(25):6327-6330.
[2] Qian Xiao-hong, Zhang Dong-min. Parameterization of inner rotor in cycloidal pump [J]. Modern Manufacturing Engineering,2007,(01):135-137.
[3] LIU Ling-tao, WANG Jian-hua. Parametric design of whole cycloid gear based on Pro /E [J]. Journal of Beijing Information Science and Technology University,2011,(06):82-85.
[4] GONG Xiaxia, SUN Fuchun, LI Fengqin. Parametric Design on Cycloid Gear with UG / OPEN API [J]. Modern Manufacturing Technology and Equipment,2010,(01):63-64+73.
[5] KANG Limao,XU Lingbin Parametic Design in UG for Cycloidal Gear[J].Modern Machinery,2013,(04):49-51.
[6] SUN Fuchun , JIANGJiawang , WEI Yong , TANGYan. Parametric Design of Whole Cycloid Gear Based on SolidWorks API [J]. Journal of Chengdu University(Natural Science Edition),2012,(04):382-384.

IOP Conf. Series: Materials Science and Engineering **250** (2017) 012069 doi:10.1088/1757-899X/250/1/012069

Research of Design Method for Key Parts of Automotive Interior Based on NX Software

Enguang Zhang [1, a] and Li Wang[1,b],

[1]Zhuhai College of Jilin University, Zhuhai, 519041, China
[a] teamcenter@163.com [b] wangli00019@163.com

Abstract. Take the car door-handle as an example, propose a design method for key parts of automotive interior based on NX software, which is to do concept design for interior first, secondly, carry out framework design for the motion plan, then do structure design, shape design and topology optimization design through kinematics and dynamics analysis, and finally verify the reliability of the parts by finite element strength analysis and fatigue life analysis. This method can guarantee the unification of product data, shorten the product design life cycle, and guarantee the quality of product design.

1.Introduction

According to incomplete statistics, in the first 4 months of 2017, China's passenger car sales have reached more than 7 million 700 thousand, which is expected to reach more than 23 million 100 thousand vehicles throughout the year. With the increasing demand for cars, the car has been developing from the initial rapid running to be comfortable, safe, intelligent, reliable, beautiful and so on. Consumers begin to have higher and higher requirements for automobile performance, interior and exterior decoration. Compared with the important parts such as the engine and gearbox, the design and manufacture of automotive interior parts are easier to get started for the PRD enterprises which are eager to improve the added value of products. Take the car door-handle as an example, as shown in figure 1, it is the most touched one of the automotive interior parts except the steering wheel. People now have more and higher requirements for its beauty and strength. Most of the door-handle material is PC+ABS plating, which can not guarantee the strength well. It often occurs issues such as pull, deformation and each other. Therefore, it is imperative to study an effective method for the design and reliability analysis of automobile door-handle.

2. Current Status and Research Methods

At present, the foreign automobile companies have rich experiences in the design of automotive door-handle assembly due to their long history and strong technology. They have great advantages in material, structure and inspection. However, most domestic manufacturers do not have their own designed products. They are basically using foreign customers' drawings for processing, production and assembly. Although some domestic manufacturers and researchers use reverse scanning and other methods for reverse design[1-6], or even copy foreign related products, they can not form their own brands, and the product added value is low. Therefore, in the country's strong advocacy of "made in China" environment, from the perspective of long-term economic efficiency of enterprises, it has become the inevitable trend of enterprise sustainable development to design beautiful, practical and reliable car door-handle independently.

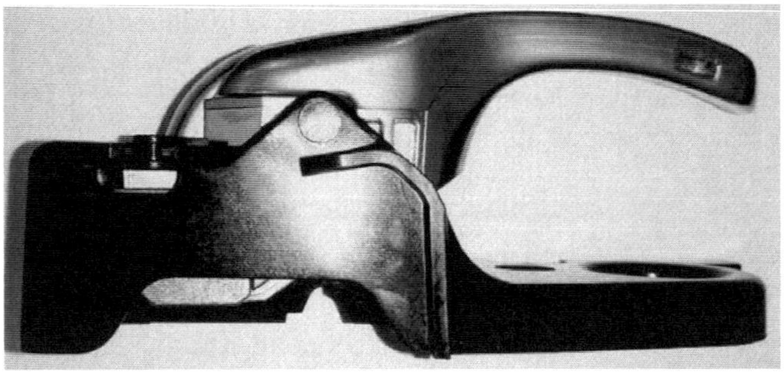

Figure 1. An automotive door-handle assembly

3. Research Scheme

Design of the car door-handle assembly usually need to meet the following requirements: a、Meet its functional requirements: switch and anti-theft lock function. b、Meet its aesthetic requirements: innovative modelling fashion. c、Meet its strength and life requirements. d、Meet cost control requirements, such as manufacturing process cost, assembly cost, etc.

The design of the car door- handle should take into account of the above problems, and these problems interact with each other. For example, in the design process of automotive door-handle assembly, how to solve the contradiction between shape and strength, life strength and cost, etc. are needed to consider. Therefore, this paper prepared to explore an effective and far-reaching automotive interior parts design method which is to speed up the design, shorten the product design cycle, ensure the quality of product design by virtual design.

There are two important simulation analysis in the design process. They are kinematic analysis of skeleton model for automotive door-handle assembly and finite element strength analysis of automotive door-handle. Therefore, the requirement of editability and parameterization of the digital model is relatively high. To solve the above problems and ensure the unity of data, the method proposed in this paper is intended to work under NX software in each section. Figure 2 shows the automotive door-handle assembly under NX software. All the processes under the NX software include skeleton design, motion simulation, finite element analysis, topology optimization design and structural detail design. The technical route is shown in Figure 3.

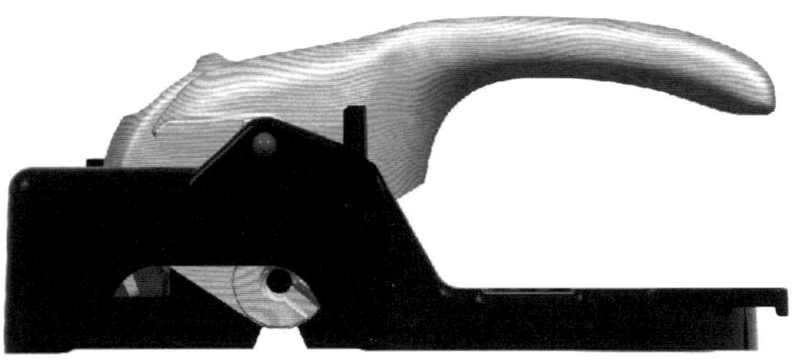

Figure 2. an automotive door-handle assembly designed by NX software

4. About NX Topology Optimization

NX10.0 is a three-dimensional powerful digital software which assembles CAD/CAE/CAM. Its feature covers the whole process of product life cycle, including the design, assembly, graphics, advanced simulation, motion simulation and processing and other modules. The NX advanced simulation module is used to realize the finite element analysis and topology optimization analysis of the mechanism. NX topology optimization uses the Tosca topology optimization solver from FE DESIGN Company. Optimization determines the optimal material distribution by adjusting the material density of units in the area of design. The material of the optimized model is distributed in block-shaped. Then the software creates a smooth transition for the model, so that the distribution of the material is more continuous. The smoothed model (STL file) can be imported into the NX modelling to be a reference for optimization design.

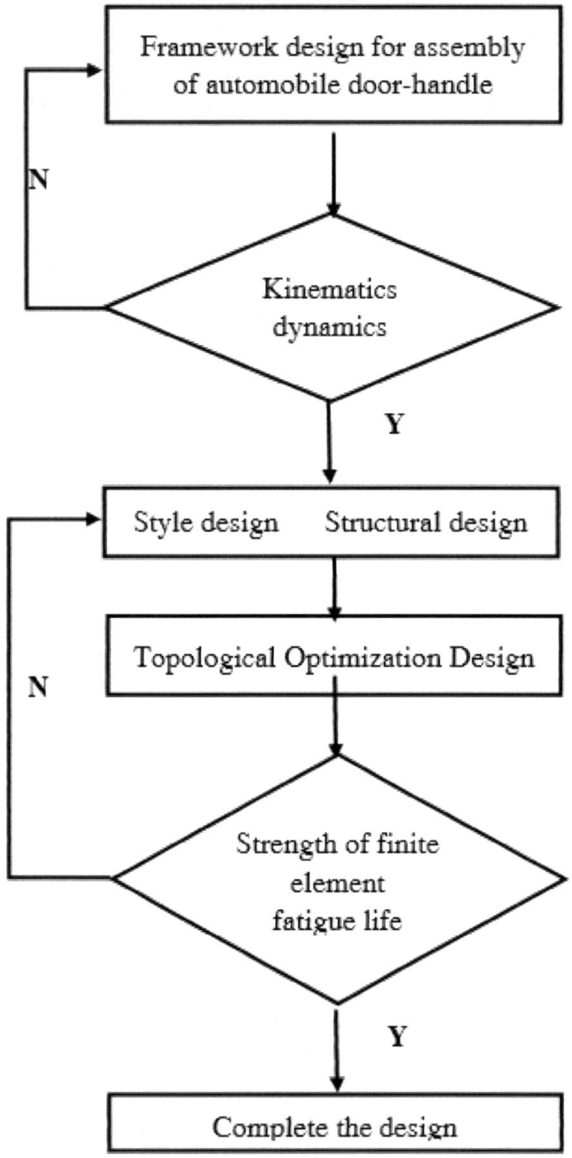

Figure 3. technique flow chart

Topology optimization technology can seek the best material distribution in a given design space, and define the material flow pattern by the two kinds of methods of Homogenization (equalization) and Density (Density). For density method, the material density of each element is directly used as a design variable, with the continuous variation between 0 ~ 1. 0 and 1 representing respectively empty or real .The process of topology optimization is shown in Figure 4[7].

Figure 4. process of topology optimization
(1) The design area to be optimized
(2) Optimized model (3) Smoothed model
(4) Final CAD design.

5. Conclusions
Thus, that use the PLM software NX10.0 to do product design can effectively integrate CAD\CAE or even CAM together, so that the product data is unified, it also can shorten the product development time, and ensure the quality of product design.

6. Acknowledgement
This paper is supported by the project "Design and reliability analysis of key parts in automobile interior", which is a innovative cultivation project in Zhuhai College of Jilin University.

7. References
[1] XU Zhenshan, LI Luxing. Reverse Design of Automobile Door Handle based on Geomagic/ [J]. China Educational Technology & Equipment,2017,(02):39-40+48.
[2] LIU Bin,JIANG Huibo,FAN Junhong. Research on reverse modeling and NC machining technology of automobile door handle [J]. Electronic Technology & Software Engineering,2015,(11):182-183.
[3] LAI Xiao, GUO Cheng. Reverse Design and Machining of Doorknob Based on Geomagic DesignX [J]. Mechanical Engineer,2017,(04):64-66.
[4] Chen Yu. The Design and Development of a Car Door [D]. Hunan University,2014.
[5] Zhang Ming chun.Design Procedure of All Terrain Vehicle Based On Reverse Engineering Technology [D]. Chang'an University,2007.
[6] Wu Jun.Freedom Surface Feature Rule Analyse in The Auto Styling[D]. Donghua University,2007.
[7] SIEMENS NX help documentation

IOP Conf. Series: Materials Science and Engineering **250** (2017) 012070 doi:10.1088/1757-899X/250/1/012070

Numerical Simulation and Monitoring of Surface Environment Influence of Waterless Sand Layer Shield Tunneling

Yanliang Shang [1,2], Tongyin Han[3], Wenjun Shi [2], Shouji Du [1]and Zhichao Qin [2]

[1]Department of civil engineering, Shanghai Jiao Tong University, Shanghai 200240, China; 2. School of civil engineering Shijiazhuang TieDao University, Shijiazhuang 050043, China; 3.Langfang Teachers University, Langfang ,065000 , China)
[1]E-mail:sty1980926@163.com

Abstract: The development of urban subway is becoming more and more rapid and plays an increasingly important role. The shield tunneling method has become the first choice for the construction of urban subway tunnel in the construction of urban subway. The paper takes the interval of Shijiazhuang Metro Line 3 Administrative Center Station and Garden Park Station as the engineering background. The establishment of double shield finite difference model by considering the thickness of covering soil, tunnel excavation and excavation at the same time, distance and other factors, the surface deformation, and soil thickness. The ground deformation law is obtained, the surface settlement is inversely proportional to the overburden thickness and the double line spacing, and the gradual excavation is smaller than the synchronous excavation.

1. Introduction

The numerical simulation analysis method is cheap and efficient. With the deepening of PC [1]technology and soil constitutive relation[2][3], it is more and more recognized. In order to study the law of surface deformation during shield construction.[4–9] In order to study the law of surface deformation during shield construction, this paper relies on the Shijiazhuang Administrative Center Station - Garden Expo Station Shield Tunnel Project, and establishes a three-dimensional numerical model for different working conditions. The numerical simulation method is used to establish shield tunneling .The interaction between the process and the surrounding soil, the thickness of the shield layer, the double line spacing, the excavation and so on are considered respectively. Finally, the numerical simulation results are analyzed.

2. Project Overview

Administrative Center Station ~ Garden Park Station is located in Zhengding New District, starting from the administrative center station, north along the planning of Beijing South Street (formerly known as Metro Avenue) below the laying, side wear and wear comprehensive pipe gallery to park garden station. The administrative center station to park station interval by shield construction ,starting mileage of the right line of the shield is K31+593.450, terminate the mileage for the K32+682.090, the total line length is 1088.640 m; the left line of the shield chainage K31+593.450, terminate the mileage for the K32+682.090, the total line length of 1088.640 m.

IOP Conf. Series: Materials Science and Engineering **250** (2017) 012070 doi:10.1088/1757-899X/250/1/012070

The soil layers in the exploration scope of the project are divided into 4 layers: artificial accumulation layer (Qml), recent sedimentary layer (Q4al), quaternary Holocene flushing deposit (Q4al+pl), and quaternary Pleistocene flush alluvium (Q3al+pl).

3. Establishment of Three - Dimensional Numerical Model of Shield Crossing Construction

In view of the deep groundwater depth in Shijiazhuang, the influence of groundwater is not considered; In accordance with the Garden Expo station section of EPB shield machine administrative center station, selection of construction parameters. In this paper, FLAC3D software is used to model the construction process of shield tunneling. Su Xiaokun [10] through numerical simulation of the multi cavity span, it is considered that in general, the net distance of double track tunnel is 3-5 times of the span of the tunnel, and can be modeled as a single hole which is independent and independent of each other. For the model, the upper boundary is the free surface, the rest is set to the normal constraint, and the model size is 48 m × 48 m × 48 m. The initial stress calculation only considers the soil weight.

The sight lining is a seamless torus. Segment material and grouting material are considered as linear elastic material. The surrounding rock follows the Mohr-Coulomb criterion.

3.1 Calculate Parameter Selection

The parameters of the material are mainly based on the results of the geological survey of the administrative center station - garden garden station. The parameters of each material are shown in Table 1 .

Table 1. Parameters of the stratum of the tunnel

material type	γ/(kN·m-3)	E/MPa	μ	c/kPa	φ/°
Miscellaneous soil	16.5	5.36	0.35	0	8
Fine sand	19.0	26.7	0.32	0	20
Loess-like silty clay	20.0	25.7	0.3	18.3	25
Silty clay	20.5	26.6	0.3	22.9	17

3.2 Establishment of Calculation Model for Shield Tunnel with Different Overburden Thickness

Respectively, the shield depth of 14m, 16m, 18m model, each model grid a total of 43 360 units, 43 excavation steps. The completed model is shown in Figure 1 to figure 3.

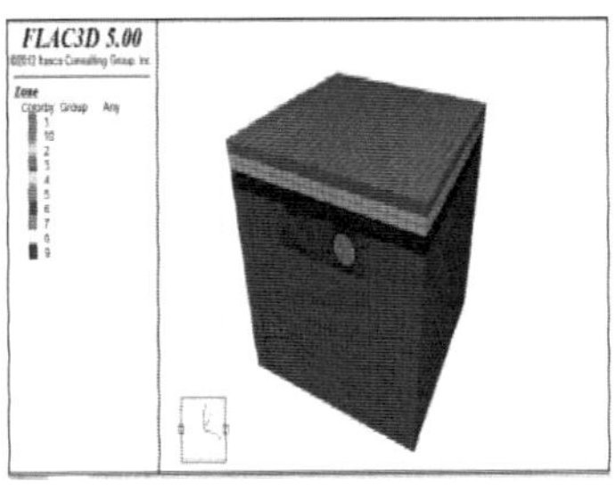

Figure 1. Shield depth (14 m)

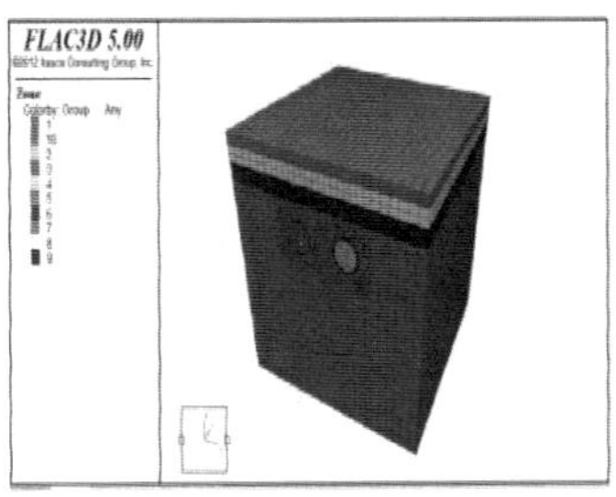

Figure 2. Shield depth (16 m)

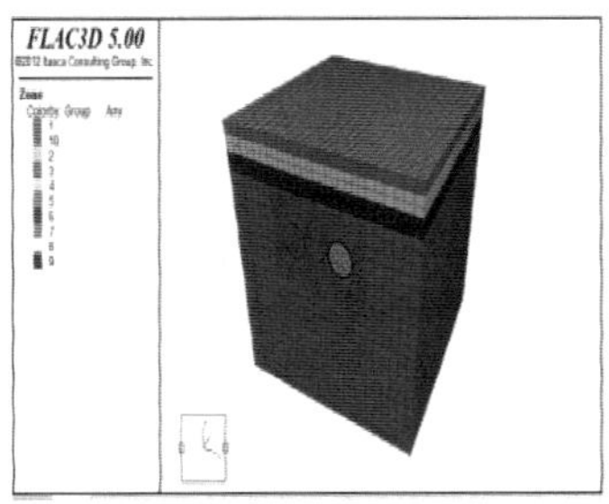

Figure 3. Shield depth (18 m)

3.3 Establishment of Shield Tunnel Calculation Model with Different Horizontal Spacing

Respectively, the establishment of the tunnel spacing of 12m, 14m, 16m numerical model, each model grid a total of 43 360 units, 43 excavation steps. The built model is shown in figure 4 to figure6.

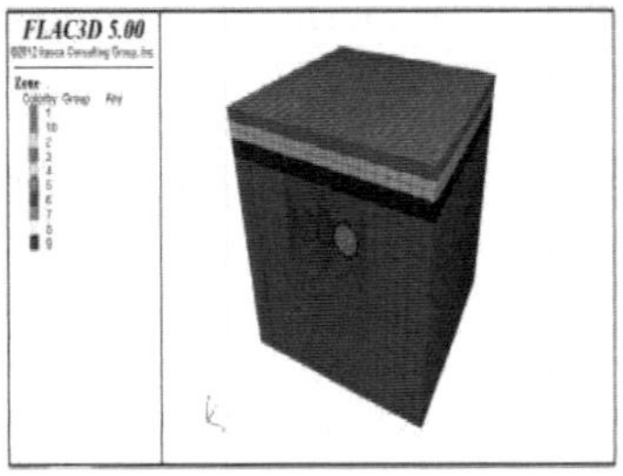

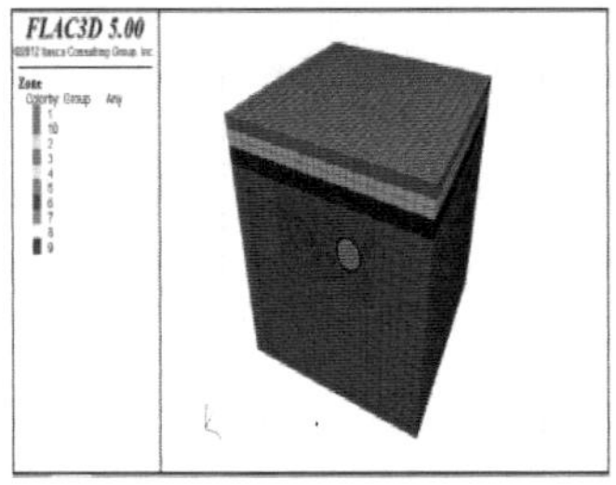

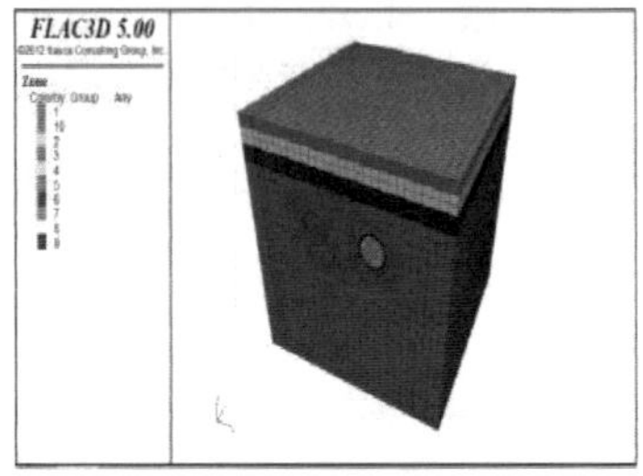

Figure 4. Shield spacing (12 m) **Figure 5.** Shield spacing (14 m) **Figure 6.** Shield spacing (16 m)

4. Calculation Results and Analysis

Based on calculation results, the paper analyzes the law of surface deformation and overburden thickness, double line spacing and synchronous excavation and step excavation during shield tunnel construction

4.1.Analysis on Influence of Shield Tunnel Construction on Different Overburden Thickness

Figure 7 shows the vertical displacement of the soil after the completion of shield tunneling with a depth of 18 m. It can be seen from the figure7 that the maximum settlement of the surface is 16.1 mm due to the construction of the shield, the settlement of the formation and the formation of the settling tank. Figure 8 shows the horizontal displacement of the soil after the depth of the shield is 18 m. As can be seen from the figure8, the horizontal horizontal displacement of the surface is 5.39 mm. Figure 9 shows the vertical displacement of soil after completion of shield tunneling with a depth of 16 m. It can be seen from the figure9, due to shield construction, formation settlement, the formation of sedimentation trough, the maximum settlement of the surface is 20.3 mm.

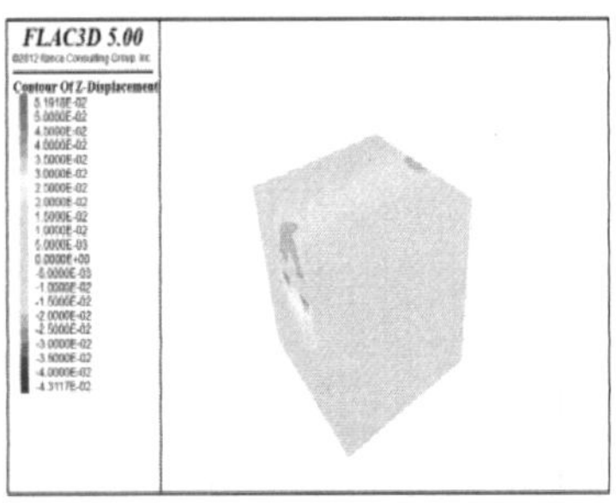

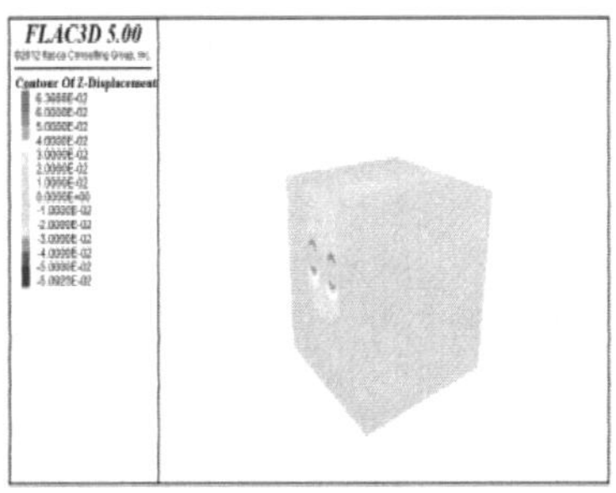

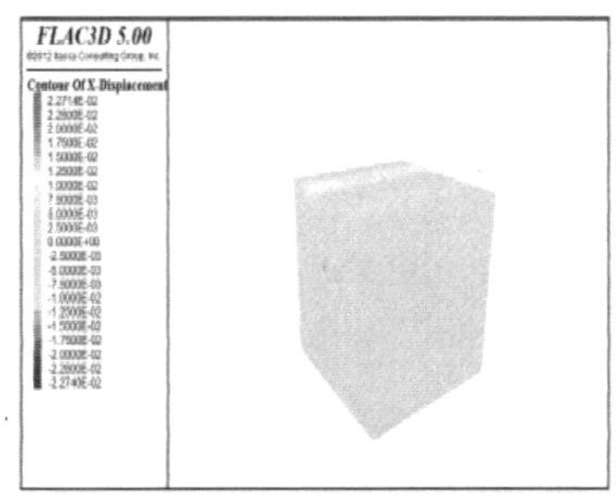

Figure7. Vertical displacement (18m) **Figure8.** Horizontal displacemen (18 m) **Figure9.** Vertical displacement (16 m)

Figure 10 shows the horizontal displacement of the soil after the completion of the shield tunneling. As can be seen from the figure10, the horizontal horizontal displacement of the surface is 7.33 mm. Figure 11 shows the vertical displacement of the soil after the completion of shield tunneling with a depth of 14 m. It can be seen from the figure11, due to shield construction, formation settlement, the formation of sedimentation tank, the maximum settlement of the surface of 21.7 mm. Figure 12 shows the horizontal displacement of the soil after the excavation of the shield is completed. As can be seen from the figure12, the horizontal horizontal displacement of the surface is 6.57 mm.

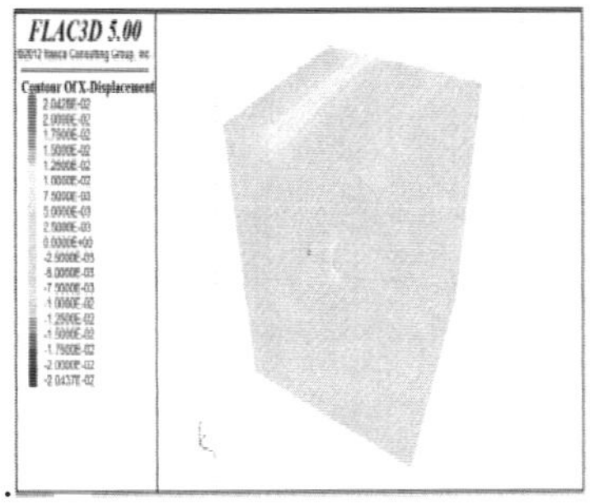

Figure10. Horizontal displacement (16 m)

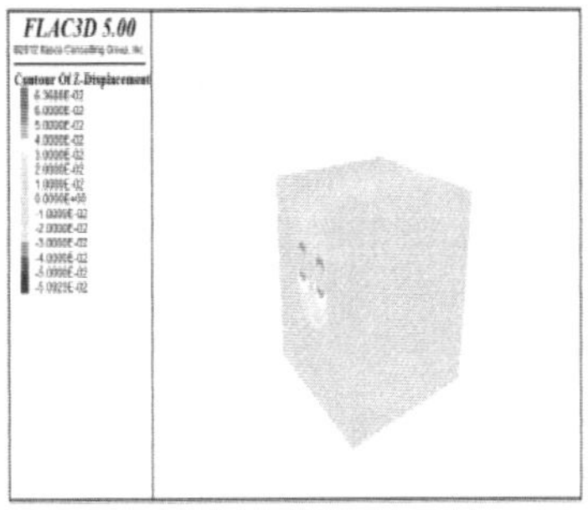

Figure11. Vertical displacement (14 m)

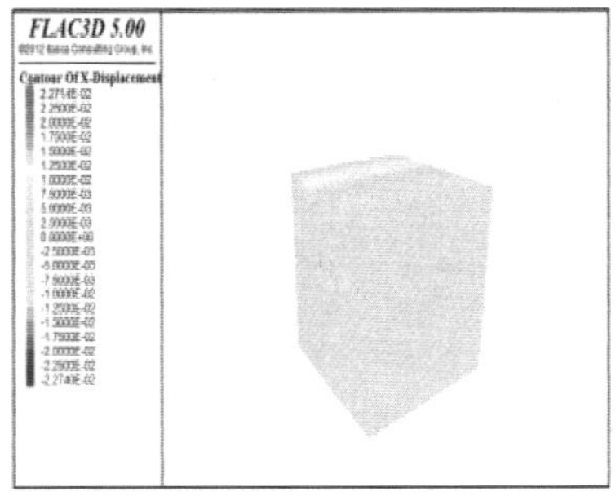

Figure12. Horizontal displacement (14 m)

According to the pre-arranged monitoring point, you can get the top of the tunnel center vertical displacement along with the excavation step curve, the left line and the right line settlement the same. The larger the buried depth, the smaller the settlement, the monitoring points on both sides due to the boundary effect of the model, the vertical displacement is rising.

4.2.Analysis on the Influence of Shield Tunnel Construction with Different Tunnel Spacing

Figure 13 shows the vertical displacement of the soil after the completion of 12 m boring distance. It can be seen from the figure13, due to shield construction, formation settlement, the formation of sedimentation trough, the maximum settlement of the surface is 26.16 mm. Figure 14 shows the horizontal displacement cloud diagram of the soil after the boring distance of 12 m. As can be seen from the figure14, the horizontal horizontal displacement of the surface is 11.6 mm. Figure 15 shows the vertical displacement of the shield after the completion of the shield tunneling distance of 14 m. It can be seen from the figure15, due to shield construction, formation settlement, the formation of sedimentation tank, the maximum settlement of the surface is 22.18 mm.

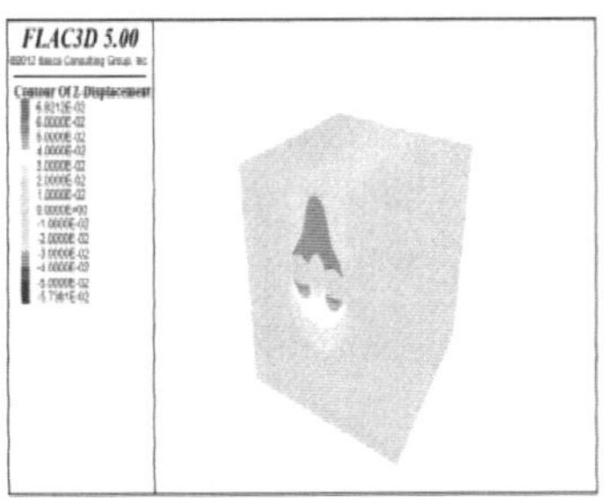

Figure13. Vertical displacemen (12m)

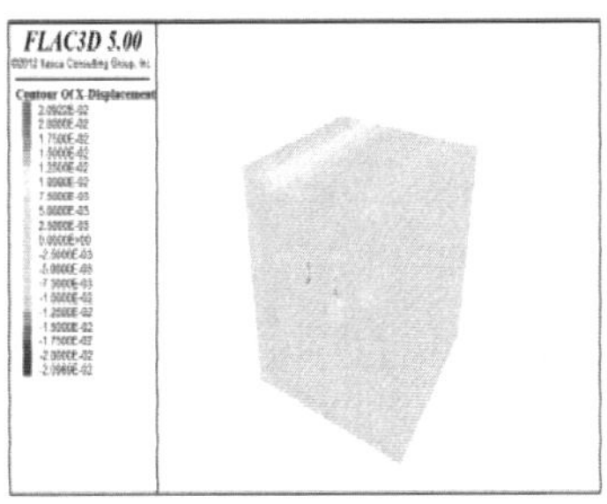

Figure14. Horizontal displacement (12m)

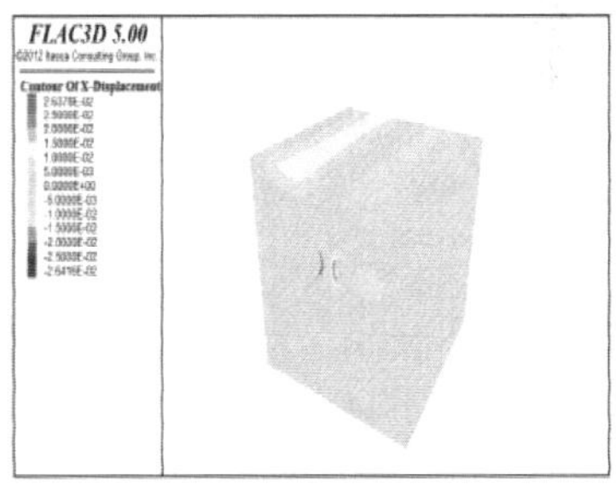

Figure15. Vertical displacement(14 m)

Figure 16 shows the horizontal displacement of the double-line shield at a distance of 14 m. It can be seen from the figure16 that the horizontal horizontal displacement of the surface is 9.6 mm. Figure 17 shows the vertical displacement of the shield after the completion of the shield tunneling distance of 16 m. It can be seen from the figure17, due to shield construction, formation settlement, the formation of sedimentation trough, the maximum settlement of 16.68 mm. Figure 18 shows the horizontal displacement of the double-line shield at a distance of 16 m. As can be seen from the figure18, the horizontal horizontal displacement of the surface is 8.1 mm.

IOP Conf. Series: Materials Science and Engineering **250** (2017) 012070 doi:10.1088/1757-899X/250/1/012070

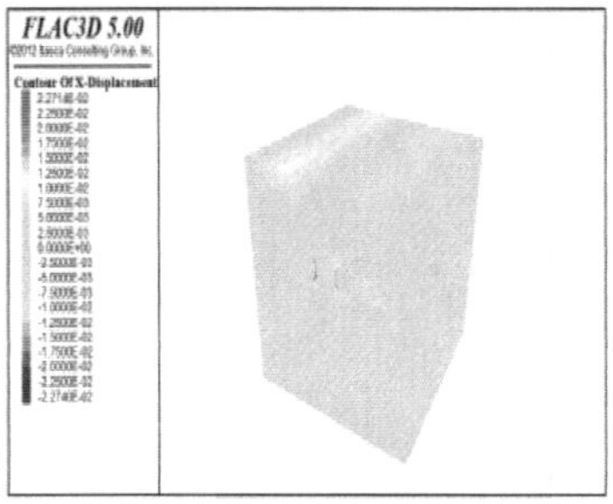

Figure16. Horizontal displacement (14 m)

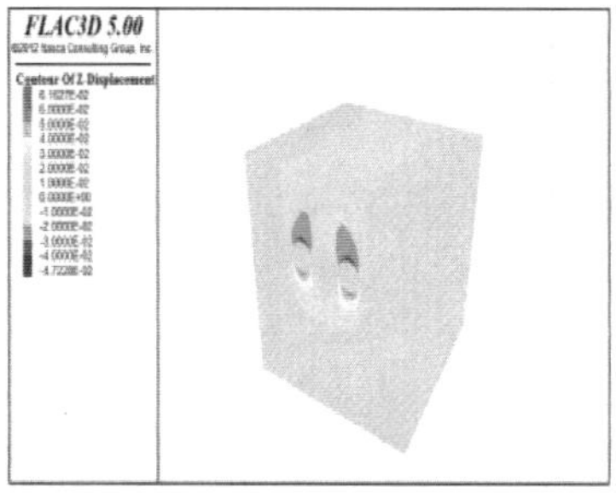

Figure17. Vertical displacement (16 m)

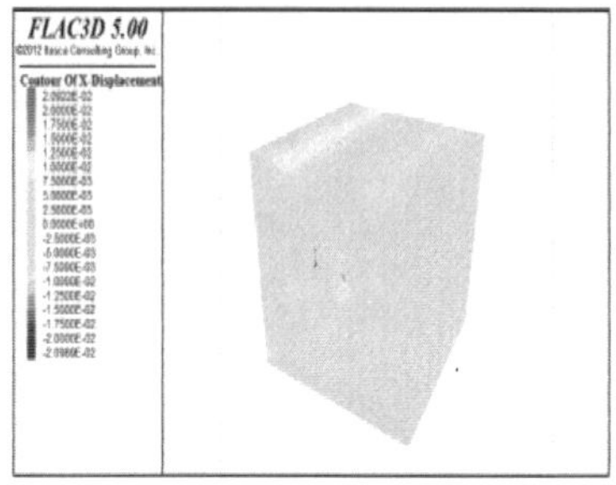

Figure18. Horizontal displacement (16 m)

According to the monitoring point laid in advance, the vertical displacement of the surface of the tunnel above the center of the tunnel will change with the excavation step, because the settlement of the left line and the right line will be the same. As can be seen, the greater the spacing between the two lines, the smaller the settlement

5. Conclusion

This paper mainly through the establishment of three-dimensional numerical simulation (FLAC3D) method to simulate and analyze the influence of shield tunneling on the deformation of the central area of the Administrative Center Station ~ Garden Park Station.And through the numerical simulation, the thickness of the upper cladding layer Spacing and shield synchronous excavation and asynchronous excavation under the conditions of the impact of the law, the main conclusions are as follows:

(1) the surface subsidence is inversely proportional to the thickness of the shield soil. When the thickness of the overlying soil is 14 m, the maximum surface subsidence is 21.7 mm.

(2) the surface subsidence is inversely proportional to the spacing of the two tunnels. When the tunnel spacing is 12 m, the maximum surface subsidence is 26.16 mm.

6. Acknolegdements

This study is subsidized by the Science and Technology Project of Hebei Province [Grant No. 16215408D],and by Hebei Provincial Disaster Prevention and Mitigation Collaborative Innovation Center of Large Infrastructure Projects . These financial supports are gratefully acknowledged.

7. References

[1] Xu Shoutian, Pu gang. Design of shield ground data acquisition and monitoring system based on [J]. *OPC tunnel construction*, 2009, (**06**): 674-677.

[2] Ren Qingyang. The research and application of the constitutive relation of numerical modeling of[A]. China society of rock mechanics and engineering geotechnical. Second national Geotechnical Engineering Conference (2) [C]. *Chinese: Journal of rock mechanics and engineering*, 2006:6.

[3] Yang Xiaoyan. Some thoughts on numerical analysis of geotechnical engineering [J]. *Sichuan cement*, 2017, (**01**): 300.

[4] Qu Kaisheng. Study on ground surface settlement caused by shield tunneling in soft soil [D]. *Zhejiang University of Technology*, 2014

[5] Liang Rongzhu, Xia Tang Dynasty, Lin Cungang, Yu Feng. Shaped Propulsion Causes Surface Deformation and Horizontal Displacement Analysis of Deep Soil [J / OL]. *Chinese Journal of Rock Mechanics and Engineering,* 2015,34 (**03**): 583-593. (January 9, 2015) [August 18, 2017]. Http://kns.cnki.net/kcms/detail/42.1397.o3.20150115.1810.027.htmlDOI:10.13722/j.cnki.jrme.2015.03.016

[6] Xiao Li, Zhang Qinghe. Numerical simulation of ground subsidence caused by track shield construction [J]. *Journal of Tongji University (Natural Science Edition)*, 2011,39 (**09**): 1286-1291. [2017-08-18]

[7] Li Hongan. Beijing Metro Line 10 shield under the bridge pile pile settlement alculation formula application [A]. China Civil Engineering Society. Underground construction and harmonious development of the environment - the fourth China International Tunnel Engineering Symposium [C]. *China Civil Engineering Society*: 2009: 5.

[8] Li Yigang. Finite element simulation analysis of shield tunnel under the tunnel [J] .*Guangdong Building Materials*, 2011, (**03**): 24-26.

[9] Li Song, Yang Xiaoping, Liu Tingjin. Effect of Shield Tunneling on Pile Foundation of Viaduct Bridge in Guangzhou Metro [J]. *Railway Construction*, 2012, (**07**): 74-78.

[10] Su Xiao-kun.Study on the Range of Surrounding Rock Boundary of Numerical Simulation of Tunnel Excavation [J]. *Journal of Railway Engineering Society*, 2012, (**03**): 64-68.

IOP Conf. Series: Materials Science and Engineering **250** (2017) 012071 doi:10.1088/1757-899X/250/1/012071

Spare Parts Demand Analysis Method Based on Field Replaceable Unit

MinZhu, ZijianXu,MingGuoand BiaoLi

Naval Aeronautical University ,No.704 Department of Teaching and Research, No.188,No.2 Road,Zhifu District,Yantai , Shandong ,China
Email:min0zhu@163.com

Abstract. It is based on the reliability of spare parts optimization modeling, and the influence of failure rate, fixability, importance and availability on spare parts reserves is added to the model, and the weight of each spare part is calculated by fuzzy analytic hierarchy process.

1. Introduction

In the reliability analysis of the field replaceable unit, we found that it is very important to impbe able to get the replacement in time when the replacement of the replaceable unit components in the field is very important. Therefore, based on the analysis of the reliability of the field replaceable units, this paper analyzes the spare parts reserve of the field replaceable units..

2. Calculation Model of Spare Parts Requirement

On the optimization of the spare parts system [1] ,some important literatureof relevant domestic and foreign mostly use Barlow's theory and methods [2,3].It considers components and their spare parts as cold storage subsystems, and the entire spare system is considered as a system of tandem systems of each subsystem. The reliability of the reserve-series system is the protection rate of the spare parts system. The optimization of the spare parts system is attributed to the minimum cost of the system under the constraints of the given system reliability. The model is as follows:

$$\begin{cases} \min \sum_{i=1}^{n} c_i x_i , i = 1,2, L, n \\ \prod_{i=1}^{n} R_i(x_i, t_i) > R_0 \end{cases} \tag{1}$$

Where n is the type of spare part system; x_i indicates the number of type i spare parts; c_i indicates the spare parts on unit price of type i; $R_i(x_i, t_i)$ is the reliability of the type i spare parts equipped with x_i spare parts in $(0, t_i)$ time. R_0 is defined as the protection rate for the entire spare parts system. The problem with this computational model is that there are many types of spare parts, and if we use the evaluation function of the series model $\prod_{i=0}^{n} R_i(x_i, t) \ge R_0$ as the evaluation criterion, even if the reliability of each component reaches 0.9, the reliability of the 30 parts is only 0.424, which is not in line with the actual situation . In general, the protection rate of the spare parts system [4] is defined by the system service level:

$$\text{Spare parts protection rate} = \left(1 - \frac{\text{The number of missing parts in time}}{\text{The number of received parts within a certain period of time}}\right) \times 100\% \quad (2)$$

From the formula (2) we see the protection rate of spare parts is equivalent to the average number of single component protection rate, which is very different from the evaluation function in the series reliability model. An evaluation function is constructed from the concept of spare rate of spare parts. It can be seen from the formula (2) that the total spare rate is considered to be the probability that the spare parts can be provided in time. $N_i(t),(i=1,2,\cdots,n)$ indicates the number of times the type i component has failed during the period $(0,t)$ .Thus, the protection rateof the type i spare parts is:

$$R_i = P\{N_i(t) \le x_i\} \quad (3)$$

Spare parts system total security rate R_s is:

$$R_s = P\{N_1(t)+N_2(t)+\cdots+N_n(t) \le (x_1+x_2+\cdots+x_n)\} \quad (4)$$

The formula (4) is a reliability model constructed according to the basic definition of the spare rate of spare parts, and continues to derive its intrinsic form from this basic definition.

The observed value of the statistic $\hat{R}_i = \dfrac{x_i^0}{N_i(t)}$ is taken as the estimate of R_i ,in formula:

$$x_i^0 = \begin{cases} x_i, N_i(t) \ge x_i \\ N_i(t), N_i(t) < x_i \end{cases} \quad (5)$$

For the entire spare system, the observed value of $\hat{R}_s$ is taken as an estimate of the total guaranteed rate R_s.

$$\hat{R}_s = \frac{\sum_{i=1}^{n} x_i^0}{\sum_{i=1}^{n} N_i(t)} = \sum_{i=1}^{n} \frac{N_i(t)}{\sum_{i=1}^{n} N_i(t)} R_i \quad (6)$$

We introduce λ_i as the probability of type i component's failure, and by the update process theory we know: $\lambda_i = E(N_i(t))$ .For parts of the service subject to the distribution of the situation, it is precisely established; for the non-exponential distribution of component life, it is asymptotically established .Therefore, the entire spare system protection rate can be expressed as

$$R_s = \sum_{i=1}^{n} \frac{\lambda_i}{\sum_{i=1}^{n} \lambda_i} R_i \quad (7)$$

We make $\rho_i = \dfrac{\lambda_i}{\sum_{i=1}^{n} \lambda_i}, i=1,2,\cdots,n$, so

$$R_s = \sum_{i=1}^{n} \rho_i R_i \quad (8)$$

ρ_i represents the weight of the protection effect of the type i component on the total protection rate of the system in formula (8) . Obviously, in the same protection rate of each type of component , high failure rate has a great impact on the overall protection rate of the spare parts system. Thus, the reliability model is established based on the derived formula:

2.1. Protection Model Based on A Certain Amount of Follow-up Spare Parts

The objective function of the model is:

$$R = \max \sum_{i=1}^{N} \rho(i) R(i) \quad (9)$$

In the formula: R is the system protection rate, $\rho(i)$ is the weight of a single component, $R(i)$ is the guarantee rate of the individual components.

The constraint function is:

$$C = \sum_{i=1}^{N} C(i) X(i) \le C(0) \quad (10)$$

$$R(01) \le R(X(i), T) \le R(02), i = 1, 2, \cdots, N \quad (11)$$

$$R(00) \le R(X(i), T), i = 1, 2, \cdots, k, R(01) \le R(00) \le R(02) \quad (12)$$

In the formula (10), $C(i)$ is the unit price of the i -thspare part, which represents the sum of the product of the number and unit price of each component and makes the it less than or equal to the total cost of the system $C(0)$; formula (11) indicates that the protection rate of each component in the order cycle is greater than or equal to the minimum protection rate, less than or equal to the maximum protection rate; formula (12) indicates that a number of particularly important spare parts protection rate is greater than the protection rate $R(00)$, where $R(00)$ isgreater than or equal to the minimum protection rate, less than or equal to the maximum protection rate.

2.2. Protection Model Based on the Protection Rate of Follow-up Spare Parts

The objective function of the model is:

$$C = \min \sum_{i=1}^{N} C(i) X(i) \quad (13)$$

Constraint function:

$$R(01) \le R(X(i), T) \le R(02) \quad (14)$$

$$\sum_{i=1}^{N} \rho(i) R(i) \ge R(0) \quad (15)$$

$$\sum_{i=1}^{J-1} \rho(i) R(i) + \sum_{i=J+1}^{N} \rho(i) R(i) + \rho(J) \times R(X(J) - 1, T) \le R(0) \quad (16)$$

J is any value in $1,\cdots,N$,indicates that the total guaranteed rate of the system R is less than R (0) when the J spare part is missing.

3. Model Optimization of Spare Parts Requirement

In practice, spare parts are also affected by many factors, not only the impact of spare parts wear and tear, but also its importance in the equipment-critical impact. In addition, we need to consider the impact of economic and so on. The key to the component is the piece in the equipment system and the role of the impact on the size of the system performance. Parts of the wear and tear refer to the size of the degree on it, mainly related with the inherent reliability of parts, the use of environmental conditions and conditions. Some of these factors are not easy to quantify, and their relationship is a vague relationshipwith the spare parts, so it is suitable for fuzzy comprehensive evaluation method in the weight of the analysis.

In the calculation of this paper, we mainly consider the influence of the failure rate, maintainability, the importance and availability of components in the four aspects, and use the fuzzy analytic hierarchy process [5,6] to obtain the weight score of the component, and add up the weight score r_i of all the components to obtain a total weight fraction G , $G=\sum_{i=1}^{n} r_i$,so the weight of the component is $\rho_i=\frac{r_i}{G}$.

3.1. Construction Judgment Matrix

Before the weight analysis, we must first establish a hierarchical structure to solve the problem, as shown in Figure 1.

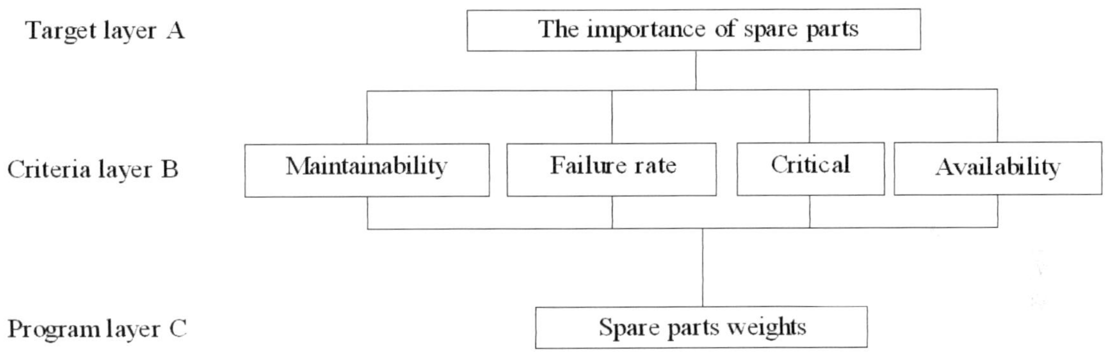

Figure 1. Weight Hierarchy Diagram

Judgment matrix is the basic information of the analytic hierarchy process. It is an element value of the above layer as the criterion of judgment, and the two elements of the next layer compared to determine the element value of the matrix. The element a_{ij} in the judgment matrix A indicates that the relative importance of the element i to the element j from the viewpoint of the judgment criterion ,which satisfies:

$$a_{ij}=\frac{1}{a_{ji}}, a_{ii}=1,(i,j=1,2,\cdots,n)\ (17)$$

Judgment scale use AHP (Scale 1-9), with 1,3,5,7,9, to construct the fuzzy judgment matrix $A^{(k)}$,respectively comparing element i to element j of the same, significant, important, more important, most important. For experts evaluation, they use fuzzy triangular number to express fuzzy judgment, with level 3 scale of fuzzy number . Experts have a certain degree of confidence δ in the degree of importance of subjective judgments , when the confidence were "very sure", "more sure", δ were the value of 0.5,1,1.5. $A^{(k)}=(a_{ij})_{n\times n}$ expressed as:

$$a_{ij}=\left[a_{ij}^{-},a_{ij}^{+}\right]=\begin{cases}[a_{ij}-\delta,a_{ij}+\delta](a_{ij}\geq 1,i\neq j)\\ \left[\dfrac{1}{1/a_{ij}+\delta},\dfrac{1}{1/a_{ij}-\delta}\right](a_{ij}\leq 1,i\neq j)\\ [1,1](i=j)\end{cases} \quad (18)$$

a_{ij} is an element in the fuzzy judgment matrix, so the judgment matrix $A^{(k)}$ is a positive reciprocal matrix.

3.2. Calculate the Normalized Weight Vector

We calculate the weight determined by the k -th expert $w^{(k)}$, $w^{(k)}=(\alpha x^{-}+\beta x^{+})/2$,

where x^{-} and x^{+} are the normalized eigenvectors corresponding to the largest eigenvalues of the matrix A^{-} and A^{+}. α , β for the value of:

$$\alpha=\left[\sum_{j=1}^{n}\frac{1}{\sum_{i=1}^{n}a_{ij}^{+}}\right]^{1/2},\quad \beta=\left[\sum_{j=1}^{n}\frac{1}{\sum_{i=1}^{n}a_{ij}^{-}}\right]^{1/2} \quad (19)$$

We make a normalization of the calculated weight vector $w^{(k)}$,and getthe normalized weight vector $w_0^{(k)}$. In order to ensure the reliability of decision-making, we need to carry on the consistency judgment to the fuzzy judgment matrix. The consistency test is carried out with the difference between $\lambda_{\max}$ and n , that is,

$$C.I.=\frac{(\alpha\lambda_{\max}^{-}+\beta\lambda_{\max}^{+})/2-n}{n-1} \quad (20)$$

If $C.I.\leq 0.1$, it is proved that the judgment matrix is consistent.

3.3. Calculate Decision Weights

In order to reduce the risk of decision-making, the assessment process has a number of experts to participate,while each expert's credibility is not the same, so we make each expert's own weight y_k , weight $w_0^{(k)}$, and we get decision-making weight

$$w=\frac{1}{\sum_{k=1}^{k}y_k}\sum_{k=1}^{k}y_k w_0^{(k)} \quad (21)$$

3.4. The Final Weight of the Component

We make the score of the failure rate, fixability, importance and availability of each component separately, and the scores of the four items are summed to obtain the weight fraction of the component, and then sum the weight scores of all components up ,draw a total weight score $G=\sum_{i=1}^{n}r_i$. The final weight of the component is $\rho_i=\dfrac{r_i}{G}$.

4. Concluding Remarks

Through the fuzzy theory, the reliability of the field replaceable parts, the importance of the installation, the fixability and the availability of the weight in the outfield were weighted to obtain the weight of the components , we build up the optimization model based on the reliability of the spare parts requirements , and this method is in accordance with the practice of spare parts protection.

5. References

[1] Qu Hongchun, Sun Chunlin. The Reliability of Model Based on the Guarantee Rate of Spare Parts System[J].Spare Parts Management,2004,3:50-51
[2]Jiang Bosong, Sun Chunlin. An Improved Optimization Model and Solution of Spare Parts System [J].Journal of China Civil Aviation University, 2003, 21(6):17-20
[3]Ronald H.Barlow. Business Logistics Management: Planning Organizing and Controlling the Supply Chain[M]. Pearson Education, Inc, Upper Saddle River, New Jersy,2004
[4]Fan qinghe. Research on the Optimization Model of Follow-up Spare Parts[J].Science & Technology and Engineering,2007,7(12):2763-2766
[5]Cui Nanfang, Luo Xue. Repair Spare Parts Based on AHP Classification model of ABC[J].Industrial Engineering Management,2004(6):33-36
[6] Ma Li, Chen Donglin. The Application of Fuzzy Comprehensive Judgment Theory in the Supply

Elliptic Curved Component Macro-Programming and Its Application

Zhibo Yang, Junchen Hu, Kaiqiang Li, Shiyu Zhang and Aiju Liu

School of Mechanical and Power Engineering, Henan Polytechnic University, Jiaozuo 454000, China

E-mail: Yangzhibo2001@163.com

Abstract. Most conventional numerical control systems do not have the function of noncircular curve interpolation instruction. Manual programming is extremely challenging, automatic programming by computer-aided manufacturing software is highly sophisticated, and processing parameters cannot be easily modified. Therefore, macro-programs, which possess powerful parametric programming, are applied for the processing of noncircular curved components. The values of arguments were determined using transfer and loop statements (IF and WHILE), and elliptic curved macro-programs were achieved using normal and parameter equations in this study. The elliptic curved components were fitted using micro-sized segments or arcs. The numerical control machining tests verified the validity and viability of the macro-programs, and elliptic curved components were processed. The results indicated that the elliptic curved components processed using macro-programs met the design requirements.

1. Introduction

In FANUC numerical control systems, the assignment of macro-programs by arguments is achieved using arguments, and the versatility and applicability of the programs can be enhanced using macro programming. Currently, most FANUC numerical control systems employ Category B macro-programs for programming[1,2]. In conventional programming, designated constants are used, and the program is executed according to the designed sequence with no skip allowed. The macro-programs, however, employ arguments that can be assigned and operated, thus enabling skip. For the processing of series components (with size variation but no shape change), conventional programs require re-programming for each component, while macro-programs need changes in argument values only[3]. Using macro-programs, noncircular curved components, such as conic curved ones, can be fitted using micro-sized lines or arcs to meet the precision requirements[4].

2. Macro-programming for Conic Curved Components

2.1. Macro-programming Mechanism

In the numerical controlled processing of elliptic curved components, the elliptic curve was divided into several tiny segments, which were fitted using micro-sized line segments or arcs. In this case, the processing precision was directly related to the quantity of segments divided[5]. For components with sophisticated configurations, only mathematical equations and algorithms were keyed in manually, and node coordinates were obtained and processed by macro-programs, thus reducing efforts for programming.

2.2. Macro-programming Format

In the first format, conditional transfer instructions (IF statements) were used:

IF [conditional expression]
GOTO m;

In this case, execution came before judgment. In other words, the program went to the mth section if the condition was satisfied; otherwise, the program went to the next section.

In the second format, loop instructions (WHILE statements) were used:

WHILE [conditional expression]
DO m;
...
END m;

In this case, judgment came before execution. In other words, the section from DO m to END m was executed if the condition was satisfied; otherwise, the section after END m was executed[6,7]. Herein, m could only be 1, 2, or 3; otherwise, alert was triggered[8].

3. Mechanism of Macro-Programming for Elliptic Curved Components

In the lathe workpiece coordinate system *xoz*, the *z*-and the *x*-axes intercepts were assumed to be *a* and *b*, respectively, and coordinates of Point p on the ellipse as *(z, x)*, as shown in Figure 1:

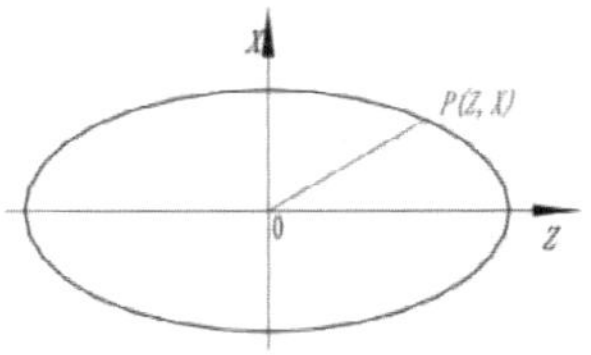

Figure 1. Elliptical mathematical model.

The standard equation of ellipse was as follows:

$$\frac{z^2}{a^2}+\frac{x^2}{b^2}=1 \quad (1)$$

Where *a* and *b* were the semi-major axis and semi-minor axis, respectively, and a > b > 0.
The parameter equation of ellipse was as follows:

$$\begin{cases} z = a\cos\alpha \\ x = b\sin\alpha \end{cases} \quad (2)$$

Where *a* was the polar angle.

If the standard equation was used for the macro-programming of ellipse, coordinates of p_0, p_1, …, p_n with *z* or *x* as the argument were calculated, and the elliptic curve was processed by pointwise interpolation[9]. If the parameter equation was used for the macro-programming of ellipse, coordinates of p_0, p_1, …, p_n with angle a as the argument were calculated, as shown in Figure 2.

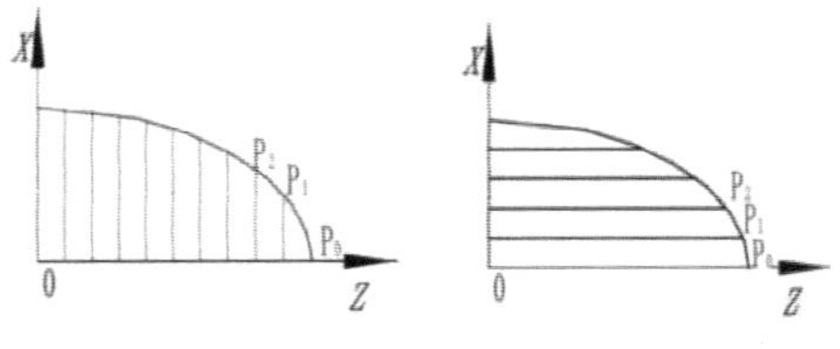

Figure 2. Elliptical Schematic of different processing method argument.

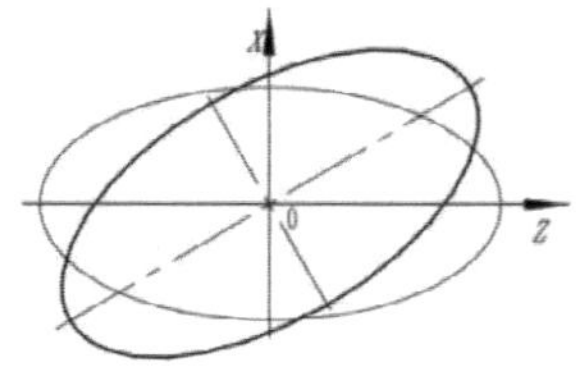

Figure 3. Mathematical model of inclined elliptic curve.

In the macro-program, arguments #24 and #26 were assumed to be x and z coordinates, respectively. The macro-program equation for the x coordinate in the standard equation was as follows:

$$\#24 = {}^{b}\!/_{a} * SQRT[[a*a]-[\#26*\#26]] \tag{3}$$

Assuming argument #10 in the parameter equation as angle a, the macro-program equation was as follows:

$$\begin{cases} \#26 = a\cos[\#10] \\ \#24 = b\sin[\#10] \end{cases} \tag{4}$$

For the macro-programming of inclined elliptic components, coordinates were usually transformed[10], as shown in Figure 3.

As an inclined ellipse was obtained by the tilting of ellipse by θ, the ellipse equation could be transformed using the rotation transformation matrix:

$$\begin{cases} z' = z\cos\theta - x\sin\theta \\ x' = z\sin\theta + x\cos\theta \end{cases} \tag{5}$$

Where (x, z) and (x', z') were the coordinates before and after rotation, and θ was the rotation angle.

The simultaneous equations (5) and (1), the equation of inclined ellipse was as follows:

$$\begin{cases} z' = z\cos\theta - \dfrac{b}{a}\sqrt{a^2 - z^2}\sin\theta \\ x' = z\sin\theta + \dfrac{b}{a}\sqrt{a^2 - z^2}\cos\theta \end{cases} \tag{6}$$

With θ as an argument, the equation of inclined ellipse could be obtained from equation (2):

$$\begin{cases} z' = a\cos\alpha\cos\theta - b\sin\alpha\sin\theta \\ x' = a\cos\alpha\sin\theta + b\sin\alpha\cos\theta \end{cases} \tag{7}$$

In the macro-program, arguments #1 and #2 were set as x' and z' coordinates of Point P. The parameter equation of the macro-program was as follows:

$$\begin{cases} \#2 = \#26*\cos[\theta] - \#24*\sin[\theta] \\ \#1 = \#26*\sin[\theta] + \#24*\cos[\theta] \end{cases} \tag{8}$$

Where $\#24 = {}^{b}\!/_{a} * SQRT[[a*a]-[\#26*\#26]]$.

4. Application of Macro-Programming for Elliptic Curved Component

The component shown in Figure 4 was processed by employing macro-programming. The blank is a round bar ($\Phi = 50 \times 100$ mm^2), and the intersection point of its central axis with the right end face of the elliptic curved component is defined as the origin for programming. The programs for rough and finish machining of elliptic curved components were developed based on the G73 fixed shape cutting cycle. Herein, the overall withdrawal amounts along x- and z-axes were both 2 mm (radius) and the cycle index was 14. T0101 was a rough turning tool with the main shaft speed = 500 r/min and feeding rate = 0.3 mm/r in the case of rough machining. During finish machining, T0202 was used and the allowances for finish were 0.5 mm and 0.1 mm along *x*-axis and *z*-axis, respectively. The feeding rate was 0.1 mm/r.

Two elliptic curves were observed on the component, and the origins from right to left were $\frac{z^2}{20^2}+\frac{x^2}{15^2}=1$, $\frac{z^2}{20^2}+\frac{x^2}{14^2}=1$. In the ellipse equation $\frac{z^2}{20^2}+\frac{x^2}{15^2}=1$, the argument #26 was z in the ellipse equation, the initial value was #4 = 20, the final value was #5 = 0; the argument #24 was x in the ellipse equation, the initial value was 0, and the increment was #6 = 0.2. The processing lists for elliptic curve macro-arguments based on standard and parameter equations were shown in Table 1 and Table 2 respectively.

According to the coordinate system in the scheme and the coordinate system for programming, the central line of the ellipse and the central line of the component were located on 20 mm away from the program origin in the negative *z* direction. Hence, the coordinates of any point in the new system were x (2*#24), z (#26-20).

In the ellipse equation $\frac{z^2}{20^2}+\frac{x^2}{14^2}=1$, the argument #26 was z in the ellipse equation, the initial value was #4 = 14, and the final value was #6 = −14; the argument #24 was x in the ellipse equation, and the increment was #6 = 0.2. The processing lists of macro-arguments in elliptic curves developed using standard and parameter equations was Table 3 and Table 4, respectively.

According to the coordinate system in the scheme and the coordinate system for programming, the central lines of the ellipse and component were parallel to each other. The origin in the ellipse coordinate system was 25 mm away from the program origin in the *x* direction and 34 mm away from the program origin in the negative *z* direction. Hence, the coordinates of any point in the new system were *x* (50-2*#24), *z* (#26-34).

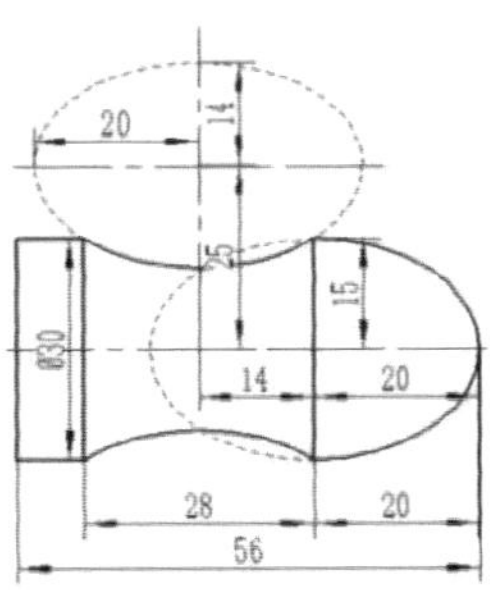

Figure 4. Elliptic Curved Component.

Table 1. Standard equation elliptic curve processing program variable table.

Variable selection	Variable representation	Macro variable	Macro variable
Independent variable	z	#26	#26
Domain	$[20,0]$	$[\#4,\#5]$	$[\#4,\#5]$
Dependent variable	$x=\frac{15}{20}\sqrt{20^2-z^2}$	#24 = 15/20 * SQRT[[20*20]-[#26*#26]]	#24 = 15/20 * SQRT[[20*20]-[#26*#26]]

Table 2. Parameter equation elliptic curve processing program variable table.

Variable selection	Variable representation	Macro variable
Independent variable	α	#10
Domain	$[0°,90°]$	$[\#4,\#5]$
Dependent variable	$\begin{cases} z=20\cos\alpha \\ x=15\sin\alpha \end{cases}$	$\begin{cases} \#26=20\cos[\#10] \\ \#24=15\sin[\#10] \end{cases}$

Table 3. Standard equation elliptic curve processing program variable table.

Variable selection	Variable representation	Macro variable
Independent variable	z	#26
Domain	$[14,-14]$	$[\#4,\#5]$
Dependent variable	$x=\frac{14}{20}\sqrt{20^2-z^2}$	#24 = 14/20 * SQRT[[20*20]-[#26*#26]]

Table 4. Parameter equation elliptic curve processing program variable table.

Variable selection	Variable representation	Macro variable
Independent variable	α	#10
Domain	$[54.3°,-54.3°]$	$[\#4,\#5]$
Dependent variable	$\begin{cases} z=20\cos\alpha \\ x=14\sin\alpha \end{cases}$	$\begin{cases} \#26=20\cos[\#10] \\ \#24=14\sin[\#10] \end{cases}$

Figure 5 shows the flow chart of macro-program for an elliptic curve. The program flow chart was developed according to the logical flow, followed by the development of processing procedures according to the program flow chart. The developed program was imported to the FANUC numerical control lathe. First, the developed program was simulated to avoid damages to lathe or cutting tools caused by program errors. Figure 6 shows the cutting tool route on the numerical control panel; the

IOP Conf. Series: Materials Science and Engineering **250** (2017) 012072 doi:10.1088/1757-899X/250/1/012072

validity of the developed program could be verified according to the cutting tool route shown. Figure 7 shows the component processed by the proposed approach. The specifications of this component were demonstrated to satisfy design requirements. In the processing of series components (with size variation but no shape change) by macro-programs, only changes in argument values were needed. In this way, standardization, modularization, and versatility of numerical control processing were realized. Using macro-programs, noncircular curved components, such as conic curved ones, could be fitted using micro-sized lines or arcs to meet the precision requirements. In the processing of elliptic curved components with varying parameters (e.g., initial and final points, angle, semi-major axis, and semi-minor axis), one general macro-program with these parameters as arguments allowed processing of various components without re-programming. Additionally, this study provided references for the process of other nonlinear curved components such as hyperbolic and parabolic curves.

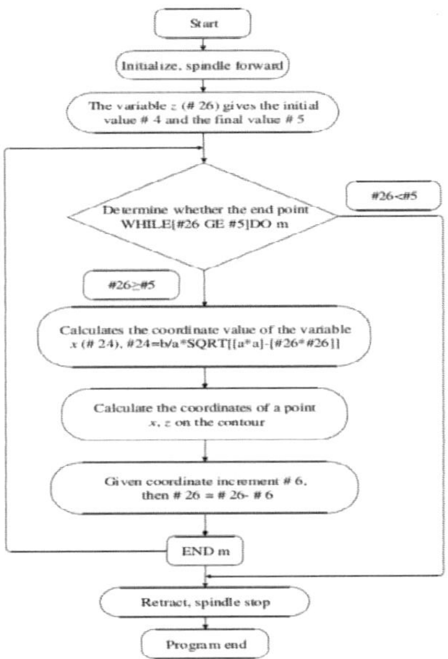

Figure 5. The flow chart of macro-program for an elliptic curve.

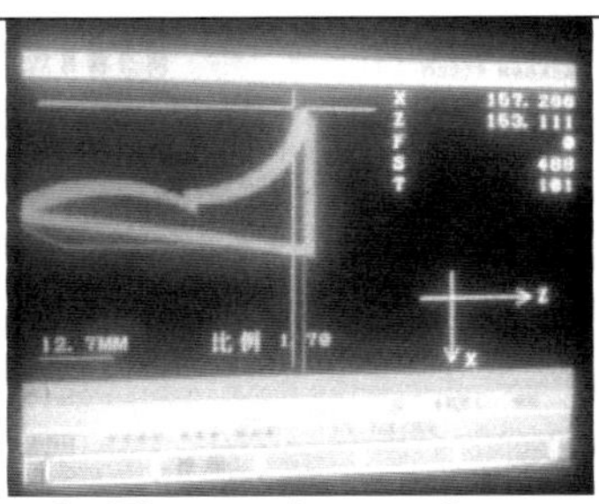

Figure 6. The cutting tool route on the numerical control panel.

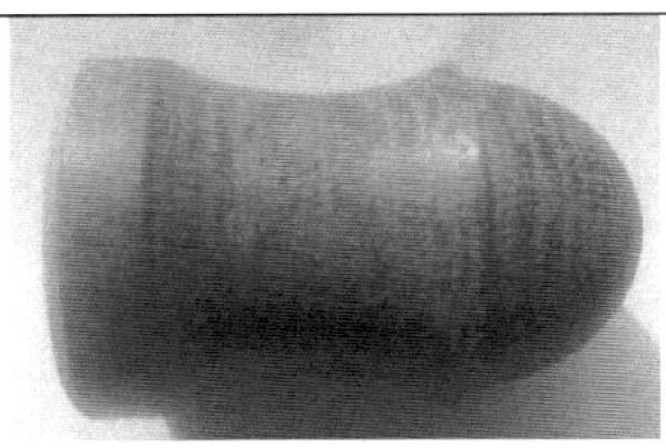

Figure 7. The component processed by the proposed approach.

5. Conclusions

(1) Owing to its mathematical calculation function, loop statements, and conditional transfer instructions, macro-programming was employed for numerical control processing of elliptic curved components. The programming procedures were reduced, while the versatility and applicability of the programs were enhanced.

(2) The elliptic curves were fitted using micro-sized segments or arcs. Rough and finish machining of elliptic curved components were achieved using the G73 fixed shape cutting cycle instructions. The

argument was defined using IF and WHILE statements. Elliptic curved components were processed using standard or equation–based macro-programs to eliminate issues of ellipse configuration in processing by numerical control lathe. In this way, tedious node calculations were avoided, and manual programming work required was significantly reduced.

(3) Additionally, this approach provided references for macro-programming for other noncircular curves. Parameter By increasing the quantity of difference points, the elliptic curve was fitted by micro-sized segments, and the processing precision was significantly enhanced.

6.Acknowledgments

This research was supported by the financial sup-ports from the National Natural Science Funds Fund (U1204517), Natural Science Fund by the Education Department of Henan Province P.R. China (12B460011), and the Opening Project of Key Laboratory of Precision Manufacturing Technology and Engineering, Henan Polytechnic University.

7.References

[1] Li P, Miao FX and Zhang M 2014 Application of fanuc cnc system macro-program in elliptic parts turning *Mech. Eng. Automat.* (3) 59-60

[2] Xu H 2013 *Fanuc cnc system with my manual programming* (Beijing:Chemical Industry Press) pp 62–63

[3] Wang MY 2015 Exploration into the macro programming of cnc lathes ellipses *Appl. Mech. Mater.* 727-728 823-826

[4] Zhang M, Li P and Miao FX 2015 Applying macro programming to turn non elliptic curve parts by fanuc system *Mech. Eng. Automat.* (2) 196-197

[5] Liu XC 2016 Study on the macro programming of nc machining convex ellipse and concave ellipse *Mech. Elecl. Eng. Technol.* 45 (3) 60-62

[6] Huang QG 2011 Application of user macro-program processing non-circular curve parts in cnc turning process *Manuf. Automat.* 33 (9) 49-51

[7] Liu YL, Jia T 2009 Research on how to write the macro program of ellipse parts and its application *Manuf . Automat.* 31 (7) 92-94

[8] Ling WC 2012 CNC programing of the quadratic curve *Mech. Elecl. Eng. Technol.* 41 (7) 76-78

[9] Li GY, Feng YH and Wang LL 2012 Macro programming skills and practice for ellipse part *Coal Mine Mach.* 33 (8) 137-138

[10] Li YX, Li XS and Li XL 2015 Application of macro program processing non-circular curve profile using NC lathe *Henan Sci.* 33(2) 204-208